World Sustainability Series

Due to its scope and nature, sustainable development is a matter which is very interdisciplinary, and draws from knowledge and inputs from the social sciences and environmental sciences on the one hand, but also from physical sciences and arts on the other. As such, there is a perceived need to foster integrative approaches, whereby the combination of inputs from various fields may contribute to a better understanding of what sustainability is, and means to people. But despite the need for and the relevance of integrative approaches towards sustainable development, there is a paucity of literature which address matters related to sustainability in an integrated way.

Notes on the quality assurance and peer review of this publication

Prior to publication, the works published in this book are initially assessed and reviewed by an in-house editor. If suitable for publication, manuscripts are sent for further review, which includes a combined effort by the editorial board and appointed subject experts, who provide independent peer-review. The feedback obtained in this way was communicated to authors, and with manuscripts checked upon return before finally accepted. The peer-reviewed nature of the books in the "World Sustainability Series" means that contributions to them have, over many years, been officially accepted for tenure and promotion purposes.

Johannes Kronenberg ·
Edith T. Lammerts van Bueren
Editors

On the Earth We Want to Live

Anthropsophy's Contributions to Sustainable Development

Springer

Editors
Johannes Kronenberg
Section for Agriculture
Goetheanum—Freie Hochschule für Geisteswissenschaft
Dornach, Switzerland

Edith T. Lammerts van Bueren
Wageningen University and Louis Bolk Institut (Emeritus)
Utrecht, Netherlands

ISSN 2199-7373 ISSN 2199-7381 (electronic)
World Sustainability Series
ISBN 978-3-031-98757-1 ISBN 978-3-031-98758-8 (eBook)
https://doi.org/10.1007/978-3-031-98758-8

This work was supported by Allgemeine Anthroposophische Gesellschaft.

This Springer imprint is published by the registered company Springer Nature Switzerland AG
The registered company address is: Gewerbestrasse 11, 6330 Cham, Switzerland

Foreword

Whether or not you are interested in anthroposophy as a school of thought or as a practice, this book raises important questions on the present and the future of humanity, and proposes original approaches to create relevant responses. Not being myself a member of the Anthroposophical Society, I feel nonetheless the strength of a vision that is also an intention: that of developing a holistic synthesis of many different threads of knowledge, consciousness and creativity, ancient and modern, for the co-creation of desirable futures.

While our backgrounds and sense of identity are built in a very different way, that intention resonates well with the purpose of The Club of Rome. To start with, there is a common recognition that the modern and globalised civilisation we live in today incubates an existential poly-crisis in its most tragic forms. Climate change, loss of biodiversity, social inequalities and fractures, geopolitical tensions, wars and genocides, you name them: these are multiple dimensions of a self-inflicted existential threat which simply seems too big to grasp. Not the least, by continuously legitimising militarisation and the industrialised demise of both fellow humans and ecosystems, the most powerful countries (self-entitled as the "developed" world) are directly accelerating the chances of our own extinction, and flirting again with the unthinkable, a fully-fledged nuclear confrontation.

In this context, all efforts to make feasible and widespread the oxymoron of "sustainable development" are welcome, especially when they have so much to show as this book. A most interesting part is the impressive compilation of close to 30 "Sustainable Natives—Stories of Sustainable Development" all around the world (Part Four). Against all odds of self-destruction, those examples will not only be sources of inspiration, they could also facilitate making sense of what could happen, by highlighting grounded realities with unconventional responses, potentially leading to regenerative societies where humans make peace among themselves and with life at large. It is not a coincidence if anthroposophy, since its inception, puts so much emphasis on the reinvention of agriculture, which is precisely the testbed to make that reconciliation possible, by providing both food security and a deeper sense of reconnection with nature, from which we crazily pretend to be separate.

After the shock created in 1972 by the publication of The Limits to Growth, with its warning that "progress" could lead to the collapse of human civilisations and the realisation that by itself the warning would not be sufficient, The Club of Rome proposed seven years later an original formulation of the civilisational challenge: in No Limits to Learning, it advocated for realising the full potential of humans to learn in order to "bridge the human gap", i.e. the difference between our capacity to act and transform the reality we live in, and our capacity to understand and deal with the consequences of our actions.

The "human gap" is a descriptive manner of invoking something terrible and scary, which was already outlined in The Limits to Growth: not only progress could lead to collapse, but our mental frameworks could be guilty of the suicide of the species. For all the incredible capacities that we have demonstrated, are we so stupid that the history of Life on Earth could only continue without us? This is also where anthroposophy comes in. It proposes a radical bet on the integral development of human capacities (as recalled in Part One), which again resonates well with the bet of No Limits to Learning: that of the development of new learning modes, that could be anticipatory, participatory, integrative and innovative enough to address our existential challenges in a positive way.

But this bet goes beyond the transformation of pedagogy and the creation of educational institutions suited to our times, as important as these could be, and indeed have been in the history of anthroposophy since the times of Rudolf Steiner. What we need to envision is not less than a whole civilisational shift. This means asking much deeper questions and, as well exemplified in the Third Part, mobilising the best capacities across domains as diverse, and usually as disconnected from each other as philosophy, economics, psychology, pedagogy, biology, agronomy, medicine, and so on. So much so that this mobilisation leads to proposing the definition of an "integral sustainable development" (Part Two) to overcome the limitations of more mechanistic and reductionistic approaches to sustainability.

For anthroposophy, as for other schools of thought, this implies the conception of humanity making a whole with the Earth, and the Earth itself being a whole living being. Needless to say, this connects not only with certain worldviews sustained by modern science but also with ancient wisdoms and extraordinary intuitions being now rediscovered with the tools of experimental sciences. There is something magic in this and in many other aspects of life, and then, as many feel in the need of tools to navigate uncharted seas, spirituality, artistic expressions and their influence on culture seem to provide a natural grounding to the whole (Part Two).

Ultimately, endeavours of the kind represented by this book or by The Fifth Element programme of The Club of Rome have, for all their differences, something in common: they use new lenses for an endless quest coming from the dawn of humanity: the two-sided search for the meaning of our presence on Earth and what makes a good life. I feel that this book, co-created by its many authors from a momentum of passion and commitment, could attract anyone concerned about humanity's future or intrigued by the fundamental questions surrounding the human existence and our ethical challenges. Not all the responses might be here, but, in resonance again with the No Limits to Hope project that The Fifth Element is developing, a strong feeling emerges: our existence does not make sense and will indeed come to an end if we are not able to honour our resonance with the infinite web of life. And then, a good life is one in which our individuality is but one face of our deep connection with others and with the history of life on Earth.

Dear reader, I hope you will enjoy the book, and that it will have a lasting impact not only on the questions you will ask, but especially on the attitude of hope with which you will go on, wisely inquiring until living the responses.

Winterthur, Switzerland

Carlos Alvarez Pereira
Secretary General of The Club of Rome
Co-creator of The Fifth Element, a Programme by The Club of Rome

Acknowledgements

Given the nature of this publication as a compendium, we would like to first express our sincere gratitude to the 75 authors who contributed their time and expertise to this work. Activists, artists, entrepreneurs, farmers, philosophers, professional practitioners, scientists, teachers, and others have engaged with the central questions of this publication, enriching its content through their diverse perspectives.

We extend our appreciation to the funders of this publication: Fonds-Goetheanum (Dornach, CH), Hamasil Stiftung (Zürich, CH), Neuguss Stiftung (Kaltenkirchen, DE), the World Goetheanum Association (Dornach, CH), and the Section for Agriculture at the Goetheanum (CH). While some of these organisations are not directly connected to anthroposophy, those that are have supported this project with their broad understanding of research and cultural initiatives related—but not limited—to anthroposophical practices. Importantly, the content and research presented here were not directed or influenced by any of these foundations beyond the intentions of the editors.

A special thanks goes to Klaas van Egmond who provided intermittent yet highly valuable support, offering his vast academic expertise in sustainable development and anthroposophy during key stages of the editing process.

From the Goetheanum, Ueli Hurter and Constanza Kaliks supported this work from its inception with unwavering confidence—without their support, the publication would not have been possible.

We are also grateful to Till Höffner, whose dedication to translation, editing, and maintaining academic quality was instrumental in refining this work. Additionally, the team at the Section for Agriculture of the Goetheanum provided ongoing encouragement and steadfast support throughout this project, for which we extend our sincere appreciation.

Finally, we thank Walter Leal as series editor, Springer Nature for facilitating the publication of this compendium in their *World Sustainability Series*, which this year will mark its 10th anniversary, to a growing body of inter- and transdisciplinary world sustainability perspectives.

Dornach, Switzerland
Utrecht, The Netherlands
September 2025

Johannes Kronenberg
Edith T. Lammerts van Bueren

Contents

Contributors

Helmy Abouleish CEO, SEKEM, Cairo, Egypt

Konstanze Abouleish Managing Director, NatureTex, El Salam City, Egypt

John Addison Newton Dee Village, Aberdeen, Scotland

Aline Amrein Montezillon, Switzerland

Christine Arlt Member of the SEKEM Future Council and Head of SEKEM-Friends Germany, SEKEM-Friends, Cologne, Germany

Maurizio Bär Pneumology and Lung Cancer Centre, Lead Climate Friendly Hospital Havelhöhe, Berlin, Germany

Ginny Bayman Newton Dee Village, Aberdeen, Scotland

Johannes Beurle Notfallpädagogik ohne Grenzen e.V., Karlsruhe, Germany

Juan Bottero Managing Director, Fundación Dinero y Conciencia, Santiago de Chile, Chile

Heike Breidenich Pneumology and Lung Cancer Centre, Lead Climate Friendly Hospital Havelhöhe, Berlin, Germany

Dirk Buchwald Pneumology and Lung Cancer Centre, Lead Climate Friendly Hospital Havelhöhe, Berlin, Germany

Trevor Button Kufunda Learning Village, Harare, Simbabwe

Siu Hou Chong Koperasi Kongsi Selangor Berhad (Kongsi Co-op), Kuala Lumpur, Malaysia

Christoph Cordes Montezillon, Switzerland

Ute Craemer Co-founder and CEO, Associação Comunitária Monte Azul, São Paulo, Brazil

Jacqueline de Kool Odin, Geldermalsen, Netherlands

Yaike Dunselman Architect, Co-founder and Director, 9graden architectuur, Amersfoort, The Netherlands

Nadia El-Hage Swette Center for Sustainable Food Systems, Arizona State University, Tempe, AZ, USA

Volkert Engelsman Founder and Former CEO Eosta, Founder Robin Food Coalition, Waddinxveen, Netherlands

Jean-Michel Florin Co-president, Biodynamic Federation Demeter International, Bonn, Germany

Christa Foppe Pneumology and Lung Cancer Centre, Lead Climate Friendly Hospital Havelhöhe, Berlin, Germany

Mathias Forster Managing Director, Bio-Stiftung Schweiz, Basel, Switzerland

Vesna Forštnerič Lesjak Co-Leader and Research Associate, Natural Science Section at the Goetheanum, Dornach, Switzerland

Nana Goebel Managing Director, Friends of Waldorf Education, Berlin, Germany

Christian Grah Pneumology and Lung Cancer Centre, Lead Climate Friendly Hospital Havelhöhe, Berlin, Germany

Christine Gruwez Faculty of Comparative Science of Religion, University of Leiden, Leiden, Netherlands

Justus Harm Managing Director, Egyptian Bio-Dynamic Association, Cairo, Egypt

Gerhard Heid Former Managing Director, Sonett GmbH, Deggenhausertal, Germany

Ning Hii Koperasi Kongsi Selangor Berhad (Kongsi Co-op), Kuala Lumpur, Malaysia

Craig Holdrege The Nature Institute, Ghent, NY, USA

Ueli Hurter Section for Agriculture at the Goetheanum, Dornach, Switzerland; Co-leader Section for Agriculture at the Goetheanum - Freie Hochschule für Geisteswissenschaft & Executive Council of the General Anthroposophical Society, Dornach, Switzerland

Claudy Jongstra Owner, Studio Claudy Jongstra, Spannum, Netherlands

Renate Keller Associação Comunitária Monte Azul, São Paulo, Brazil

Maaianne Knuth Founder and Managing Director, Kufunda Learning Village, Harare, Simbabwe

Maja Kolar Co-president, Biodynamic Federation Demeter International, Bonn, Germany

Merle Koomans van den Dries Managing Director, Odin, Geldermalsen, Netherlands

Johannes Kronenberg Research Associate Section for Agriculture, Goetheanum - Freie Hochschule für Geisteswissenschaft, Dornach, Switzerland

Inke Kruse Managing Director, Hans Stockmar GmbH & Co. KG, Kaltenkirchen, Germany

Edith T. Lammerts van Bueren Emeritus Professor, Wageningen University and Louis Bolk Institute, Organic Plant Breeding, Utrecht, Netherlands

Jia Hao Lau Koperasi Kongsi Selangor Berhad (Kongsi Co-op), Kuala Lumpur, Malaysia

Salvatore Lavecchia Professor, Dipartimento di Studi Umanistici e del Patrimonio Culturale, Università degli Studi di Udine, Udine, Italy

Margareta Leber Pedagogy, Anthropology, Methodology and Didactics of the Class Teacher, Freie Hochschule Stuttgart, Stuttgart, Germany

Chiou Woan Leng Koperasi Kongsi Selangor Berhad (Kongsi Co-op), Kuala Lumpur, Malaysia

Flore Lutters PhD candidate, VU Graduate School of Humanities, Free University Amsterdam, Amsterdam, Netherlands

David Martin Professor, Medical Theory, Integrative and Anthroposophic Medicine, Institute of Integrative Medicine, University of Witten/Herdecke, Witten, Germany

Jon McAlice The Nature Institute, Ghent, NY, USA

Dan McKanan Senior Lecturer, Harvard Divinity School, Cambridge, MA, USA

M D Isa M. Saffiuddin Koperasi Kongsi Selangor Berhad (Kongsi Co-op), Kuala Lumpur, Malaysia

Jesse Mulder Assistant Professor, Faculty of Humanities, Department of Philosophy and Religious Studies, Utrecht University, Utrecht, Netherlands

Tina Müller CEO, Weleda, Arlesheim, Switzerland

Chee Yuan Ng Koperasi Kongsi Selangor Berhad (Kongsi Co-op), Kuala Lumpur, Malaysia

Beate Oberdorfer Former Managing Director, Sonett GmbH, Deggenhausertal, Germany

Jasmin Peschke Research Associate, Section for Agriculture at the Goetheanum, Dornach, Switzerland

Jörg Probst Pneumology and Lung Cancer Centre, Lead Climate Friendly Hospital Havelhöhe, Berlin, Germany

Wolodymyr Radysh Newton Dee Village, Aberdeen, Scotland

Matthias Rang Co-Leader, Natural Science Section at the Goetheanum, Dornach, Switzerland

Susanne Charlotte Rotermund Associação Comunitária Monte Azul, São Paulo, Brazil

Bernd Ruf Notfallpädagogik ohne Grenzen e.V., Karlsruhe, Germany

Stefan Ruf Managing Director, Maeander Youth Health Center, Berlin, Germany

Max Ruhri Managing Director, Freie Gemeinschaftsbank, Basel, Switzerland

Siao Hoong Sam Koperasi Kongsi Selangor Berhad (Kongsi Co-op), Kuala Lumpur, Malaysia

Nadine-Hélène Santiago CSO, Weleda, Arlesheim, Switzerland

Guillermo Scallan Managing Director of Social Innovation at the AVINA Foundation, Ñuñoa, Chile

Albrecht Schad Professor, Methodology and Didactics of Upper Secondary Education, Biology, Geography, Freie Hochschule Stuttgart, Stuttgart, Germany

Friedemann Schad Pneumology and Lung Cancer Centre, Lead Climate Friendly Hospital Havelhöhe, Berlin, Germany

Christopher Schümann Bio-Stiftung Schweiz, Basel, Switzerland

Ryan Shea The Nature Institute, Ghent, NY, USA

Marjan Sikkel Newton Dee Village, Aberdeen, Scotland

Meinhard Simon Emeritus Professor Aquatic Microbial Ecology, Carl von Ossietzky University Oldenburg, Oldenburg, Germany

Georg Soldner Medical Section, Goetheanum - Freie Hochschule für Geisteswissenschaft, Dornach, Switzerland

Johannes Steinke Voelkel GmbH Fruit and Vegetable Juices, Pevestorf, Germany

Paw Lan Tai Koperasi Kongsi Selangor Berhad (Kongsi Co-op), Kuala Lumpur, Malaysia

Jia Yi Tan Koperasi Kongsi Selangor Berhad (Kongsi Co-op), Kuala Lumpur, Malaysia

Chen Yee Teh Koperasi Kongsi Selangor Berhad (Kongsi Co-op), Kuala Lumpur, Malaysia

Ha Vinh Tho Eurasia Learning Institute for Happiness and Wellbeing, Palézieux, Switzerland;
Gross National Happiness Center Bhutan, Thimphu, Bhutan;
Traning International Committee of the Red Cross, Geneva, Switzerland

Klaas van Egmond Emeritus Professor, Copernican Institute for Sustainable Development, Utrecht University, Utrecht, Netherlands

Boris Voelkel Managing Director, Voelkel GmbH Fruit and Vegetable Juices, Pevestorf, Germany

Nathaniel Williams Head of the Youth Section at the Goetheanum - Freie Hochschule für Geisteswissenschaft, Dornach, Switzerland

Roland Zerm Pneumology and Lung Cancer Centre, Lead Climate Friendly Hospital Havelhöhe, Berlin, Germany

Kees Zoeteman Emeritus Professor, Sustainable Development, Tilburg University, Tilburg, Netherlands

Amadeus Zschunke Managing Director, Sativa AG, Rheinau, Switzerland

Introduction

Johannes Kronenberg and Edith T. Lammerts van Bueren

Abstract

This compendium explores the contributions of anthroposophy to the evolving field of sustainable development, addressing the urgent need for a more integral and culturally rooted response to the socio-ecological crisis of the Anthropocene. While mainstream sustainability discourses often adopt technocratic and one-dimensional frameworks, this publication foregrounds the neglected spiritual and cultural dimensions as critical to meaningful transformation. Tracing the roots of the crisis beyond the Industrial Revolution to the colonial expansion post-1492, it positions anthroposophy—a spiritual-scientific worldview initiated by Rudolf Steiner—as a historically grounded and practically applied approach to sustainable development. Through interdisciplinary contributions from 75 authors worldwide, the compendium highlights anthroposophical practices in agriculture, education, medicine, economics, and the arts, and introduces the concept of "Integral Sustainable Development," which integrates inner and outer dimensions of human and planetary development. Emphasising appropriate anthropocentrism and a renewed Earth-human relationship, this work presents anthroposophy as both a theoretical framework and a practical path toward a cosmopolitan, ecologically resilient future. It aims to foster further research, dialogue, and cooperation within the global sustainability movement, while acknowledging the complexity and unfinished nature of its subject matter.

J. Kronenberg (✉)
Research Associate Section for Agriculture, Goetheanum - Freie Hochschule für Geisteswissenschaft, Dornach, Switzerland
e-mail: johannes.kronenberg@goetheanum.ch

E. T. Lammerts van Bueren
Wageningen University and Louis Bolk Institute, Organic Plant Breeding, Utrecht, Netherlands
e-mail: edith@bioacademy.nl

J. Kronenberg and E. T. Lammerts van Bueren (eds.), *On the Earth We Want to Live*,
World Sustainability Series, https://doi.org/10.1007/978-3-031-98758-8_1

Keywords

Anthroposophy • Sustainable development • Socio-ecological crisis • Anthropocene • Cultural-spiritual dimension • Integral sustainability • Rudolf Steiner • Appropriate anthropocentrism • Spiritual science • Systemic worldview • Biodynamic agriculture • Inner development goals • Global transformation • Earth-human relationship

1 About This Publication: Its Scope and Relevance

Overcoming the socio-ecological crisis of the Anthropocene demands new narratives and paradigms of thought, social reciprocity, and inclusive action. While sustainable development has gained global traction, its often technocratic and one-dimensional focus has the risk to neglect the cultural-spiritual dimensions essential for meaningful transformation. Also, the history and cause of this crisis is still attributed to the industrial revolution and the Great Acceleration (Steffen et al., 2015), though the colonial impact after 1492 seems to be a more precise start of global social injustice and the destruction of vital earth organisms (Jhagroe, 2024).

For over 100 years, anthroposophical approaches and practices have engaged deeply with questions concerning human, Earth, and cosmic development. Many of these approaches have found pioneering practical applications across continents and cultures, including Biodynamic Agriculture (as first organic farming method), Waldorf education, complementary medicine, social inclusive development and the arts (McKanan, 2017). In 2025, the centenary of Rudolf Steiner's death will be commemorated. Steiner, as initiator of anthroposophy, described anthroposophy as "a path of knowledge to guide the Spiritual in the human being to the Spiritual in the universe" (Steiner, 1973, p. 13). Building on traditions such as German Idealism, Steiner advanced the exploration of integrating sensory and supersensory realities, and applying his hypotheses to the understanding of human nature, the Earth, and cosmic development as a spiritual science (see Part 1).

What are the contributions of anthroposophy towards sustainable development in this respect? What are its perspectives and practical approaches to addressing socio-ecological challenges in an integral and systemic way? These questions form the foundation of this compendium and highlight a research gap that warrants further exploration (McKanan, 2017).

Since 1924, the Goetheanum, as the School of Spiritual Science and headquarters of the General Anthroposophical Society, has been dedicated to developing, researching, teaching, and training anthroposophical contributions through its 12 faculties and related disciplines. These include the Section for Agriculture, General Anthroposophical Section, Section for Literary Arts and Humanities, Mathematical-Astronomical Section, Medical Section, Natural Science Section,

Pedagogical Section, Section for Social Inclusive Development, Section for the Performing Arts, Section for Social Sciences, Section for Visual Arts, and the Youth Section. This publication was initiated by the Section for Agriculture as part of its ongoing efforts to engage with global sustainable development.

The aim of this compendium is to bring together a broad collection of multi- and transdisciplinary contributions from the anthroposophical discourse on sustainable development and to introduce these to the wider public and professional communities. Furthermore, this publication aspires to serve as a foundation for future research and educational programmes for those seeking a culturally and spiritually enriched approach to socio-ecological challenges and sustainable development. At the same time, it seeks to foster constructive dialogue with partner organisations within the sustainable development movement on these topics.

Over the past century, anthroposophy has been both critiqued and praised. Critical self-reflection remains a central part of its discourse (see, e.g., Van Baarda et al., 2000; Ebert et al., 2024; Selg et al., 2024). Notably, global recognition has been accorded to initiatives such as the United Nations Environment Programme's "Champion of the Earth 2024" laureate, SEKEM in Egypt, for their pioneering contributions to biodynamic agriculture as a solution to environmental crises.

In addressing the central question of this publication, we acknowledge that the breadth and complexity of subjects such as socio-ecological crises, sustainable development, and anthroposophy across disciplines (e.g., agriculture, education, medicine) make it impossible to cover their full depth and entirety. Nevertheless, we see this collection of perspectives as a starting point for continued dialogue and the potential for new collaborations, building discourse in the field of sustainable development and anthroposophy's contributions.

2 Content of This Compendium

This publication is structured as a compendium and comprises five parts, featuring 45 texts written by 75 authors from across the globe. Activists, artists, entrepreneurs, farmers, philosophers, professional practitioners, scientists, and teachers have contributed to this inter- and transdisciplinary exploration of the subject matter.

In this chapter, the Introduction, discusses the aim of this publication and provides background on how sustainability challenges are perceived and addressed from an anthroposophical perspective.

Part 1, *Anthroposophy in the context of Sustainable Development*, includes contributions that introduce some fundamental anthroposophical principles and concepts, particularly regarding the relationship between human beings and the Earth in the face of current socio-ecological challenges. Additionally, Dan McKanan of Harvard Divinity School shares his research findings on the contributions of anthroposophy to the environmental movement.

Part 2, *Sustainable Development: Definitions and an Expansion Towards Integral Sustainable Development*, examines the evolution of various definitions of sustainable development. It argues that the socio-ecological crisis is, in essence, a cultural-spiritual crisis that calls for a renewed and constructive anthropocentrism. This section also proposes expanding the established definition of sustainable development—which typically focuses on environmental, social, and economic dimensions—by adding a fourth dimension informed by the anthroposophical worldview and practices, thereby deepening the concept of Integral Sustainable Development. This remains still an incomplete and first sketch, that needs to be explored further.

Part 3, *Worldview: Perspectives from Anthroposophy*, presents an integral and holistic worldview based on anthroposophy and its relevance to socio-ecological challenges and sustainable development. Contributions from 19 authors provide insights and perspectives on various key elements of this theme.

Part 4, *Sustainable Natives: Stories of Sustainable Development*, features essays from 26 organisations worldwide, illustrating how they have contributed to sustainable development through anthroposophical practices in fields such as medicine, education, social development, agriculture, and food processing. Many of these organisations have been pioneers in their respective sectors, working towards sustainability for decades and, in some cases, for over 100 years. We introduce these organisations as *Sustainable Natives,* acknowledging their pioneering roles. As a disclaimer, we emphasise that this compendium does of course not claim to offer "the" solution to the complex challenges facing our world but instead presents contributions and stepping stones towards solutions.

Part 5, *Concluding Remarks and Outlook*, concludes the compendium with a call for further dialogue and collaboration.

This publication is aimed at a diverse readership, including anyone interested in sustainable development, professional practitioners seeking integral perspectives for their work, and the academic sustainable development community.

We are honoured to be part of Springer's *World Sustainability Series,* which, since 2015, has fostered interdisciplinary and integral perspectives on sustainable development, its definitions, and its significance. We hope this compendium will contribute to the growing body of knowledge and applications, offering new perspectives and furthering the development of sustainable development itself.

3 Evolving Sustainable Development and Anthroposophy

Every day, more people and institutions contribute to the effort of overcoming the global socio-ecological polycrisis (Deloitte, 2024). This publication seeks to affirm and reinforce these sustainability efforts, recognising that such challenges will accompany global society for many decades to come. In this context, sustainable development is understood and applied as a transformative tool to address socio-ecological challenges, working towards a resilient, cosmopolitan society in equilibrium, rooted in the vast diversity of landscapes, cultures, local identities, and human and non-human beings. Sustainable development, in this sense, may be seen as a grand attempt to reconcile freedom with responsibility—from the personal and local level to the global, earthly, atmospheric and even cosmic dimensions.

However, 40 years after the Brundtland Report of the United Nations World Commission on Environment and Development and more than 50 years after *The Limits to Growth,* published as a report to the Club of Rome, sustainable development remains predominantly a technically focused, one-dimensional strategy. This approach still permits "organised irresponsibility" (Scientific Council for Integral Sustainable Agriculture & Nutrition, 2012) on the part of Big Agribusiness, Big Finance, Big Pharma, Big Oil, and others under the banner of "Green Growth" and a decoupled economy (Weizsäcker et al., 2018; Dixson-Declève, 2022; European Union, 2022). As a response, protests such as those organised by Extinction Rebellion have become a near-daily occurrence, while the global rise of nationalism and conservatism deprioritises socio-ecological challenges, dismissing them as moralising left-wing ideologies (United Nations, 2024; The Guardian, 2025). Moreover, one can also observe parts of the sustainability agenda of the West as an exclusive pursuit of the elite, perpetuating issues such as the proliferation of electric SUVs, greenwashing philanthropy, justifications for continued exponential economic growth, and the ongoing exploitation and colonialisation by the West and Global North (Jhagroe, 2024; Krenak, 2023).

Amid these dynamics, sustainable development itself is evolving. To remain relevant and effective, it must adapt to societal developments (Ametepey et al., 2023). In our view, sustainable development must become more than a monitoring framework or certification labyrinth—it must inspire action and hope. As demonstrated by the initiatives described in Part 4, sustainable development can serve as a narrative of mutuality, reciprocity, and a shared future for all. While critiques of sustainable development as a tool of Western neo-colonialism exist (Jhagroe, 2024), we argue strongly for "reclaiming" sustainable development as a potential meta-discipline, rather than dismissing it as corrupt or obsolete. It is a relatively young discipline, still maturing, and holds great potential. This motivates the contributors to this publication to explore how anthroposophy can contribute to a more integral understanding of and practical response to the ecological crisis in the context of sustainable development (see Part 2).

By introducing anthroposophical principles and worldviews in this context, this publication aims to showcase an established, research- and practice-based approach to addressing the socio-ecological crisis and sustainability challenges (see Part 3). Building on these experiences, a sketch of an Integral Sustainable Development approach emerges, addressing the cultural-spiritual dimensions of society alongside sustainability practices. This perspective explores the Earth as a living organism and humankind as a potential co-creator in its development, aligning with the concept of Inner Development Goals (Eleftheria & Campos Suarez, 2024) as prerequisites for outer transformation (see Part 2).

Anthroposophy, as such, offers both a worldview and a set of practices that can be recognised as a path of knowledge, integrating mainstream science, spiritual science, and the arts as drivers of societal, human, and earthly development (see Part 4).

4 Our Current World

Numerous analyses describe the dynamics of our contemporary era, this publication provides a brief perspective to offer an overview and context for its specific contribution.

Beneath the present *hyperreality,* in which we appear to have been living for some time (Baudrillard, 1994), a new, unknown world is slumbering. Following the proclaimed "death of God" (Nietzsche, 1883–1885), the abyss of humanity in the twentieth century, the "end of art" (Hegel, 1955), the "end of history" as the culmination of the ideal state (Fukuyama, 1989), and the ecological destruction of the Earth, much seems to have been laid bare. Yet, many challenges continue to be approached through "one-way" or "single-issue strategies," with a truly integral approach still largely absent. This absence has led to an accumulation of "organised irresponsibility" (Scientific Council for Integral Sustainable Agriculture & Nutrition, 2012). Nevertheless, the dramatic climate and atmospheric changes are propelling global society into action. The demand for new steps, inspirations, and horizons is urgent.

As a specific example of grappling with this momentum, Bruno Latour, in his "Protective Measures" at the onset of 2020, called on society during the coronavirus pandemic not to revert to the old dominant production model in the so-called post-crisis phase. He proposed six pragmatic questions to guide the next steps in addressing the deeper causes of contemporary challenges, which he framed as an "ecological mutation" (Latour, 2020). Latour urged everyone to reflect on which practices and behaviours should be abandoned and which could support building the future we wish to inhabit. In essence, he highlighted how our current exposed world is increasingly shaped by the daily actions and behaviours of individuals, regardless of their occupation or location. This growing emphasis on individual responsibility transforms society from an abstract concept into a concrete, personal reality: what you do—or fail to do—is society itself (Latour, 2020).

This raises a pressing question: what are we doing that defines our current time and society? One might observe a *Departing Humanity,* where humankind increasingly distances itself from the Earth. The "technological human," or transhuman, seems to turn toward other worlds, seeking a "multiplanetary life" (Musk, 2020). Is this an escape from Earth to never return, shattering humanity across the cosmos? Digital planets and virtual worlds are being constructed, while Mars is imagined as the next destination. Will Earth itself become a nature conservation zone—a relic of the Old World where we once lived? Increasingly, dystopian visions such as *The Matrix* seem to edge closer to reality, as individuals spend an average of 6.5 h online daily (Statista, 2024), with the yet-unknown impacts of artificial intelligence likely to amplify this trend.

This compendium seeks to reintroduce a terrestrial turn—a "Down to Earth" approach that calls for rediscovery of the Earth as Gaia, as a living organism rather than a broken machine in need of repair. It advocates an understanding of humanity as a responsible actor and co-creator of Earth's substance. It calls for reflection on what it means to live "on the Earth we want to live."

This paradigm shift—this transition, transformation, metamorphosis, socio-ecological crisis, or however it might be termed—is clear, universal, and global. It heralds a new form of globalisation: a cosmopolitanism where all unite as Earthlings, while remaining rooted in their own places and localities. It demands new competencies, perspectives for action, and ways of living and working together—a renewed sense of fraternity, sisterhood, or siblinghood. It calls for a revitalised morality and togetherness (Jonas, 1979), transcending differences in politics, economic status, or ideological divides in mono- or multipolar worlds.

What seems impossible must, and will, happen: the birth of a new world. But where are the beginnings? Where are the modest turns and initial steps? Where is the future already emerging? What worldviews and images of humanity are involved? How will the Earth and humanity's course continue to evolve, and how can we co-create and shape its substance?

5 Appropriate Anthropocentrism

Current socio-ecological events increasingly appear to culminate in dramatic scenarios: the Earth is breaking, and human dignity is being widely violated. Increasing awareness of the global dimension of this situation has existed for several centuries (von Humboldt, 1801) and has been on the political and scientific agenda for the past 50 years (e.g., Carson, 1962; Meadows et al., 1972; Brundtland, 1987; Steffen et al., 2015). However, only recently has it become a central topic in public debate, amplified through global media platforms and warnings of António Guterres for instance at the COP25 in 2019 or Greta Thunberg in 2019. Reflection on this socio-ecological crisis now spans all dimensions, areas, and professions, encompassing research disciplines as well. The issue is overarching, transdisciplinary, and universal in nature.

The Anthropocene, defined as the human era in which human activity has a lasting influence on the Earth's vital ecosystems, has been proposed as a framework to sharply define and clarify the time in which we live (Crutzen et al., 2000). While the term remains contested and classified as geologically imprecise by some, other scientists (Steffen et al., 2015) pinpoint the start of the Anthropocene as the *Great Acceleration*—a period of exponential growth in the economy, population, and resource consumption that began in the mid-twentieth century. According to this interpretation, the Anthropocene commenced on Monday, 16 July 1945, marked by the detonation of the first atomic bomb in the New Mexico desert, USA. Steffen et al., (2015, p. 93) describe this moment as follows:

> Setting the start date of the Anthropocene at the beginning of the Great Acceleration makes it possible to specify the onset of the Anthropocene with a high degree of precision. On Monday 16 July 1945, about the time that the Great Acceleration began, the first atomic bomb was detonated in the New Mexico desert. Radioactive isotopes from this detonation were emitted to the atmosphere and spread worldwide, entering the sedimentary record to provide a unique signal of the start of the Great Acceleration, a signal that is unequivocally attributable to human activities.

Since this point, human influence and destructive potential have transitioned from localised impacts to a global scale, signalling the definitive onset of the Anthropocene. Current discussions whether 1492 might be the more accurate date, or the since thousands of years taught virtue ethics, its meaning and striving for a "good live" are ongoing.

The "atomic" interpretation of the Anthropocene offers a dramatic characterisation of the human age, but it could—and, in our opinion, should—be approached differently, particularly when considering the question of what a sustainable future might look like. Depicting humanity solely as a destructive force provides a partially accurate record of time but fails to unlock humanity's full potential or open pathways to societal resilience and viability.

A fundamental question emerges: What is, and what could be, the trajectory of human development beyond one dominant narrative? Is it an irreversible path towards destruction, or can it evolve into a journey toward resilience, responsibility, and sustainable progress? Such progress, we propose, would necessitate worldviews that regards the Earth as a living organism and humanity as a potential co-creator of the planet as an integrated whole.

This brings forth the concept of "appropriate anthropocentrism" (see Sect. 5, in Chap. 6), which anthroposophy identifies as a significant potential and contribution to the environmental movement (McKanan, 2017). By reimagining humanity's role not as a destroyer but as a responsible and creative participant in the Earth's future, anthroposophy offers an alternative perspective—one that balances human development with the health and integrity of the planet.

6 Contributions from Anthroposophy

By exploring the Earth as a living organism and humankind as a co-creator (Wachsmuth, 1945/1980; Zoeteman, 1989, 2009; Steiner, 1924/2022), anthroposophy offers a complementary perspective on sustainable development and environmentalism (McKanan, 2017). On the one hand, anthroposophy has fostered a renewed worldview and concept of the human being, drawing on Rudolf Steiner's (1861–1925) "spiritual science" (*Geisteswissenschaft*). On the other hand, the anthroposophical movement has contributed practical solutions to socio-ecological challenges over the past century, pioneering efforts in already named fields such as organic (biodynamic) agriculture, ecological and ethical banking, ecological medicine, Earthbound pedagogy, associative economics, and the arts. Notable examples include the founding of Weleda in 1921, which produces natural cosmetics, self-care products, and medicines, as well as the GLS Bank in 1974 and Triodos Bank in 1980, which were among the first sustainable initiatives in economic life with ties to medicine, art, education, and broader social and cultural organisation (McKanan, 2017).

One notable pioneer of anthroposophy, Guenther Wachsmuth, published *Earth and Humankind—Their Formative Forces, Rhythms and Life Processes* in 1945, presenting the Earth as a living organism. In this work, he advocated for an integral and systemic view of the Earth as a living whole. In the foreword, Wachsmuth noted:

> The fact that today it has become more important and urgent to understand the interrelationships in the total system of the Earth, is due to the fact that humankind, by penetrating the higher layers of the atmosphere, is now more than ever making the whole of the Earth his own and subjecting it to his influence. As a result of the rapid development of technology, whereby, for example, aircraft flying at supersonic speeds, by generating explosions in connection with nuclear tests, through rocket projectiles up to high atmospheric layers, etc., we are daily intervening more and more in every hitherto unexplored area of the atmosphere, we are intervening in processes of the Earth as a whole that we do not have a nearly sufficient overview of. We know from the processes in the smaller organisms what importance is often attached to the finest effects in the developmental process, but what it means in the so unstable and reactive high layers of the atmosphere when we now begin to arbitrarily intervene in their structure and processes remains to be explored. (Wachsmuth, 1945/1980, pp. 11–12)[1]

Wachsmuth's call for a systemic understanding of the Earth as a living organism, based on anthroposophy, emphasised the importance of understanding the Earth's rhythms and life processes before undertaking (global and atmospheric) human interventions. His perspective aligns with other integral worldviews, such as the Gaia hypothesis (Lovelock, 1972), systems thinking (Meadows, 2001, 2008; Capra, 2002), and Bruno Latour's *Down to Earth* (2018), which stress the interconnectedness of all living and non-living dimensions on Earth.

Earlier contributors to this integral worldview include Alexander von Humboldt (1769–1859), who, in collaboration with Johann Wolfgang von Goethe (Wulf, 2015, pp. 25–39), which laid some of the conceptual groundwork for anthroposophy. Humboldt described nature as a living unity, stating:

> For the thinking observer, nature is a unity in multiplicity, the combination of the manifold in form and mixture, the epitome of natural things and natural forces, as a living whole. The most important result of sensible physical research is therefore to recognise the unity in diversity, to embrace from the individual all that the discoveries of the latter ages offer us, to examine the details and yet not to succumb to their mass, mindful of the sublime destiny of humankind, to grasp the spirit of nature, which lies concealed under the cover of phenomena. In this way our endeavours reach beyond the narrow world of the senses, and we can succeed in comprehending nature, in mastering the raw material of empirical perception through ideas, as it were. (von Humboldt, 1845–1862, vol. 1, p. 5)[2]

This perspective can be seen as part of the groundwork for anthroposophy's spiritual-scientific approach, which transcends sensory perception and seeks to grasp the essence of nature and life through non-sensory, idea-based thinking. Rudolf Steiner, as the first editor of Goethe's *Naturwissenschaftliche Werke*, built on these ideas, establishing the foundations of anthroposophy (Steiner, 2021; 1884–1897).

Anthroposophy views development as central—not only for humankind but for the Earth itself, including the mineral, plant, and animal realms, and their cosmic interrelations (Steiner, 1904/2013a, 1919/2019). In the context of today's Anthropocene, humanity's role in Earth's development is increasingly critical.

---

[1] Editors translation from German.

[2] Editors translation from German.

Anthroposophy explores how human beings can act as co-creators with the Earth, fostering a shared developmental path. Steiner captured this sentiment in 1909: "Our mother Earth has hardened through. Our mission is to spiritualise her again, by reworking her through the power of our hands into a spirit-filled work of art" (Steiner, 1993, p. 113).

Despite their pioneering nature, anthroposophical contributions remain underrepresented in broader socio-ecological discourse (McKanan, 2017). This compendium seeks to address this gap by further elaborating on these contributions in subsequent chapters.

In the face of today's challenges, this publication advocates for a broader, more differentiated worldview and actions that transcend the often reductionist, technical solutions prevalent in sustainability efforts. Reports such as *Come On!* by the Club of Rome (Wijkman & von Weizsäcker, 2018) argue that the socio-ecological crisis also has a philosophical and spiritual dimension that must be addressed to enable systemic solutions. Similarly, longstanding environmentalists such as Paul Kingsnorth (2018) and initiatives such as the *Earth Charter* (2000) have called for integrating cultural and spiritual dimensions into sustainable development. Van Egmond and colleagues (2011, 2025) echo this perspective, informed by anthroposophy and German Idealism, and call for an integral worldview.

This compendium serves as an appeal to individuals and institutions engaged in culture, philosophy, and the humanities, as well as those shaping civil society through economic and political life. As mentioned before, it of course does not claim to offer definitive answers but rather suggests directions for thought and action—a hypothesis to be tested and further developed.

7 Expansion of Sustainable Development

Through the contributions of anthroposophy to sustainable development presented in this publication, we aim to expand the understanding and definition of sustainable development as a potential meta-discipline. The integration of inner and outer dimensions of sustainable development is increasingly a topic of debate at the United Nations level (Patterson, 2025). The absence of a cultural-spiritual dimension in mainstream approaches to sustainable development, as recently been highlighted by e.g. the Earth Charter Commission (2000), reflecting the growing recognition of the attention for such an expansion.

This expanded perspective may offer a much-needed synthesis or a step beyond the dualistic polarisation of "left" and "right" extremisms, both of which often lack a clear vision for the future where all are part of in reciprocity and dignity. A cosmopolitan world in equilibrium, rooted in the vast diversity of landscapes, cultures, local identities, and human and non-human beings—both individually and collectively—remains in our view a key aspiration for sustainable development.

References

Ametepey, S. O., Aigbavboa, C. O., Ansah, S. K., Gyadu-Asiedu, W., & Boamah, L. (2023). *Evolution, principles, and future of sustainable development: A systematic review. Preprints*, 2023020158. https://doi.org/10.20944/preprints202302.0158.v1

Baudrillard, J. (1994). *Simulacra and simulation* (S. F. Glaser, Trans.). University of Michigan Press. https://doi.org/10.3998/mpub.9904

Brundtland, G. H. (1987). *Our common future*. Oxford University Press.

Capra, F. (2002). *The hidden connections: A science for sustainable living*. Doubleday.

Carson, R. (1962). *Silent spring*. Houghton Mifflin.

Crutzen, P. J., et al. (2000). The 'Anthropocene'. *Global Change Newsletter, 41*, 17–18.

Deloitte. (2024, July). *2024 Sustainability action report: Survey findings on ESG disclosure and preparedness*. Deloitte. https://www2.deloitte.com/content/dam/Deloitte/us/Documents/audit/2024-sustainability-action-report.pdf

Dixson-Decleve, S., Gaffney, O., Ghosh, J., Randers, J., Rockstrom, J., & Stoknes, P. E. (2022). *Earth for all: A survival guide for humanity*. New Society Publishers.

Earth Charter Commission. (2000). *The Earth Charter*. Earth Charter Initiative. https://earthcharter.org/read-the-earth-charter/

Ebert, J., zur Nieden, S., & Pieschel, M. (2024). Die biodynamische Bewegung und Demeter in der NS-Zeit: Akteure, Verbindungen, Haltungen. Metropol Verlag.

Eleftheria, E., & Campos Suarez, M. (Eds.). (2024). *Inner development goals: Stories of collective leadership in action*. De Gruyter.

European Union. Directive (EU) 2022/2464 of the European Parliament and of the Council of 14 December 2022 Amending Regulation (EU) No 537/2014, Directive 2004/109/EC, Directive 2006/43/EC and Directive 2013/34/EU, as Regards Corporate Sustainability Reporting. Official Journal of the European Union, 16 Dec. 2022, L 322/15. EUR-Lex. https://eur-lex.europa.eu/legal-content/EN/TXT/?uri=CELEX%3A32022L2464

Fukuyama, F. (1989). The end of history? *The National Interest, 16*, 3–18.

Hegel, G. W. F. (1955). *Lectures on aesthetics* (F. Bassenge, Ed.). Felix Meiner Verlag.

Jhagroe, S. (2024). *Voorbij duurzaamheid: Op weg naar een ecorechtvaardige samenleving*. Mazirel Pers.

Jonas, H. (1979). *Das Prinzip Verantwortung: Versuch einer Ethik für die technologische Zivilisation*. Suhrkamp.

Krenak, A. (2023). *Interview: Ausbeutung im Amazonas. Indigener Philosoph hört auf, die Welt zu verschlingen* [Interview]. SRF. https://www.srf.ch/kultur/gesellschaft-religion/ausbeutung-im-amazonas-indigener-philosoph-hoert-auf-die-welt-zu-verschlingen

Latour, B. (2020). *What protective measures can you think of so we don't go back to the pre-crisis production model?* Unpublished

Lovelock, J. E. (1972). Gaia as seen through the atmosphere. *Atmospheric Environment, 6*(8), 579–580.

McKanan, D. (2017). *Eco-alchemy: Anthroposophy and the history of nature*. University of California Press.

Meadows, D. H. (2001). *Dancing with systems*. The Sustainability Institute. Retrieved from http://donellameadows.org/archives/dancing-with-systems/

Meadows, D. H. (2008). *Thinking in systems: A primer*. Chelsea Green Publishing.

Meadows, D. H., Meadows, D. L., Randers, J., & Behrens, W. W. (1972). *The limits to growth: A report for the Club of Rome's project on the predicament of mankind*. Universe Books.

Musk, E. (2020, May 30). *Interview*. NASA.

Nietzsche, F. (1883–1885). *Also sprach Zarathustra: Ein Buch für Alle und Keinen*. E. W. Fritzsch.

Patterson, L. (2025, January 17). Why do we collectively produce results that nobody wants? [LinkedIn post]. LinkedIn. https://www.linkedin.com/feed/update/urn:li:activity:7285694341707149312/

Scientific Council for Integral Sustainable Agriculture and Nutrition. (2012). *Towards an integral approach to sustainable agriculture and healthy nutrition: Vision of the Scientific Council for Integral Sustainable Agriculture and Nutrition.* https://www.pe-rc.nl/ridlv/Publication%20RIDLV_UK_March%202012.pdf. Last accessed January 19, 2025.

Selg, P., Gross, S. H., & Mochner, M. (2024). *Anthroposophie und Nationalsozialismus: Die anthroposophische Ärzteschaft*. Schwabe Verlag.

Statista. (2024). *Daily time spent online worldwide from 2011 to 2023.* https://www.statista.com/statistics/1380282/daily-time-spent-online-global/

Steffen, W., et al. (2015). The trajectory of the Anthropocene: The great acceleration. *The Anthropocene Review, 2*(1), 81–98.

Steiner, R. (1973). *Anthroposophical leading thoughts: Anthroposophy as path of Knowledge.* Rudolf Steiner Press.

Steiner, R. (1992). *Bilder okkulter Siegel und Säulen. Der Münchner Kongress Pfingsten 1907 und seine Auswirkungen* (3. Aufl., GA 284). Rudolf Steiner Verlag.

Steiner, R. (2013a). *Theosophie: Einführung in übersinnliche Welterkenntnis und Menschenbestimmung* (33rd ed., GA 9). Rudolf Steiner Verlag.

Steiner, R. (2013b). *Die Geheimwissenschaft im Umriss* (31st ed., GA 13). Rudolf Steiner Verlag.

Steiner, R. (2019). *Allgemeine Menschenkunde als Grundlage der Pädagogik* (10th ed., GA 293). Rudolf Steiner Verlag.

Steiner, R. (2021). *Die Philosophie der Freiheit: Grundzüge einer modernen Weltanschauung* (17th ed., GA 4). Rudolf Steiner Verlag.

Steiner, R. (2022). *Geisteswissenschaftliche Grundlagen zum Gedeihen der Landwirtschaft* (9th ed., GA 327). Rudolf Steiner Verlag.

The Guardian. (2025, January 8). *Why is Donald Trump talking about annexing Greenland?* Retrieved January 9, 2025, from https://www.theguardian.com/world/2025/jan/08/why-is-donald-trump-talking-about-annexing-greenland

United Nations. (2019, December 2). *Secretary-General's remarks at the opening ceremony of the UN Climate Change Conference COP25* [Delivered]. United Nations. https://www.un.org/sg/en/content/sg/statement/2019-12-02/secretary-generals-remarks-opening-ceremony-of-un-climate-change-conference-cop25-delivered

United Nations. (2024). *The sustainable development goals report 2024.* United Nations. https://unstats.un.org/sdgs/report/2024/

Van Baarda, T. (2000). *Antroposofie en het vraagstuk van de rassen.* Antroposofische Vereniging in Nederland.

Van Egmond, N. D., & de Vries, H. J. M. (2011). Sustainability: The search for the integral worldview. *Futures, 43*(8), 853–867. https://doi.org/10.1016/j.futures.2011.05.027

Van Egmond, N. D., & Mulder, J. M. (2025). The narrative of a new renaissance [Manuscript under peer review].

von Humboldt, A. (1845–1862). *Cosmos* (Vol. 1, p. 5f.). Digitised version and full text in the German Text Archive.

VPRO Tegenlicht. (2018, December 16). Paul Kingsnorth: Portrait of a recovering environmentalist [Video]. VPRO.

Wachsmuth, G. (1980). *Erde und Mensch: Ihre Bildekräfte, Rhythmen und Lebensprozesse* (4th ed.). Verlag am Goetheanum.

Wijkman, A., & von Weizsäcker, E. U. (2018). *Come on!: Capitalism, short-termism, population, and the destruction of the planet*. Springer.

World Commission on Environment and Development. (1987). *Our common future.* Oxford University Press.

Wulf, A. (2015). *The invention of nature: Alexander von Humboldt's new world.* Knopf.

Zoeteman, K. (1989). *Gaiasofie: Anders kijken naar evolutie, ruimtelijke ordening en milieubeheer.* VBK Media.

Zoeteman, K. (2009). *Gaia Logica.* Zeist.

Anthropsophy in the Context of Sustainable Development

Introduction

Anthroposophy is closely linked to the concept of development and sustainable development. This chapter aims to establish an initial connection between anthroposophy and sustainable development by integrating and building upon McKanan's work (2017). Following this, three additional authors will explore key aspects of anthroposophical principles, addressing fundamental questions about human nature and the Earth from an anthroposophical perspective.

It is important to note that neither anthroposophy nor sustainable development will be fully portrayed in this chapter. Instead, this section serves as an introduction to key aspects that bridge the two disciplines. Chapter 3 will further explore the concept of sustainable development, its evolution, and its connections to anthroposophy from different perspectives.

Both anthroposophy and sustainable development encompass a vast body of literature, research, and practical applications embedded in formal and informal learning institutions worldwide. Since the introduction of anthroposophy over a century ago by Rudolf Steiner, extensive research, publications, and study programs have been conducted, written, and developed. As primary literature in this field, the Rudolf Steiner Verlag, in collaboration with the Rudolf Steiner Nachlassverwaltung and archive, has nearly completed publishing the so-called "Collected Works" (CW, or "Gesamtausgabe/GA") of Rudolf Steiner's work, comprising 354 volumes (and up to 50 extra volumes concerning the arts), including over 30 written works and 6500 lectures. Furthermore, many scholars and practitioners worldwide have contributed to secondary literature, further research, and practical applications in relation to anthroposophy.

Sustainable development, as a discipline, is much younger, with its institutionalised conception emerging in the 1970s and 1980s. A substantial body of literature has since been developed in this field. A recent systemic review identified approximately 200 comprehensive definitions of sustainable development (Ametepey et al., 2023), highlighting both the diversity, the relative novelty and future directions of this discipline. The following chapter will serve as an entry point into these disciplines, their connections and exploring their potential as emerging meta-disciplines.

Anthropoposophy's Four Gifts for Sustainable Development

Dan McKanan

Abstract

For more than a century, anthroposophy has brought spiritual vitality to the global environmental movement through its influence on organic agriculture, anti-pesticide campaigns, green banking, holistic education, and intentional community. The unique gifts of anthroposophy—gifts I identify as "cosmic holism," "homeopathic social change," "appropriate anthropocentrism," and "planetary transmutation"—can deepen our shared pursuit of the model of sustainable development outlined in the United Nations' 17 Sustainable Development Goals (SDGs).

Keywords

Anthroposophy • Sustainable development • Biodynamic agriculture • Waldorf education • Intentional communities • Environmental activism • Cosmic holism • Homeopathic social change • Appropriate anthropocentrism • Planetary transmutation • United Nations sustainable development goals (SDGs) • Ecological sustainability • Holistic education • Green banking • Social therapy

1 Introduction

For more than a century, anthroposophy has brought spiritual vitality to the global environmental movement. The biodynamic movement is certainly the most visible expression of anthroposophy's care for the more-than-human world. But environmental ideals inform virtually every anthroposophical initiative, from

D. McKanan (✉)
Harvard Divinity School, Cambridge, MA, USA
e-mail: dmckanan@hds.harvard.edu

J. Kronenberg and E. T. Lammerts van Bueren (eds.), *On the Earth We Want to Live*, World Sustainability Series, https://doi.org/10.1007/978-3-031-98758-8_2

Waldorf schools to socially responsible banks to "social therapy" intentional communities, including the Camphill movement, that support adults and children with intellectual disabilities. What's more, nearly every current of environmental activism has been enriched in small or large ways by anthroposophical spirituality. The practice of organic certification, the anti-pesticide campaign of the 1960s, community-supported agriculture, and dozens of green businesses all emerged because of the creative interchange between deeply committed anthroposophists and environmentalists with other spiritual commitments.

In my book *Eco-Alchemy* (2018), I encapsulated anthroposophy's contributions to environmentalism in four "gifts." The gift of *cosmic holism*, I wrote, challenges environmentalists to expand their imaginations and attend to the widest possible circles of interconnection. Anthroposophy's *homeopathic model of social change* suggests that even the smallest acts of care for the planet can have a big impact as their effects ripple through the environment. *Appropriate anthropocentrism* affirms that human beings need not be a threat to other species but can enrich the ecological whole by bringing our full selves into mutual relationship with our neighbours. And *planetary transmutation* allows us to embrace the evolutionary processes that are constantly changing planet Earth, while resisting unbalanced and one-sided changes, such as the rapid pumping of carbon dioxide into the atmosphere.

In this essay, I will identify some of the ways these gifts, as expressed through anthroposophical initiatives, can help the world community attain the 17 "Sustainable Development Goals" (SDGs) articulated and tracked in 2015 by the United Nations (United Nations Department of Economic and Social Affairs).

2 Sustainable Development

The concept of sustainable development was first articulated on behalf of the United Nations by the 1987 Brundtland Commission, which defined it as "meeting the needs of the present without compromising the ability of future generations to meet their own needs" (World Commission on Environment and Development, 1987). Many interpreters elaborate this definition by speaking of the three "pillars" of environmental, social, and economic sustainability. The SDGs represent a still more elaborate rendition of the same idea, accompanied by a host of specific metrics and targets. In whatever form, "sustainable development" is not a perfect concept. The word "development" often implies that all societies will follow the same path of economic growth pioneered by the wealthy Western nations, yet such development was clearly not sustainable insofar as it came at the expense of non-Western cultures and more-than-human nature. For this reason, some of the economic targets used by the UN are probably incompatible with the environmental goals. By foregrounding the balance of present and future human generations, the Brundtland Commission obscured the equally important balance between humans and other species, as well as the possibility that we might also be accountable to past generations. And many spiritual leaders—but also increasingly academic scholars—argue that we will not make progress toward the SDGs unless we also work inwardly on such "transformational skills" such as "being, thinking, relating, collaborating, and acting" (Inner Development Goals, 2024). Nevertheless, the goals in their current form may be the best approximation to a consensus statement of what a positive human and planetary future might look like. As such, they can help us understand anthroposophy's potential contribution to such a future.

3 Cosmic Holism

The breadth of that contribution is aptly suggested by what I call anthroposophy's "cosmic holism." Rudolf Steiner gave classic expression to this idea in the Agriculture Course (Steiner, 1924), when he declared that "in great Nature, everything, everything is connected" (lecture 7). "We must extend our view to the whole Cosmos," he went on, and "summon all the Universe into our counsels" (lecture 1). In practice, this has shaped anthroposophical environmentalism in two distinct ways. First, it has inspired anthroposophists to consider forms of connection that might not occur to other environmentalists. All organic farmers are attentive to the connections between plants and animals and between wild and domestic organisms. But biodynamic farmers are much more likely than their counterparts to ask how their farms might connect to the empowerment of people with intellectual disabilities, to community arts initiatives, and to school curricula. It is easy for biodynamic farmers to act on these vital connections, because all anthroposophical initiatives exist within a shared ecosystem that includes farms, schools, banks, clinics, intentional communities, and arts organisations. Attention to the interplay of economic, social, and ecological dimensions is inescapable for anthroposophical initiatives.

In this sense, cosmic holism resonates perfectly with the vision of the UNs Sustainable Development Goals. The goals place such seemingly diverse items as "zero hunger" (goal #2), "quality education" (#4), "climate action" (#13), and "life below water" (#14) into a single framework, ensuring that people with a primary commitment to just one of the goals will be challenged to consider how their work might foster or impede the remaining goals. Just as anthroposophy has inspired many Camphill intentional communities to pursue "clean water and sanitation" (goal #6) with biologically based water treatment systems or "affordable and clean energy" (goal #7) with biomass or biogas electric production, so too the United Nations is challenging the urban planners responsible for "sustainable cities and communities" (#11) to also work to "reduce inequalities" (#10) and for politicians committed to "decent work and economic growth" (#8) to also consider the "good health and well-being" (#3) of workers.[1]

Yet, "cosmic holism" suggests more than simply connecting education and health and farming and finance. It also implies connecting all of these things

[1] For more information on the environmental commitments of the Camphill network of intentional communities supporting people with intellectual disabilities, see McKanan (2018, pp. 153–174) and McKanan (2020, pp. 170–175).

to a *spiritual* dimension of reality that extends from the most distant corners of the cosmos to the depths of the human heart. By this measure, engagement with anthroposophical initiatives can widen and deepen the implementation of the Sustainable Development Goals. Waldorf educators, for example, have long known that one way to open students to the spiritual dimension of reality is by incorporating the arts into every lesson plan. But arts education does not appear at all among the targets and indicators the UN uses to flesh out the goal of "quality education" (#4). Similarly, holistic and integrative medical practices are not mentioned at all in the UN's description of the goal of "good health and well-being" (#3).

The spiritual dimension of cosmic holism has special relevance in our current moment of global peril. The past few years have witnessed a global upsurge of "populist" politics, often driven by antipathy to the "neoliberal" economic policies foisted upon the world in the late twentieth century by the likes of Margaret Thatcher, Ronald Reagan, Bill Clinton, and Tony Blair. Neoliberalism seeks to measure human well-being in exclusively economic terms. By that measure it has made a real contribution to sustainable development. Since the onset of neoliberalism, the number of people living in extreme poverty has been cut in half (Development Initiatives). But too often, higher incomes go hand in hand not only with environmental destruction but also with the hollowing out of traditional communities and lifeways. Populism recognises that humans cannot live by bread alone, but the forms of populism that are currently surging seek to foster meaning and identity by scapegoating immigrants, racial minorities, and people who live beyond traditional gender roles. Cosmic holism, as fostered by anthroposophy and other spiritual traditions, offers a more loving path to transcendent meaning.

4 The Homeopathic Approach to Social Change

The *homeopathic approach to social change*, as I define it, is the practice of trusting that even small actions can catalyse larger processes of social change. Biodynamic agriculture, of course, involves homeopathy in the literal sense: the biodynamic preparations introduce minute quantities of key substances into the farm organism. The expectation is that these will encourage, rather than overwhelm, the lifegiving forces that are already present on the farm. Analogously, the biodynamic farmers at Temple Wilton Community Farm deliberately chose *not* to "scale up" their cooperative enterprise, but to preserve a farm community where all the farmers and all the consumers could sustain face-to-face relationships. This catalysed broader change in two ways. On the one hand, Temple Wilton's CSA model was available to other farmers who could scale it up to serve thousands of customers, or adapt it to meet other needs. One CSA in my neighbourhood, for example, is actually a network of more than a dozen tiny farms run by new immigrants to the United States. On the other hand, Temple-Wilton's abiding commitment to its original ideals means that those ideals will still be available to the larger CSAs when they discover that growth has caused them to sacrifice too many ideals.

Like cosmic holism, social homeopathy has special relevance right now. I am writing this in the immediate aftermath of the 2024 presidential election in the United States, which guaranteed that the US federal government will take no positive action on climate change until at least 2029. Given the structure of the US Congress, even after 2029 legislative action will likely be impossible, unless an economic or environmental catastrophe causes a total realignment of US politics. And so long as the United States is unwilling to be part of the solution, other nations may choose to castigate the US rather than devoting enough resources to offset the American shortfall. This was the dispiriting outcome of the 2024 United Nations Climate Change Conference, or COP29.

What this means is that environmentalists will need to imagine more local solutions to our global challenges. If nations won't act, what about cities, villages, or schools? What if even the tiniest organisations committed to taking homeopathic action on each of the seventeen Sustainable Development Goals? Many Waldorf schools and Camphill communities are already, in effect, doing just that (McKanan, 2018).

5 Appropriate Anthropocentrism

Anthroposophy's third gift, appropriate anthropocentrism, is rooted in anthroposophy's distinctive cosmology, which holds that planet Earth represents just one stage in a longer, cosmic process of human evolution. For environmentalists inclined to believe that the Earth is the only available home for a multi-species community of persons, this cosmology is extremely disconcerting! It is easy to worry that such a cosmology instrumentalises both the planet and the creatures that inhabit it. In practice, however, most anthroposophical initiatives invite participants into deeper symbiotic partnership with plants, animals, and fungi. Though human evolution plays a key role, it is not possible without the collaboration of beings at other evolutionary stages. For this reason, the biodynamic movement seeks to increase the number of people with a personal connection to farming. Its farming methods, such as the use of homeopathic "preparations," are inherently labour-intensive, and community-supported farms also invite consumers to participate in the life of the farm through regular visits and economic support. Biodynamic farms also host cultural festivals that help people connect their spiritual lives to the well-being of other organisms.

This human-centred approach to farming stands in contrast to global trends in both agriculture and conservation. The share of the human population involved in farming has declined precipitously over the past century, driven down by "Green Revolution" techniques that have expanded yields per farmer, at the expense of both biodiversity and rural community life. Meanwhile, the conservationist ethic of "leaving no trace" has justified the expulsion of Indigenous and pastoralist communities from national parks, and persuaded many idealistic young people that *they* are the greatest threat to the well-being of the planet.

In a broad sense, the Sustainable Development Goals also stand in opposition to these trends. The very phrase "sustainable development" implies that economic and cultural "development" is compatible with environmental "sustainability," just as the word "anthroposophy" implies that humans carry wisdom that may be healing for the cosmos. Yet in the current articulation of the goals, sustainability and development sit awkwardly alongside one another—at best in creative tension, at worst in outright contradiction. For example, the targets and indicators used to measure "sustained, inclusive, and sustainable economic growth" (goal #8) include gross national product, a measure that can reflect practices that are devastating for ecosystems. What is lacking in the goals is a clear articulation of the sorts of economic activity that will heal the planet, and the sorts of environmental activity that will enhance human well-being.

Such articulation is a central task of anthroposophical banking, as exemplified by such initiatives as GLS Bank in Germany, Triodos Bank in the Netherlands and United Kingdom and the Freie Gemeinschaftsbank in Switzerland, along with other participants in the Global Alliance for Banking on Values. Rather than measuring their success in purely numerical terms, anthroposophically inspired banks use money to build relationships—by, for example, inviting lenders and borrowers to set interest rates collaboratively after hearing one another's stories. Though one might imagine that a "green bank" would invest exclusively in things like renewable energy or recycling, the named banks are equally committed to affordable housing, human health, and cultural initiatives including schools and religious congregations. As a direct consequence, they bring together exactly the mix of perspectives needed to redefine sustainable development.

This is not work that can be done once for all. Sustainable development might mean one thing in 2025, when both human population and carbon emissions are still rising, and something quite different at the end of the century, when both may be declining quickly. Already in 1987, the Brundtland Commission affirmed that "sustainable development is not a fixed state of harmony, but rather a process of change" (p. 18). Anthroposophy's fourth gift, *planetary transmutation*, provides an intriguing framework for embracing the process of change.

6 Planetary Transmutation

Like cosmic holism, planetary transmutation is rooted in cosmological understandings that were articulated by Rudolf Steiner, but may not be fully embraced by most participants in anthroposophical initiatives. Steiner taught that traces of human evolution have already occurred on multiple pre-phases of the Earth and will potentially transform planet Earth into a "spirit-filled work of art" (Steiner, 1907 (GA284), p. 113). But he also insisted that such transformation would be healthy only if involved a truly collaborative relationship between humans and other species. Though he warned that the "human will" was connected to "the forces of dissolution and destruction inherent in our planet," he also promised that "the constructive, up-building forces of the planet earth also lie in humanity itself" (Steiner, 1919, lecture 5).

This promise speaks powerfully to the present moment, when environmentalists are confronting the fact that humanity has crossed all the bright lines marked out by the first generation of climate activists. A few decades ago, scientists projected that an atmospheric carbon level of 350 parts per million would trigger a "tipping point" after which climate change would accelerate, as shrinking forests and melting permafrost would augment human-caused carbon emissions. By the time "350.org" was organised, the world had far exceeded that target level; today, the level stands at 421 parts per million. NASA (2024) continues to argue that it "may not be too late to avoid or limit some of the worst effects of climate change," but such words ring hollow in villages that have already experienced total desertification, inundation, or destruction by fire or hurricane.

Climate change, in other words, is not a future threat but a present reality. Saving the world from mass extinction is, sadly, not an option. The current cycle of mass extinction began before we were born and will continue at least for the remainder of our lives. The concept of planetary transmutation acknowledges this but also invites us to think about how we'd like to contribute to the wave of evolutionary creativity that will surely come after the mass extinction. Do we want the new species of the future to be genetically engineered to maximise corporate profits? Or do we want them to emerge from a symbiotic conversation in which humans are no longer apex predators, but consummate cooperators?

In order to answer such questions, humanity needs to foster millions of spaces where people engage in face-to-face democratic deliberation, informed by direct encounters with the other species that are evolving along with us. Such spaces will include ecovillages, cooperative businesses, religious communities, and local governments. They will also include thousands of biodynamic farms, Waldorf schools, and social therapeutic communities. Anthroposophical initiatives, in other words, try to exemplify the social forms that humanity needs to cultivate if we are to pass through the great changes of this century with more connection, more resilience, and more love.

7 Conclusion

Anthroposophy offers profound insights for achieving sustainable development through its gifts of cosmic holism, homeopathic social change, appropriate anthropocentrism, and planetary transmutation. These principles bridge spiritual vitality with practical solutions, providing pathways for tackling global challenges. Thousands of anthroposophical initiatives are already implementing these practical solutions, and a deeper partnership between those initiatives and other social and environmental projects would enable more powerful resistance to the dominant paradigms of economic growth that stand in the way of development that is deeply ecological and fully human. Through conscious evolution and intentional community-building, anthroposophy invites us to envision and enact a world in which sustainable development becomes a shared reality.

References

Development Initiatives. (2024). *Economic poverty trends: Global, regional and national.* https://devinit.org/resources/poverty-trends-global-regional-and-national/#:~:text=Poverty%20rates%20rebounded%20slightly%20in,less%20than%20%246.85%20a%20day

Inner Development Goals. (2024). https://innerdevelopmentgoals.org/

McKanan, D. (2018). *Eco-Alchemy: Anthroposophy and the history and future of environmentalism.* University of California Press.

McKanan, D. (2020). *Camphill and the future: Spirituality and disability in an evolving communal movement.* University of California Press.

NASA. (2024). *Is it too late to prevent climate change?* https://science.nasa.gov/climate-change/faq/is-it-too-late-to-prevent-climate-change/

Steiner, R. (1907). *Images of Occult Seals and Columns.* Trans. Steiner Online Library.

Steiner, R. (1919). *The influences of Lucifer and Ahriman.* Trans. D. S. Osmond. GA 191.

Steiner, R. (1924). *Agriculture course.* Trans. G. Adams. GA 327.

United Nations Department of Economic and Social Affairs. (2024). *The 17 goals.* https://sdgs.un.org/goals

World Commission on Environment and Development. (1987). *Our common future.* http://www.un-documents.net/our-common-future.pdf

Anthropoposophical Principles and Sustainable Development—The Human "I" as Dialogical Encounter with the World

Salvatore Lavecchia

Abstract

This chapter will concentrate on the I as essential constituent of anthroposophical principles. The notion of the "I" presupposed by anthroposophy could be in fact an inexhaustible source of productive, both theoretical and practical implications regarding any discourse on sustainable development. In the horizon of anthroposophy, the I is perceived as an eminently and unrestrictedly dialogical principle, transcending any perspective or unilateral identification with the self, and thus capable of manifesting the human being as a protagonist of impulses not rooted in egoism or narcissism, but in openness and self-givingness toward every other being and the world.

Keywords

Anthroposophy • Sustainable development • Human "I" • Dialogical principle • Transperspectival consciousness • Human dignity • Ecological ethics • Epistemology • Rudolf Steiner • Perception and knowledge • Freedom and consciousness • Holistic science • Relational ontology

1 Introduction

More than 100 years ago, anthroposophy initiated a worldwide movement encompassing agriculture, education, medicine, the arts, humanities, natural sciences, and more, driven by an impulse towards spiritual, cultural, and societal renewal. This

S. Lavecchia (✉)
Dipartimento di Studi Umanistici e del Patrimonio Culturale, Università degli Studi di Udine, Udine, Italy
e-mail: salvatore.lavecchia@uniud.it

J. Kronenberg and E. T. Lammerts van Bueren (eds.), *On the Earth We Want to Live*, World Sustainability Series, https://doi.org/10.1007/978-3-031-98758-8_3

impulse was based on a coherent vision concerning the human being and human development in relation to the future of both the Earth and the whole cosmos.

Since anthroposophy is a manifold, transdisciplinary meta-discipline, inaugurated by Rudolf Steiner and further developed by many persons from all backgrounds and cultures, it will be necessary to limit this short contribution to a specific subject, intended as a productive threshold towards the exploration of as many fields of present and future research as possible. This threshold will be constituted by a concentration on the I as a central principle of anthroposophy.

2 The I as a Central Principle of Anthroposophy

Anthroposophy is "the wisdom that is spoken by the human being" (Steiner, 2001/09, p. 18). But the human being *is* the I,[1] which authorises to regard the I as the true essence of the human being (Steiner, 1994, p. 49). And again: "The I is all beings/All beings are the I" (Steiner, 1999/1908, p. 36). These three concise formulations paradigmatically emphasise the centrality of the I as principle of anthroposophy. This centrality is connected with the spiritual nature of the I: in the context of a threefold anthropology, like that presupposed by anthroposophy (Steiner, 1994), the human being is constituted by body, soul, and spirit (Heusser, 2016; Heusser & Weinzirl, 2014; Weger & Wagemann, 2021; Weger et al., 2021), whereas the I, that is the spiritual individuality of the human being, transcends body and soul, and therefore manifests itself as the original fundament regarding all dimensions of the human constitution.

While body and soul are in their true essence oriented towards the I, whereas the I lives in body and soul, the I is oriented towards the spirit, so that the spirit can pervade the I, living in the I, and herewith bestowing to the I the character of eternity (Steiner, 1994, pp. 50–51). Through the presence of the spirit, the I is capable of participating consciously and individually in the reality which, in the horizon of anthroposophy, is the origin and fundament of all forms of consciousness and being, which explains the aforementioned identification of the I with all beings.

3 The Intrinsic Openness and Self-Givingness of the I

The just emphasised centrality of the I can be perceived as contradictory with any discourse regarding sustainability if the reality of the I is identified with the current representation of the I, corresponding to the image connected with solipsism as characterised by Ludwig Wittgenstein: "The I in solipsism shrinks to an extensionless point and there remains the reality co-ordinated with it" (Wittgenstein,

[1] I prefer the translation of "das Ich" as "the I" to the translation as "the ego" – since "das Ego" has in German univocal negative connotations pointing to egoism or narcissism—or "the self"—since the notion of self presupposes a reference to reflexivity and frontality.

1922, § 5.64). This notion coincides with the representation of the I as constituting "the limit—not a part of the world" (Wittgenstein, 1922, § 5.641), that is as confined into the limits of the first-person perspective. In our encounter with the world, we habitually tend to, in fact, represent the perceiving, thinking, and understanding I as a point localisable within our body, whose interiority we understand as separated from the exteriority of the world. Consequently, we tend to associate the modality of perception actualised by our I with the perspective of a tunnel vision (Metzinger, 2009), which hopelessly splits the first-person perspective from the second- and third-person perspectives, separating a postulated mere subjectivity of the perceiving consciousness from a postulated purely objective being. The contrastive dualism between the I and the world is thus originated, which usually determinates and configurates every field of knowledge and action. Exactly this dualism is contradictory with any discourse regarding the notion of sustainability, since it presupposes an I encapsulated in its own subjectivity and thus striving towards univocal self-affirmation. The here intended solipsistic I is therefore constitutionally incapable of an unrestricted openness towards "meeting the needs of the present without compromising the ability of future generations to meet their own needs" (United Nations General Assembly, 1987, Chap. 2, para. 1). However, some exemplary passages from Steiner's work evidently show that the reality of the I presupposed in anthroposophy points to the contrary of the solipsistic notion of the I. Steiner emphasises namely that the true nature of the I is intimately connected with love:

> You have to lose your earthly I in order to get a sight of your real I. And the person who does not develop [...] self-givingness cannot approach this true I. [...] The true I does not want to be sought, if it is to become manifest, if it is to reveal itself; and it hides itself if it is sought. For it is only found within love. And love is the self-givingness of one's own being to the extrinsic being. As a consequence, the true I has to be found like an extrinsic being. (Steiner, 1986, lecture of 22. April 1923, p. 142, transl. S. Lavecchia)

This implicate that the terrestrial I is in itself capable of transcending the first-person perspective tending to solipsism; otherwise, it would be intrinsically undisposed towards substantiating the unrestricted self-givingness leading to the manifestation of the true I. This innate disposition towards self-ginvingness is implicitly affirmed by Steiner through some characterisations regarding the human sensory organism. According to Steiner, said organism is inherently oriented towards the perception of another, extrinsic I:

> It can be recognised from the I-experience that the human being forms an organism [...] which can make present in itself the image of an equal extrinsic I. What shapes itself as such an organism can be regarded as "typus" of a perceptive organ. (Steiner, 2002, p. 186, transl. S. Lavecchia)

This means that already the activity of the embodied (the "earthly") I involved in every current sensory perception can be truly understood not on the basis of the determination through the first-person perspective, but, on the contrary, only presupposing with regard to the perceiving I a constitutional disposition to unrestrained openness and self-givingness towards an extrinsic being. Steiner suggests that this disposition is quintessentially realised by the activity of the perceiving I in the encounter with another I, so that in this encounter the whole human organism manifests itself as the archetype, that is as the original form of every sensory organ (Lavecchia, 2022, pp. 37–43). The aforesaid activity is connected by Steiner with the "sense of the I" (Lavecchia, 2022, pp. 79–84; Lavecchia, 2024), which consists precisely in the perception of another I:

> this is the relationship that exists when one person faces the other, perceiving the I: self-givingness towards the perceived person (HIngabe)/inner defence (Abwehr). [...] I am not talking about the emotional life, but only about the perceptive relationship. The soul vibrates; it vibrates sympathy/antipathy [...] As sympathy develops, you sleep into the other person; as antipathy develops, you wake up [...]. This is a very short-lived alternation between waking and sleeping in vibrations [...]. (Steiner, 2019, lecture of 29. August 1919, pp. 134–135, transl. S. Lavecchia)

As Steiner affirms in the first appendix to the 1918 edition of *The Philosophy of Freedom* with reference to the I perceiving a thinking person, the here presupposed self-givingness implicates that "the other thinking takes the place of my thinking," which in turn means that "the separation between the two spheres of consciousness is actually eliminated," so that I "experience my own consciousness just as little as I experience it in dreamless sleep;" only, no unconsciousness occurs through the perception of the other I, on the one hand because the other content of consciousness takes the place of my own content of consciousness, on the other hand because "the alternating states between extinguishing and relighting of the consciousness of myself follow each other too quickly to be usually noticed" (Steiner, 1994, pp. 260–261, transl. S. Lavecchia). The here intended elimination of the separation between the spheres of consciousness—which Steiner emphasises also concerning the perception of a speaking human being through the perceiving I (Steiner, 2002, pp. 196–197)—explains why the I is in general capable of attaining true knowledge of an exterior being, which capacity would be absent if the I would not be able of letting the other being *dwell* in the own sphere of consciousness, thus revealing itself as potential identity with all beings (see Steiner, 1999, p. 36).[2]

On this basis, it can be asserted that in the horizon of anthroposophy an unrestricted openness and self-givingness—which makes possible any gesture of authentic attention, care, and love—is intrinsically connected with the I not only with regard to its reality transcending the embodiment, but also with respect to its embodied dimension, manifesting itself in sensual perception as well as in the

[2] In *Theosophy* Steiner affirms that "the I receives its essence and meaning from that with which it is connected" (Steiner, 1994, p. 50). However, this is not a concession to radical determinism, since the most essential characteristic of the I is the faculty of determining through the own free will the orientation of its spiritual, psychical, and physical connections.

possibility of attaining knowledge. As we will see shortly, this reveals that the I, as intended in the context of anthroposophy, is an intrinsically, eminently dialogical principle, whose presupposition enables to build a solid foundation for any discourse on sustainability capable of involving the human being as its co-creative actor.

4 A Selfless "I": on the I as a Dialogical, Transperspectival Principle

Steiner's characterisation of the perceiving I as intrinsically associated with an unrestricted openness is indeed consonant with any unprejudiced observation regarding our embodied experience. In fact, the above considered image of a localisable point constrained in the first-person perspective contradicts even the most elementary consideration concerning perception: in every perception we are actually unable to localise the perceiving I at any place in space or moment in time, neither within a supposed merely subjective interiority, nor in a supposed objective exteriority (Scheurle, 2013). Additionally, both modalities of localisation would involve absurd consequences: in the case of identifying the I with a merely subjective interiority—with a point shrunk in itself—, perception would be a mere projection of our subjectivity, without any authentic encounter with the perceived being; in the case of reducing the perceiving I to external determining factors, the I would be an illusionary no-one (Metzinger, 2003), and we would have no reason to be certain of any evidence manifested by this no-one (Lavecchia, 2014d). Consequently, the most consistent image of the perceiving I is that of a non-localisable, non-physical, dimensionless (i.e., *spiritual*!), and thus *selfless* point—or, more precisely, *centre of attention* (Ganeri, 2017)—which, transcending through its unrestricted openness and self-givingness any first-, second-, and third-person perspective—both singular and plural—, instantaneously generates an unlimited *sphericity* illuminating the perceived being, offering to every being the possibility of manifesting itself, whereas the point and the sphere build an instant unity, not a temporal succession (Lavecchia, 2014a, 2014b, 2014c).

The herewith presupposed eminently *dialogical and trasperspectival* reality of the I, which is perfectly consonant with the suggestions concerning the I coming from anthroposophy—especially from Steiner's characterisation of the "sense of the I"[3]—, preserves both the indispensable openness of the I towards the perceived being and the equally indispensable remaining centred in itself characterising the perceiving I—which in fact mirrors the rhythmic polarity to which Steiner points

[3] The characterisation contained in Steiner, 2019, pp. 134–135 excludes that the "sense of the I" is to be intended as a "sense of the You" or "sense of the We"—which would restrict it to the second person singular or first person plural perspectives—, since for the perception of the other I Steiner creates here the verb ichen, which unequivocally indicates that on the one side the other I is perceived as I, on the other side this perception transcends every perspective (otherwise the reciprocal interpenetration of the involved spheres of consciousness would be impossible).

in the above quoted characterisation regarding the "sense of the I" (Steiner, 2019, pp. 134–135). In the absence of the first, the I would perceive and act as a solipsistic subject; without the second, the I would dwindle till to the status of an inessential no-one.

In effect, like Steiner's characterisations regarding the perceiving I, an unbiased examination of the perceptive experience shows that already our embodied I, as we already saw, has to be considered as an intrinsically selfless reality, that is as transcending any unilateral identification with a self, and therefore contradicting any tunnel vision. On the contrary, the image of a tunnel vision, presupposing the identification of the I with the image of a localisable point, implicates a hopeless separation of the human being from any possibility of attaining a knowledge really significant both for the human self and for the world, whereas the I would thus reveal itself as an eminently undialogical, subjectivistic principle, incapable of an unprejudiced encounter with phenomena in whatever dimension of life.

The image of the perceiving I as non-localisable, dimensionless and selfless centre/sphere, to which Steiner's suggestions implicitly point, instantaneously manifests an intrinsic, unrestricted *dialogicity* in the encounter with every kind of phenomena, that is in the encounter between interiority and exteriority, subiectivity and objectivity, individual and community. According to this premise, if not conditioned through conceptual prejudices or preconceived representations—like the habitually internalised representation of the localisable point shrunk in itself—, the I encountering the world reveals itself as constitutionally capable of preserving in its consciousness both its own autonomous self-awareness/-transparency and the possibility for the perceived world of autonomously manifesting its own transparency. This constitutionally dialogical and transperspectival I, which is *the* central principle of anthroposophy, can be assumed as a consonant fundament for any reasonable notion of sustainability, which notion is necessarily oriented towards a fecund harmony between the I and the world, neither tending to collectivistically annihilating the individuality of the human being nor affirming any narcissistic form of individualism. The elaboration of a conception of the human being relying on such a fundament was an essential concern in Rudolf Steiner's research, since a dialogical, trasperspectival I is the indispensable presupposition for the elimination not only of any dichotomy between interiority and exteriority, subjectivity and objectivity, individuality and relationality/universality, but also between perception and knowledge on the one side, ethics on the other side. Exactly the elimination of the last-mentioned dichotomy has constantly built the most profound intention in Steiner's work—as already Steiner's earliest philosophical writings evidently testify (Traub, 2011)—, and it may be still nowadays regarded as the intention which builds the fundament for any productive impulse of anthroposophy. For this reason the last considerations presented in this contribution will be dedicated to this theme, briefly concentrating on the conception of science presupposed in anthroposophy.

5 Preserving the Dignity of Both the Human Being and the Earth

Precisely the one-sided identification of the I with the first-person perspective has caused the dualistic dichotomy between I-consciousness and the world, on the basis of which the current conception of science strives towards excluding the operativity of the I from all scientific activity. In the horizon of anthroposophy the overcoming of said dichotomy—which implies the dichotomy between perception and concept—can take place through the unrestricted trust in thinking as a potentially vigilant and free activity which, acting beyond subjectivity and objectivity, stimulates the I to practise its constitutional self-givingness towards extrinsic being and, consequently, towards the world, whereas thinking reveals itself herewith as a spiritual modality of love (Steiner, 1995). The practice of thinking helps the I to develop a modality of consciousness transcending the aforementioned dichotomy, which is associated exclusively with the physical organisation of the human being (Steiner, 1995, Chapter V). The I is, on the contrary, in itself free from that dichotomy, since its nature transcends the physical organisation, being inherent to the dimension of reality which originates the principles governing the world of phenomena (Steiner, 1984, p. 135–144); thinking is in turn the activity capable of connecting the physical organisation with that dimension of reality, bridging the gap between the embodied I and the I as reality which intrinsically transcends and funds subjectivity and objectivity.

Building on these premises, authentic science is only present when scientific activity takes the thinking and knowing I seriously as its own essential and inexhaustibly generative source. The here regarded science will always, in other words:

> first recognize the only point at which perception and concept coincide and at which we can directly grasp reality in its completeness. This point consists in the perception of the I in pure thinking, the only point at which perception and concept coincide and at which we can directly grasp reality in its completeness. This point consists in the perception of the I in pure thinking, whereby perception itself consists of thinking the logical activity of thinking itself. In this experience, the dichotomy between thinking and perception, i.e. between subject and object, is overcome. (Borghi & Nani, 2024, pp. 66–67, transl. S. Lavecchia)

Science—not excluding the knowing I, hereby eliminating itself as a consequential concept!—can receive through this premise a truly immovable foundation. Through the here intended science, which recognises the centrality of the human experience (Heusser, 2016)—thus being consonant with some recent tendencies concerning the notion of science (Frank et al., 2024)—, the human being manifests itself as an:

> organ through which nature reveals its secrets, because an ideal reality reveals itself [...] when it [scil. the human being] observes a fact. This view of the world, which recognizes the essence of things in ideal realities and understands knowledge as a living into the essence of things [...], does not regard objective truth as something given in the external world,

> but as something that can really be grasped within the interiority of the human being. The opposite world view places the fundaments of things behind the phenomena, in a realm beyond human experience. This world view can either indulge in a blind faith in these fundaments, which receives its content from a positive religion, or it can formulate intellectual hypotheses and theories about the nature of this otherworldly realm of reality. (Steiner, 1987, pp. 334, 338–339, transl. S. Lavecchia)

The here presupposed notion of science would not lead to the revival of an unsustainable or destructive anthropocentrism; on the contrary, it would eradicate from the self-perception of the human being that representation of the I—the image of a localisable point constrained in the first-person perspective—producing the existing negative anthropocentrisms. In other words, it would manifest the human being as constitutionally characterised by an unegoistic orientation towards a dialogical, transperspectival self-transcendence, that is by unprejudicial, generative openness towards harmonious—and thus sustainable—relations with every being.

The conception of the human being on which the here intended science is grounded reveals that the fundament for sustainability is intrinsically present already in the embodied I of *every* human being, manifesting that every human being, transcending any separation of knowledge and ethics, *is* constitutionally a transperspectival, and thus eminently *both dialogical and ethical* being, whose striving for true knowledge and authentic well-being cannot be separated from a harmonious, intrinsically *ethic* relationship with the world. This would constitute a relevant premise for genuine creativity in coping with present crises. It would namely encourage confidence in the fact that *every* human being is intrinsically able to autonomously generate sense and striving for sustainability. Crises would be encountered much more efficiently if individual awareness of the aforesaid premise as well as confidence in free individual initiative would be fostered, instead of implementing more and more standardising and collectivising dynamics. Univocally reducing the human I to the first-person perspective, those dynamics presuppose namely that individual well-being is in itself egoistically conflicting with the well-being of the Earth. Predominant approaches to crises are, in fact, not based on confidence in free individual initiative, since they have no confidence in the reality of the I. However, if taken seriously, this lack of confidence leads to manifest absurdities: why should we trust in any conceptual elaboration, epistemic or ethical decision, or economic enterprise of a non-real I? Why should we, in general, trust in someone or something, since only a real and free I can decide to trust or not to trust?

6 Conclusion

The absurdities which the elimination of trust in the reality of the perceiving, thinking, understanding, deciding, acting, enterprising I implies, lead to the irreducibility of freedom (Ebersbach et al., 2019; Esfeld, 2020; Tewes, 2017). They further lead to the irreducibility of consciousness to any form of being, and consequently, to any form of standardisation, algorhythmisation, and computation (Bayne, 2010; Bitbol, 2014; Chalmers, 1996; Dainton, 2011).

Is the radicality of present crises maybe intrinsically associated with an increasing mistrust in the fact that in a world governed by peace and harmony—that is in a world characterised by true sustainability—consciousness determines being, and not the contrary? Would genuine sustainability be possible if we would not perceive the untranscendible irreducibility of human consciousness and freedom? Recognising, acknowledging, and affirming in every field of activity the irreducibility and freedom of the human I, as well as its intrinsically dialogical and transperspectival character, as *the* own central principle, anthroposophy manifests itself radically consonant with the anthropological and epistemological premises founding a notion of sustainability, integrating the human being and the human dignity as its most essential protagonists. Preserving the human dignity as central value, the here intended notion of sustainability is the only one authentically capable of preserving the dignity of the Earth. In fact, how could it be possible to preserve the dignity of the Earth while neglecting the presence and dignity of the only being capable of consciously and creatively encountering the Earth through the unrestricted, loving self-givingness of a dialogical principle?

References

Bayne, T. (2010). *The unity of consciousness*. Oxford: Oxford University Press.

Bitbol, M. (2014). *La conscience a-t-elle une origine? Des neurosciences à la pleine conscience: une nouvelle approche de l'esprit*. Paris: Flammarion.

Borghi. L., Nani, D. (2024). *Aurora und Zentaur. Wirklichkeit und Beschreibung in der Ära der Technik* (pp. 66–67). Dornach: Verlag am Goetheanum

Chalmers, D. J. (1996). *The conscious mind. In search of a fundamental theory*. New York, Oxford: Oxford University Press.

Dainton, B. (2011). *The phenomenal self*. Oxford: Oxford University Press.

Ebersbach, R., Kettner, M., & Weger, U. (eds.). (2019). *Freiheit?! Freiheitsbewusstsein—Neurowissenschaftliche Tatsachen—Gesellschaftliche Bedeutung*. Würzburg: Könighshausen und Neumann.

Esfeld, M. (2020). *Science and human freedom*. Berlin: Palgrave Macmillan /Springer.

Frank, A., Gleiser, M., & Thompson, E. (2024). *The blind spot: why science cannot ignore human experience*. Cambridge, Mass: Cambridge University Press.

Ganeri, J. (2017). *Attention, not self*. Oxford: Oxford University Press.

Heusser, P., Weinzirl, J. (Eds.). (2014). *Was ist Geist?* Würzburg: Königshausen&Neumann.

Heusser, P. (2016). *Anthroposophie und Wissenschaft. Eine Einführung*. Verlag am Goetheanum, Dornach. English: *Anthroposophy and science*. Frankfurt a.M.: Peter Lang.

Lavecchia, S. (2014a). Agathological realism: Searching for the good beyond subjectivity and objectivity or on the importance of being platonic. *Etica & Politica, 16*, 533–549 (https://www.openstarts.units.it/bitstream/10077/10742/1/LAVECCHIA.pdf). In G. De Anna, & R. Martinelli (Eds.), *Moral realism and political decisions* (pp. 29–50). Bamberg: University of Bamberg Press 2015.

Lavecchia, S. (2014b). Das Ich und das Gute. Ansätze einer Licht-Philosophie in Anknüpfung an Novalis und Platon. *Perspektiven der Philosophie, 40*, 9–46.

Lavecchia, S. (2014c). Trasparenza o illusione dell'autocoscienza? Metafisica della luce e neurofilosofia. In F. Fabbro (Ed.), *Neuroscienze e spiritualità. Mente e coscienza nelle tradizioni religiose* (pp. 74–91). Roma: Astrolabio-Ubaldini Editore.

Lavecchia, S. (2022). *Ich als Gespräch. Anthroposophie der Sinne.* Stuttgart: Freies Geistesleben.

Lavecchia, S. (2024). Ich-Sinn und Ich-Bild. Wahrnehmen jenseits von Innen und Außen. *STIL. Goetheanismus in Kunst und Wissenschaft, 46–4*, 71–77.

Metzinger, T. (2003). *Being no one. The self-model theory of subjectivity.* Cambridge, Mass: MIT Press.

Metzinger, T. (2009). *The ego tunnel: The science of mind and the myth of the self*. New York: Basic Books.

Scheurle, H.J. (2013). *Das Gehirn ist nicht einsam. Resonanzen zwischen Gehirn, Leib und Umwelt.* Stuttgart: Kohlhammer

Steiner, R. (1984). Die psychologischen Grundlagen und die erkenntnistheoretische Stellung der Anthroposophie. In R. Steiner (Ed.), *Philosophie und Anthroposophie. Gesammelte Aufsätze 1904–1923* (Gesamtausgabe 35) (pp. 111–144). Dornach: Rudolf Steiner Verlag.

Steiner, R. (1986). *Was wollte das Goetheanum und was soll die Anthroposophie?* (Gesamtausgabe 84). Dornach: Rudolf Steiner Verlag.

Steiner, R. (1987). *Einführungen zu Goethes naturwissenschaftlichen Schriften* (Gesamtausgabe 1). Dornach: Rudolf Steiner Verlag.

Steiner, R. (1994). *Theosophie. Einführung in übersinnliche Welterkenntnis und Menschenbestimmung* (Gesamtausgabe 9). Dornach: Rudolf Steiner Verlag.

Steiner, R. (1995). *Die Philosophie der Freiheit. Grundzüge einer modernen Weltanschauung* (Gesamtausgabe 4). Dornach: Rudolf Steiner Verlag.

Steiner, R. (1999). *Mantrische Sprüche. Seelenübungen, Band II, 1903–1925* (Gesamtausgabe 268). Dornach: Rudolf Steiner Verlag.

Steiner, R. (2001). *Anthroposophie—Psychosophie—Pneumatosophie* (Gesamtausgabe 115). Dornach: Rudolf Steiner Verlag.

Steiner, R. (2002). *Anthroposophie. Ein Fragment* (Gesamtausgabe 45). Dornach: Rudolf Steiner Verlag.

Steiner, R. (2019). *Allgemeine Menschenkunde als Grundlage der Pädagogik* (Gesamtausgabe 293). Dornach: Rudolf Steiner Verlag (Taschenbuchausgabe).

Tewes, C. (2017). *Libertarismus, Willensfreiheit und Verursachung*. Frankfurt: Vittorio Klostermann.

Traub, H. (2011). Philosophie und Anthroposophie. Die philosophische Weltanschauung Rudolf Steiners. Grundlegung und Kritik. Stuttgart: Kohlhammer.

United Nations General Assembly (20 March 1987). Report of the world commission on environment and development: Our common future. Retrieved on January 1, 2025. From https://sustainabledevelopment.un.org/content/documents/5987our-common-future.pdf

Weger, U., Wagemann, J. (2021). Soul, spirit, and consciousness in psychology and philosophy: Traditions, current views, perspectives. *European Psychologist, 26*(2), 71–72.

Weger, U., Sparby, T., & Edelhäuser, Fr. (2021). Dualistic and trichotomic approaches in psychological enquiry. *European Psychologist, 26*(2), 85–95.

Wittgenstein, L. (1922). *Tractatus Logico-Philosophicus.* Transl. Fr. Plumpton Ramsey. London: Kegan Paul, Trench, Trubner & Co.

Who Is the Human Being? Exploring a Question Without an Answer

Christine Gruwez

Abstract

Only recently has the question, 'who is the human being?' become a part of human reflection and self-understanding. There is no answer to it! Raising the question is already answering it. For human being is the answer, in the same way as he is the question. Ensuing the first, a second question came up. No longer human being as ontologically one and undifferentiated in itself, no longer 'who is the human being', but 'what is it?' What human being is, was answered in successive narratives, from mythologies, where the Gods gave the human being its place in the different epochs, where individualities appeared as landmarks in a continuing process of self-exploration. In the Western narrative, individualities as Aristotle, Pico della Mirandola, Kant, and others contributed to this inquiry, each opening a window on how to make those two questions (who-what) meet. It was Rudolf Steiner who introduced the faculty of human being to think the possible as a new category of human cognition. In our times, the movement gathering around the Anthropocene, for the first time surpassed the boundaries of nations and people. Are we at the brink of a new narrative?

Keywords

Anthroposophy • Human I • Self-understanding • Anthropocene • Human dignity • Logos • Ethical individualism • Potentiality • Becoming • Perception and cognition • Rudolf Steiner • Philosophical anthropology • Intuition • Transdisciplinary inquiry • Enlightenment and self-awareness

C. Gruwez (✉)
Faculty of Comparative Science of Religion, University of Leiden, Leiden, Netherlands
e-mail: christine.gruwez@gmail.com

J. Kronenberg and E. T. Lammerts van Bueren (eds.), *On the Earth We Want to Live*, World Sustainability Series, https://doi.org/10.1007/978-3-031-98758-8_4

1 Introduction

Some websites areas ask you to prove that you are human. Most of the times you have to identify some distorted characters from the alphabet or find a missing element in a given picture. Last time, when it happened to me, I had to recognise a particular detail in some four different pictures. I could not do so! I could not discover that one single detail,—dots in a circle-, in any of the 4 examples. Nowhere was it to be found. To my surprise this proved to be the correct answer.

Who is the human being and how do you prove that you are a human? There is some irony in the fact that I have attested my being human by not being able to answer a simple question posed to me. And that my failure to do so, was the only proof possible! But this proof is merely the answer to quite another question: what the human being is.

Something similar happens when one is asked: 'who is the human being'? Also there, I must confess that I fail to come up with an answer. And that this makes me enter in all kinds of descriptions, for instance of what the human being is capable of or is failing to enact. Only, in going that way, without being so aware, I have already changed the initial question. The question is no longer who the human being is, but it is now: what the human being is. My attention has switched from the essence of being human, towards the faculties that can be connected to this essence, the so-called 'accidentia', all what can be attributed to an essence. And it is this second question that has turned out to become more and more central to research in various disciplines till today, where the environmental science is becoming different movements as well. For instance, "Dan McKanan has created a most practical and informative guide for activists of all faiths and persuasions seeking to determine the options available for guiding the future relationship between humanity, the living planet and the cosmos" (Hutchinson, 2018, Para. 4). A continuing investigation that found its expression in successive and different narratives: from myths, theology, philosophical parables and reflection, to sociology, psychology and anthropology.

In dealing with the so-called Anthropocene, where Anthropos means human, it seems little surprising that this fundamental question is showing up more and more. If human activity is the source of Earth destruction and social injustice, the search for who the human being is, is almost not negatable any longer.

2 Paradise Lost?

Narratives about human being belong to all people and cultures around the globe as they have unfolded in cosmogonies and mythic tales. Many present themselves as a kind of 'paradisiacal landscapes' on Earth, created by divine powers in which human being has been installed in a place, determined by his creators. In those surroundings, humankind appears as an integral part of what is around him and he has to act in accordance with the place that had been given to him. Who is he? It is this place, destined by higher powers, which had to tell him who he is. And in realising which place the Gods had given him, there was an immediate understanding of who the human being is.

The most compelling narrative with its many ramifications in Western thinking and self-understanding is that of the garden of Eden, the paradise as a state of 'Edenic' being, in which humankind lived in harmony as a creature among all other creatures (e.g., Agamben, 2020). And of which he was chased away, a moment of harsh awakening as the once given harmony had broken down. For "Paradise lost" the title of the epic poem by John Milton (1608–1674) at the dawn of Modernity, means to have lost at the same time the place once granted to human being:

> Man is the living being that has been expelled from his own dwelling place, who has lost his originary place. He is on earth doubly peregrinus: not only because his eternal life will be in the celestial paradise, but also and above all because he has been exiled from his Edenic homeland. (Agamben, 2020, section 1.7)

Aristoteles defined human being as a ζώον (zoo-ön), a living being, apt to live in the polis, as in a community with its own intrinsic order, hence 'zoo-ön politikon' (Bini, 2009; Aristotle, Politeia, [Sect. 1252b-1253a]). Living being, ζώον (zoo-ön) is often translated by 'animal' although the original meaning underlines the element of aliveness, an alive being (Bini, 2009). Interestingly, in Greek language till today, there are two words in use for 'life': zoè and bios. Bios means natural life, while zoè is life in its intrinsic quality. But what according to Aristotle distinguishes humankind from all other living beings, such as the bee, is that he disposes of the gift of speech. Aristotle, as quoted by Tricot clearly states that the human being possesses the word, he is able to speak, - logos-: 'λόγον ἔχον, logon echon' and thus he is able to distinguish between moral qualities such as good and evil (Tricot, 1970, p. 29). It is this capacity of distinction, which raises man above the status of the animal reign.

> (…) but speech is designed to indicate the advantageous and the harmful, and therefore also the right and the wrong; for it is the special property of man in distinction from the other animals that he alone has perception of good and bad and right and wrong and the other moral qualities, (…) . (Aristotle, Politeia, [Section 1252b-1253a]).

Taken out of its general context, the definition of Aristoteles could be understood as an approach answering this most fundamental question. Yet the initial question to Aristoteles was not: who is human being, but: who is worth of living in a 'polis'? The answer lies in this 'special property of man' and points to a description of the human ability to distinguish, not to the being itself. In that way, at least regarding the development of civilisation in the Western world, a mindset for further interpretations was introduced. Not what unites humankind with his surroundings, but what distinguishes him from them, will become the new criterion. A criterion later to be followed by what separates humankind from his surroundings. Those criteria belong to the vast and changing landscapes human beings build around them, as gardens where to read and to earn their own story. Only: new interrogations followed, some with far reaching consequences. Does the human being own his surroundings? Also the living beings which are part of them? And as an owner, does he have the right to dispose of them?

3 Paradise Found?

At the end of the fifteenth century in Europe, a new narrative appeared, suddenly lighting up the sky as comets do. It was one of those moments, brief and intense, when the opacity of mind was momentarily lifted, and the burden of heart went illuminated from within. In the midst of the whole creaturely world, the human being was given a radical new place: a non-place, but not as an ou-topos, a utopia! The young Giovanni Pico della Mirandola (1463–1494) was the first to address this special position of humankind in an unrivalled way in his speech On the Dignity of Man (Mirandola, 1486). A speech that he had intended to give on January 1487 at the age of 23. With this speech, he wanted to open a symposium organised by him in Rome, to which representatives from all religions or worldviews known at the time, had been invited. He had written 900 theses for this congress, organised by topic, with each thesis formulated from a different point of view. In the chapter about his own opinions, he advances the thesis that one should not believe an article of faith, because it pleases him (Farmer, 1998, pp. 430–431).

At the turn of the modern era, this project must be understood not only as a sum of all previously developed worldviews, but also as a first contribution to a newly emerging self-understanding: What does it mean to be human and what is the place human being is occupying in the world? What do all people have in common with each other in such a way that this can be seen as a fundamental principle of their being human? The answer is the most stunning: what is common to all human beings is that they have no pre determinate place. No place has been given to them. And in this not having a place, the dignity of all men is anchored.

Humankind is the only being who can raise the question who he is, in other words: who the human being is. And there is no other one to whom this question can be addressed. The human being has only himself to ask about himself. Even when I am submitting this question to a friend, I am directing my question to him as a human being. Exactly in the same way as if I were asking it to myself.

No factor external to his own being, no instance overviewing humankind and its history, no higher nor lower authority, are entitled to replace man's own questioning. Let alone answering to it. Not even man's own acting and the manifested results of it, can offer him an answer. For all possible answer that does not take its origin in man's being, can only be an answer to the question what human being is. Yet, man does not 'have' a question and an answer. He is the question, just as he is the answer. When it comes to the initial 'who', that is to the being, the being itself is the question and the answer in one. At the centre of his message,

Pico della Mirandola places the unlimited potential of the human being together with his inherent dignity. In all human beings, without any exception, dignity is the principle they have in common.

In the innermost sense, dignity is linked to the fact that every human being disposes of the potential to realise what is already at disposition in him. Dignity is rooted in this possibility, which must be understood as an essential potency, remaining completely independent of what a person realises—or not—from this potential. For his being as human is the receptacle of his dignity. They are one, in the same way as question and answer concerning who the human being is, are one.

In view of his audience and his adversaries, Giovanni Pico della Mirandola had to clothe his understanding of 'who the human being is' in a theological parable, a narrative so to say. Yet he stresses his being different from other narratives, when after praising the human being as the greatest miracle of creation, Pico della Mirandola addresses the question of how such a statement could be justified:

> A great miracle, Asclepius, is man. As I pondered the meaning of these sayings, I was not satisfied with all that is said by many about the excellence of human nature. In his discourse, Pico ultimately comes to the point where God the Father, as creator and artist, has not provided a place for man in the whole of creation, has not given him a purpose, but has given him the opportunity to determine his own place. Man is called to be the faber sui, the sculptor of himself:
>
> At last, the supreme artist decided that he to whom he could give nothing of his own would have a share in all that the individuals had each had for themselves. So he was satisfied with man as a creature of indeterminate form, placed him in the center of the world and addressed him thus: "We have given you no fixed abode, Adam, no appearance of your own, nor any special gift, so that you may have and possess the abode, appearance and gifts that you yourself choose, according to your desire and decision. The nature of the other creatures is firmly determined and limited within the laws prescribed by us. You shall determine your own without any restriction and narrowness, according to your own judgement, to which I have entrusted you. I have placed you in the center of the world so that from there you can look around more comfortably at what there is in the world. We have created you neither heavenly nor earthly, neither mortal nor immortal, so that you, like your own creative sculptor, freely deciding in honour, can shape yourself into the form you prefer. You can degenerate into the lower, the animal; but you can also be reborn into the higher, the divine, if your soul so decides. (Pico della Mirandola, as cited in Gruwez, 2023, pp. 169–171).

The opening speech Pico intended to deliver, as well as the Congress, never came to realisation. Immediately before, Pico was banned from Rome as the orthodoxy of the theses had to be verified. Suspected of heresy, he had to flee! His theses were condemned in a bull by Pope Innocent VIII.

Dora Baker in concluding her portrait of this exceptional destiny, declares Pico to be our contemporary (Baker, 1983, p. 144). Yet too soon the comet had to leave our sky. But some comets do return.

4 A Door for Ever Closed?

In the words of Giovanni Pico della Mirandola, no place, neither in heaven nor on Earth is owned by the human being, unless he, as the 'sculptor of himself', creates it. This could suggest that humankind can rise himself above all other living creatures with the aim of putting them at his own service. But to become the sculptor of himself is to 'carve away' what still is obscuring the essence of your humanity. It is an exercise in limiting oneself out of a free decision, even then when it comes to accept the boundaries of higher cognition (Baker, 1983, pp. 141–142).

At the time Immanuel Kant was founding his system, the notion of human as 'faber sui', together with the inherent concept of creativity, had been replaced by a system of concepts, the twelve categories, already present in the mind and thus not derived from sensorial experiences (Kant, 1982, pp. 116–118). Therefore, they were called 'pure' and were to be applied to make the world and human being 'understandable'. They were necessary to think along the ratio, and by doing so, they allowed to judge what was there outside in the world (Kant, 1982, pp. 116–118 What at the time of Pico, opened new horizons, was soon lost after the wave of the Kantian revolution engulfed the outposts of thinking in the western world. The paradigm of progress had entered. (Gruwez, 2013, pp. 209–211). Together with the idea of progress, all immediacy in experiencing the world was excluded. Any intuitive access to the world was blocked and what by Kant was named 'Anschauung' was reduced to an immediate grasping of a content, transmitted by the senses, without any guarantee that the true reality of the world could be cognised. What remained was the homelessness of the human being in an immense cosmos, where his intimate experiences could never tell him who he was, and the limits imposed on cognition could not reveal where he was.

The conflict between experience and cognition, engraved by Kant on the unrelenting tablets ('Tafel der Kategorieën') of the ontological divide between the empirical—what we can observe and experience by the senses—and the not observable transcendent, only to be found in the a priori rational concepts, without the senses interfering (Kant, 1982, p. 118). This has permeated our understanding of who we are and has been engraved in our soul. The question "who the human being is," was considered by Kant as the most important of all, and still he did not enter in a direct answer (Frierson, 2013). Instead, in his Critique of Pure Reason, Kant transformed the question to the well-known three key questions: "What can I know?", "What ought I do?", and "What can I hope for?" Each question finding an answer on a different level. As for the last question about hope, in Kantian terminology, hope must meet the criteria for being an a priori principle. This means to understand the world independently of experience, by reason alone (Danziger, 2020, pp. 1–14).

In an interview, asked about the human being, the French philosopher Jean-Luc Nancy (1940–2021) testifies how with his fellow philosophers, for instance Jacques Derrida, he came to the understanding that there is no paradise. *'Neither on earth nor in heaven!'* (Nancy, 2020). In other words, we cannot know! Nancy, who was part of the movement of post-structuralists, also called deconstructivists in France, declares his having finally understood that there is no paradise. Paradise has not been lost. It never existed (Nancy, 2016). After Kant, Hegel and Heidegger, at the end of the twentieth century there are no 'big stories' to be told anymore, no history as the story of human progress, no presence in the world of a subject, not even the presence of human being to himself. And no narrative about who human being is. (Nancy, 2020, Para. 2–3).

Part of the crisis we are traversing is the still not recognised mourning at our loss of our once belonging to what is greater than ourselves. We are being confronted to a 'scientific worldview' based on the mechanistic point of view that was introduced by the rationalistic and materialistic orientation of the Enlightenment period and strongly actualised in the turn of the 20th and the beginning of the twenty-first century (Beaulieu, 2017). The step-by-step untying of the bond with nature as being part of whom we are, has left us bereaved and powerless when it comes to a self-understanding able to respond to the actual state of crisis we are living in. In Western societies a kind of implicit 'taboo' imposes itself, declaring that, instead of becoming the artisan of ourselves, we herald the idea of an unlimited growth, be it at the cost of all life, including our own. The legacy of having been severed from the place humankind once owned, continues till today. It manifests itself in an almost subconscious and dreamy atmosphere of nostalgia, in which the lost paradise is mirrored in a picture calling for a regressive return 'into the wild,' fascinating, but eventually fading out the human being.

5 Anthroposophy and the Anthropocene, and Their Encounter in the Future

Since two decades ago, the term 'Anthropocene' has been going around. As a noun it has been composed of 2 parts: the first being'anthropos' (human being), the second one is '-cene', a suffix ending, that refers to the word endings of terms for geological period. A definition is for instance: "anthropocene is the proposed name for a geological epoch in which humans have become the major force determining the continuing 'livability' of the earth" (Tsing et al.,2017, p. G1). What it represents is a break with all geological periods that have preceded it in the history of Earth and of humankind. And despite the suffix, it largely transcends what is meant by a geological period. The name could suggest that the human being is back! That he has created out of himself a place, where a new balance between what surrounds him and himself is on the way to be found, the first condition being that he develops awareness for his disturbing and damaging 'footprints'.

Or is it nature that is finally back? An understanding of nature that has broken free from the refrainments in which epistemology and science held it? The term, promoted by various 'ecological' movements from all over the world-, points to a new epoch in which the focus is on the negative impact humans made on the planet. At the same time, there is also a growing recognition of human dependence and precariousness. Is it turning a page? Or the 'great turnaround' as Bruno Latour and Nikolaj Schultz have called it, when the limitless stepping up of production will be understood as an aberration' (Latour & Schulz, 2022)? A new nature? And by this, a new narrative, starting with the opening of the Anthropocene (Raffnsøe, 2016)? A narrative that is not the property of one single fraction, society, group, or culture? But a narrative belonging to all living beings, to the ones that 'have speech' and to the ones who do not have it? At least not in the same way! A worldwide new landscape, so that everywhere gardens can be cultivated. Words such as landscape or garden, are used in a different way. Or new words have made their entrance, 'liveability' for instance. And as for some, Earth is a living being, with at the centre, not one, but two actors, both humankind and nature. Nature and humankind and the focus on their mutual relationship, new thinking, not just new thoughts will have to be developed. And the burning question: how not to let this core centre disappear once more, transforming it into yet another branch of the sciences already at hand? Or mingle it with other narratives, be it political, socio-cultural or religious? Is the momentum preparing to go beyond the necessity for a narrative, by lack of finding an adequate answer?

What then are the possibilities of self-understanding before a narrative comes in? Is there an opportunity for humankind to approach his own essence, suspending for a moment the urgent need to already fill it in with a description of qualities derived from that essence? Or to free itself from the imposed Kantian limits, restraining his own potentiality? In order to come to self-imposed restraint? And this not only on the individual level, but also on the level of the family of the 'human being', of humanity. For being human is not an individual quality, albeit the human being must enact it as an individual. Only on this condition, my being born as a human can reach its actualisation and becomes true.

All of us are born in a given world, a world that is already there. Growing up, different modes to relate to this given world unfold. Exactly there sensing, feeling, as well as willing and thinking enter. By relating, gradually I constitute my identity. But it remains an identity that is based on the past. To this kind of relating to the world belonging to the past, Rudolf Steiner added a new category: I can also relate to what is not there yet as actualised, but what is there in the modality of the possible. And I can think the possible because my essence as human is potentiality. A potentiality asking me to become realised, be it with small steps. The keyword for the potential I am carrying in me is 'to become.' This aligns with what Steiner (2016/1894, pp. 141–142) describes as 41'ethical individualism.' I can become human, and this includes all what I can become in my biographical journey as an individual (Gruwez, 2023, pp. 50–53).

Thinking myself as possibility to become, does not need a prior narrative. If still so, the narrative can only come afterwards. First comes the act to think, in which the possibility is already present. It is new every time, from the very first movement, in the sense of not having been there, present beforehand! Within the act of thinking the possible, I become. I have entered a process of actualisation, where who I am and what I am may converge. Because it is no longer neither a question nor an answer addressed only to what part of the past is, the divide between experience and cognition is suspended. And the faculty of speech the human being is endowed with, has turned its face towards the possible, to the future. From now on, its name is Logos, and it includes all beings (Gruwez, 2024, pp. 91–103). There is no other place for the human being then to stay awake amidst the becoming, daring to think the possible. This means daring to become a witness to his essence in becoming who he is. Then possible is not the same as a possibility. And it has nothing to do with probability. It is the experience, when in the thinking as a self-induced activity a content appears—which is no longer bound to the sense world—and holding this pure activity going on, before any content configurates. The content comes out of the activity. Both possibility and probability have already a content from the start. By offering this view both in 'The Philosophy of Freedom' (2006/1894) and in 'Leading Thoughts of Anthroposophy' (1982/1925) and in the fullness of what was created in between, Rudolf Steiner (1861–1925), in defining intuition as a self-sustaining spiritual activity, in which percept and concept are no longer separated, has restored the potential of the intuition, reduced as it had been by the Kantian 'Enlightenment' (Steiner, 2016, p. 121). An Enlightenment, similar to a light that within its limits, could possibly reflect the given world, but on the condition of extinguishing the inner light in the thinker himself.

How long will we still keep mourning for what has not been lost? And shall never extinguish?

References

Agamben, G. (2020). *The kingdom and the garden* (A. Kotsko, Trans.). Seagull Books. https://www.seagullbooks.org/the-seagull-blog/the-kingdom-and-the-garden-by-giorgio-agamben-excerpt/

Baker, D. (1983). *Giovanni Pico della Mirandola*. Verlag am Goetheanum.

Beaulieu, P. (2017). Review of Peter Heusser's *Anthropposophy and science: An introduction. Bio-cosmology–Neo-Aristotelism, 7*(1), 100–105.

Bini, M. (2009). Fourth post: Ζώον λόγον έχον, on the relationship between philosophy and language—How language should be seen from a philosophical perspective. https://maxbini.wordpress.com/2009/06/16/fourth-post-ζώον-λόγον-έχον/

Danziger, M. (2020). Kant: The critique of pure reason (A805/B833). *Contemporary Studies in Kantian Philosophy, 5*(1), 1–14.

Eliot, T. S. (2001). *Four quartets*. Faber and Faber.

Farmer, S. A. (1998). *Syncretism in the West: Pico's 900 theses (1486): The evolution of traditional religious and philosophical systems*. Medieval and Renaissance Texts and Studies.

Frierson, P. R. (2013). *Kant's questions: What is the human being?* Routledge.

Gruwez, C. (2013). *Meditaties over onmacht*. Via Libra.

Gruwez, C. (2023). *Die Wunde und das Recht auf Verletzlichkeit*. Urachhaus.

Gruwez, C. (2024). In het teken van de Logos. In *Het labyrinth van de Logos* (pp. 91–103). Kolisko.

Hutchinson, F. (2018). Sustainability, eco-alchemy. *Art of Education (Erziehungskunst)*. https://www.artofeducating.org/article/eco-alchemy

Kant, I. (1982). *Kritik der reinen Vernunft* (Vol. 1, W. Weischedel, Ed.). Suhrkamp Taschenbuch.

Latour, B., & Schultz, N. (2022). *On the emergence of an ecological class: A memo* p. 18. Polity Press.

Milton, J. (2005). *Paradise lost*. Penguin Classics.

Mirandola, G. P. (1990). *De hominis dignitate. On the dignity of man* (A. Buck, Ed.). Felix Meiner Verlag.

Nancy, J.-L. (2016, December 26). Im Gespräch mit Michael Magercord. https://www.deutschlandfunk.de/philosophie-der-mensch-ist-die-frage-was-ist-der-mensch-100.html
Nancy, J.-L. (2020, March). Exégèse. *Philosophie Magazine*. https://www.philomag.com/articles/jean-luc-nancy-avec-derrida-jai-compris-que-le-paradis-nexiste-ni-sur-terre-ni-au-ciel
Raffnsøe, S. (2016). *Philosophy of the Anthropocene: The human turn*. Springer. https://doi.org/10.1007/978-3-319-09885-2
Steiner, R. (1964). *The philosophy of freedom: Chapter IX, the idea of freedom*. https://rsarchive.org/Books/GA004/English/RSP1964/GA004_
Steiner, R. (2016). *Die Philosophie der Freiheit*. Rudolf Steiner Verlag.
Steiner, R. (1982). *Anthroposophische Leitsätze*. Rudolf Steiner Verlag.
Tricot, J. (1970). *Aristote, la politique: Introduction, notes et index*. Vrin.
Tsing, A. L., Bubandt, N., Gan, E., & Swanson, H. A. (Eds.). (2017). *Arts of living on a damaged planet: Ghosts and monsters of the Anthropocene*. University of Minnesota Press.

The Earth as a Living Organism

Albrecht Schad

Abstract

Until around five hundred years ago, it was commonly accepted to perceive the Earth as a living entity. Since then, many authors, including Rudolf Steiner in the last century, have expressed this idea. Contemporary empirical scientific research also identifies characteristics and properties of organisms that are specific to life processes and cannot be reduced to inorganic principles. This section describes certain distinctive features of life—such as autonomy and balance, sensitivity, and self-activity—which we recognise not only in ourselves but also in the Earth. Life itself alters its environment to better serve its needs. For millennia, humans have shaped landscapes in ways that enhance their diversity and richness—a remarkable cultural achievement. The Earth is a highly developed organism that requires our assistance and cooperation.

Keywords

Earth as a living organism • Gaia hypothesis • Rudolf Steiner • Planetary homeostasis • Cosmic balance • Ecological systems • Anthropogenic impact • Biodiversity • Cultural landscapes • Planetary metabolism • Homeothermy • Environmental sustainability • Geobiology • Planetary evolution • Human stewardship

A. Schad (✉)
Methodology and Didactics of Upper Secondary Education, Biology, Geography, Freie Hochschule Stuttgart, Stuttgart, Germany
e-mail: schad@freie-hochschule-stuttgart.de

J. Kronenberg and E. T. Lammerts van Bueren (eds.), *On the Earth We Want to Live*, World Sustainability Series, https://doi.org/10.1007/978-3-031-98758-8_5

1 Introduction

On 7 December 1972, Apollo 17 was launched as the eleventh and final manned mission of the Apollo programme. During their journey to the Moon, the Apollo crew captured what has become the most iconic full-image of Earth: the Blue Marble. Since then, no human has ventured far enough from Earth to see it as a whole. This blue jewel is unique in the cosmos. A multitude of finely balanced conditions must be met for life to exist on our home planet (Fig. 1).

One of the fundamental prerequisites for life is water. Shortly after the formation of our solar system, water arrived on Earth as a "cosmic gift" from a region beyond Mars (Lesch, 2013, p. 44). Water can only sustain life within a very narrow temperature range. If Earth were just 1% farther from the Sun, all water would freeze. Conversely, if Earth were 5% closer to the Sun, all water would evaporate, subjecting Earth to a fate similar to that of Venus or Mercury. Mercury experiences daytime temperatures of up to 430 °C and nighttime lows of – 180 °C, while Venus endures an average temperature of 460 °C.

Fig. 1 Blue Marble, 7 December 1972. Taken by Harrison Schmitt or Ron Evans (NASA, 1972)

Earth's geological history reveals that its temperature has always remained remarkably stable. On average, it has not been colder than + 5 °C nor hotter than + 50 °C. This is an astonishingly narrow temperature corridor compared to other planets. Earth exhibits a characteristic we usually associate with the most highly developed animals, such as birds and mammals: it is homeothermic, maintaining a consistent internal temperature (Pflug, 1984). How is this possible? These mild and stable temperatures alone would not suffice to retain water on Earth. The situation on Mars provides insight: its surface clearly bears marks left by water, yet today, no water remains. It has entirely disappeared. Why?

Moreover, even the presence of water on a planet does not guarantee the possibility of life on its surface. Harsh cosmic radiation would prevent life from flourishing. What protects Earth from cosmic radiation, and how is this protection achieved? We shall return to these questions in due course (Figs. 2 and 3).

But that is not all. If Earth were smaller, its atmosphere would be thinner, similar to what we observe on the Moon. It would be far more exposed to bombardment from space, such as by meteoroids, and there would be much greater temperature fluctuations. However, even the atmosphere alone would not suffice to protect us from meteoroids. Fortunately, we have Jupiter acting as a "bodyguard." Its mass,

Fig. 2 Okavango, 30 December 2016 (Schad)

Fig. 3 Namib Desert, 29 December 2015 (Schad)

which is 300 times greater than Earth's, enables it to attract most meteoroids from space, preventing them from reaching Earth.

If Earth were larger, like the giant gas planets Jupiter and Saturn, there would be no solid surface, as their thick atmospheres generate extremely high temperatures. Thus, not only are Earth's size and its distance from the Sun conducive to life, but so is the presence of other planets in our solar system. The Earth's axial tilt of 23.5° facilitates the existence of seasons, while its 24-h rotation establishes the rhythm of day and night. The Moon, which also arrived as a "cosmic gift" to Earth, stabilises the planet's axis, ensuring that it only wobbles slightly and does not tilt excessively.

2 The Three Cosmic "Stroke of Luck" Events

Astronomers tell us that during the formation of our solar system, the Sun emerged, surrounded by a massive gas and dust disk. Through an initial process of organisation, this gave rise to the four hot, rocky inner planets and the cold, gaseous outer planets, such as Jupiter and Saturn (Lesch, 2013). The gas and dust disk surrounding the Sun thus differentiated into a hot inner region and a cold, peripheral outer region. Here, we observe an initial formation of balance—one that aligns with the conditions necessary for life. In this sense, Earth owes its existence to the formation of the other planets, as their emergence caused matter to organise into the Sun-adjacent hot planets and the distant cold planets. This was the first cosmic stroke of luck for Earth. Earth's position is also remarkable. It is not among the cold planets like Jupiter and Saturn, nor among the hot inner planets like Mercury and Venus. It is also not precisely at the boundary between hot and cold but slightly on the inner side, within the orbit of Mars. Thus, all extremes are avoided.

For a long time, many large and small rocky fragments wandered through the cosmos. One particularly large fragment struck Earth at a glancing angle. As a result of this event, shortly after the formation of our solar system, the Moon was created. We can consider the Moon a bringer of good fortune. Without its influence, Earth would tumble chaotically through space. "Its axis of rotation would fluctuate so dramatically that one side would always face the Sun, while the other would freeze in complete darkness. Instead, the Earth-Moon system is stable, and Earth's axis only wobbles slightly, allowing for the formation of different seasons" (Lesch, 2013, pp. 44–45).

The Moon has a second effect: tidal forces that cause ebb and flow. These forces not only churn Earth's water but also knead its interior. As a result, Earth's rotation around its axis was slowed from an initial 5–6 h to the familiar 24-h day over an extended period. Without this deceleration, with Earth spinning in just 5–6 h, surface winds would reach 300–500 km/h. No living being could stand upright under such conditions. We therefore owe our ability to stand upright to the Moon. The Moon is Earth's second cosmic stroke of luck.

Originally, there was no water on Earth. Water arrived with ice-laden asteroids from the solar system beyond Mars. In Earth's primordial atmosphere, water vapour formed and

> […] condensed into a global rain of unimaginable proportions. Over tens of thousands of years, this rain washed carbon dioxide and methane out of the air, dissolving it in the oceans and storing it in rocks. This spared Earth from Venus's fate, where a runaway greenhouse effect caused by carbon dioxide has led to surface temperatures of 460°C. On Venus, it never rained; apparently, it was not supplied with asteroids. On the other hand, Earth's primordial atmosphere initially prevented it from freezing entirely. In the early solar system, the young Sun was about 25–30% less luminous than it is today. Without a sufficient amount of greenhouse gases, Earth would still be a white sphere, reflecting nearly all solar radiation back into space. However, Earth evidently had just enough methane, water vapour, and carbon dioxide to escape the fate of global glaciation. (Lesch, 2013, pp. 44–45)

The "gift" of water represents the third cosmic stroke of luck. Life avoids extremes. All three cosmic strokes of luck enable the formation of balance. At this point, we can already discern an organismic character.

3 Earth—A Living Organism

Over the past 500 years, various authors have commented on the idea that Earth is a living being. Up until the time of Johannes Kepler (1571–1630), it was quite common to understand Earth as a living body (Göpfert, 1999, p. 7). In his significant work Harmonices Mundi, Kepler stated: "The globe of the Earth is a body that belongs to a living being" (Kepler, 1619, Book IV, p. 7). In 1785, the Scottish naturalist James Hutton compared the circulation of nutrients in soil and the movements of the oceans to the Earth's own circulation system (Scarpula, 2015, p. 5). In 1841, the natural philosopher Carl Gustav Carus published 12 Letters on the Life of the Earth (Carus, 1986). Alfred Wegener's groundbreaking publication in 1915 (Wegener, 1915) on the horizontal movement of continents, which we now know as plate tectonics, made it clear that Earth is highly dynamic and constantly undergoing transformation. "This dynamism is largely self-regulating. Self-regulating processes are typically associated with life processes" (Schad, 2023a, p. 25).

Rudolf Steiner considered Earth to be the final seed of life in a dying cosmos. From Earth, the dead cosmos would be revived (Steiner, 1982, pp. 197–200). He frequently spoke of Earth as being alive:

> Today, people mostly speak nonsense. When they see a plant, they tear it out of the earth and study botany, knowing nothing about the matter. If I were to pluck a hair and describe it, that would be nonsense, because a hair can only grow on a living being—it cannot exist independently. You cannot place a hair on a lifeless stone and expect it to grow. There must be a living entity underlying it. Well, plants are the hairs of the Earth because the Earth is alive. And just as humans need air to live, Earth needs the starlight and its spirituality; it breathes this in so it can live. And just as humans move about on Earth, Earth moves about in the cosmos. It lives in the entire cosmos. The Earth is a living being. (Steiner, 1924, pp. 153–154)

Building on Steiner's ideas, Günther Wachsmuth published a book in 1945 titled *Earth and Humanity*, in which he wrote:

> The phenomena of the Earth organism demonstrate most clearly, upon closer examination, that it has not only receptive but also active organs. Its structure and segmentation, nourishment and excretion, respiration and circulation, the impulses and rhythms active within it, and the developmental factors it exhibits all belong to a living being [...]. (Wachsmuth, 1945, p. 45)

In 1986, geophysicist Raymond Siever stated:

> We must increasingly view Earth as a highly differentiated organism: a system so complex that its marvellous diversity and richness in detail, observable on its surface and inferred within its interior, cannot be explained by simple forces and mass compositions. (Siever, 1986, p. 18)

In 1997, geomicrobiologist Wolfgang Krumbein wrote: "Yes, Earth is alive. It is a living being, a species not unlike microorganisms, plants, and animals" (Krumbein,

1997). The renowned Heidelberg zoologist and senior professor at the University of Heidelberg also remarked:

> Organisms are characterised by a set of attributes unique to them. Their elemental building blocks do not reflect the distribution of elements in the abiotic world. (…) Furthermore, they undergo evolution, changing over time, often in collaboration with partners—for instance, in symbioses. (Storch, 2014, p. 32)

The idea of a living Earth has a long tradition and remains deeply fascinating. Every new idea encounters criticism. This is not only normal but desirable, as criticism contributes to refining ideas. Developing such an idea requires a cross-disciplinary perspective—something the current academic system is often hesitant to encourage. Therefore, everyone is invited to contribute to the further development of the idea of a living Earth.

James Lovelock and Lynn Margulis have already made significant contributions in this regard—James Lovelock with his Gaia hypothesis (Lovelock, 1992, 1993), and the wonderful Lynn Margulis published her important book Symbiotic Planet in 2017. There she wrote: "The non-biological natural sciences are insufficient to explain the environmental conditions on Earth's surface. When one considers the diverse functions of living beings, which produce gases and alter temperatures, the apparent discrepancy disappears" (Margulis, 2017, pp. 164–165). Earth is thoroughly alive—not only because of its living beings but as a whole. This will be explored further in the following discussion.

4 The Relationship Between Earth and the Cosmos

The Sun, like all stars, moves through the cosmos. As Earth orbits the Sun, this means that its yearly orbit is not merely an ellipse but a kind of elliptical spiral through the cosmos—a helical path through space. In this way, Earth demonstrates something on a cosmic scale that is familiar to all living beings: all substances formed by life exhibit a spiral structure, which biochemists call chirality. "All movements of the planets, including Earth, are chiral within the fixed star sky" (Schad, 2023b, p. 77).

On the map of the Milky Way, one can see the four arms of our home galaxy (Fig. 4):

> two larger, brighter arms (the Perseus Arm and the Scutum-Centaurus Arm) and two smaller, fainter arms (the Norma Arm and the Sagittarius Arm). Our solar system is neither in the bright centre of the Milky Way nor in its dark periphery. It is also not located on a bright arm or in the darkness between two spiral arms. Our solar system is situated between the two bright arms on a weaker arm (the Sagittarius Arm), and there on the brighter side. There is no greater compensation of opposites, no more perfect balance. From this perspective, the cosmic position of our solar system, with the planet Earth, leads us to conclude: Earth is alive. The planetary system and the cosmos display structures that only life can bring forth. (Schad, 2023a, p. 65)

Fig. 4 Depiction of Our Home Galaxy: The Milky Way (NASA, 2008)

Before we turn our attention to Earth itself, let us first take a look at the human being.

5 Our Early Nursery

After the fertilisation of a human egg cell, it divides into two cells. One of these cells becomes the child and its attachments, which biologists refer to as the embryoblast. The other cell forms the membranes of the developing child, known as the trophoblast. Both are tissues belonging to the child. The embryoblast initially pauses its development until the trophoblast has fully formed. Once the trophoblast, which we call the chorion, has developed, the embryoblast can continue its growth under the protection of these membranes (Figs. 5 and 6).

Initially, the chorion features protrusions all over, called villi. It then differentiates into the serosa (without villi) and the placenta (with villi). Through the placenta, the child is nourished throughout pregnancy. At first, these villi are distributed everywhere—what one might describe as peripheral. Later, the villi concentrate in the central placenta. The child first creates a membrane, a periphery. This periphery establishes an environment in which the child can continue to

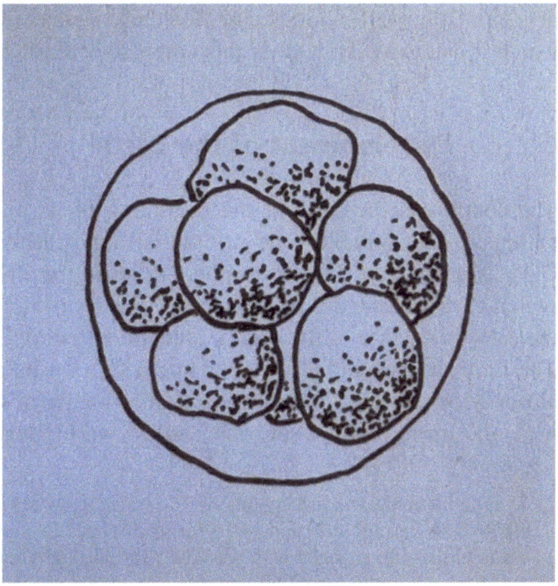

Fig. 5 8-cell human embryo (Schad)

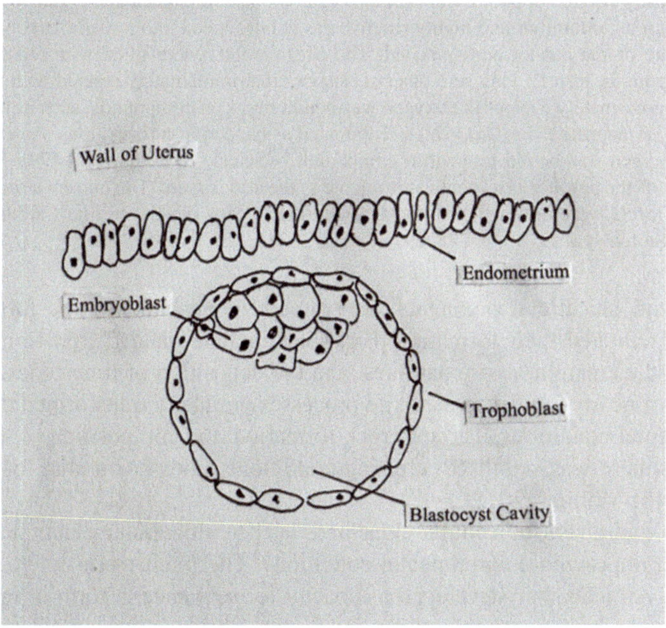

Fig. 6 Human Embryo 6 days old (Schad)

develop. Life itself—in this case, the child—creates a surrounding environment in which it can exist. How does this compare to Earth?

6 Development of the Earth

The conditions on Earth at the beginning of its existence must be imagined as extremely inhospitable. The primordial atmosphere was filled with toxic gases, CO_2, ash, and soot. This atmosphere dimmed the already weak sunlight. The eerie twilight was repeatedly illuminated by flashes of light caused by rock debris from the early formation of the solar system bombarding Earth over a prolonged period. The Earth rotated rapidly, completing a turn on its axis in just 5–6 h. The much closer Moon exerted intense tidal forces on Earth's water masses, pulling them back and forth. It was a feverish, restless world, devoid of balance.

> Despite these nightmarish conditions, life began to take hold of the physical substances on Earth. How exactly this happened remains shrouded in a mysterious darkness. The earliest material that life engaged with was likely viroids and viruses. With the emergence of the first life on Earth around 4 billion years ago (4,000 million years), bacteria also appeared. Evidence of this can be traced back to approximately 3.8 billion years ago. These bacteria were already capable of photosynthesis and began producing molecular oxygen. What happens when an unfamiliar and highly reactive gas is introduced into a world largely devoid of it? The drama cannot be overstated. The ocean waters were filled with reduced sulfur compounds, iron(II) ions, and other substances that immediately reacted with the new gas. 'Approximately 57% of the oxygen was bound into iron compounds, such as hematite (Fe_2O_3) or magnetite (Fe_3O_4), which sedimented as banded iron formations. Around 38% of the oxygen was bound into sulfur compounds.' (Schad, 2019. pp. 178–179). Initially, only 4% of the produced oxygen was available to the atmosphere. The current atmospheric oxygen levels were only reached about 350 million years ago during the Carboniferous period (Storch et al., 2013, pp. 88–89). (Schad, 2023a, 2023b. p. 55)

Bacteria and unicellular organisms also directly contributed to the formation of rocks or facilitated their formation. Examples include banded iron formations in Australia, the creation of stromatolites, and the deposition of limestones. Over the vast span of nearly 3 billion years, life processes gradually transformed the Earth's environmental conditions. Through rock formation, the composition of the oceans and atmosphere was stabilised, creating conditions more favourable for life (see: Kegel, 2015; Mölling, 2015).

In a second phase, unicellular algae and, later, multicellular plants further contributed to improving environmental conditions. The plant world internalised life processes. Animals then developed a capacity for sentience, a form of inner experience. Human beings, capable of mastering this sentience, have their own dignity and individuality. Life itself transforms its environment. It creates a living envelope, a periphery for Earth, just as the life of an embryo forms its trophoblast—a periphery. One could also say: peripheral life reshapes the environment to enable centralised life to exist (Figs. 7, 8, 9 and 10).

Fig. 7 Stone (Aquamarine) (Schad)

Fig. 8 Plant (Water Lily) (Schad)

Fig. 9 Animal (Elephant) (Schad)

Fig. 10 Human Being (Schad)

7 Characteristics of Earth's Life

Generally, we believe that the Earth is an inanimate object on which life somehow emerged. Life is usually understood as a combination of processes that we can observe in the inanimate world. "However, empirical natural scientific research has now advanced to the point where it can describe properties and characteristics of organisms that are specific to life processes. These specific characteristics and properties of life cannot be reduced to inorganic principles. That is to say, inanimate systems and objects typically do not exhibit these properties" (Schad, 2023a, p. 68). Biologist and physician Bernd Rosslenbroich identifies 13 specific characteristics of life (Rosslenbroich, 2023). Of course, we are familiar with these traits in ourselves. Below, some of these characteristics are briefly introduced, followed by an exploration of whether they can also be found in Earth.

Autonomy and Balance

Every living being constructs a body that is distinct from its environment. The organism is considered dead when this "being different" from its surroundings ceases—when the creation of autonomy stops. At that point, it disintegrates and returns to the external cycles of matter. However, as long as the organism is alive, it maintains its own internal environment in contrast to the external world. A chemical imbalance is sustained between the organism and its environment, and this imbalance is consistently regulated. Biologists refer to such regulatory processes as homeostasis.

From the Earth, we know of many homeostases that sustain the autonomy of life. Two examples of this are: the constant salinity of the world's oceans and the respiratory cycle.

Example 1: Constant salinity of the world's oceans

The solid crust of the Earth is broken into several plates that move. On a spherical body like the Earth, this movement creates zones where plates collide (subduction zones) and zones where they drift apart (dilatation zones). Where plates drift apart, mid-ocean ridges are formed. At these rifts, phenomena such as black smokers and white smokers occur. These are hot springs where water at temperatures exceeding 400 °C emerges. The water released by these vents contains dissolved salts. In black smokers, these are primarily sulfides and other salts of iron, manganese, copper, and zinc, whereas white smokers primarily release sulfates like anhydrite and gypsum, along with carbonates and silicon dioxide. When the hot water encounters the cold water of the seabed, fine mineral clouds precipitate, forming the characteristic chimneys and depositing minerals in the surrounding area.

Where does the water from these springs originate? Ultimately, it is seawater from the nearby or distant surroundings, drawn into the ocean sediments. As it passes through the sediments, it is purified of dissolved ions, much like a kidney. Since rivers continuously carry small amounts of salts into the sea, the salinity of the oceans would otherwise increase. The process described above significantly contributes to maintaining the constant salinity of the world's oceans over vast periods. This stability is crucial for most marine life and represents a true "life achievement" of the Earth organism in sustaining this homeostasis.

Example 2: The respiratory cycle

There is the well-known respiratory cycle between plants and animals. In this global regulatory loop, plants absorb CO_2 and produce oxygen through photosynthesis. Conversely, animals and humans take in oxygen and release CO_2. However, this cycle is only complete when the Earth as a whole is included—in this case, the oxygen-consuming processes of rock formation. The high oxygen concentration of about 20% in our atmosphere represents a chemical imbalance (a labile dynamic equilibrium) that is continuously maintained by the processes described above. Such a condition does not exist in a world without life.

Sensitivity: Maintaining Constant Body Temperature

Every living being possesses the ability to perceive the relationship between its internal environment and the external environment (sensitivity). This allows it to respond to changing environmental conditions and maintain its own living conditions, such as body temperature in warm-blooded animals.

Over the past four billion years, the Sun has become approximately 30% hotter, which should have caused Earth's temperatures to rise. However, evidence shows that Earth has been able to maintain its "body temperature" within a narrow range for about four billion years. This has been made possible primarily through the regulation of CO_2 levels. Carbon dioxide was repeatedly removed from the atmosphere through processes such as the formation of limestone, coal, oil, and natural gas, with the result that Earth's temperatures could remain stable. Today, we are partially reversing these regulatory processes that maintain Earth's internal environment by burning lime and fossil fuels. This reversal underpins our current prosperity. To sustain this prosperity, we urgently need to change our lifestyle.

Thus, Earth has been able to respond to changes in the cosmic environment and maintain its internal environment—an impressive achievement of life. We see this capability today only in the most highly developed vertebrates: birds and mammals.

Self-activity

Earth travels around the Sun in one year. Its 24-h rotation on its axis can also be regarded as self-activity. This rotation has significant effects on processes within Earth's interior. In Earth's plastic outer core, convective currents occur. Due to the heat of the inner core, material rises, cools at the boundary with the lower mantle, and then sinks back down. The Earth's rotation deflects these currents through the Coriolis force, forcing them into spiral trajectories. This movement generates Earth's magnetic field (Seyfried, 2005).

The magnetosphere is crucial for life on Earth. It shields life from deadly cosmic radiation and solar winds by diverting ionised matter from space and the Sun around the planet. Only near Earth's magnetic north and south poles, where the magnetic field lines meet the Earth's surface, can some of these particle showers penetrate the atmosphere. This phenomenon is known as the auroras. The outer field lines of Earth's magnetic field remain stationary, while the electromagnetic circulation systems closer to Earth rotate along with the planet. These near-Earth rhythmic waves and vortices arise from the interaction between solar wind and Earth's living processes (see Figs. 11, 12 and 13). These vortices are created when solar particles encounter Earth's magnetosphere. As these particles enter the atmosphere, the auroras light up. In this sense, the shape and form of the auroras are an expression of Earth's life.

Protective Function: Ozone Layer

The Earth's magnetic field alone is not sufficient to protect life from cosmic radiation. The ozone layer is absolutely essential for this purpose. Without oxygen, ozone cannot form. The primordial life process, photosynthesis, produces molecular oxygen. At an altitude of about 25 km, there exists an optimal ratio of oxygen and cosmic radiation, which splits oxygen molecules (O_2) to form ozone (O_3). The ozone layer protects water from being split by ultraviolet radiation into hydrogen and oxygen. Without this protection, the lightweight hydrogen would escape into space, while the highly reactive oxygen would combine with various elements (e.g., iron and sulfur) and disappear from the atmosphere. This would eventually

Fig. 11 Magnetic field lines of Earth

Fig. 12 Aurora, Hammerfest, February 1998 (Schad)

Fig. 13 Aurora, Hammerfest, February 1998 (Abrecht Schad)

lead to the drying up of Earth's ocean basins. We can observe this fate on Mars. The fact that Earth has not suffered the same destiny as Mars is due to its status as a living planet. Life itself creates the ozone layer that protects it.

If we examine ourselves, we notice that we are organisms that regulate our internal environment and maintain a dynamic chemical imbalance relative to the outside world. We regulate our temperature within a narrow range of fluctuations, and we create protective membranes, such as our skin, which not only separate us from the environment but also preserve our autonomy. We can observe these same qualities in Earth. It maintains material cycles in a constant balance, even though non-living processes—such as the influx of salts into the oceans—would lead to destabilisation. Earth has been capable of maintaining a stable "body temperature" for over four billion years. Moreover, it actively generates protective layers, such as the magnetic field and the ozone layer, which act like a skin, enabling life and contributing to Earth's autonomy. Earth is alive. It is a living being, comparable to bacteria, a lily of the valley, an earthworm, an elephant, or a human being.

8 Life Changes the Environment

Life adapts to environmental conditions, a fact that biology has studied extensively over the past hundred years. There are countless examples demonstrating this process. At the same time, however, life also transforms its environment, shaping it to better suit its needs. This aspect has been increasingly recognised in contemporary research. We already noted this in Sect. 6: Development of the Earth. Here, we will explore this idea further with additional examples.

Deer Gardens

When we walk through the forest, we might notice small "bonsai-like beeches" that have been nibbled by deer. In specific areas, deer repeatedly browse young, regenerating beeches. Essentially, the deer are creating a kind of "garden" that provides them with what they need. Deer are classified as concentrate selectors, meaning they prefer high-energy food with easily digestible nutrients such as sugars, starches, and proteins. These are primarily found in the buds and fresh leaves of the browsed beeches, which conveniently grow at the exact height the deer can reach. Additionally, deer enjoy eating flowers, herbs, and acorns. In their self-made "deer garden," the deer create an environment that is more favourable for them. This behaviour also reduces pressure on other areas of the forest, which remain less intensively browsed. Thus, the deer actively reshape the forest, adapting it to provide better living conditions for themselves.

Beaver Ponds

Beavers also transform their environment by building dams. These dams slow the flow of water, extending the shoreline where more willows and poplars can grow. The stumps of felled trees, whose young shoots and twigs are eaten, sprout new growth. In this way, beavers create a plantation of fresh saplings for their needs.

Fig. 14 Elephants as landscape shapers, 28 August 2018, Kwando, Namibia (Schad)

Here, too, beavers modify their environment to make it more suitable for themselves. At the same time, this "gardening" activity has a positive effect on the water balance of the landscape, benefiting many other living beings.

Living beings are not solitary actors. They work together with other living beings as part of the landscape, much like organs cooperate within an organism. Additionally, we can observe that life creates the environment it needs to thrive. This idea will be explored further with one final example.

Elephants

Ecologists regard elephants as a keystone species because they act as landscapers. Their influence on shaping the landscape, and consequently on the habitat as a whole, cannot be overstated (Parker, 2017).

The Kwando River, a tributary of the Zambezi, partially marks the border between Namibia and Botswana. Its banks and the marshes between its tributaries are densely covered with papyrus and elephant grass, forming impenetrable thickets in places. Elephants not only graze on these plants but also plough pathways through the vegetation, opening up areas and creating new habitats for other species such as hippos, waterbucks, sitatunga antelopes and red lechwe antelopes. These pathways enable other wildlife to move through the forested areas (see Fig. 14).

During the rainy season, elephants use their large feet to create new small ponds, which are expanded further in the following rainy seasons (see Fig. 15). In this way, elephants create water sources that can be used by many other wild animals. Elephants are often accused of destroying the landscape. While this can happen, it is typically only the case where there is a significant overpopulation or where the animals cannot migrate, such as in fenced national parks.

One visible effect of elephants is their tendency to break or fell trees to access leaves and twigs. While this may appear destructive it actually benefits many other organisms. A felled tree is not necessarily dead; it often sprouts new growth. These new shoots and buds are more nutrient-rich than older parts of the tree. Additionally, smaller animals can now browse on tree leaves that were previously out of reach. Black rhinos and other browsers such as kudu antelopes rely on these fresh shoots, especially during the dry season, for vital sustenance—an opportunity they would not have without the activities of elephants (see Fig. 16). By felling trees

Fig. 15 New waterhole forming, 27 August 2017 (Schad)

Fig. 16 Altered bushes and trees, 27 August 2023, Kwando, Namibia (Schad)

elephants also open up the landscape locally, allowing grass to grow which benefits numerous grazing animals.

Elephants consume large quantities of fruits and seed pods. As they travel long distances—up to 80 km per day—they act as major seed dispersers. Seeds in their dung are not only spread over wide areas but are also fertilised by the dung itself. Elephants spread more plant seeds than any other animal. They do not just live in the landscape; they actively cultivate it like gardeners.

Trees felled by elephants may eventually die. Dead trees, however, provide significant benefits to the landscape and the organisms living within it. They enrich the soil with nutrients, prevent erosion by wind and water and protect young seedlings from being grazed too early. Dead trees also provide shade for plants that need shelter to grow, such as certain tree and grass species.

In areas where elephants have disappeared the landscape often becomes overgrown and impenetrable thickets form. This has significant consequences for many other species. Elephants, by keeping the landscape open, create diverse habitats over decades and centuries. With their feet, trunks and tusks they craft unique ecosystems even on a small scale. For instance, a partially damaged tree can become a network of refuges and microhabitats for countless insects and reptiles.

Elephants consume 150–180 kg of plant material daily and produce about 100 kg of dung per day. This translates to approximately 37 tons of dung per square kilometre for every elephant in that area. The landscape is fertilised on a massive scale. Elephant dung contains many undigested plant seeds which not only enable new plants to grow but also provide food for other animals such as baboons who enjoy the seeds as a snack. Undigested leaves from grasses and trees are eagerly consumed by termites which use them to feed their underground fungus gardens. In turn termites themselves are prey for many other species.

Dung beetles also play a significant role in processing elephant dung. Africa is home to approximately 1800 species of dung beetles which bury the dung and fertilise the soil. Elephants, along with their dung-specialised beetles, are crucial contributors to maintaining and enriching the landscape.

Elephants fertilise the land on a large scale. Their inefficient digestion leaves behind food for other species in their dung. It is also important to note that in African savannas, and indeed wherever hoofed animals exist, landscapes are shaped and fertilised by their presence and activity.

Now we can ask ourselves: Who are the "elephants" in our landscapes? Who shapes and cultivates the environment so that all life forms can coexist more harmoniously? One could argue that humans, with their cultural and creative abilities, are the "elephants" of our world. Humanity has the potential to make the world more diverse and guide all living beings toward a shared evolutionary future. Humanity is not only a destructive force but also has a constructive side. This constructive side will be the focus of the discussion to follow.

9 Cultural Landscapes as Ecological Organisms

We have now explored this remarkable ability of life to transform an inhospitable environment into one that is more conducive to life. Throughout Earth's geological history, extinction events have repeatedly occurred, drastically "pruning back" the progress achieved in the development of life. Each of these events has prompted life to renew itself and evolve further, achieving ever greater independence from the environment and increasing autonomy (see also Rosslenbroich, 2014). This natural process, which has recurred throughout evolution, has been amplified by humanity's collaboration with nature and, consequently, with life itself—a cultural phenomenon (Fig. 17).

Central Europe was originally covered by forest, from sea level to the tree line in the high mountains. This primeval forest was species-poor, both in terms of plants and animals. The locations where our ancestors established villages were chosen with remarkable and fascinating care, often in transitional areas between wet and dry conditions. The forest had to be cleared first, of course. In the immediate vicinity of the village, vegetables were primarily cultivated. The soil was intensively fertilised with composted manure. Plants grown in this innermost circle around the village were bred specifically for leaf production, such as cabbages. Flowering was undesirable. In the second circle around the village, cereals were cultivated on the drier side.

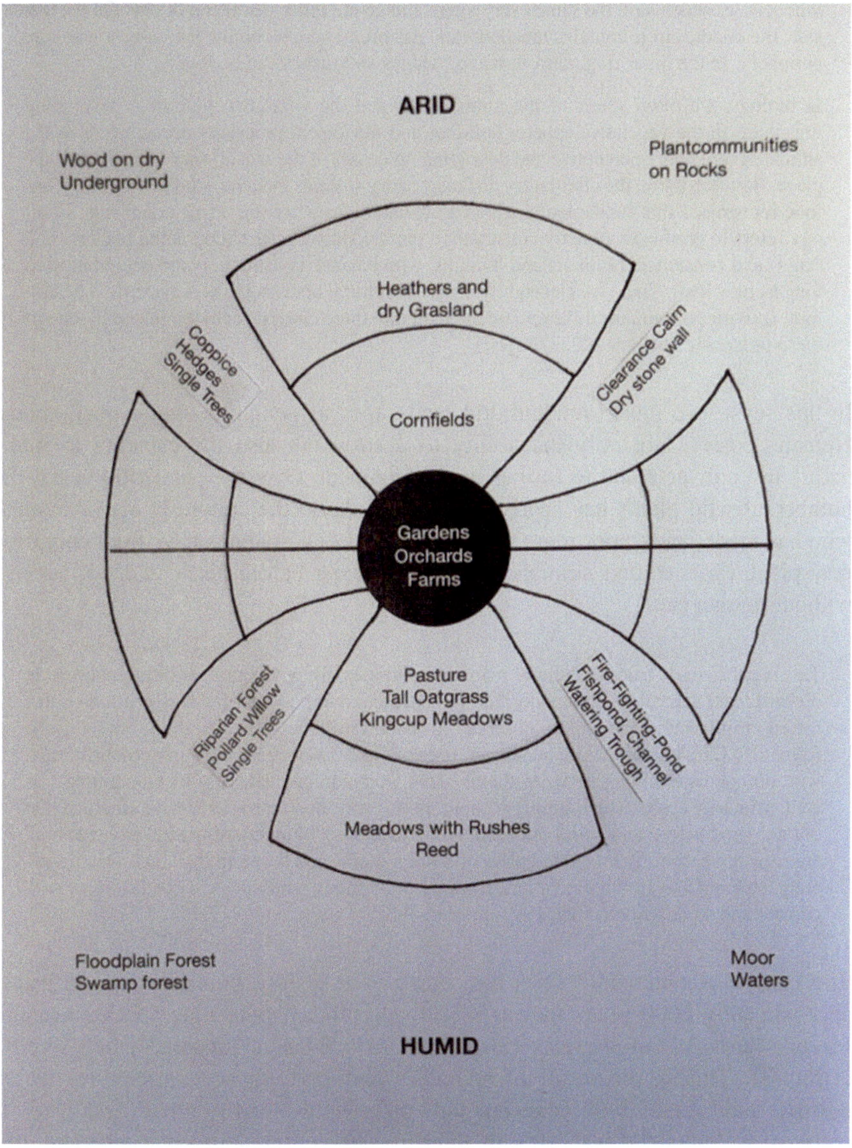

Fig. 17 Ideal representation of the Northwestern German Cultural Landscape (after Vahle, 1991)

With the drying of cereal stalks, the bright blossoms of the accompanying flora—such as poppies, cornflowers or field thistles—become more visible. This tendency increases in the third circle. On sheep pastures with their nutrient-poor grasslands, a small-leaved vegetation appears, enriched with an abundance of blossoms. The intense nutrient extraction in this outer circle causes the vegetation to respond very sensitively to even the smallest differences in site conditions. This is not the case with the vegetation-dominated plants of the

> inner circle, which look the same everywhere and do not reflect local conditions. On the wet side, the middle circle contains meadows and pastures where blooming increases toward the periphery. In the outer ring, narrow-leaved sedges and rushes can be found.
>
> In biology, we often speak of the animal pole and the vegetative pole of a developing organism. In the vegetative sphere, building and movement processes occur, while in the animal sphere, more perceptive and degrading processes of the sensory-nervous system take place. Between these, the circulatory and respiratory systems mediate. Classical ecology has long recognised this fundamental tripartite division characteristic of all organisms. Ecology refers to producers (mostly constructive plants), decomposers (degrading bacteria and fungi) and consumers (animals and humans, who balance construction and degradation as long as they live). Thus, we can rightly describe cultural landscapes as organisms. The cultural activities of humans differentiate the land into three circles, each dominated by one of these processes. (Schad, 2023a, p. 167)

In this sense, we must view cultural landscapes as genuine cultural phenomena. Humans possess not only the ability to destroy but also the capacity to make nature more diverse and to further its development. Over the past 7000 years, the number of wild plants has nearly doubled thanks to the culturally active human, who has made landscapes more diverse. Beyond that, humans have bred countless new plant varieties and animals that did not exist before and could not survive without human care.

> This is particularly true for cultural plant communities, the association of different crops in orchards and vegetable gardens, or, as was once common, also in fields. Even today, this mixed-crop farming can still be observed in 'backward' areas, such as the foothills of the Romanian Carpathians. Maize and beans, potatoes and Swiss chard grow in a colourful mix, with pumpkins climbing between them. These communities naturally include animals as well, primarily cattle, which supply manure. In this way, new ecosystems have emerged that did not exist before and could not exist in the unaltered, human-free natural environment. Humans are both fully integrated parts of these systems and far more than that—the generating, shaping and sustaining factor, naturally in harmony with the local conditions of soil, climate and so on. (Suchantke, 1993, pp. 111–112)

The Earth needs humanity. Over the course of its history, humanity has cultivated approximately 6000 plant species worldwide. Today, only nine species account for two-thirds of all harvests (Strobel, 2021). What an unprecedented impoverishment. This is the result of so-called high-performance agriculture, which destroys biodiversity with large amounts of pesticides and synthetic fertilisers. If we were to factor the real costs of industrial agriculture—its environmental, climatic and health impacts—into its prices, it would quickly become clear: this is not something we can afford.

10 Earth as a Living Organism

Through its cultural activity, humanity has the ability to collaborate with the life processes of landscapes and living beings. Just as the human head is the centre of sensory organs and nervous organisation, humanity can emphasise and develop corresponding life processes in the landscape, as we observed with nutrient-poor grasslands. Similarly, as the body's building processes are linked to the abdominal organs and movement processes to the limbs, humanity can also cultivate these aspects of life processes in the landscape, for example, through intensive vegetable cultivation. In the human chest, we find the centre of breathing processes (the lungs) and circulation (the heart). In agriculture, this corresponds to cereal cultivation, while in the landscape, it aligns with standing and flowing bodies of water.

For Earth, we have identified corresponding characteristics of life. The ability of sensitivity enables Earth to "perceive" changes in its cosmic environment, such as the Sun becoming hotter. In response, Earth has managed to keep its surface temperatures stable over immense spans of time; it is homoiothermic. It has also maintained a constant salinity in the oceans and, more recently, a steady oxygen concentration in the atmosphere. Earth shows self-activity and movement. It orbits the Sun and rotates on its axis. Through convection currents in its outer core, it generates a magnetic field that protects it from lethal cosmic radiation. It also forms an ozone layer, which can be seen as part of its metabolic processes. Oxygen production—a true life process—leads to the creation of ozone. Both serve as a form of skin formation.

The global currents of water and air can be seen as Earth's breathing and circulatory systems. Specifically, during the Northern Hemisphere summer, the atmosphere heats and expands, while in the Southern Hemisphere winter, it contracts. Thus, Earth "breathes" in and out in a half-yearly rhythm. Cold oceans hold more oxygen, while warm oceans hold less—much like the relationship between arterial and venous blood in our bodies.

Earth is a highly developed organism that needs our help and cooperation. And we, in turn, need the living Earth. As far as we know, it is the only seed of life in the unimaginably vast cosmos. It is the sole place in the universe where we can exist. We need the living Earth and other human beings to realise our life intentions (Fig. 18).

Fig. 18 Little Blue Dot. Taken on 14 February 1990 by the Voyager 1 spacecraft from a distance of 6 billion km or 40.5 astronomical units. No other photo of Earth has ever been taken from a greater distance

11 Conclusion

Our solar system is part of the Milky Way galaxy, which is itself just one among countless other galaxies in the unimaginably vast cosmos (see Fig. 19). From this perspective, one might conclude that the Milky Way, our solar system, and especially Earth, are insignificant.

However, as shown, the position of our solar system within the Milky Way exhibits structures that are characteristic of life. Similarly, Earth's position in the solar system and the three cosmic "fortunate coincidences" reflect organismic qualities. The movement of all planets, including Earth, is chiral within the fixed star sky, and chirality is a hallmark of life.

For Earth itself, we have demonstrated that it possesses attributes typically associated with living beings, such as autonomy and balance, sensitivity, self-activity, and even the formation of a protective "skin." Remarkably, Earth also maintains

Fig. 19 Each point in this image represents the position of a galaxy as it was 6 billion years ago. The image covers approximately one-twentieth of the sky and displays 48,741 galaxies. Grey patches indicate small regions with no data. The colour represents the distance from Earth, ranging from yellow for galaxies on the near side of the disc to purple for those on the farther side

a constant body temperature, a trait otherwise known only in birds and mammals. We can therefore rightfully regard Earth as a living being that requires our collaboration.

Humans have the ability to shape landscapes in ways that are full of life and to enrich them through cultural activity. This benefits humanity, other living beings, and Earth itself. We should commit to caring for and further developing this unique blue jewel in the cosmos.

References

Carus, C. G. (1986). *Zwölf Briefe über das Erdenleben* (H. Meffert, Ed.). Frommann-Holzboog.

Göpfert, C. (1999). *Das lebendige Wesen der Erde*. Verlag Freies Geistesleben.

Kegel, B. (2015). *Die Herrscher der Welt*. DuMont Buchverlag.

Kepler, J. (1619). Harmonices mundi (Vol. 4).

Krumbein, W., & Levit, G. (1997). Die Erde – ein Lebewesen. *Einblicke: Forschungsmagazin der Carl von Ossietzky Universität Oldenburg, 25*, 12–15.

Lesch, H. (2013). Unter dem Diktat der Physik. In V. Storch, U. Welsch, & M. Wink (Eds.), *Evolutionsbiologie* (pp. 44–60). Springer-Verlag.

Lovelock, J. (1992). *Gaia: Die Erde ist ein Lebewesen*. Scherz Verlag.

Lovelock, J. (1993). Das Gaia-Prinzip: Die Biographie unseres Planeten. Insel Taschenbuch.

Margulis, L. (2017). *Der symbiotische Planet: Wie die Evolution wirklich verlief*. Westend Verlag.

Mölling, K. (2015). *Supermacht des Lebens*. C.H. Beck.

NASA, (1972). https://www.nasa.gov/multimedia/imagegallery/image_feature_329.html
NASA, (2008). http://www.eso.org/public/images/eso1339e/
Parker, P. (2017). *Understanding elephants.* Random House.
Pflug, H. (1984). *Die Spur des Lebens.* Springer-Verlag.
Rosslenbroich, B. (2014). *On the origin of autonomy: A new look at the major transitions in evolution.* Springer.
Rosslenbroich, B. (2023). *Properties of life: Towards a theory of organismic biology.* The MIT Press.
Scarpula, L. (2015). *Gaia oder die beste aller möglichen Welten: Reflexionen über die Erde* (Master's thesis). Freie Hochschule Stuttgart.
Schad, A. (2019). Der Geographieunterricht an Waldorfschulen und sein Verhältnis zur modernen Wissenschaft. In A. Hüttig (Ed.), *Wissenschaften im Wandel* (pp. 149–184). Berliner Wissenschaftsverlag.
Schad, A. (2023a). *Vom Leben unserer Erde.* Verlag Freies Geistesleben.
Schad, W. (2023b). *Vom Geist der Erde: Unsere Welt als lebendiges Wesen.* Rosenkreuzer Verlag.
Seyfried, H. (2005). *Ein Planet organisiert sich selbst* (pp. 71–105). Jahrbuch aus Lehre und Forschung der Universität Stuttgart.
Siever, R. (1986). Die Dynamik der Erde: Bewegungen, Strukturen, Wechselwirkungen. Spektrum der Wissenschaft.
Steiner, R. (1924). Die Geschichte der Menschheit. In *Gesamtausgabe* (Vol. 353, pp. 153–154). Rudolf Steiner Verlag.
Steiner, R. (1982). *Anthroposophische Leitsätze* (Vol. 26, pp. 197–200). Rudolf Steiner Verlag.
Storch, et al. (2013). Evolutionsbiologie: 88f.
Storch, V. (2014). Die Entstehung des Lebens auf der Erde. In P. Rothe, V. Storch, & C. von See (Eds.), *Lebensspuren in Stein: Ausflüge in die Erdgeschichte Mitteleuropas* (pp. 10–25). Wiley-VCH Verlag.
Strobel, Jacob y Serra (2021). Die hausgemachte Selbstzerstörung. Interview mit Felix Prinz zu Löwenstein. In: Frankfurter Allgemeine Zeitung, 31.7.2021.
Suchantke, A. (1993). *Partnerschaft mit der Natur: Entscheidung für das kommende Jahrtausend.* Verlag Urachhaus.
Vahle, Hans-Christoph (1991). Die Idee der Kulturlandschaft am Beispiel Nordwestdeutschlands. In Die Drei, Heft 7/8, S. 581 ff.
Wachsmuth, G. (1945). *Erde und Mensch.* Archimedes Verlag.
Wegener, A. (1915). *Die Entstehung der Kontinente und Ozeane.* Vieweg & Sohn.

Sustainable Development—Definitions and an Expansion Towards an Integral Sustainable Development

Sustainable Development—Definitions and an Expansion Towards an Integral Sustainable Development

Johannes Kronenberg and Edith T. Lammerts van Bueren

Abstract

Sustainable development (SD) has undergone continuous evolution since its institutional emergence approximately 50 years ago. In light of persistent global challenges, there is increasing recognition of the need to expand the prevailing SD paradigm to include a cultural-spiritual and inner dimension. From this perspective, the current socio-ecological crisis may be understood as a manifestation of a deeper cultural-spiritual crisis. This chapter proposes an Integral Sustainable Development approach informed by anthroposophy. It begins by outlining key definitions and historical developments in the discourse on sustainable development. It then explores the implications of incorporating a cultural-spiritual dimension, which calls for a renewed understanding of human nature in relation to the environment. Anthroposophy, developed by Rudolf Steiner, offers a relevant framework through its emphasis on three interconnected human faculties—thinking, feeling, and willing—which shape individual and collective activity. Steiner's concept of the Threefold Social Order is also examined, distinguishing three interdependent yet autonomous spheres of society: the cultural/spiritual, the legal/political, and the economic. These are associated respectively with the guiding principles of freedom (Liberté), equality (Égalité), and solidarity (Fraternité). Within this context, the chapter

J. Kronenberg (✉)
Research Associate Section for Agriculture, Goetheanum - Freie Hochschule für Geisteswissenschaft, Dornach, Switzerland
e-mail: johannes.kronenberg@goetheanum.ch
URL: https://www.sektion-landwirtschaft.org/en/

E. T. L. van Bueren
Wageningen University and Louis Bolk Institute, Organic Plant Breeding, Utrecht, Netherlands
e-mail: edith@bioacademy.nl

J. Kronenberg and E. T. Lammerts van Bueren (eds.), *On the Earth We Want to Live*, World Sustainability Series, https://doi.org/10.1007/978-3-031-98758-8_6

introduces the idea of appropriate anthropocentrism embedded in a holocentric ethic, forming the foundation of Integral Sustainable Development. This approach integrates the cultural-spiritual dimension into the established pillars of sustainability—ecological, social, and economic—through the concepts of world-viewing, world-sensing, and world-acting. The chapter concludes with four case studies of anthroposophically inspired enterprises that exemplify the practical application of Integral Sustainable Development principles.

Keywords

Sustainable development • Integral sustainable development • Cultural-spiritual dimension • New narratives • Socio-ecological crisis • Anthroposophy • Threefold social order • Appropriate anthropocentrism • World-viewing • World-sensing • World-acting • Planetary boundaries • Inner development goals • Regenerative agriculture • Biodynamic agriculture • Paradigm shift • Human and earth dignity • Systems thinking • Living organism thinking • Social threefolding • Economy of love • True cost accounting

1 Introduction

Our global society has been aware of the sustainability challenges at least since the publication of the Club of Rome's 1972 report on "The Limits to Growth" (Meadows et al., 1972). It expected that, over the course of the current century, human, industrialised society would no longer be supported by the "earth systems," leading to a "path of overshoot and collapse." The continuous exponential growth—both of the population and of the economic system—cannot be sustained in a finite world. Steffen et al. (2015), in "The trajectory of the Anthropocene: The great acceleration," stressed this inconvenient situation empirically in a dramatic way, showing the exponential growth of human population and of human activity and their environmental impact from 1945 onwards. A few years after "The Limits to Growth," Fred Hirsch published *Social Limits to Growth* (1977). In it, he demonstrated how, once basic material needs are satisfied, a new and more fundamental scarcity will arise for "positional goods." Under the new circumstances, goods such as high-quality housing and higher education will inevitably remain scarce, which will give rise to increasing social tensions. The history and cause of this crisis is still mainly attributed to the industrial revolution and the Great Acceleration, though the colonial impact after 1492 seems to be a more precise start of social injustice and the destruction of vital earth organisms (Jhagroe, 2024).

Now, more than half a century after the increased awareness of the state of our world, the anticipated socio-ecological crisis is manifesting itself strongly on a global scale. The global climate is changing as predicted by the IPCC (1990); in 2024, the average global temperature reached the highest level since records began (World Meterological Organization, 2025). Extreme weather events are becoming more common around the world. Global biodiversity is constantly decreasing. In terms of social conditions, inequality in the world is widening, especially the financial gap between the rich and poor (Oxfam International, 2024). Human basic needs are not being met, malnutrition is not being solved, and two billion people will be living in water-scarce areas by 2025 (Earth.Org, 2024). Geopolitical security has been declining since Russia's invasion of Ukraine and the recently intensified war between Israel and Hamas. Worldwide, social justice is being violated and working conditions are being compromised, with the most severe impacts observed in manufacturing countries. Human action generates brutal resource extraction, discard-consumption patterns, poisoning and pollution of land, water, and air from conventional agriculture and industry (Wijkman & von Weizsäcker, 2018). The prevailing economic system is still—after 50 years of institutionalised sustainability efforts—focused on profit maximisation, exponential growth, and "short-termism" (Wijkman & von Weizsäcker, 2018; Dixson-Declève et al., 2022). In the name of progress and development for all, the very 'habitability' of the Earth is at stake (Latour & Schultz, 2022).

Continued economic growth is still seen as a sustainable path (Jackson, 2009), referring to the myth of 'decoupling' growth from environmental impact through technology. This assumes that economic growth can be achieved without a net increase in resource use or negative environmental consequences. But such a required absolute decoupling has not yet been achieved anywhere. Only relative decoupling has been achieved, with growth being realised with less sharply increasing—but still negative—environmental impacts.

On the other hand, with the same technological euphoria, one could also argue that in the last 50 or even 200 years, life expectancy has increased, while early infant mortality, malnutrition, water scarcity, poverty, inequality, and injustice have decreased enormously (Rosling et al., 2018). But the core problem is that this has been realised at the expense of the "habitability" of the Earth and by crossing planetary boundaries as written above. The "development" achieved is apparently very fragile and its benefits still accrue predominantly to Western countries.

Over the past 50 years, it has become increasingly clear that the causes of this socio-ecological crisis are not solely technical or physical and, therefore, cannot be resolved through materialistic technical solutions alone (Dresner, 2008). The 'systemic change' called for in the "Limits to Growth" report is gradually becoming more fundamental and profound. The 1992 update of the report stated that "the sustainability revolution will have to be, above all, a collective transformation that makes it possible to express and nurture the best of human nature, rather than the worst" (Randers et al., 1992, p. 181) In 1999, Donella Meadows, one of the lead authors of the "Limits to Growth" report, went even further, pointing out in her hypothesis about system change "that the least effective leverage points for a system change are changing parameters, numbers, the rules of a system, changing structures, or even changing the goals. The most radical and effective lever to change a system is the so-called "transcendence of paradigms" as such" (Meadows, 1999, pp. 19). This is the point at which one might say that an "ethical universalism"[1] satisfies the call for individual responsibility and development in accordance with the greater whole. In line with these evolving views, the Club of Rome's more recent 50th anniversary report also calls for "changing the narrative", emphasising the need to include the "spiritual dimension" as a path to a moral viewpoint in a new narrative for humanity and the Earth (Wijkman & von Weizsäcker, 2018, p. 67).

In this chapter, we will first will discuss evolving and more widely spread views on 'sustainable development' which followed up to the initial definition of the (1987) UN Brundtland Commission. Given the call for a new narrative considering the current socio-ecological crisis as a cultural-spiritual crisis, we will then discuss the implications of a cultural-spiritual dimension and the later proposals made in this direction. As pointed out by Meadows (1999, p. 19), and discussed above, we have to take into account that such an expansion of the current definition of Sustainable Development is far from random. It requires a vision of "human nature" in order to derive an "ethical universalism." Such an integral view of human nature, encompassing the cultural-spiritual dimension, is an important approach and here we can learn from anthroposophy as developed by Rudolf Steiner (1861–1925) and the subsequent anthroposophical movement as presented in Part 4 with the examples of the Sustainable Natives. Steiner (1886/2003, 1894/2021) pointed out the importance of a thinking practice that goes beyond mere reductionist intellectual thinking. In this, the thinking practices are directly linked to our actions and their results, touching on what Meadows called "transcendence of paradigms" (Meadows, 1999).

Because we ultimately have to formulate a vision of "human nature" in relation to its surroundings, the Earth, and beyond, our approach starts necessarily anthropocentrically but aims to integrate all life forms. Since this can be seen as contrary to the ecocentrism currently advocated, we explain this position in a separate section. This brings about the concept of "appropriate anthropocentrism", providing a constructive interpretation and presentation of the Anthropocene itself, triggered by anthroposophical principles and practices over the past 100 years (McKanan, 2017). We consider altering the narrative of the Anthropocene as essential in order to advance in the sustainability effort.

Finally, we sketch a proposal for a concept of integral sustainable development. This is to be understood as an instrument and tool to face and overcome the culminating 'global socio-ecological crisis,' as overarching terminology for the *polycrisis*, including the social, ecological, economic, and spiritual dimensions.

[1] Ethical Individualism is one of the core principles of anthroposophy regarding human development stated by Rudolf Steiners. Ethical individualism emphasises the idea that moral action should arise from the individual's inner sense of ethical intuition rather than external rules, traditions, or social norms.

2 Evolving Views on Sustainable Development

In this section, we explore current definitions of Sustainable Development, that knows over 200 definitions (Ametepey et al., 2023). Herewith we connect to the understanding that Sustainable Development is a rather new science and discipline, one that is still evolving and gradually advancing towards a completely new meta-discipline (Ametepey et al., 2023; Mihelcic et al., 2003). In recent decades, Sustainable Development (SD) has become the central political, academic and pragmatic instrument for tackling socio-ecological challenges, in the endeavour of creating a world where human dignity and planetary health are cornerstones for the wellbeing of all inhabitants of the earth. SD tries to offer concrete solutions and knows many frameworks, agreements, and also ambivalences in its concepts, interpretations, and applications

As already indicated, the first global call for a 'sustainable development' originated from the United Nations World Commission on Environment and Development from the UN summit in Norway in 1987 (Mulligan, 2015). Chaired by Gro Harlem Brundtland, the Commission published the report 'Our Common Future', which can be seen as a political response to the scientific assessment which was earlier given by the 'Limits to Growth' report. The Commission formulated the first and most common 'Brundtland-definition' of SD: "Sustainable development is development that meets the needs of the present without compromising the ability of future generations to meet their own needs." (Brundtland, 1987, p. 43) This definition contains the concept of 'needs', in particular the essential needs of the world's financially poor, to which overriding priority should be given. This definition prioritises social sustainability, the fulfilment of which requires the basic needs of all people, starting with the poorest in the world, to be met as soon as possible, without, however, preventing this from remaining possible in the future. But the 'needs' of present and future generations are not only to be understood as social, but also as ecological and economical. The development thus is 'sustainable' as long as the developments within the social, the ecological, and the economic pillars or dimensions of society can be maintained within certain specified boundaries, together forming the "carrying capacity" of the respective system or society (Ametepey et al., 2023). A crucial point here is whether a sustainability approach can be categorised as "weak" or "strong" sustainability.[2] "Weak" here means that the three dimensions or aspects can substitute each other. Following this way of thinking, for example, when the number of pollinators is decreasing the function of pollinating insects in nature could be taken over by the development of specialised robots. The human-made dimension or capital 'economy,' from which this technical innovation originates, should thus equalise or completely take over the dimension or capital of<nature>. In "strong" sustainability, this substitution is not possible; on the contrary, natural and human-made dimensions are even

[2] Following 'Our Common Future', the United Nations Development program (UNDP) initiated the Human Development Index, which monitors the trends in education, health and per capita income in order to determine human wellbeing.

complementary (Daly, 1999; Ott, 2003; Ott & Döring, 2004). Strong sustainability therefore recognises the interdependence of the three dimensions or aspects of sustainability.

A variation of the Three Pillar model of SD is the introduction of a "fourth pillar of sustainability," which identifies environment, economy, social and culture as the basis of Sustainable Development (Hawkes, 2001). Jon Hawkes argues in favour of this division, as this model assigns a more differentiated role to the "social" dimension in that people can find and live out their personal and social value within the "culture" area, which differs enough from other indicators of social sustainability (such as the satisfaction of basic needs) to justify a separation into the "social" and "culture" pillars.

Subsequently, much research has been devoted to the specification of the respective carrying capacities within the three domains. With respect to the ecological dimension, this research was summarised in the "Planetary Boundaries" (Rockström et al., 2009). In the so called "Doughnut Economics" (Raworth, 2017), (sustainable) development requires not only to stay below the ecological ceiling of the planetary boundaries but also above a social foundation which specifies basic living conditions and social justice.

In order to translate the carrying capacity further into concrete goals, the UN Member States adopted in 2015 a shared blueprint for peace and prosperity for people and the planet, now and into the future. At its heart are the 17 Sustainable Development Goals (SDGs), which are an urgent call for action by all countries—developed and developing—in a global partnership. These countries "recognise that ending poverty and other deprivations must go hand-in-hand with strategies that improve health and education, reduce inequality, and spur economic growth—all while tackling climate change and working to preserve our oceans and forests" (United Nations, 2015). The individual goals range from 'no poverty', via health, education and environment to 'decent work and economic growth' (United Nations, 2015). Apparently, the underlying assumption is that reduction of the impacts on the environment can be achieved in combination with continued economic growth.

However, the contradiction of decoupling the unlimited economic growth from the earth's carrying capacities aggravate the challenge of a SD, making it urgent and existential (Harribey, 2002). Nevertheless, the European Commission has made the same assumption in their new European Green Deal.[3] However, as discussed earlier, the required absolute decoupling of materialist economic growth and environmental pressure has nowhere been achieved or demonstrated (Jackson, 2009). Exponential growth is maintained, instead of fundamentally moving towards a steady-state economy or even a degrowth economy (Daly, 1999; Jackson,

[3] The European Green Deal and European Parliament and of the Council of 14 December 2022 Amending Regulation (EU) No 537/2014, Directive 2004/109/EC, Directive 2006/43/EC and Directive 2013/34/EU, as Regards Corporate Sustainability Reporting. Official Journal of the European Union, 16 Dec. 2022, L 322/15. EUR-Lex, https://eur-lex.europa.eu/legal-content/EN/TXT/?uri=CELEX%3A32022L2464.

2009). This reflects itself in the ambivalence between a pro-growth developmentalist perspective on SD and no-growth environmentalist concerns (Mitcham, 1995).

Furthermore, critiques from the Global South towards SD agendas are present, arguing that it remains a privileged endeavour since hundreds of millions of people cannot even cover their basic human needs on the short term, let alone doing this in a sustainable manner. Or even more prominent and recent is the viewpoint that SD allows the Western capitalist imperialism to continue neo-colonial behaviour and dynamics within a new myth of 'green growth' (Corinne, 2023; Jhagroe, 2024).

And so, the debate on SD goes on and knows a dense character, starting to become more and more a global discussion. In the 1990s and 2000 when SD became more part of mainstream business the Triple Bottom Line of People, Planet and Profit were introduced (later "Profit" being transformed in "Prosperity" and also "Purpose"). The Triple Bottom Line are critiqued as corrupted interpretation of Sustainable Development and allowing Green Washing and a diluted SD understanding (Wayne and McDonald, 2004). This is certainly a problematic issue, and simultaneously, can also be seen as contributions towards a growing consciousness that SD is not merely a nice-to-have or reporting directive, but the very backbone of our common human and planetary future. In order for SD to become part of the very backbone of societies' common future, its deeper and integral layers need to be further discussed and explored.

3 Socio-ecological Crisis as a Cultural-Spiritual Crisis

There is a deeper layer that goes beyond the symptoms of the economic and ecological dimensions. It is what Pope Francis calls the move towards a *"bold cultural revolution"*, unveiling an "ethical, cultural and spiritual crisis of modernity" (Francis, 2015, p. 117; p. 89). In this cultural-spiritual dimension—cited earlier from the 50st anniversary report of the Club of Rome—there lay the axioms of our global society, our paradigm about human and earth development. These symptoms of what we could call the signatures of our time, interlocking crises, or polycrisis, are asking for a systemic change where the most effective leverage point is to transcendent not only the current paradigm, but paradigms in general (Meadows, 1999).

> There is yet one leverage point that is even higher than changing a paradigm. That is to keep oneself unattached in the area of paradigms, to stay flexible, to realize that no paradigm is "true", that everyone, including the one that sweetly shapes your own worldview, is a tremendously limited understanding of an immense and amazing universe that is far beyond human comprehension (Meadows, 1999, p.19).

In 2018 Paul Kingsnorth as well-known (ex-) climate activist, writer and philosopher called out for a "new philosophy for human and nature" where he argues that the ecological crisis beyond its symptoms needs to be understood as a spiritual crisis (VPRO Tegenlicht, 2018).

> [To] work on yourself, your norms and values. What human being you want to be? And how to bring in your talents [...] I would never have used the word spirituality publicly. But this is the conclusion to which I have arrived. The crisis which we experience, the by ourself created downfall of nature, and the likely collapse of our modern civilisation in this century, is in essence a spiritual crisis. Because this must be the first civilisation in history, where nothing is sacred anymore. There is no central faith, but moreover: no notion anymore from something greater than ourselves. Something to have reverence for. According the myth of progress, we have transcended the outdated notion of faith and sacredness we find primitive. And this, I think, this lack of the divine is cause of the destruction we create. We have put ourselves in the centre. We have become individuals that devour the world. And my conclusion, or maybe it's only a step, is that if we want to have future for our current culture, it needs to be found in the sacredness of nature. We need to go back, or forward to a reverence for the divine in all. The divine in the sea, the rock and spirits in the wind, or however you might want to call it. If you only think that the web of life where we are part of, is solely a resource to be used, you are lost. (Kingsnorth as cited in VPRO Tegenlicht, 2018)

Kingsnorth points out hat each action comes from a certain way of thinking, from a faculty of knowledge that can be seen as part of the cultural-spiritual dimension of SD. Currently, in modern sciences, the "spiritual dimension" can only be something to be studied by theologists, or even anthropologists, sociologists, or philosophers, but barely be involved as faculty of knowledge in inquiry or research (Wilber, 2000). However, the work of Steiner—building on for instance Goethe and German Idealism—does approach and integrate the spiritual dimension of reality as faculty of knowledge, being, and origin (Heusser, 2016).

Steiner introduces an ontology and anthropology including a threefold approach to the human being in addressing the spirit (also "I"), soul (or psyche), and body. These build a first part of a spiritual or holistic anthropology (Steiner, 1904/2013a). The soul (or psyche) is introduced as mediator between our bodily, or physical sensorial reality on the one hand, and the spirit as non-sensorial reality on the other. The soul is the receptor where sensorial experiences through our senses meet non-sensorial experiences. Through the constitution of our soul, being organised with a *thinking, feeling,* and *willing* faculty, the human being is supported to perceive and understand the world in its manifold. If, for instance, the human being would not have a (non-sensorial) thinking faculty,[4] all empirical perceptions perceived through our senses would be chaos (Steiner, 1886, 1894/2021, 1904/2013a, 1904/2022a). The thinking organises our experiences into concepts and therefore creates reference points. Both the spiritual and physical realms of life influence our state of the soul, that needs to navigate through it. Thus, the human being can be seen as being of two worlds, a sensorial one and a non-sensorial one. The spirit in the human being is the directive faculty, that operates with the soul as mediating faculty into the physical reality. Therefore, human behaviour and action can be

[4] With *faculty* we mean a part or section of the human condition.

seen as a direct expression of both the soul and spirit (Steiner, 1904/2013a, 1904/2022a). If someone decides to cut down a rain forest for financial profit, the action can be led back to a certain worldview descending from the faculty of his or her spirit. Of course this concerns a generalisation, in reality human action is very complex, multi-layered and nuanced. However, the hypothesis that behind every action one can discover a larger background of values and worldviews, remains crucial for a deeper insight in the socio-ecological crisis.

The inner connection to the faculties of thinking, feeling, and willing as part of our psyche are getting more and more challenged. Under an increased materialistic worldview, the human being is increasingly losing the connection not only to the world (including Earth and cosmos) and each other, but ultimately also to oneself. This as well is understood as the cultural-spiritual source of the current sustainability crisis (Berry, 1999; Scharmer, 2007; Steiner, 2022). The fast increase of embodiment practices, mindfulness, yoga, and the increased use of certain drug substances testifies of a search to handle the personal crises (depression, suicides and aggression, but also the search for meaning), as well as the socio-economic crisis (increasing gap between financially poor and rich, inequalities, increasing geopolitical conflicts), and the ecological crisis (climate change, loss of biodiversity and extreme land degradation by big agro, big tech, big industry, etc.). Crafting new narratives and stories, as already mentioned, is therefore fundamental to create perspective and orientation (Eisenstein, 2018).

The disconnect to the *self*, the *other*, and the world is therefore marking the cultural-spiritual crisis of humanity and shapes part of a deeper root cause of the ecological, social, and economic challenges we face. These elements are of course not new and part of many world religions and Indigenous Knowledge and practices. The main point here is to not introduce them as something new, but to reintegrate them into an Integral Sustainable Development.

4 Evolving Integral Views on Sustainable Development

In a way, this cultural-spiritual component provides the ground that the other pillars of Sustainable Development need to stand on. It goes a decisive step further than Brundlandt's political definition, as it is no longer only about ensuring that basic needs are met but also focusses on the dignity and development opportunities of every human being as a compass for Sustainable Development. The term "development" is thus elevated from a purely existential-pragmatic to a moral-civilisational dimension. This idea can already be found in, for instance, ancient Greece, exemplified in Aristotle's "Ethica Nicomachea": the development of virtues as a basic prerequisite for a "good life". Or even above the entrance of the Temple of Apollo in Delphi, where the two essential lines already state *"Know thyself"* (Gnothi seauton) and *"nothing in excess"* (Meden agan), pointing to the inner (Know thyself) and outer (Nothing in excess) parts of a Sustainable Development.

As indicated, the cultural-spiritual component was already touched upon in the successive Limits to Growth reports and in later works of Donella Meadows in the 1990s. These early notions have found their way for instance into the *Earth Charter,* initiated during the UNCED Rio Earth Summit by Mikhail Gorbachev and Maurice Strong in 1992 and launched in 2000. The very first principles of this charter are the "Respect and Care for the Community of Life" with as first point "Respect Earth and life in all its diversity: (a) Recognize that all beings are interdependent, and every form of life has value regardless of its worth to human beings; (b) Affirm faith in the inherent dignity of all human beings and in the intellectual, artistic, ethical, and spiritual potential of humanity" (Earth Charter International, 2000). The cultural-spiritual dimension or pillar of Sustainable Development started here to become articulated in a UN context.

Inspired by these insights with respect to the relevance of human dignity, and the human intellectual, artistic, ethical, and spiritual potential, Van Egmond and de Vries (2011) studied value orientations within the Dutch population by means of a social survey amongst 40.000 people. The results of the survey appeared to correspond strongly to Steiner's worldview and 12 differentiated perspectives (Steiner, 1919/2015). This worldview was translated into a vision of 'human nature' and 'human dignity' and a notion of 'sustainability ethics'. On this basis, sustainable development was defined as identical to the preservation of human dignity, where human dignity is the measure of coherence between the values in the pattern of 'human nature.' It also implies a human purpose (telos) and herewith a 'meaning'. As such it anticipates Broman's and Robèrt's framework (Broman & Robèrt, 2017) and Schneidewind's definitions of SD (Schneidewind, 2018), which stresses the dignity and development opportunities for every human being, and harmony with the carrying capacity of the earth shapes the concept of SD.

Broman's *Framework for Strategic Sustainable Development* (FSSD) refers to eight<Sustainability Principles>, five of which are categorised in the socio-cultural dimension: health (physical and mental), influence (participation), competence (education), impartiality (no discrimination) and meaning-making (Broman & Robèrt, 2017). It is important for the FSSD that people are not prevented from realising these principles. Meaning-making was meant to be an individual sustainability principle, but gets an overarching societal dimension if posed for the whole of society, albeit that it remains unclear which meaning is made in 'meaning-making'.

Uwe Schneidewind (2018) is touching on this notion of meaning-making for the larger society. He describes Sustainable Development as "a further step in the evolution of human civilisation towards a world in which the dignity and development opportunities of people everywhere are the compass for social, political and economic action today and in the future" (Schneidewind, 2018, p. 23).[5] Also here, a deeper and broader dimension of SD are addressed.

A recent related publication concerning moral and inner development in the context of sustainability, can be found in the Club of Romes initiative "Earth4ALL", where the "inner dimensions" of sustainability and system change

[5] Translated from German by the editor.

such as the psychological, social, and spiritual factors are presented as often overlooked parts (Bristow et al., 2024). Stewarded by Harvard University, Stockholm University, and the World Business Council for Sustainable Development (WBCSD), the *Inner Development Goals (IDGs) were launched* in 2021[6] (Bristow et al., 2024). Herein *Being, Thinking, Relating, Collaborating and Action"* were introduced as inner prerequisites in order to achieves outer transformation and Sustainable Development. These IDGs serve as a complementary tool to the already existing, ecological, social and economic SDGs. As discussed above, the IDGs were introduced because achieving the SDGs is deemed impossible without inner development goals.[7] The above mentioned five key IDGs, accommodate 23 skills related to these 5 dimensions.

These five IDGs can be linked to the anthroposophical anthropology: (1) *Being* as the "I" or "Spirit", (2) *Thinking* directly linked to the faculty of "thinking" of the soul's constitution, (3) *Relating* and (4) *Collaborating* as part of "feeling" of the soul's constitution and (5) *Acting* as part of the "will" of the soul's constitution. The five inner dimensions of the human being are meant to approximate the more fundamental human constitution. A constitution that all human beings beyond their differences and uniqueness have and therefore share. This can be seen as concrete part of the spiritual dimension of (individual) SD.

Another evolving view on SD can already be found in the presented Threefold Social Order of Steiner (2019, 2022b), and their relation to the expansion of the three pillars of SD:

Anthroposophy's Threefold Social Order and the Pillars of Sustainable Development Anthroposophy's Threefold Social Order and the Pillars of Sustainable Development

After World War I, in 1919 Rudolf Steiner was approached by industry leaders to draft a concept of what then became the so-called Threefold Social Order of society. An approach in which society is organised in interconnectedly but differentiated three sovereign domains, where one domain cannot dominate the other: cultural (or spiritual) life, legal (or political) life and economic life. The three domains are supposed to serve the wellbeing and development of both humanity and the Earth with its manifold forms of life. (Steiner, 2019, 2022b; Schmelzer, 2017; also see Chap. 16).

Steiner sketches how *freedom (*Liberté*)* is attributed to cultural life (education, research, etc.), *equality (*Égalité*)* to legal life (rights, justice, and politics) and solidarity (Fraternité) to economic life (labour and production) (Steiner, 2019, 2022b). This for instance means that cultural and economic life are interconnected but separate in the sense that *freedom* should not reign in the economic but is functional in cultural life to create large diversity in worldviews. Solidarity or brotherhood

[6] IDG. Inner Development Goals, https://innerdevelopmentgoals.org/about/. Accessed 12 Sept. 2024.

[7] IDG. Inner Development Goals, https://innerdevelopmentgoals.org/about/. Accessed 12 Sept. 2024.

should be the driver of every economic activity to become altruistic and serve the greater wellbeing of society and the Earth. A remarkable view already at the beginning of the 20st century, since at that time the highly profit-maximalisation oriented free market mechanisms were not as dominant as they have been for the last three decades. Nowadays one could state that freedom slipped out of the cultural realm into the economic realm where the *Big* win on so-called free markets, which even led to:

> Oligarchs in the US, Russia, Ukraine, and around the world further their private interests by playing the nation-state system against itself, quietly benefitting from the tremendous profits won through international trade while fuelling nationalist hatred at home through their monopolistic control of media. Further, the fusion of state and economic interests represented by the military-industrial complex has turned war and the threat of war into a highly profitable enterprise [...]. (Segall, 2023, p. 246)

Simultaneously the Threefold Social Order stresses that the social realm of law and politics should create the backbone for equality, equity, and justice, not meaning a "communist" state where governments are overregulating, censoring, and ruling for instance cultural life by determining education, academic life, or freedom of the press. As of January 2025, these are still highly relevant themes since the world is shaking geopolitically on its fundaments, with (even Western) world leaders threatening to economically or militarily invade allied countries, misusing their capital and power to manipulate mass opinion and therewith undermining our democracies (The Guardian, 2025).

Furthermore, equality in the form of law or regulation should not reign cultural life, since there the freedom of plurality and expression should be supported (Perlas, 2019; Schmelzer, 2017; Segall, 2021; Steiner, 1919/2022b, 2019). The Threefold Social Order was not intended to be a one-size-fits-all ready-made concept, but a step into a direction to overcome one-sided tendencies in society:

> There are no easy solutions to resolve the conflicts that continue to lead our world down the warpath. History is a bloody ruin of failed ideologies. What social threefolding offers is not a ready-made solution but a new understanding of the problem rooted in a deeper perception of human needs, desires, and capacities. Its specific suggestions in the realms of politics, economics, and cultural life seek to balance the one-sidedness of more well-worn modern ideologies stemming from progressive and conservative orientations alike. (Segall, 2023, p. 246)

In this publication, the Chap. 16 "Earthbound Economics" by Nathaniel Williams elaborates further on the Threefold Social Order.

The Threefold Social Order in Ecological, Social and Economic Sustainable Development

Applied to the common three pillars of ecological, social, and economic sustainable development, one could recognise that the Threefold Social Order brings pioneering values and definitions: a sustainable economy not mainly directed by regulations and law (that belong in the legal realm) or free market competition and profit maximalisation, but by the principle of solidarity. Anthroposophy herewith indicates towards an expanded understanding of a sustainable economy, one not necessarily of green growth, a steady state economy, or a degrowth economy, but an economy led by solidarity and associative organisational structures carried by civil society (Perlas, 2019).

One could argue that solidarity in and for itself would, however, not guarantee the mitigation of planetary boundary violations. In this regard, Walter Johannes Stein, engaged in the Threefolding movement, emphasised that the economy cannot be seen as separate from the Earth but is inherently a part of it (Stein, 1937).

> It is the present-day tendency to be concerned only with individual nations and groups, and to regard the earth only as the stage on which the action takes place, as the mere provider of raw material. If we conceive economic life as forming part of the life of the whole earth, the complexion of the matter changes entirely, the inexorability of the economic forces largely disappears and man is thus enabled to build up his own sphere. (Stein, 1937, p. 14)

A sustainable economy, according to Stein, starts with the awareness that this very economy is the human fundament of not only individual and collective welfare and wellbeing, but also is the realm where humans' awareness and responsibility for the Earth's development expresses itself:

> Man is thus a creature, that is to say, a phenomenon in the process of evolution in the same way as are other creatures; but in him the universe becomes conscious of itself, and in the form of perceptive thinking, it is able to understand the creative powers. Man must then be recognized as the being in which al these creative powers are embodied, holding thereby dominion over all nature and living things. In awakening to his own responsibility for the whole, man becomes aware of himself as a social being. (Stein, 1937, p. 12)

For this awareness and responsibility to arise, Stein indicates that:

> to obtain a sensible view of the earth we must see it not as an inanimate clod but as a living organism whose powers are balanced, and whose life-force run into its furthest extremities, demonstrating the principle of the whole in every part, and the way in which the welfare of each such part is bound up with the welfare of the whole. It can be perceived thus, that the earth should be regarded not only as a living organism but as the producer of moral stimuli also. (Stein, 1937, p. 12)

Herewith a cultural-spiritual domain of SD out of the anthroposophical approach becomes visible: the moral stimuli to see the Earth as living organism where human activity does not only encompass fulfilling basic human needs, but potentially carries the very responsibility for the Earth as a living whole. An "Earth Dignity" can thus be attributed to ecological sustainability from the viewpoint of the Threefold Social Order.

In this line of thought, ecological sustainability is expanded with an "Earth Dignity," social sustainability with the equal rights that should reign in legal and political life (and not for instance in cultural life) and a sustainable economy is rooted as very part of the earth systems with the potential out of a solidarity principle, shaping the development of the Earth as living organism. The role of the human being on the Earth can in this context move from a destructive anthropocentrism into a possible appropriate or constructive anthropocentrism.

From here we will further explore the contribution of anthroposophy towards Sustainable Development along approaching the socio-ecological crisis as cultural-spiritual crisis and through an "appropriate" or "constructive" anthropocentrism.

5 An Appropriate Anthropocentrism

The Anthropocene, the *human era* in which human activity has a lasting influence on the Earth's vital ecosystems, is introduced to define the start of a new geological era (Crutzen and stormer, 2000). Although the term has long been disputed and classified as incorrect by geologists, some scientists (Steffen et al., 2015) define the start of the Anthropocene during the start of the so-called "Great Acceleration": the exponential growth of the economy, population, and resource consumption at all levels from the middle of the twentieth century (initiated at the beginning of the industrial revolution, or according to some already 12.000 years ago when agriculture started). This start took place on Monday 16 July 1945, when the first atomic bomb was detonated in the desert of New Mexico, USA.

> Setting the starting date of the Anthropocene at the beginning of the Great Acceleration makes it possible to specify the onset of the Anthropocene with a high degree of precision. On Monday 16 July 1945, about the time that the Great Acceleration began, the first atomic bomb was detonated in the New Mexico desert. Radioactive isotopes from this detonation were emitted to the atmosphere and spread worldwide entering the sedimentary record to provide a unique signal of the start of the Great Acceleration, a signal that is unequivocally attributable to human activities. (Steffen et al., 2015, pp. 13–14)

Since then, human influence and destructive potential appears to have penetrated from a local to a global scale, and the age of humankind, the Anthropocene, is said to have begun.

This interpretation of the Anthropocene is a dramatic categorisation and description of the age of humankind, which could—and in our opinion, should—be done quite differently following the question how a sustainable future looks like. To portray humankind as a mere destroyer appears to be partly correct, however does not show the also present constructive and creating force humanity. A fundamental question is what the human being is, what human development is, and ultimately what the responsibility of humankind as part of the Earth could look like beyond satisfying human needs, nature conservation, and climate justice.

McKanan (2017, p. 226) argues that one of anthroposophy's key contributions towards the environmental movement is "an appropriate anthropocentrism that allows us to experience ourselves as fully at home in the world [...] protects us from the temptation to respond to ecological devastation with bitter misanthropy." Central to this appropriate anthropocentrism is "the compelling hope" that humankind can live in harmony with the earth and all its forms of life. This encompasses and provokes a new ethical framework especially since the "Great Acceleration" of the twentieth century due the overgrowth and omnipresence of modern technical innovation. Hans Jonas therefore called out for an "imperative of responsibility", where humanity needs to define a new ethical framework based on the responsibility for the Earth, since the enormous destructive potential that modern technology unleashed and demands a totally revised imperative of responsibility (Jonas, 1979).

An appropriate anthropocentrism is questioning the role of humankind on Earth. Rudolf Steiner introduced a radical image of the role of humankind, related to the suffering of the Earth under human activity: "Our mother earth has hardened through pain. Our mission is to spiritualise her again, by reworking her through the power of our hands into a spirit-filled work of art" (Steiner, 1909/1992, p.113).[8] This leads to a first key proposition of an appropriate anthropocentrism, in which the human being is not a destroyer or ultimate dominator but is being described as a potential artist with both a healing (regenerating) and transformative (generating) role. Biodynamic agriculture as the first organic agriculture practice, founded as early as 1924, is one practical example in which this cornerstone of anthroposophical practice is addressing an appropriate anthropocentrism (more in Chap. 5 , 10, and Part 4).

This role of humankind as both healer and transformer, does imply that the earth not only needs to be preserved, protected, or regenerated for what was destroyed (the healing aspect), but that there is a future image where the Earth becomes a "spirit fulfilled" artwork itself (the transformative aspect). It is what McKanan calls a "Planetary Transmutation," the transmutation of the Earth, not only by random happenings but by the active participation of humanity in the Earth's development itself. One could argue that without the human factor, Earth already is an artwork, which of course we can only confirm. However, the human contribution is to introduce *culture* in its broadest potential; to *cultivate* development of the Earth itself in interdependence and interrelationship with all living organisms present. Something humankind is already doing, but momentarily in a most destructive way and thus a "negative" example of what the potential holds.

Humankind as co-creator is thereby given a healing and transforming artistic task and indeed a deeper reason for its presence on the Earth, and thus becomes part of the new story of Earth and humankind. A story that can hold "our common future" addressing each nation, culture, and individual to co-create and carry responsibility for the Earth's future collectively.

A second key proposition of anthroposophy concerning an appropriate anthropocentrism, is perceiving the Earth as living organism. This organism is placed in a cosmic composition precise enough to provide the prerequisites to carry life as

[8] Translated by the editor.

we know it. Unlike the pre-scientific assumption of modern natural sciences that the Earth's existence is a big coincidence, anthroposophy introduces the hypothesis that the Earth itself knows a moment of birth and development alike all life (Steiner, 1910/2013b; Wachsmuth, 1945/1980; Zoeteman, 1989, 2009). The human being is understood as inherent part and organ of this larger organism but also carries the possibility to change the cause of the Earth's development, which is being showed by the precise current global situation from the negative feedback loop. However, this possibility to change or alter the Earth's development path is directly linked to the first proposition of the appropriate anthropocentrism, stating that at the moment it is humankind's obligation and responsibility to heal and transform the Earth through a future proof path of development. As already stated, this goes beyond the protection, preservation, conservation and regeneration of nature, which is only one part of human's responsibility regarding the Earth's developmental path. The second part marks the task of transforming the Earth and enabling its fullest development potential precisely by positive human activity. In this sense, Sustainable Development, as a rising meta-discipline, can be seen as carrier of Inner Development Goals (IDGs), Outer Development Goals (SDGs), and even Development Goals for Humanity and Earth. Stressing development as most essential part and driver of change and transformation, this calls for a transition from System Thinking towards Living Organism Thinking. System Thinking still represents a mechanical "system dynamics" approach, whereas a Living Organism Thinking introduces the faculty of development towards a transcendental Earth destination. These attributes are part of a "cultural-spiritual" dimension of Sustainable Development, complementing SD towards an *Integral Sustainable Development.* In order to evolve this dimension to the existing pillars of SD, we will make a first sketch in the following section.

The main critique on the above could be that the anthropocentric approach is still dominant and the risk to place the human being at the top of creation remains. This is rather a more precise critique on the current exponential economic growth and neoliberal practices which descents from a "human as dominator" attitude. With an appropriate anthropocentrism, we strongly want to invite the discussion to go beyond a naïve believe that the current systems and human activity can be balanced out by regenerating and conserving nature or green growth only by making consumption "green" or "pulling back" the human being in its place before the agricultural start 12.000 years ago. It rather requires exploring and totally renewing the role and co-creative attitude and responsibility humanity can potentially unfold.

6 Introducing Integral Sustainable Development

The growing awareness of the cultural-spiritual component of sustainable development and the introduction of the IDGs challenged us to reformulate the current Brundtland definition into a more inclusive, integral concept of Sustainable Development wherein the ethical but also inner transformative factors are taken into account, as prerequisite for outer change and societal transformation towards a society acting towards both Earth and human dignity. This expansion of the current SD concepts and also new integral approaches will be referred to by us as *Integral Sustainable Development* (ISD).

The *Integral Sustainable Development* approach introduces an integral dimension that embraces and interweaves the existing ecological, social (cultural), and economic sustainability. Therefore, we will introduce the concepts of "World-viewing", "World-sensing" and "World-acting" that can be applied on the three existing pillars of SD, known as ecological, social, and economic sustainability. World-viewing, World-Sensing and World-Acting are derived from a bioethical value-clarification model (Simon et al., 1978; Keulartz et al., 2002; Verhoog et al., 2003). They are also connected to the thinking-feeling-acting interpretation of the five IDG's. They consist of the *Cognitive* (World-viewing),[9] *Emotive* (World-Sensing) and *Normative* (World-Acting) Dimensions of the human constitution, and will be linked to anthroposophical anthropology as well. These form a first description of how the spiritual dimension of Sustainable Development could be approached.

World-Viewing

The *Cognitive Dimension* or the *"World-viewing"* is the first dimension. Here lie the concepts, mental schemes and worldviews from which our actions emerge. Of course, one could also call this ontology. This concerns, for instance, our understanding of the human being, nature, Earth, and cosmos; the relationship of humankind and nature; and also different worldviews from where these concepts arise, including scientific, artistic, the civil societal, transcendental, religious aspects, etc. Complemented with the anthroposophical anthropology (Steiner, 1886/2003, 1894/2021, 1904/2013a, 1910/2013b, 1919/2019, 1924/2022c) this could be also called the *spiritual dimension* of value-clarification, which starts with the understanding of the human condition as being three-folded in body, soul, and spirit. World-viewing is the part of the human being's condition that connects the sensorial world and the non-sensorial world, with thinking as first step or example of a non-sensorial activity: thinking here interacts with observation and the creation of concepts of perceived reality (Steiner, 1886/2003).

In daily life it means that, for instance, all products, but also all matter, have an origin, a concept, a spiritual dimension behind it.[10] It implies that all we create or do, has a spiritual dimension or specific worldview preceding it. This World-viewing then needs to be explored behind the social, ecological and economic pillars of Sustainable Development before we create and implement actions: what is the concept and understanding of ecology, of nature, of the social life, and the economy? Taking the Leverage Points of Meadows, World-viewing could be also

[9] We introduce World-viewing, World-sensing and World-acting in this publication.

[10] We will not go into this fundamental part of continental philosophy from i.a., Hegel's but also Goethe on the spirit here.

seen as understanding "paradigms" or even "the power to transcend paradigms" (Meadows, 1999) by recognising the concepts and mental schemes behind how we understand ecology, society (and culture), and the economy.

Furthermore, World-viewing also includes what McKanan (2017) calls "Cosmic holism", as a contribution of anthroposophy to the environmental movement. From this cosmic holism perspective, the Earth as an organism is embedded in a broader cosmic development that needs to be taken in consideration in order to get a real "systems perspective", or as we would state here: a "post-system thinking" attitude, towards a *thinking in organisms.* This thread connects to the fundamental statement that the Earth needs to be studied, understood, and treated as living organism (instead of a mechanical system) with a past, present, and future biography. Kepler (1619) already stated in his *Harmonices Mundi* that the Earth's physical body belongs to a living being attributed with an animal-like constitution. This understanding of the Earth as living organism according to Kepler, would need centuries to be further examined. The call of Kepler, however, moved to the background after the Industrial Revolution brought a strong mechanical worldview in place, affecting human civilisation on all its dimensions (Capra, 1982; Mumford, 1934). Gunther Wachsmuth, as first director of the Natural Sciences Section at the Goetheanum,[11] reintroduced this task in his work *Earth and Human* (1945). He already warned not to alter our atmospheric layers of the Earth by combustion and other technological effects, before knowing what the actual feedback loops of these interventions are (Wachsmuth, 1945/1980, pp. 12–13). In order to overcome the dominant mechanical worldview of the nineteenth and twentieth century, he called for "building a new worldview" because "our human action is, as mentioned, fundamentally determined by the essence of our image of the Earth" (Wachsmuth, 1945/1980, pp. 14, 19). World-viewing herewith is directly addressed and understood as first part of the spiritual dimension of SD. Rudolf Steiner similarly addressed the impact of the human activity of thinking—or *World-viewing*—on the environment directly.

> And indeed, if human beings do not enliven their thoughts, if people remain stuck in merely intellectual, dead thoughts, they must break the Earth. The breaking, however, begins with the most delicate element, with warmth. (...) By further and further developing merely intellectualistic thoughts, the warmth atmosphere of the Earth will be corrupted.[12]

Before we go to human action which we call here "World-acting" as results of our thinking and World-viewing, we firstly introduce the dimension in between, which we call "World-sensing", as the emotive part of Integral Sustainable Development.

---

[11] The Goetheanum was founded in 1924 by Rudolf Steiner and the Collegium in Dornach, Switzerland as «Freie Hochschule für Geisteswissenschaft», with nowadays 12 faculties inquiring, teaching and working with anthroposophy.

[12] Translated by the editor from the original edition Steiner, Rudolf. *Die Impulsierung des weltgeschichtlichen Geschehens durch geistige Mächte.* 4. Auflage, Rudolf Steiner Verlag, 1989. Gesamtausgabe, GA 222. P108 (Steiner, 1989, p. 108).

World-Sensing

The second dimension of an ISD can be linked to the *Emotive Dimension of* the bioethical value-clarification, or what we call here "World-sensing". This dimension introduces *four fundamental attitudes.* Linked to anthroposophical anthropology the *Emotive Dimension* is equivalent to the "soul" or psyche, being defined with the faculties of *thinking, feeling,* and *willing* (Steiner, 1884/2021, 1919/2019). In anthroposophical anthropology, the thinking can be understood as bridge to the spiritual realm, and the will can be seen as bridge to the physical realm. As written before, the soul herewith mediates between the non-sensorial and the sensorial reality.

Four fundamental attitudes as described in literature address the relationship between human and nature and are presented as: (1) ruler; (2) steward; (3) partner; or (4) participant (e.g., Verhoog et al., 2003; De Groot et al. 2011). Humankind as ruler puts the human being in the centre and considers the earth merely as raw material for our produce. The human being as steward confirms the ruler attitude but puts an emphasis on exploiting the earth in such a way that following generations can also profit. The human being as partner is built on the respect for nature. And the human being as participant acknowledges reciprocity between the development of humankind and nature. The Ruler and Steward only acknowledge extrinsic values of nature (the utility) and the partner and participant also acknowledge intrinsic values of natures (the value of its own, the dignity).

This *World-Sensing* with its *four fundamental attitudes* provides a base for an integral understanding of the social, ecological, and economic pillars of Sustainable Development: from what fundamental attitude does one depart before implementing action? Does one strive for "ruling" the ecological or natural realm? Does one see social life and challenges as subject to be controlled, organised and ruled? Does economic activity rest on competition and getting the biggest market shares? Or is there a striving beyond humankind's reproduction and self-interest, and even beyond just having respect for nature, moving towards a participatory and co-creative reciprocity between humankind and the pillars of sustainable development? World-sensing therefore can expose the fundamental attitude before even starting Sustainable Development in any pillar or more specific field of work, sector, or industry. At the same time, it can be used as an analysis and evaluation indicator to understand specific challenging situations that are in need of SD.

World-sensing also includes what McKanan (2017) calls an "appropriate anthropocentrism", as described in the previous Sect. (3.3). This appropriate anthropocentrism adds a fifth fundamental attitude that we would like to call "co-creator", which can be understood as one core aspect of human development in the relationship to the Earth's development derived from Anthroposophy (Steiner, 1909, 1910/2013b). Co-creating herewith is understood as an emotive attitude towards nature, that is a prerequisite for positive and constructive action from an integral sustainability point of view.

In anthroposophical biodynamic agriculture, co-creating with nature is the ultimate task of the farmer—firstly, in *healing* the land (regeneration, preservation and conservation of soils, plants, and animals), and secondly in *transforming* and evolving the very Earth substance. Or, as Steiner indicates, the "freeing of nature" from the reductionist thinking of humankind by an "ethical individualism" (Steiner, 1893, 1924/2022c).

In order to foster this World-sensing, anthroposophy offers a very broad set of contemplative practices in order to educate the inner life or psyche's thinking, feeling, and willing faculties (Steiner, 1904/2013a, 1904/2022a, 1910/2013b). More recently, these were being reintroduced in an academic context by Arthur Zajonc's contemplative understanding (2008) or adapted in Otto Scharmer's Theory U (2009, 2013) for business and governance context and in his work regarding the transition from an ego-system to an eco-system practice. And as mentioned, the recently developed Inner Development Goals also touch the training of the inner attitude and directly link them to achieve outer Sustainable Development.

The World-Sensing—as the middle of three dimensions of the Integral Sustainable Development approach—hereby forms the fundament for an Appropriate Anthropocentrism leading to constructive "World-Acting" as third and final dimension of the integral approach.

World-Acting

The third dimension–known as the *Normative dimension* by the bioethical value-clarification–we call *"World-Acting"*. Here the consequences of the two other dimensions find their culmination in human action; our concept and understanding of the world (World-Viewing) and fundamental human attitude (World-Sensing) direct all what we do or do not do. Here, we build on four *bioethical frameworks of action* as are distinguished in the literature (see Verhoog et al., 2003): (1) *anthropocentric (only the human being is considered as ethically relevant)*; (2) *zoocentric (humans and sentient animals are considered as ethical relevant)*; (3) *biocentric (all living organisms are considered ethically relevant)*; and (4) *ecocentric (here beyond individual organisms also species or ecosystems as a whole are considered ethically relevant)*. This dimension can be linked to the anthroposophical anthropology to the "physical", "embodied" or "material" part of life, where the "will" and "action" are taking place (Steiner, 1894, 1904/2013a). However, none of these four bio-ethical frameworks alone are applicable to our notion of integral sustainable acting as co-creature in nature, respecting both the dignity of the Earth and human beings. From an appropriate anthropocentric attitude, there is *some form of human-centred stewardship needed to emphasise the role of humankind as bearer of culture* (Verhoog et al., 2003). Besides, an integration of elements of a zoocentric, biocentric and ecocentric framework of action is required to respect the integrity of plants, animals, ecosystems, and the Earth as a whole. Verhoog et al. (2003) came to the same conclusion in the context of organic agriculture and proposed a holocentric ethic as more appropriate ethos overarching and integrating the existing four bio-ethical frameworks of action.

This Holocentrism (not to be confused with the Holoscene) can be understood as crucial for an Integral Sustainable Development: an action that includes all forms of life being backed by an awareness about the driving fundamental attitude (World-Sensing) and understanding and concepts of reality (World-Viewing). The

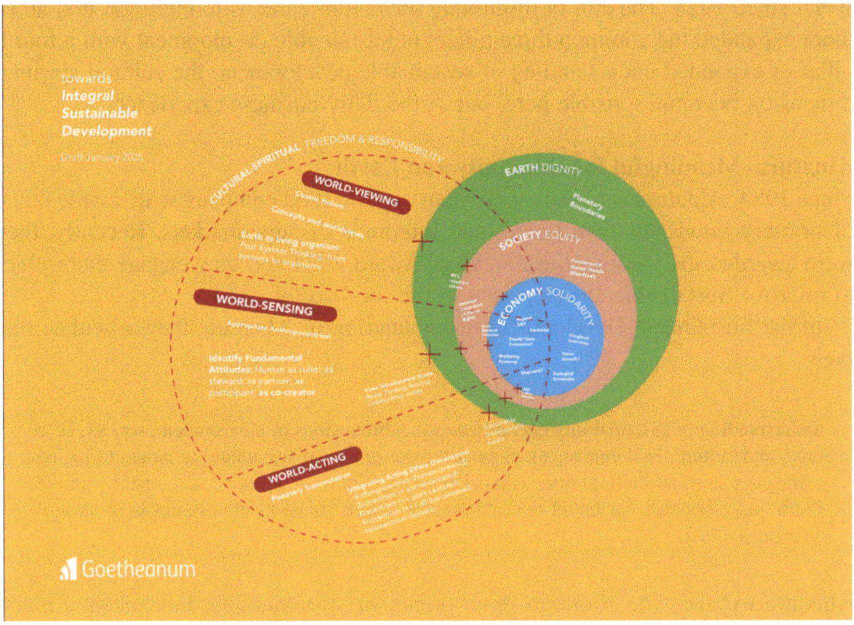

Fig. 1 Draft towards an Integral Sustainable Development. The integral approach as "cultural-spiritual dimension" of SD both embracing and interweaving with the conventional three pillars of Sustainable Development. Also added are the main values of the Social Threefold Order, solidarity in economy, equity/equality in legal/social realm and freedom in the cultural realm

Acting itself herewith is the end and beginning of an Integral Sustainable Development. This integral approach would be a possible translation and interpretation of the "spiritual dimension" that is needed to continue (sustainable) human and Earth development. The classical threefoldness of anthroposophical anthropology—of spirit, soul (thinking, feeling, willing), and body—is herewith included and indicate a way forward into an integral, holistic or spiritual approach on the very question of development in its broadest sense (Fig. 1).

7 Examples of Implementing Integral Sustainable Development

Before closing this chapter, we address the existing Integral Sustainable Development frameworks of *Alnatura*, the *Biodynamic Federation Demeter International* (BFDI), *SEKEM,* and *Eosta.* All four enterprises also contributed to part 4 *Sustainable Natives—Stories of Sustainable Development,* where their perspectives can be found in more depth. They have in common that they are all founded on anthroposophical principles and have integrated ecological, social, and economic sustainable development in their activities for decades and Demeter as part of

BFDI since 1928. The aim of discussing these four cases is to highlight that all of them expanded the common three pillars of sustainable development with a fourth pillar or expanded understanding of sustainable development: the cultural-spiritual dimension becomnes visible here, out of the daily business experience.

Alnatura–Meaningful for Humanity and Earth

Since 1984, Alnatura is a leading German organic food company with a wide range of primary goods, its own brand, and hundreds of supermarkets. Recently, they were awarded the German Sustainability Award 2024 and they engage themselves to improve Sustainable Development in the food sector.

In the Sustainable Development understanding of Alnatura, they articulate that their:

> understanding of sustainability encompasses the dimensions of environment, society, economy, and culture. In doing so, we expand the conventional sustainability model to include a unifying spiritual-cultural dimension. This reflects how we think about humanity and the Earth and embodies our belief that sustainable action stems from sustainable thinking.[13] (Alnatura, 2022)

Alnatura expands the common three pillars of sustainability and put an emphasis that all their activities should be meaningful for both humans and the Earth, including its different dimensions of life (soils, plants, animals, water, and air). Thus, the cultural-spiritual dimension introduces the notion of responsibility by directing the company's meaning to the wellbeing of humans and Earth. If this is done in a consistent manner and "we equally consider all dimensions in our thinking, sustainable action will follow."[14] The endeavour for constantly seeking a healthy balance and interaction between the interconnectedness and "inseparable" dimensions of life, is motivated by the right intentions to create an Integral Sustainable Development practice.

For Alnatura this practically means at the human level to establish "long-term partnerships with farmers and manufacturers"; "[h]igh-quality organic food" for their customers; "[m]eaningful work" for their employees, and "[r]espect for human rights throughout the supply chain". On the other hand, for the Earth they formulate to "[p]romote organic farming", "exclusive trade in organic foods, natural cosmetics, and natural goods", the "support and promotion of initiatives such as animal welfare" or the "protection of rare ecosystems", and an "environmentally friendly design of their supermarkets".

---

[13] Translated by the editor from the Alnatura website https://www.alnatura.de/de-de/ueber-uns/nachhaltigkeit-bei-alnatura/

[14] Alnatura. *Unser Beitrag für Mensch und Erde: Nachhaltigkeitsbericht 2021/22*. 2022. https://www.alnatura.de/de-de/ueber-uns/nachhaltigkeit-bei-alnatura/nachhaltigkeitsbericht/. Accessed 22 Nov. 2024. und Alnatura. "Nachhaltigkeit bei Alnatura." *Alnatura*, www.alnatura.de/de-de/ueber-uns/nachhaltigkeit-bei-alnatura/. Accessed 22 Nov. 2024. (Translated by the editor).

Seen from the Integral Sustainable Development approach, they apply and practise a clear World-Viewing, World-Sensing, and World-Acting on the pillars of SD by creating their of understanding of the cultural-spiritual dimension. World-Viewing expressed the sense that by *"consider all dimensions in our thinking, sustainable action will follow.* " World-Sensing in the fundamental attitude seeing human (and their) activity as *participatory* in relation to Earth development; and World-Acting by considering and *Ecocentric* or even *Holocentric* level of action.

SEKEM—Sustainable Development Since 1977

"Greening the desert" in Egypt with Biodynamic Agriculture, social innovation, and cultural enrichment was the founding impulse of Sekem as early as 1977. It led towards nowadays almost 50 years of Sustainable Development for thousands of co-workers and a worldwide distribution of their products. In the fall of 2024, Sekem won the "Gulbenkian Prize for Humanity",[15] after the jury including Angela Merkel and Johan Rockström selected Sekem as one of three winners of the year.

Sekem always had Sustainable Development at its centre as formulated by its founder Ibrahim Abouleish (1937–2017).

> In the midst of sand and desert I see myself standing as a well drawing water. Carefully I plant trees, herbs and flowers and wet their roots with the precious drops. The cool well water attracts human beings and animals to refresh and quicken themselves. Trees give shade, the land turns green, fragrant flowers bloom, insects, birds and butterflies show their devotion to God, the creator, as if they were citing the first Sura of the Qu'ran. The human, perceiving the hidden praise of God, care for and see all that is created as a reflection of paradise on earth. For me, this idea of an oasis in the middle of a hostile environment is like an image of the resurrection at dawn, after a long journey through the nightly desert. I saw it in front of me like a model before the actual work in the desert started. And yet in reality I desired even more: I wanted the whole world to develop. (SEKEM, 2023, p. 7)

From the very beginning Sekem introduced a Sustainable Development understanding built on the expanded definition of SD, not only including a fourth cultural dimension but also giving their own definition and goals of the ecological, social and economic sustainability expressed in their sustainability flower. The flower has 16 vision goals inspired by the Sustainable Development Goals and expanded with a holistic dimension:

> We recognize the interconnectedness of all aspects of life and our impact within this unity. This understanding drives us to carefully balance our actions, managing trade-offs inherent in sustainable development. While we adhere to standards to implement best practices, we also value open dialogue with stakeholders to avoid over-rationalization. (SEKEM, 2023, p. 7)

[15] "Gulbenkian Prize for Humanity – Impact and winners 2024." *Gulbenkian Foundation*, www.gulbenkian.pt/en/the-foundation/gulbenkian-prize-for-humanity/impact/#5edition. Accessed 22 Nov. 2024.

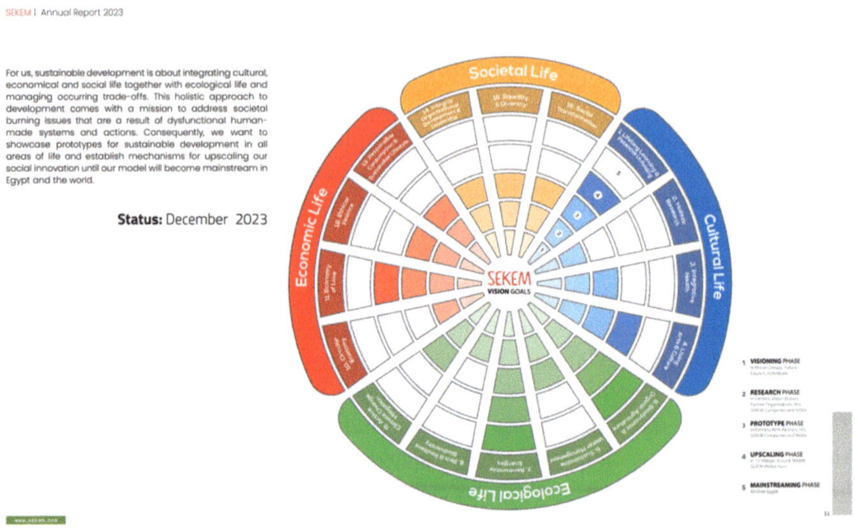

Fig. 2 Sustainability Flower of Sekem with its vision goals. Sekem. Annual *Report 2023*

Unique to Sekems sustainability flower is its total own innovation of the 16 goals divided in the four main dimensions. For instance, the "Economy of Love" (EoL) in the economic dimension is as framework created by Sekem in order to launch an alternative economic model, where profit maximalisation is transformed into solidarity and associative practices. An elaborate description can be found in Chap. 23 "Egyptian Biodynamic Association EBDA and EoL." Also in the cultural dimension of Sekems sustainability understanding, goals like "Holistic Research," "Lifelong learning & unfolding potential," or "Living Arts & culture" form a total novel and unique expansion of the conventional SD understanding (Fig. 2).

Sekem's showcase seen from the Integral Sustainable Development approach, attributes solidarity to economic life, equality aspects to social life, and a strong dignity to the ecological dimension. Also, World-Viewing expresses itself strongly in Sekem, having created a whole own understanding of each pillar of SD and enhancing it with the cultural dimension as clear driver for holistic development. World-Sensing clearly shows aspects of a co-creative attitude: the situation in which they created "green" out of the desert, but also aspects of inner development with goals such as "life-long learning & potential unfolding". In World-Acting, all ethical action dimensions (anthro-, zoo-, bio-, eco-, and holocentric) are taken in consideration. This is being expressed by the founder's vision that he aims for development for the whole of humanity and the Earth alike.

Eosta–where Ecology Meets Economy

First as a pioneer in 1990 and by now as a European market leader in organic fruit and vegetable in- and export, Eosta can be seen as pioneer in the food sector. They have been revolutionising the social, economic and cultural realm of their sector based on the social threefold order of Rudolf Steiner.

Freedom, justice, and solidarity lie at the heart of their self-developed Sustainability Flower, which can be understood as the integral or cultural-spiritual dimension of their SD understanding. As introduced by Steiner (2019, 1919/2022b), freedom belongs to cultural life, justice to social life, and solidarity to economic life. For Eosta, this manifests itself in the statement that these three baseline approaches can contribute towards the health and functioning of their six identified Earth systems: animals, energy, air, water, plants, and soil. They advocate strongly for True Pricing and therefore a True Cost Accounting. True Pricing means that the whole value chain up until the farmers and beyond back into Earth system organisms are taken into account. A correct accounting will show that, for instance, non-organic products have a higher true costs due ecosystem destruction and negative impact on workers' health and rights (more on this matter in Chap. 24. This would ultimately lead towards an economy based on solidarity: profit is only profit, if Earth and human beings are also winning (Engelsman, 2024).

The three core values of Eosta show a clear formulation of a cultural-spiritual understanding of SD:

> The potential of the entire human race is hidden in every human—that we can become who we are; human encounters encompass a wholeness that is greater than we are; that we maximise progress in collaboration; our responsibility extends across the entire planet and all of humanity—that we want to do what is right. (Eosta, n.d.)

This formulation goes beyond Brundtland or the question how business can be made more efficient, sufficient, or consistent. It touches on the very question of how we define development. The sustainability flower of Eosta embodies the attempt to offer a concrete framework to do so.

One of the many offsprings of Eosta is the Robin Food Coalition (see Chap 5.6) that, in the same spirit of Eosta and their expanded SD understanding, tries to drive change and transform the food system by more activist and political engagement. In Eosta's Sustainability Flower and understanding of SD, one can apply the Integral approach: a World-Viewing where the baseline and understanding of the human being is defined in the freedom of cultural life, justice in the social life, and solidarity in the economic life. A World-Sensing can be found in the fundamental attitude of co-creation where "our responsibility extends across the entire planet and all of humanity—that we want to do what is right [...] to make people and the earth healthier " (Eosta, n.d). And finally, World-acting outspokenly incorporates the ethics of the anthropocentric level (freedom, justice, and solidarity), the zoocentric level (animals), the biocentric level (plants and soil), the ecocentric level (water, air, and energy).

Eosta therefore embodies what we sketched and introduced as Integral Sustainable Development by expanding the three pillars of SD by a cultural-spiritual dimension and also redefining for instance economic sustainability based on

Fig. 3 Sustainability flower of Eosta

Steiner's indications, by introducing solidarity as the main economic point of reference (Fig. 3).

Biodynamic Federation Demeter International (BFDI)—100 Years in the Rhythm of Change
BFDI is the worldwide body governing the movement of biodynamic farmers, biodynamic associations, and their Demeter certification standard, the latter of which was founded in 1928 as first organic agriculture certification standard. The biodynamic impulse started just before in 1924, after eight lectures on agriculture by Rudolf Steiner in response to an increasing chemical and industrialising agriculture (Steiner, 2022). It is known as the first and most strict organic agriculture approach (Hurter et al., 2025; McKanan, 2017). BFDI is based on participatory, democratic, and collaborative principles and working with an on a peer-review based system to guarantee quality, collective learning, development, and the Demeter standard (Hurter et al., 2025). After decades of work in differentiated organisational structures, the main entities fused in 2020 to form the BFDI as worldwide umbrella.

In the heart of BFDI's vision, one can find the appeal "to take over the responsibility for the holistic development of the earth", in order to build an "Agri-culture for the Future. Inspired by intuition and responsibility" (Demeter International, n.d.). This in combination with the outer part of their vision "Cosmic and Spiritual Impact and Evolution of Earth and Mankind" can be seen as cultural-spiritual approach to a sustainable farming practices. The ecological dimension, influenced by "Social Relationships", "Economic Value Creation" and "Human Development", is embedded in this cultural-spiritual dimension, and the BFDI thus offers an integral approach for sustainable agriculture.

A strong "World-Viewing" is formulated in: a "holistic development of the earth," to "enable people to unfold their individual potential"; agriculture that "produces wholesome and health food [...] [that] nourish[es] body, soul and spirit"; an agriculture that "foster[s] [...] dignity, mutual respect, and tolerance"; and finally "embed cosmic and terrestrial forces and substances." A broad understanding of sustainable agriculture including Earth and human development is formulated. These are all the way made into concrete guidelines in the Demeter standards that enable farmers to practice Biodynamics.

A "World-Sensing" is attributed to the attitude of "mankind to take over responsibility for the holistic development of the earth", which directly addresses the "co-creator" fundamental attitude. But also, a human development indicating the very individual farmers development is addressed, which can be understood as parallel to the Inner Development Goals (2024), for instance. Biodynamic farmers are trained in Goetheanistic nature observation but also meditative introspection (Hurter et al., 2025).

The "World-acting" in BFDI's approach to sustainable agriculture was already mentioned in the core aspect of holistically developing the Earth's course. One of the Biodynamics principles is to understand the farm and individual farm organism, which means that like any organism it is a semi-closed system with different parts or organs with an interdependent and circular quality: soils, water, plants, animals, and the farmer all have an allocated place and task in the farms organism and try to produce what is needed for the wanted output.

SEKEMas enterprise working with Biodynamic Agriculture (see Chap. 5.22) is an example of how this approach "greened the dessert" in Egypt since 1977. With their work they generated many socio-ecological benefits, for which they were appointed by the United Nations Environmental Programme as "Champion of the Earth 2024" laureate (United Nations Environment Programme, 2024) (Fig. 4).

Final Remarks

Whereas many enterprises in the 1990s and 2000s were leaning towards the Triple Bottom Line of People, Planet, and Profit, being critiqued as corrupted interpretation of Sustainable Development and leading towards Green Washing and a diluted SD understanding (Wayne and Mcdonald. 2004), these four showcases tried to work in the opposite direction, where human and Earth development shape the backbone of their sustainable development understanding. An early implementation and sketches of a cultural-spiritual dimension of SD herewith become visible. These examples, like many in Chap. 5, can be seen as "Sustainable Natives" trying to envision, research, prototype, scale, and mainstream integral sustainable development approaches. Not long ago, these initiatives were seen as being niches or prototypes but they are now quickly reaching the stage of upscaling and mainstreaming. At the same pace of this success, the understanding and frameworks of

Fig. 4 Vision of BFDI

SD need to keep up: stay in movement, stay alive, and be open for needed changes to not get obsolete.

8 Conclusion

In the evolving definition of sustainable development in the past 50 years of its institutionalised existence, amid a continuous *Great Acceleration* and exponential economic growth, we see that economic growth-driven systems and thinking fail to ensure sustainability and continue to perpetuate social inequality and massively cross planetary boundaries.

There is a need for a more integral approach to SD, one that incorporates cultural and spiritual dimensions alongside technical solutions. Meaningful system change requires rethinking human nature, our relationship with the Earth, and the ethical principles guiding our actions. This perspective aligns with contributions of anthroposophy, calling for an integral view of human nature that integrates ethical individualism with sustainability practices. For SD to be evolved, it must embrace these deeper dimensions and address the global contradictions in current models. The "spiritual dimension" must be central to both individual responsibility and collective action. The long-term habitability and transformation of the Earth is herewith the addressed aim. The socio-ecological crisis is rooted not only in environmental degradation but also in a deeper cultural and spiritual crisis. Modern civilisation's disconnection from its humanity underpins this crisis. A dominant materialistic worldview has reduced nature to a resource, eroding reverence for life's interconnectedness. This spiritual void, where humans are seen as the planet's sole destroyers, drives much of the destruction we observe.

Anthroposophy and Rudolf Steiner's work contribute to understanding and addressing the relationship between human and Earth and enates with sustainable development since over 100 years. Anthroposophy's view of the human being, integrating spiritual, soul, and bodily dimensions, provides an approach for reconnecting with the deeper faculties of thinking, feeling, and willing. These faculties guide human actions, bridging the material and spiritual realms. The current crisis reflects a rupture in this relationship, as humans increasingly lose touch with both their inner selves and the Earth. Addressing this crisis requires a transformative approach that reawakens cultural and spiritual dimensions and reintegrates them into a more integral sustainable development.

We consider anthroposophy's "Appropriate Anthropocentrism" can offer a transformative vision of humanity's relationship with Earth. Rather than being seen as mere destroyers, humans are envisioned as co-creators tasked with regenerating and transforming the Earth through healing and co-creative activities. This shift from a destructive to a co-creative role requires a new ethical framework balancing human progress with planetary well-being. Anthroposophy's view of Earth as a living organism reinforces this, suggesting that humans must actively participate in the Earth's development toward a more harmonious future. Regeneration and nature conservation are key, but not enough. Ultimately, appropriate anthropocentrism calls for a rethinking of humanity's role, transcending current economic and ecological paradigms as already stated by Donella Meadows in 1999.

The introduced Integral Sustainable Development (ISD) approach provides the sketch for addressing the ecological, social-cultural, and economic pillars of sustainability by incorporating the spiritual dimension into sustainable development practices. By introducing the interrelated dimensions of World-Viewing (cognitive), World-Sensing (emotive), and World-Acting (normative), ISD emphasises the necessity of an integral and holistic perspective that links human cognition, emotion, and action with the broader context of ecological systems, socio-economic development and spiritual values.

World-Viewing encourages a deeper understanding of paradigms and worldviews that shape our perception of sustainability, drawing from anthroposophical anthropology and the recognition of Earth as a living organism. World-Sensing bridges this understanding with human attitudes, emphasising co-creation and reciprocity between humanity and nature. Finally, World-Acting translates these insights into actionable frameworks, fostering an integrative approach that encompasses anthropocentric, zoocentric, biocentric, and ecocentric ethics.

Together, these dimensions offer a pathway towards an integral approach, aligning the human condition with sustainable development. This alignment represents a shift from mechanistic thinking to a holistic paradigm that integrates spiritual, ecological, and social perspectives, contributing to a more profound and enduring evolution of both humanity and the Earth.

In the four examples of enterprises that took anthroposophy as the source of inspiration, we see how the elements of Integral Sustainable Development emerge from practice and how they are already implemented. This provides a solid base for further discussion and development of the concept of Integral Sustainable Development in the future. Chapter 4 will now introduce a manifold of perspectives from different disciplines on the questions of what contributions anthroposophy's worldview provides to sustainable development.

References

Alnatura. (2022). *Unser Beitrag für Mensch und Erde: Nachhaltigkeitsbericht 2021/22*. Alnatura. Retrieved November 22, 2024, from https://www.alnatura.de/-/media/Alnatura/B2C/Dateien/ueber-uns/Alnatura-Nachhaltigkeitsbericht-2021_2022.pdf

Ametepey, S. O., Aigbavboa, C. O., Ansah, S. K., Gyadu-Asiedu, W., & Boamah, L. (2023). *Meaning, evolution, principles, and future of sustainable development: A systematic review* (Preprint No. 2023020158). Preprints. https://doi.org/10.20944/preprints202302.0158.v1

Berry, T. (1999). *The great work: Our way into the future*. Bell Tower.

Bind, R., & Hurter, U. (2025). *From the agriculture course to sustainable farming: 100 years of the biodynamic movement*. Floris Books.

Bristow, J., Bell, R., Wamsler, C., Björkman, T., Tickell, P., Kim, J., & Scharmer, O. (2024). *The system within: Addressing the inner dimensions of sustainability and systems transformation* (Earth4All: Deep-Dive Paper 17). The Club of Rome. https://www.clubofrome.org/publication/earth4all-bristow-bell/

Broman, G., & Robèrt, K.-H. (2017). A framework for strategic sustainable development. *Journal of Cleaner Production, 140*, 17–31. https://doi.org/10.1016/j.jclepro.2015.10.121

Brundtland, G. H. (1987). *Our common future*. Oxford University Press.

Capra, F. (1982). *The turning point: Science, society, and the rising culture*. Simon and Schuster.

Crutzen, P. J., & Stoermer, E. F. (2000). The "Anthropocene." *Global Change Newsletter, 41*, 17–18.

Daly, H. E. (1999). *Ecological economics and the ecology of economics: Essays in criticism*. Edward Elgar.

Demeter International. (n.d.). *History*. Demeter. Retrieved January 8, 2025, from https://demeter.net/about/history/

Dixson-Declève, S., Gaffney, O., Ghosh, J., Randers, J., Rockström, J., & Stoknes, P. E. (2022). *Earth for all: A survival guide for humanity*. New Society Publishers.

Dresner, S. (2008). *The principles of sustainability* (2nd ed.). Earthscan.

Earth Charter International. (2000). *The Earth Charter*. The Earth Charter Initiative. Retrieved February 3, 2025, from https://earthcharter.org/discover/the-earth-chart

Earth.Org. (2024). Why global water security matters in 2024. https://earth.org/why-global-water-security-matters-in-2024/

Eisenstein, C. (2018). *Climate: A new story*. North Atlantic Books.

Eosta. (n.d.). *Manifest*. Retrieved January 8, 2025, from https://eosta.com/node/3814/

European Parliament and Council of the European Union. (2022, December 16). Directive (EU) 2022/2464 of the European Parliament and of the Council of 14 December 2022. *Official Journal of the European Union*.

Guardian. (2025). Why is Donald Trump talking about annexing Greenland? *The Guardian*. Retrieved January 9, 2025, from https://www.theguardian.com/world/2025/jan/08/why-is-donald-trump-talking-about-annexing-greenland

Harribey, J.-M. (2002). Ein nachhaltiger Widerspruch. *Le Monde diplomatique*. Retrieved February 3, 2025, from https://monde-diplomatique.de/artikel/!1073087

Hawkes, J. (2001). *The fourth pillar of sustainability: Culture's essential role in public planning*. Common Ground.

Heusser, P. (2016). *Anthroposophie und Wissenschaft: Eine Einführung*. Verlag am Goetheanum.

Hirsch, F. (1977). *Social limits to growth*. Routledge & Kegan Paul.

Inner Development Goals. (2024). *About*. Inner Development Goals. Retrieved September 12, 2024, from https://innerdevelopmentgoals.org/about/

Jackson, T. (2009). *Prosperity without growth: Economics for a finite planet*. Earthscan.

Jhagroe, S. (2024). *Voorbij duurzaamheid: Op weg naar een ecorechtvaardige samenleving*. Mazirel Pers.

Jonas, H. (1979). *Das Prinzip Verantwortung: Versuch einer Ethik für die technologische Zivilisation*. Suhrkamp.

Kepler, J. (1619). *Harmonices mundi* [The harmony of the world].

Kingsnorth, P. (2017). *Confessions of a recovering environmentalist and other essays*. Graywolf Press.

Latour, B., & Schultz, N. (2022). *Zur Entstehung einer ökologischen Klasse: Ein Memorandum*. Suhrkamp Verlag.

McKanan, D. (2017). *Eco-alchemy: Anthroposophy and the history and future of environmentalism*. University of California Press.

Meadows, D. H. (1999). Leverage points: Places to intervene in a system. *The Sustainability Institute*.

Meadows, D. H. (2008). *Thinking in systems: A primer*. Chelsea Green Publishing.

Meadows, D. H., Meadows, D. L., Randers, J., & Behrens, W. W., III. (1972). *The limits to growth: A report for the Club of Rome's project on the predicament of mankind*. Universe Books.

Mihelcic, J. R., Zimmerman, J. B., & Ramaswami, A. (2003). Sustainability science and engineering: The emergence of a new meta-discipline. *Environmental Science & Technology, 37*(23), 5314–5324. https://doi.org/10.1021/es034605h

Mitcham, C. (1995). The concept of sustainable development: Its origins and ambivalence. *Technology in Society, 17*(3), 311–326. https://doi.org/10.1016/0160-791X(95)00008-F

Mulligan, M. (2015). *An introduction to sustainability: Environmental, social and personal perspectives*. Routledge.

Mumford, L. (1934). *Technics and civilization*. Harcourt.

Nietzsche, F. (1978). *Thus spoke Zarathustra: A book for all and none* (W. Kaufmann, Trans.). Penguin Books (Original work published 1882).

Ott, K. (2003). The case for strong sustainability. In K. Ott & P. Thapa (Eds.), *Greifswald's environmental ethics* (pp. 1–20). Steinbecker Verlag Ulrich Rose.

Ott, K., & Döring, R. (2004). *Theorie und Praxis starker Nachhaltigkeit*. Metropolis.

Oxfam International. (2024). *Inequality Inc.: How wealth and corporate power are intensifying global inequality*. Oxfam International.

Perlas, N. (2019). *Shaping globalization: Civil society, cultural power, and threefolding*. Temple Lodge.

Pope Francis. (2015). *Laudato Si': On care for our common home* [Encyclical letter]. Vatican Press.

Prinz, P. (2023). *Ausbeutung im Amazonas—Indigener Philosoph: «Hört auf, die Welt zu verschlingen!»* SRF Kultur, Schweizer Radio und Fernsehen (SRF). https://www.srf.ch/kultur/gesellschaft-religion/ausbeutung-im-amazonas-indigener-philosoph-hoert-auf-die-welt-zu-verschlingen

Raworth, K. (2017). *Doughnut economics: Seven ways to think like a 21st-century economist*. Chelsea Green Publishing.

Rockström, J., Steffen, W., Noone, K., Persson, Å., Chapin, F. S., Lambin, E., Lenton, T. M., Scheffer, M., Folke, C., Schellnhuber, H. J., Nykvist, B., de Wit, C. A., Hughes, T., van der Leeuw, S., Rodhe, H., Sörlin, S., Snyder, P. K., Costanza, R., Svedin, U., … Foley, J. (2009). A safe operating space for humanity. *Nature, 461*(7263), 472–475. https://doi.org/10.1038/461472a

Rosling, H., Rosling, O., & Rosling Rönnlund, A. (2018). *Factfulness: Ten reasons we're wrong about the world—and why things are better than you think*. Flatiron Books.

Scharmer, C. O. (2009). *Theory U: Leading from the future as it emerges*. Berrett-Koehler Publishers.

Scharmer, C. O., & Kaufer, K. (2013). *Leading from the emerging future: From ego-system to eco-system economies*. Berrett-Koehler Publishers.

Schmelzer, A. (2017). *The threefolding movement, 1919: A history*. Rudolf Steiner Press.

Schneidewind, U. (2018). *Die Große Transformation: Eine Einführung in die Kunst gesellschaftlichen Wandels*. Fischer Taschenbuch.
Segall, M. D. (2023). The urgency of social threefolding in a world still at war with itself. *Cosmos and History: The Journal of Natural and Social Philosophy, 19*(1), 229. https://cosmosandhistory.org/index.php/journal/article/view/1069/1723
Segall, M. T. (2021). The urgency of social threefolding in a world still at war with itself. *Cosmos and History: The Journal of Natural and Social Philosophy, 17*(1), 240–257.
SEKEM. (2023). *Sustainability report 2023*. SEKEM. Retrieved November 22, 2024, from https://sekem.com/en/about/sustainability/sustainability-reports/
Steffen, W., Broadgate, W., Deutsch, L., Gaffney, O., & Ludwig, C. (2015). The trajectory of the Anthropocene: The great acceleration. *The Anthropocene Review, 2*(1), 81–98. https://doi.org/10.1177/2053019614564785
Steiner, R. (1989). *Die Impulsierung des weltgeschichtlichen Geschehens durch geistige Mächte* (4th ed., GA 222). Rudolf Steiner Verlag.
Steiner, R. (1992). *Bilder okkulter Siegel und Säulen* (3rd ed., GA 284, S. 113). Rudolf Steiner Verlag.
Steiner, R. (2002). *Nationalökonomischer Kurs: Geisteswissenschaftliche Grundlagen zum sozialen Fragenkomplex* (6th ed., GA 340). Rudolf Steiner Verlag.
Steiner, R. (2003). *Grundlinien einer Erkenntnistheorie der Goetheschen Weltanschauung, mit besonderer Rücksicht auf Schiller* (8th ed.). Rudolf Steiner Verlag.
Steiner, R. (2013a). *Theosophie: Einführung in übersinnliche Welterkenntnis und Menschenbestimmung* (33rd ed., GA 9). Rudolf Steiner Verlag.
Steiner, R. (2013b). *Geheimwissenschaft im Umriss* (31st ed., GA 13). Rudolf Steiner Verlag.
Steiner, R. (2015). *Der menschliche und der kosmische Gedanke* (7th ed., GA 151). Rudolf Steiner Verlag.
Steiner, R. (2019). *Allgemeine Menschenkunde als Grundlage der Pädagogik* (10th ed., GA 293). Rudolf Steiner Verlag.
Steiner, R. (2021). *Die Philosophie der Freiheit: Grundzüge einer modernen Weltanschauung* (17th ed., GA 4). Rudolf Steiner Verlag.
Steiner, R. (2022a). *Wie erlangt man Erkenntnisse der höheren Welten?* (26th ed., GA 10). Rudolf Steiner Verlag.
Steiner, R. (2022b). *Die Kernpunkte der sozialen Frage in den Lebensnotwendigkeiten der Gegenwart und Zukunft* (7th ed., GA 23). Rudolf Steiner Verlag.
Steiner, R. (2022c). *Geisteswissenschaftliche Grundlagen zum Gedeihen der Landwirtschaft* (9th ed., GA 327). Rudolf Steiner Verlag.
Steiner, R. (2024). *Anthroposophische Leitsätze: Wahrheiten und Aufgaben der Anthroposophie—Ergebnisse aus der Arbeit der anthroposophischen Bewegung* (12th ed., GA 26). Rufolg Steiner Verlag.
United Nations. (n.d.). *The 17 goals. Sustainable development*. United Nations Department of Economic and Social Affairs. Retrieved February 3, 2025, from https://sdgs.un.org/goals
United Nations Environment Programme. (2024). *SEKEM: Champion of the Earth 2024*. United Nations Environment Programme. Retrieved January 8, 2025, from https://www.unep.org/championsofearth/laureates/2024/sekem
van Egmond, N. D., & de Vries, H. J. M. (2011). Sustainability: The search for the integral worldview. *Futures, 43*(8), 853–867. https://doi.org/10.1016/j.futures.2011.05.027
Verhoog, H., Matze, M., Lammerts van Bueren, E. T., & Baars, T. (2003). The role of the concept of the natural (naturalness) in organic farming. *Journal of Agricultural and Environmental Ethics, 16*(1), 29–49. https://doi.org/10.1023/A:1023343610533
VPRO Tegenlicht. (2018). *Paul Kingsnorth: Portrait of a recovering environmentalist* [Video]. VPRO.
Wachsmuth, G. (1980). *Erde und Mensch: Ihre Bildekräfte, Rhythmen und Lebensprozesse* (4th ed.). Verlag am Goetheanum.
Wijkman, A., & von Weizsäcker, E. U. (2018). *Come on!: Capitalism, short-termism, population, and the destruction of the planet*. Springer.

Wilber, K. (2000). *A theory of everything: An integral vision for business, politics, science, and spirituality*. Shambhala Publications.
World Meterological Organization. (2025). *WMO cconfirms 2024 as warmest year on record, about 1.55°C above pre-industrial level*. Retrieved February 21, 2025.
Zajonc, A. (2008). *Meditation as contemplative inquiry: When knowing becomes love*. Lindisfarne Books.
Zoeteman, K. (1989). *Gaiasofie: Anders kijken naar evolutie, ruimtelijke ordening en milieubeheer*. VBK Media.
Zoeteman, K. (2009). *Gaia Logica*. Christofoor.

Worldview: Perspectives from Anthroposophy

Introduction

This chapter introduces twelve texts, each presenting a distinct perspective rooted in anthroposophical principles, practices, and worldview, as well as their applications in professional fields and broader multidisciplinary contexts. We acknowledge that the term "worldview" may appear problematic or simplistic to some. However, we have chosen to use it as it effectively conveys the overarching perspective, background, and values of anthroposophy. Furthermore, in early anthroposophical texts, such as *The Philosophy of Freedom* (Steiner, 1894), the concept of a worldview played a central role. In part 2 the notion of "world-viewing" has already been introduced as a potential integral approach to sustainable development.

The contributors to this section were invited to write academic essays exploring anthroposophy's contributions to sustainable development from their respective disciplines and perspectives. This collection does not aim to provide a definitive or exhaustive answer to the question but rather to offer initial insights and directions for further exploration.

Among the authors are both scholars and practitioners engaged in anthroposophy, its applications, and sustainable development. Some are or have been professors of sustainable development, while others are newer to the discipline. This diversity has resulted in a range of perspectives and understandings regarding anthroposophy's contributions to sustainable development within its multidisciplinary context.

Each text is designed to be readable independently. However, we recommend first exploring parts 1 and 2 for a comprehensive entrance in the content.

Sustainable Development and the Anthroposophical Worldview

Klaas van Egmond and Jesse Mulder

Abstract

Anthroposophy provides profound insights into the challenges of sustainable development. In Steiner's vision of human nature, sustainability requires maintaining a delicate balance in the human value orientation between the fundamentally opposing forces that determine the human condition: the materialistic versus the spiritual and the particular (individual) versus the universal (collective). When this holistic awareness of the human condition is lost, one-sided value orientations dominate, leading to the erosion of both human dignity and sustainability. By organising society in a threefold social order, according to the three constituent (cultural, political, and economic) parts of the human condition, Steiner aimed to support a balanced, "sustainable" value orientation with organisational means. This balanced and thus humane view of human nature holds the promise of a meaningful life in a healthy society.

Keywords

Sustainable development • Anthroposophy • Threefold social order • Human condition • Virtue Ethics • Human dignity • Value orientation • Cultural-spiritual dimension • Macro-history • Systems thinking • Ethical framework • Worldviews • Freedom-equality-solidarity • Political economy • Economic life • Cultural life • Financialisation • Steward ownership • Biodynamic agriculture

K. van Egmond (✉)
Copernican Institute for Sustainable Development, Utrecht University, Utrecht, Netherlands
e-mail: n.d.vanegmond@uu.nl

J. Mulder
Faculty of Humanities, Department of Philosophy and Religious Studies, Utrecht University, Utrecht, Netherlands
e-mail: j.m.mulder@uu.nl

J. Kronenberg and E. T. Lammerts van Bueren (eds.), *On the Earth We Want to Live*, World Sustainability Series, https://doi.org/10.1007/978-3-031-98758-8_7

1 Introduction

Over the past 50 years, since the Meadows' Club of Rome's report on the limits to growth and the UN policy response by Brundtland's Our Common Future, the sustainability of the physical Earth and the global socioeconomic system has gotten weaker, not stronger. Climate change and biodiversity loss evolved from scientific expectations to observable facts. At the same time, philosophers and politicians conclude that we no longer have a genuinely moral framework. The growing gap between rich and poor is fuelling socio-political tensions. The constant concentration of economic power weakens the democratic functioning of Western societies in particular. These democracies were already threatened in their functioning by the financial crisis of 2008, after which the population in general lost trust in social institutions, resulting in a massive shift to right-wing populist political parties (Wolf, 2024). Meanwhile, sustainability policies have been limited to setting inconsistent long-term objectives such as 'green growth.' It assumes that the effects of continuous physical growth are compensated by technological innovations, without addressing the underlying socio-economic structure.

After 50 years, however, it has become clear that sustainable development is not a merely technical issue, and that the various observations we just raised are tied up in an overarching crisis of human dignity. In essence, the sustainability issue highlights our overall failure in unfolding a 'good', meaningful life. In this essay, we explore how the anthroposophical view of humankind, Earth, and world can offer the outlines of an answer to this deep crisis in our very humanity. We develop this outline in the form of a renewed ethical framework, based on a conceptualisation of human nature and human value orientations. By further articulating this integral vision in the threefold social structure of cultural, political, and economic life, a moral framework and concrete policy measures can be formulated that promise to help us overcome the current socio-ecological crisis.

2 The Anthroposophical View on Earth, World and Humankind

How do we attain knowledge of the natural world around us, and of our own nature as human beings? From an anthroposophical point of view, the human being is first and foremost a spiritual being, who incarnates into a physical body, and as such lives in the 'materialistic' environment of the physical world. Bringing sensation and thinking together, we come to understand the world, or 'reality,' and our own place within it. In his *Human and Cosmic Thinking*, Rudolf Steiner (1914) distinguishes twelve different directions to view the world which human existence on Earth provides. These twelve 'worldviews' (Weltanschauungen) are shown in Fig. 1.

The worldviews appear in pairs of opposites of individual orientations. The vertical pair represents the main contrast in anthroposophical thought; the symmetry between mind and matter. The horizontal axis represents the contrast between

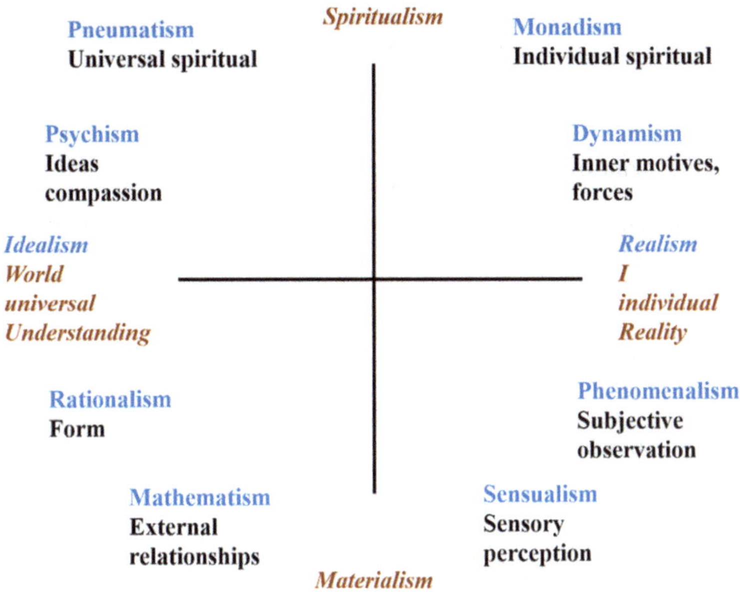

Fig. 1 Steiner's 'circle of worldviews', comprising twelve individual worldviews along with characteristic qualities (Rudolf Steiner, 1914)

realism and idealism. Realism perceives reality as given through direct, subjective observation of the surrounding world, whether spiritual or materialistic in nature. Idealism represents the Platonist (or Hegelian) position in which the world is seen as the manifestation of universal Ideas (or as the dialectical unfolding of Geist). The other eight individual ways of seeing the world can be characterised as transitions between the four worldviews given by the vertical and the horizontal axes/contrasts (Fig. 1).

Steiner remarks that different issues require different worldviews and warns against the tendency towards one-sidedness in which a single worldview pretends to be the only true worldview. We will come back to this warning later, as it appears that sustainable development is frustrated precisely when this happens.

The circle of worldviews ultimately allows us to relate the worldview to which our own individual thinking tends to this overarching whole. Where we might individually tend to take our own worldview to be that overarching whole, the circle helps us realise that we need the fullness of all twelve worldviews. The ultimate truth both about the world and about ourselves is made up of a series of contrasts or opposites, between which we must apparently maintain a certain balance. We thus see the world and ourselves as spanning the six-fold opposites by which the worldviews are defined. Apparently, then, this circle of worldviews defines the 'human condition' on the level of thought, within which we position ourselves intellectually, both individually as well as collectively (on various scales). Depending on where one experiences a foothold in reality (in which of the mentioned qualities),

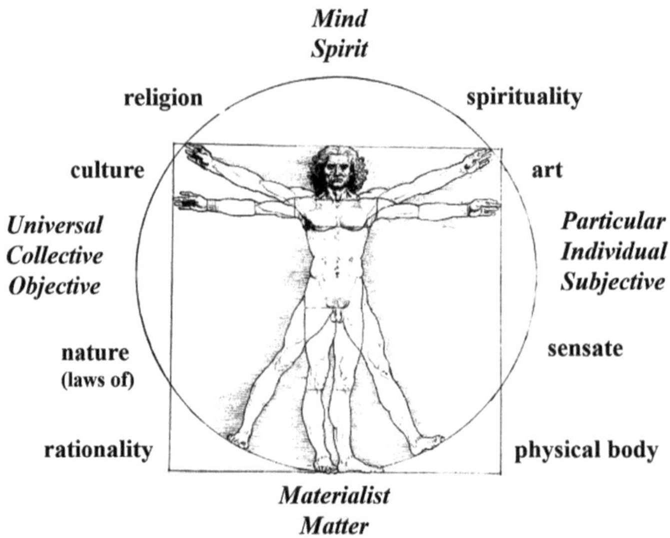

Fig. 2 The human condition, superimposed on Da Vinci's Vitruvian Man

one leans towards one of the twelve viewpoints, which brings with it the danger of excluding the others.

Human Nature: The Human Condition and Value Orientation

Although Steiner presents his 'circle of twelve' as encompassing the fundamental orientations our human thinking can take, it is possible to position more comprehensive facets of human cultural life within it. Figure 2 presents the more general conception of the human condition that we have thus lifted from Steiner's circle of worldviews.

Our basic orientation in life can be viewed as an assignment of values to the many qualities and capacities which the human condition encompasses. For example, we may consider material objects more important than cultural life, or vice versa. This implies that our consciousness tends towards one part of the pattern, which brings with it a danger of losing sight of the whole. Since, psychologically speaking, we are the values we have (Hitlin, 2003), our 'value orientation' on the human condition can be seen as an expression of our individual identity. We thus have an individual-subjective appreciation of the universal-objective human condition. For our purposes here, we can view 'human nature' as these two taken together. The pattern of the human condition can be compared very well to the image of the human body. That is why in Fig. 2 the human condition is depicted against the background of Da Vinci's Vitruvian Man. He stands with his arms and legs both in the square and in the circle, representing the finite physical world and the infinite spiritual world, respectively, but also the microcosm and macrocosm of Renaissance thought.

Now, Steiner (1919a, 1919b, 1919c) describes the human organism as consisting of three systems, which work side by side in organic interdependency, i.e., they can be clearly distinguished but not separated: the nervous and sensory system, more or less centred in the head; the rhythmic system, including breathing and circulation, centred in the heart area; the metabolic system, which includes all the processes related to the transformation of matter, concentrated in the lower part of the human body, specifically the abdomen.

In Figure 2, Da Vinci's Vitruvian Man indicates where these three systems find their counterpart within the circle: we have the upper spiritual part, the intermediate level between spirit and matter in the central part, and the lower, matter-oriented part of the human body. The systems also correspond to the fundamental faculties of the soul: thinking, feeling and willing.

As will be discussed below, Steiner connected the distinction between these three with his understanding of society in terms of a threefold social order in which the cultural system, the system of public rights and the economic system function similarly interdependently. We will then be able to explore in more detail what shape the 12-fold conception of the human condition takes on, not just on the level of cultural life, but also 'further down' on the level of public rights and on the level of economic life.

Macro-History

As indicated, Steiner warned against one-sided appreciations of the human condition, as this could lead to the predominance of singular, one-sided worldviews. The dangers of this become apparent when one studies such one-sided worldviews gaining dominance during certain historical episodes. We can connect this to a view of history in which what happens on the 'outer stage' of the public world can be 'symptomatologically' regarded as expressive of underlying spiritual realities (Steiner, 1914). Taking the historical studies of Sorokin (1957) and Toulmin (1990) into account, a macro-interpretation of history appears to be meaningful, and can be meaningfully related to the pattern of human nature as described (van Egmond, 2014). To illustrate the general developments in Western societies, for the four quadrants:

The period of early Christianity can be seen as dominated by the individual-spiritual value orientation of the upper right quadrant of the human condition in Fig. 2.

Around the year 400, a process of universalisation was brought about by the establishment of the Roman Catholic (i.e. universal) church, implying a claim to the one and only truth as it manifested on the upper left side of the human condition. Supported by the brutal Inquisition, the exaggeration of the now more universal, but still spiritual, value orientation resulted in theocracy and the subsequent domination of the political and economic system.

During the Renaissance, the rediscovery of Greek philosophy led to a more encompassing dominant value orientation nearer the centre of the human condition, resulting in tolerance, harmony, and creativity in all areas of society, in science, art, governance, and economics.

In the 'modern' age since, particularly during the Enlightenment, the universal but increasingly materialistic values of the lower left quadrant were on the rise. Here, the state still dominates the cultural and economic system. Initially, this resulted in the industrial revolution, later in (universal) state communism and Nazism with catastrophic consequences.

The increasingly materialistic value orientation towards the lowest part of the human condition caused hyper-consumption and -production and thus the ecological crisis (climate change, loss of biodiversity).

Over the past 50 years, the dominant value orientation has shifted to the lower right quadrant, where individualistic and materialistic value orientations began to dominate. As a result, politics and culture are overruled by an unregulated financial-economic sector, while postmodern attitudes, with their emphasis on the subjective and their scepticism regarding 'objective' truth, question the former universalism of science.

As a general conclusion, value orientations tend to become one-sided at first and later turn into obsessive caricatures. Such recurring shifts to the unilateral periphery of the human condition provoke catastrophes, such as institutional terror by church (inquisition) and state, or wars that later magnify into world wars. The empirical-historical analysis thus reflects Steiner's observation that the shift to very one-sided, caricatural value orientations resulted again and again in human and social catastrophes. Apparently, centrifugal shifts to the periphery of the human condition should be seen as malignant, with negative ethical connotations. Conversely, centripetal shifts to the centre are expected to be virtuous and thus ethically positive, because they avoid such catastrophes. This naturally leads us to the principles of virtue ethics.

Virtue Ethics

In virtue ethics, the 'good' is a means between two evils. Aristotle considers a virtue to be "a state of character, a disposition, which is concerned with what we value, desire, realize, or promote, and which lies in an average, determined by a rational principle. Virtue, then, is a middle way between two vices, that which depends on excess and that which depends on lack, between too much and too little." (Aristotle, 1999, Book II, p. 33). In terms of the circle of worldviews, then, 'the good' is situated among the many one-sided opposites that form the circle periphery. More precisely, virtues can, in our presentation of human nature, be viewed as centripetal forces that shift value orientations to the centre, increasing the coherence between all human values. Aristotle calls the most important, overarching virtue 'friendship' (philia), because it involves empathy, respect for 'the other side' of the human condition and thus a more or less equal appreciation of all diagonally opposed elements. Accordingly, in Christianity, the most important virtue is love. Love, we might say, relates to any of the aspects or parts of the human condition (the other, oneself, art, nature, ...) in such a way that it simply cannot become an obsession. It is by nature a centripetal force, directing appreciation to the centre of the human condition, which can then be seen as an ultimate human goal (or 'telos'). Aristotle describes the state of the centre as eudaimonia: human flourishing, the good life, happiness.[1]

Vices, on the other hand, are centrifugal forces, which bring about fragmentation rather than coherence. Once the predominant value orientation shifts to the periphery of the human condition, the respective values become so fragmented that any connection with the neglected parts of the human condition is lost altogether. Single, isolated values begin to dominate obsessively. Human functioning is no longer based on, or related to, the whole range of interrelated values and abilities (within the circle) that makes people human. Aristotle's virtue ethics thus conforms to our findings that historical orientations outside the circle are destructive. Obviously, the same basic thought of human virtue as the 'middle road' is present in the anthroposophical understanding of 'holding the right balance' between the spiritual beings called Lucifer and Ahriman, who represent the two extremes, and yet are also necessary so that there is a middle path for us to find.

Interestingly, similar thoughts can be found in the work of psychologist C.G. Jung (1953, p. 244), who warns that "when the separation (of values) is carried to such an extent that the complementary opposite pole is lost sight of, the result is one-sidedness, which is then compensated by the unconscious without our help." The result will be psychosis, obsessive preoccupation with very one-sided value orientations. Jung thus refers to the domain outside the circle as 'the unconscious'. The 'conscious', on the other hand, corresponds to the appreciations of (parts of) the human condition. As such, it 'slides' over the human condition in psychic processes (Jung, 1946, p. 408). This conscious part also represents the 'ego,' "which retains its integrity only if it does not identify with any of the opposites, and if it understands how to maintain balance between them" (Jung, 1946, p. 425). This can only be done if it remains aware of both. By bridging the opposites, consciousness inevitably spreads more and more over the whole base (of the human condition), so that the middle comes to be experienced as self-consciousness. Thus, Jung interprets the pattern of the human condition as the representation of the human 'Self', whose integrity and internal coherence must be maintained and strengthened. It requires a restless balance between the many opposites.

But as the conscious situation moves further away from a certain equilibrium point, the unconscious content struggling to restore equilibrium becomes more powerful and correspondingly more dangerous. The farther away from the centre, the greater the decrease in consciousness and the increase in the unconscious. This gives the process a self-reinforcing character. As a result, "the individual is generally so unconscious that he doesn't even know his own ability to make decisions. That's why, in all their desperation, people continue their anxious search

[1] The term 'eudaimonia' literally means *good* (eu) *daemon* (daimon); this leads us back to Socrates's awareness of his *daimonion*, which Steiner characterises as "the spiritual force which leads the soul" (*Die Rätsel der Philosophie*, Rudolf Steiner Verlag, Dornach 1985, p. 68).

for external rules and regulations to hold on to" (Jung, 1961 p. 333). To this end, they identify with the one-sided, and ultimately destructive, periphery, after which 'human dignity' is lost.

Human Dignity

As should be obvious by now, human dignity lies in the coherence of value orientations within the circle periphery of opposites. The preservation of human dignity therefore means the preservation of the values that make us human. Outside the circle boundary, human functioning is no longer based on, or related to, the whole range of interrelated values and faculties (within the circle), after which 'human dignity' is lost. The resulting narrow-mindedness makes catastrophe nigh inevitable. Human dignity is therefore closely linked to 'sustainability.' With the loss of human dignity, sustainability is also lost, given the disruptive consequences of straying too far into the periphery.

Sustainable Development

We are now in a position to characterise sustainable development. Briefly put, it comes down to maintaining value orientations in dynamic equilibrium within the periphery of the human condition, thereby avoiding the catastrophic discontinuities of the periphery. This applies to sustainable development in a socio-cultural, financial-economic and ecological sense. The ability to maintain that balance depends very much on the further increase of human consciousness. As Jung (1957, p 575) puts it: "if only there could be a global awareness that all division and all cleavage are due to the splitting of opposites in the psyche, then we would know where to begin."

Differentiation: Threefold Social Order

Where the balance within the circle of the human condition is lost, sustainable development comes under serious pressure. The underlying cause is that people are unaware of the position they take in relation to the whole circle, and thus do not realise their ability to take genuine control of their lives by developing virtuous value orientations. As already briefly indicated, the macro-historical consequence is that there have always been one-sided value orientations that dominate all others. The Middle Ages were unilaterally dominated by the Catholic-universal Church, the (early) modern period was dominated by states, up to state communism and Nazism. The gradually increasing exaggeration of materialistic values in the lowest part of the human condition led to the domination of the economic system and the current ecological crisis. The further shift towards individualistic orientations led to the financialisation of society and the subsequent crises.

In order to maintain the ability to balance the respective levels of the human condition, Steiner characterised a threefold understanding of social reality that would maintain the balance based on its own organisational structure. To this end, in the human condition three separate levels are distinguished; on top the main cognitive- 'spiritualist' system, in the centre the heart system as the 'soul' and at the bottom the metabolic- 'materialist' system. By differentiating these levels into independent yet interdependent organisations, the balance between them could be better maintained. Socially, these levels correspond to the cultural, the political (juridical) and the economic spheres. Steiner argued that greater autonomy for these three spheres would prevent single spheres from dominating the others, as they had often done in the past. With this, the threefold social order is not a ready-made solution, but a new understanding of the problem, rooted in a deeper perception of human needs, desires, and abilities (Segall, 2023). The separation in the three spheres is shown in Fig. 3.

Cultural *life* is the sphere of freedom (liberté), where every individual capacity can be developed in order to enrich and develop society as a whole. It is the realm of circle of worldviews we started with, encompassing education, art, science, and religion (Goetheanum, 2024). Although this is the domain of the human psyche, the free 'sliding' movement of human consciousness, and therefore human appreciation, can reach all parts of the human condition. It establishes the virtue ethical notion of maintaining the balance between the fundamental opposites and increasing human consciousness through their reconciliation.

The *political life* of the state, as the middle sphere of the threefold social order, is exclusively concerned with rights and duties and their enforcement. It includes administrative law, constitutional law and statutory law. Rights and duties laws

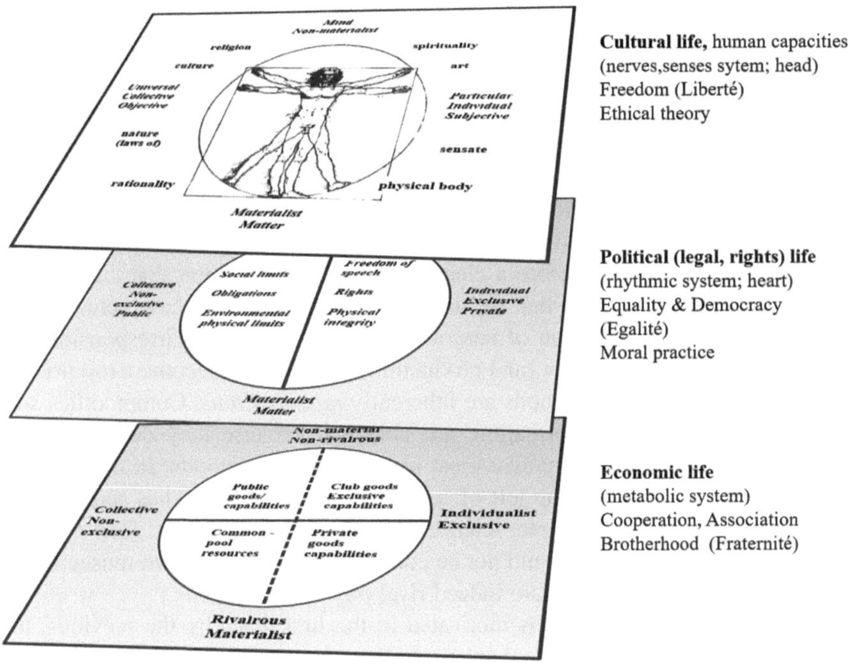

Fig. 3 Threefold social order

are in line with the core principle of equality (egalité): they do not discriminate between people. The laws can be seen as the practical moral translations of theoretical-ethical insights acquired in the cultural sphere. The laws that are formulated in this area must be formulated independently of economic interests and power. This means that economic resources should not play a role in determining people's rights and duties. The same applies to religions or ideologies that originate from the cultural domain. They are not allowed to control the levers of the state. Conversely, the state should not dictate the culture of how people think, learn, or worship.

Economic life, too, must form an independent, organic branch within the social organism. It is concerned with everything that has to do with commodity production, circulation, and consumption (Steiner, 1919a, 1919b, 1919c). The principle that governs this sphere is solidarity (fraternité), because the abilities and activities of the one fulfil the human needs of the other in mutual dependence. Economic life operates outside the state within the legal and regulatory limits, including labour laws, set by the democratic state. The basic rights and necessities of human life take precedence over all other economic imperatives. The economy is expected to function through cooperative forms of capitalism, such as those sometimes referred to today as steward ownership and stakeholder capitalism.

As indicated, the threefold social order, as shown in Fig. 3, should be seen as a differentiation of the human condition (Fig. 2) along its vertical mind-matter axis, where, as one would expect, the full circle reappears on each level. Economic life is the most 'materialistic'/physical. In general, physical economic goods are *'rivalrous'* in the sense that their consumption or use by one excludes others (we cannot both eat the same sandwich whole). This applies not only to private goods, such as a house or a bicycle, but also to 'ordinary' goods that are 'owned' by the collective community as a whole, e.g. natural areas, natural resources and public roads (lower in economic life; Fig. 3). It is true that everyone has access to these common goods, but overuse will lead to overcrowding of areas or depletion of resources. There is therefore a clear problem of distribution that calls for the mentality of solidarity. Now that the limits of (material) growth have actually been reached and further depletion of natural resources has become irresponsible, the problem of rival consumption (and production) patterns has become a top priority.

Non-physical economic goods are inherently *non-rivalrous.* Commodities such as culture, insights, or information, are non-rival because they do not run out by increasing consumption, unlike most physical-material goods. In the non-rival 'common' quadrant at the top left of 'economic life,' everyone has access to the cultural heritage of, for example, science, music and the visual arts. (The non-rival nature of cultural heritage should not be confused with overcrowded museums and historic city centres, as these are indeed rival physical commons.)

The threefold social order is motivated in the first place by the previous, historical domination of the cultural sphere in theocracy, the political sphere in state socialism, and the economic sphere in irresponsible capitalism. These destructive developments all point to the domination of the respective spheres as collective wholes, corresponding to the left side of the human condition (in Fig. 2). However, more recently (1970's, 80's) the dominant value orientation has shifted to the right, where individual, but still materialistic orientations prevail. Extreme individualism not only further strengthened the dominance of the economic system but also caused the "financialisation" of society as a whole. Based on an individualistic-hedonistic "greed is good" attitude, the new dominant sphere was the (global) financial system. This raises the question of how the individualistic side of the human condition feeds into the threefold social order. According to Brüll (1984), there are two opposing forces at work, which have everything to do with the horizontal, social interaction between people.

The A-Social and the Anti-social

In the cultural-spiritual sphere, we tend to be *a-social*, i.e., "out of contact with human society" (Brüll, 1984, p. 41). Think about people striving for knowledge: they must, to some extent, detach themselves from the immediate social context in order to dedicate their entire being to the search for the unknown. Whoever has found insight, however, also has the drive to spread the news. At the end of the day, people tend to bestow or even impose their truths on all of humanity. The process is confirmed by a social survey (van Egmond, 2014) in which spiritual value orientations were associated with 'giving' as a basic attitude. In the event that 'giving' becomes 'imposing', this leads to intolerance. In the human condition, this equates to a shift to the (evil) outside of the upper left quadrant. The transition from the early Christian faith to the universal Catholic Church is an example of this. Intolerance later resulted in the Inquisition there.

The nature of *anti-social forces*, on the other hand, finds its home in the bodily-material aspect of our being, which is the home of economic life. Anti-social behaviour is a universal phenomenon because the natural urge to stay alive always concerns rivalrous goods (to put it in terms of our above classification), and thus potentially at the expense of others. Where nature is abundant, no problem directly arises – picking some apples from the tree still leaves many for others. Where we start 'scaling up,' appropriating more than our basic needs, or where our biological drives grow into psychological drives for, e.g., dominion over others in some form or other for the satisfaction of our drives, the anti-social nature of these forces shows its true face: a shift to the (evil) outside in the lower right quadrant.[2]

According to Brüll, the task of the threefold social order is to prevent the anti-social forces in economic life and the a-social forces in the life of the mind from flowing into social areas where they have adverse effects. This is achieved by placing the political sphere (rights and duties) between the spiritual sphere and the

[2] Combined with the anti-social shift to the left in the upper left quadrant, this explains the counterclockwise movement of the "Zeitgeist" through macro history.

economic sphere. "The political sphere helps control this anti-social urge by keeping it within the proper confines of personal consumption. Therefore, the economy requires an order within which production and distribution no longer have the capacity to be used for exploitation" (Brüll, p. 182).

The political-rights sphere controls the a-social and anti-social forces by maintaining the balance between the individual-private and the collective-public orientation along the horizontal axis. On the right in Fig. 3, there are laws that protect physical integrity and limited private property, as well as the right to freedom of expression. Obligations have been created on the left, such as the obligation to refrain from imposing absolute truth.

To complete the picture: at the level of the economic sphere, the vertical distinction between rivalrous (physical) and non-rivalrous (non-physical) goods combines with the horizontal distinction in individual versus collective, i.e., public versus private. Individualism tends to make goods exclusive, often through appropriation. Given the increasing predominance of individualistic value orientations, the problem of appropriation becomes significant, all the more so since it is facilitated by an immoral financial system. The goods circulating in economic life must therefore be distinguished vertically with regard to their physical versus non-physical nature and horizontally with regard to their social nature. This results in the four basic categories (quadrants) of economic goods:

- Private goods are rival and exclusive and can be considered "prostheses of the human body," such as housing (shelter), clothing, and telephones (bottom right) (Munzer, 1990);
- Common goods are rival, but not exclusive; they can be considered as community property; natural areas, drinking water, fresh air, etc. (bottom left);
- Public goods are neither rival nor exclusive, and are available to everyone without limits; cultural heritage, intellectual/scientific insights (such as the Pythagorean theorem), (top left);
- Club goods (top right) are not rival, but they are exclusive, for example intellectual property such as patents or trademarks. (It is named 'club goods' after the classic economic example of the exclusive membership of a golf club.)

Interactions Between the Cultural, the Political and the Economic Life

So far, we have supplemented the threefold social order, as a vertical (mind-matter) differentiation of the human condition, by horizontal social relations to capture the increasing individualistic value orientations over the past 50 years. As a result, the four quadrants (resulting from the vertical and horizontal axes) can be translated into political life and economic life. This leads to the following conclusions:

- Also in the now supplemented system, the cultural, political, and economic life indeed function largely on their own, as was the intention of the threefold social order.

- The virtue ethical framework, as derived from human nature (via the appreciation of the human condition) is translated into the moral practical framework on which the political sphere or state legitimises its legal system. It implies that at all levels (spheres) the 'good' is to be found in the centre. As discussed above, a society based on such a virtuous ethical system is also a sustainable one. In it, a 'theoretical' balance is maintained between the three spheres, but increasingly also between the public–private dividing line. In this respect, the ethical framework also provides for the centred equilibrium value orientation as a universal-individual social objective (telos).[3]
- Laws, created in the political sphere, are aimed at maintaining the above-mentioned balances, thus preventing the caricatural consequences of the a-social character of the cultural sphere and the anti-social character of the economic sphere (Brüll, p. 182). The first is to prevent bigotry and obsessive claims to one absolute truth; the second, the continued appropriation of the physical world by an ever-shrinking economic and financial elite.
- Limitations on the carrying capacity of the Earth's physical system must be examined by the cultural, scientific sphere and imposed on the economic sphere, for example in the form of environmental standards.

The middle sphere of political and legal life thus plays a crucial role in mediating between the three spheres. How exactly the details of these processes could or should take shape, is a topic that lies beyond the scope of the present article.

3 Implications for the Societal Spheres

In the 100 years since Steiner formulated anthroposophy, the urgency for social reform in the proposed direction has increased. In particular, the obvious inadequacy of the currently dominant financial and economic system in light of worsening sustainability problems calls for action. That is not to say that nothing has been achieved in these 100 years; on the contrary. Later chapters in this book bring out many successful initiatives that have emerged from the anthroposophical movement in the fields of agriculture, ecology, economics and finance, education, health and medicine, art, and (spiritual) science, among others. Apart from these more 'gradual, opportune changes,' however, Steiner (1920a, 1920b) saw the need for systemic change at the level of society as a whole: "of all the measures on a small scale, one really cannot hope for much today. After all, today we should learn to see that there is essentially nothing else to be achieved than by dealing with things on a large scale" (Steiner, 1920a, 1920b). Understandably, change is

[3] This corresponds to the Bhutanese concept of Gross National Happiness, which is close to Aristotle's *eudaimonia*.

likely to occur simultaneously through both bottom-up and top-down initiatives. We will close this article by summarising the proposed top-down system changes, and by highlighting the great importance of the many already thriving bottom-up initiatives.

The Economic Life

In the threefold social order, the economic domain and the rule of law are highly differentiated (if not separated). On the one hand, "the legislative and administrative bodies in the economic field will emerge from the forces of economic life", suggesting a pronounced autonomy. This also seems to apply to bodies that set the conditions nature provides from herself. On the other hand, economic life is seen as "dependent on those legal relations established by the state between individuals and groups engaged in economic work" (Steiner, 1919a, 1919b, 1919c, Chapter II). Apparently, the political sphere (rule of law) applies only to the production factor labour.

However, given the currently dominant materialist-individualist value orientation in the lower right corner of the human condition, economic life cannot be left to its own devices. Despite half a century of increasing understanding of both the physical and social limits to growth, human needs have grown into human desires, at odds with ethical considerations about "the good life," sustainability, and the attainment of human happiness. Given the current circumstances, the political sphere, which is fully subject to the democratic process, must define the playing field for economic life according to the four quadrants in Fig. 3. The individualistic orientation is thus limited by public–private relations and the materialist orientation by ecological sustainability.

The political sphere, in turn, must be partially constrained in this at the constitutional level by the ethical notions that must be suggested by the cultural sphere. These notions amount to discouraging one-sided value orientations, including extreme materialistic consumption patterns, and inevitably resuming the cultural and political discussion of finite needs versus infinite desires. In our finite world, it will also reopen the debate about a well-founded industrial policy.

As early as 100 years ago, Steiner (1920a, 1920b) advocated cooperatives and associations in economic life, rather than companies owned by shareholders. He argued that the shareholder who has income, without working, gives an anti-social aspect to the entire economic body. Ownership should belong not only to shareholders on the basis of the factor of production of capital, but also to the factors of labour, nature-environment and culture (science, technology). This has become all the more topical as legal cases pile up where shareholders legally deny their responsibility for sustainability issues and claim direct payouts of company profits as dividends. The first such case was multinational Unilever which sought to improve their palm oil plantations in Indonesia, but lost the case against its short-sighted shareholders.

Within the proposed framework, the political sphere can redefine 'private property,' by amending the Shareholders' Directive (of the European Union). In doing so, the shareholder would be replaced by all the interested parties involved and capitalist ownership would be transformed into steward ownership. In accordance with the threefold social order, and as has been constantly debated, the ownership of land can also be questioned.

The Financial System

The gradual takeover of the economic system by the financial system ('financialisation') dates back to the nineteenth century but has accelerated in the last 40 to 50 years (Lagoarde-Segot, 2017). It is the most pronounced caricature of our time. Steiner (1922) was aware of this development and distinguished three phases of money in the economic cycle:

- purchase money that is used to purchase the perishable goods of agriculture;
- loan money to invest the wealth of the community through its intelligent use by industry to produce new products, and gift money, used to reinvest surpluses in cultural activities. The political sphere receives support from the cultural sphere in formulating (and evaluating) social objectives, including an underlying ethical framework as proposed. The constitutional nature of these very fundamental principles (human dignity, etc.) narrows the democratic playing field.

Conversely, the political sphere translates the ethical framework into moral practice, resulting in a limitation of:

- economic life vis-à-vis the anti-social forces by limiting economic activities to the four balanced quadrants (Fig. 3)
- cultural life vis-à-vis the a-social, intolerant forces by imposing restrictions on the publicly expressed claims to absolute truth (extreme fanaticism).

The Cultural, Spiritual Life

Freedom for the cultural sphere is unproblematic for the arts but already complicated for different types of science. To what extent, for example, should science provide answers to unasked questions and non-existent needs? Brüll (1984, p. 221) quotes Steiner arguing that in the absence of needs, not even truth is created (1919, CW 190). The interaction is even more problematic for (free) education. According to Segall (2023, p. 240), the idea that "state-controlled education also fosters a situation in which both corporate interests and parents with different cultural views vie for the political power to impose their worldview on curricula, pushing teachers into passive, prescribed roles." But without the democratic effort to keep education programmes within the bounds of what we have defined as "human dignity," the situation would get worse. Countless polarised pockets of very pronounced "truths" or even absolute truths would emerge.

The general problem related to this is that the contribution of cultural life to political and economic life depends on the dominant world view. Thus, the current dominance of the materialist-individualist worldview is fully reflected in political and economic life, further reinforcing the one-sidedness of the worldview. This emphasises the need to transcend the ever-changing wanderings of the worldview towards one-sided orientations and the associated fashionable worldview. The current 'postmodern' worldview, as given by the lower right quadrant, then turns out to be only part of the bigger picture of the human condition. Thus, assessing an ethical framework based on human nature only makes sense when we broaden our scope to centuries or even millennia.

Anthroposophy: The Reintegration into the Coherent Whole

Following the threefold social order, we differentiated the human condition in Fig. 2 (the ground plan of human nature) into the three levels of Fig. 3. Now that the role and functioning of cultural, political and economic life have been explored further, these three constituent parts must be reintegrated into a coherent whole. Following Jung, this process of differentiation and integration, which is similar to the integration/reconciliation of opposites, brings about a higher level of consciousness, namely the ability to oversee one's own totality. It is like the master pianist Horowitz, who argued that good music can only be realised by differentiating a composition into individual notes, which are then integrated again to form a new whole on a higher level.

It is noteworthy that this integration takes place mainly in the various fields of practice of anthroposophical work. In all cases, the approach is holistic; humankind, the earth and the cosmos are seen as a coherent whole.

Biodynamic agriculture is the 100-year-old solution to the now urgent problem of soil mineralisation, eutrophication, weakened ecosystem, a deterioration in food quality, increased antibiotic resistance, and increasing effects of pesticide use on human health.

Education is primarily focused on the healthy development of body, soul and spirit, rather than meeting the ever-changing performance targets demanded by industry or society.

Anthroposophical *medicine and health care* is highly integrative and complementary to regular medical education and practice. The diagnostic and therapeutic tools of mainstream medicine are complemented by self-healing through the stimulation of life forces through a health-promoting lifestyle. Strikingly, but not surprisingly, health is seen as related to human nature, as discussed above. Just as human happiness is associated with a more integral appreciation of the human condition, this is also true for health. According to Antonovsky (1987), health is defined by developing an identity that is based on the entirety of the human condition, rather than a part. It thus not only includes the physical body, but also the physical and social environment, and the mental-spiritual structure. Health and well-being then seem to be part of 'sustainability' as an overarching concept.

4 Conclusion

Anthroposophy offers a profound understanding of the question after sustainable development. It recommends approaching sustainable development in terms of maintaining the balance of the human value orientation between the fundamental opposing forces that define the human condition: the materialistic versus the spiritual and the particular (individual) versus the universal (collective). We derived our conception of the human condition from Steiner's circle of worldviews that represent the fundamental orientations of our human thinking. Wherever the awareness of the human condition as a whole is lost, a single orientation begins to unilaterally dominate and the balance between these opposites is lost. What initially had its proper place within the whole circle then quickly devolves into its own obsessive caricature, whereby both human dignity and sustainability are lost from view. From the 'circle of world views' a (virtue) ethical framework thus can be derived.

In support of the apparent societal goal of maintaining a certain balance, Steiner characterised a threefold understanding of the human condition and its social counterpart that would maintain equilibrium based on its own organisational structure. To this end, he differentiated the human condition into three fairly independent spheres; in social reality these spheres correspond to cultural life, political (legal) life and economic life. By seeing these differentiated spheres as an integral whole at the same time, the respective organisational structures could be designed in such a way as to maintain the balance between the spheres, as we attempted to illustrate by way of the example of economic life.

As we have seen time and time again, the quest for sustainable development boils down to striving for a higher level of awareness of our human condition. By taking a more holistic and integral view of that condition, one-sided valuations are avoided along with their caricatural identifications and catastrophic obsessions. It is undeniable that anthroposophy has shown promise when it comes to a truly humane view of human nature, and thus of the ability to support a meaningful life in a healthy society.

References

Antonovsky, A. (1987). *Unraveling the mystery of health.* Jossey-Bass.

Aristotle. (1999). *Nicomachean Ethics (Book II).* Batoche Books.

Brüll, D. (1984). The Mysteries of Social Encounters: The Anthroposophical Social Impulse. Hudson, NY: Waldorf Publications. (Note: The publication year appears to be 1984, but the publisher's date is listed as 2002. Please verify the correct publication year.)

van Egmond, N. D., & Mulder, J. M. (2025). The narrative of a new renaissance. [Manuscript under peer review].

Goetheanum. (2024). Department of Social Sciences. Retrieved from https://social.goetheanum.ch/en/fields-of-work/threefolding

Hitlin, S. (2003). Values at the core of personal identity: Making connections between two self-theories. *Social Psychology Quarterly, 66*(2), 118–137.

Jung, C. G. (1946). On the nature of the psyche. In H. Read, M. Fordham, & G. Adler (Eds.), The Collected Works of C. G. Jung (Vol. 8, pp. 408, 425). Princeton, NJ: Princeton University Press.

Jung, C. G. (1953). Psychological reflections. London, England: Routledge. (Also in The Collected Works of C. G. Jung (Vol. 14, Part 2, p. 470).
Jung, C. G. (1957). Present and future. In H. Read, M. Fordham, & G. Adler (Eds.), The Collected Works of C. G. Jung (Vol. 10, p. 575). Princeton, NJ: Princeton University Press.
Jung, C. G. (1961). Memories, Dreams, Reflections. New York, NY: Random House. (Original work published in German as Erinnerungen, Träume, Gedanken.)
Lagoarde-Segot, T. (2017). Financialization: Towards a new research agenda. *International Review of Financial Analysis, 51*, 113–123.
Munzer, S. R. (1990). *A theory of property.* Cambridge University Press.
Segall, M. D. (2023). The urgency of social trinity in a world still at war with itself. *Cosmos and History: The Journal of Natural and Social Philosophy, 19*(1), 240, 246.
Sorokin, P. (1957). *Social and Cultural Dynamics.* Transaction Publishers.
Steiner, R. (1919b). The threefold social order: Meeting social needs. In The Threefold Social Order (CW 23, Chapter II). Dornach, Switzerland: Rudolf Steiner Verlag.
Steiner, R. (1920a). The triple order of the social body: Study series II. In On the Propaganda of the Triple Social Order (CW 337a II).
Steiner, R. (1920b). Aufsätze über die Dreigliederung des Sozialen Organismus und zur Zeitlage.
Steiner, R. (1914). *Human and cosmic thought (CW 151).* Rudolf Steiner Press.
Steiner, R. (1918). *Geschichtliche symptomatologie (GA 185).* Rudolf Steiner Verlag.
Steiner, R. (1919a). *The threefold social order (CW 23).* Rudolf Steiner Verlag.
Steiner, R. (1919c). *The spiritual background of the social question (GA 190).* Rudolf Steiner Verlag.
Steiner, R. (1922). *Nationalökonomischer kurs: world economy (GA 340).* Rudolf Steiner Verlag.
Toulmin, S. (1990). *Cosmopolis: The hidden agenda of modernity.* University of Chicago Press.
van Egmond, N. D. (2014). *Sustainable civilization.* Palgrave Macmillan.
van Egmond, N. D., & de Vries, B. J. M. (2024). Reform of the Eurozone financial system: A system dynamic approach. *International Review of Financial Analysis, 93*, Article 103192.
Wolf, M. (2024). The crisis of democratic capitalism. United Kingdom: Penguin Random House.

Constructive Anthropocentrism—A Possibility for Rebalancing the Current Eco-philosophical Discourse

Flore Lutters

Abstract

This article explores the concept of "constructive anthropocentrism" as a potential pathway to rebalance eco-philosophical discourse in the Anthropocene and transition toward the Symbiocene. In contrast to destructive anthropocentrism, which prioritises unbounded human dominance over nature, and ecocentric perspectives that decentralise humanity entirely, constructive anthropocentrism seeks a middle ground. It emphasises reciprocal participation, where human creativity aligns with ecological and cosmic systems to create benefits for both. By examining eco-philosophical discourse, the evolution of human consciousness, and successful biodynamic initiatives Apricot Lane Farms and SEKEM, the article explores how constructive anthropocentrism can foster a mutual beneficial, reciprocal relationship between humanity and nature.

Keywords

Constructive anthropocentrism • Anthropocene • Symbiocene • Eco-philosophy • Environmental ethics • Biodynamic agriculture • Reciprocal participation • Evolution of consciousness • Gaia theory • Threefold social order • Human-nature relationship • Ecological regeneration • Sustainable development • Earth system science • Climate philosophy

F. Lutters (✉)
VU Graduate School of Humanities, Free University Amsterdam, Amsterdam, Netherlands
e-mail: lutters@denieuweschool.amsterdam

J. Kronenberg and E. T. Lammerts van Bueren (eds.), *On the Earth We Want to Live*,
World Sustainability Series, https://doi.org/10.1007/978-3-031-98758-8_8

1 Introduction

1.1 Introduction: The Biggest Little Farm

In 2011, John and Molly purchased a 234-ha plot of land. The soil had been completely depleted after 50 years of intensive industrial farming in the surrounding, rendering it barren and nearly incapable of supporting life. John often jokes that they "bought a bank that had been robbed" (Apricot Lane Farms, n.d.). Their dream was to create a thriving farm that would benefit both people and the environment. The gap between their vision and the initial reality was vast. To bridge this gap, they sought out a mentor figure, eventually connecting with Alan York, an expert in biodynamic agriculture.[1,2]

In the documentary The Biggest Little Farm, viewers witness the transformation of the depleted land into a flourishing, highly biodiverse farm, with over two hundred varieties of fruits and vegetables, alongside a diverse range of animals, including sheep, chickens, cows, pigs, dogs, owls, and more.

After 8 years, John and Molly's vision has succeeded. Apricot Lane Farms is now "working in harmony (or a comfortable level of disharmony) within a dynamic ecosystem" (Apricot Lane Farms, n.d.). John explains why, after years of experimentation, they are increasingly able to achieve this relative balance: "The trick is to balance the needs of the farm with those of the wild, something Alan encouraged from the beginning" (Apricot Lane Farms, n.d.).[3] So Apricot Lane Farms stands as a prime example of how humans and nature can work and coexist in a relatively beneficial harmony.

1.2 Constructive Anthropocentrism

Dan McKanan, Senior Lecturer at Harvard Divinity School, explores biodynamic agriculture and other ecological practices rooted in Rudolf Steiner's anthroposophy in his book Eco-Alchemy: Anthroposophy and the History and Future of Environmentalism, referencing successful implementations like those seen at the

[1] Alan York was a landscape and garden designer, global vineyard advisor and lecturer, former president of the Biodynamic Association, and editor of the quarterly Biodynamics magazine (Nigro, 2014).

[2] Biodynamic farming is a holistic method that recognises the intrinsic interconnections among soil, plants, animals, and humans. Initiated by Rudolf Steiner in 1924 and further developed into a successful agricultural practice by others, biodynamic so represents a farming model for integrating ecology with farming.

[3] John and Molly's approach exemplifies four core biodynamic principles. First, they view the farm as a living ecosystem in which every plant, animal, and microbe plays a unique role. Second, they focus on maximizing the diversity of crops and animals, promoting mutual and beneficial interactions. Third, they make extensive use of compost and organic fertilization to enhance soil health. Finally, they consider the cycles and natural rhythms of nature.

Biggest Little Farms. He writes: "Though I continue to appreciate the contributions of mainstream science, the Left, and Gaian spirituality, my environmental commitments have been rebalanced by the alchemy of anthroposophy. I believe the environmental movement as a whole might benefit from a similar rebalancing" (McKanan, 2017, p. 225).

McKanan outlines several factors contributing to this "rebalancing by the alchemy of anthroposophy," one of which he describes as "constructive anthropocentrism" (McKanan, 2017, p. 234).[4] This approach, he explains, reflects a capacity to achieve "harmony through creative work with the polarities of human and nature", enabling one "to balance artistic and cultural commitments with reverence for nature" (McKanan, 2017, p. 13).[5]

1.3 Rebalancing the Eco-philosophical Discourse

This article will explore whether, and if so, how constructive anthropocentrism could contribute to a "rebalancing" of the broader green movement, as McKanan suggests. As my expertise as a PhD scholar is in environmental philosophy, I will narrow this question to the possibility of rebalancing the current eco-philosophical discourse.

To address the research question, I will first explore the current ecological challenges, particularly the 'Anthropocene,' and assess the eco-philosophical responses and strategies related to it. This will map the current discourse and highlight any need for "rebalancing." Next, I will contextualise the main environmental beliefs within the framework of 'the evolution of consciousness' to uncover their roots. Finally, I will attempt to define "constructive anthropocentrism" conceptually and compare it with prevailing environmental perspectives to evaluate its potential role in rebalancing eco-philosophical discussions.

2 Face to Face with Gaia in the Anthropocene

2.1 The Anthropocene

Contemporary eco-philosophy is largely shaped by the concept of the 'Anthropocene,' a term that gained prominence in the twenty-first century. Will Steffen, Executive Director of the International Geosphere-Biosphere Programme (IGBP), recalls the moment when atmospheric chemist and Nobel laureate Paul Crutzen introduced the term. Frustrated, Crutzen exclaimed to paleontologists at a conference who continued to refer to the current geological epoch as the "Holocene":

[4] The others are a homeopathic model of social change, planetary transmutation and cosmic holism (McKanan, 2017, p. 225).

[5] It appears that Alan, the biodynamic expert, taught John and Molly this 'constructive anthropocentric' so they could find a balance between the needs of the farm and the wild.

"Stop using the word Holocene! We're no longer in the Holocene. We're in the... the... the... [searching for the right term]... the Anthropocene!" (Crutzen as cited in Angus, 2018, p. 28).

In geology, the Holocene refers to the recent epoch, which began around 11,700 years ago after the Pleistocene. This Holocene epoch was characterised by a stable Earth, a period conducive to human flourishing, in contrast to the large temperature fluctuations of the preceding Pleistocene. These stable conditions were so ideal that some referred to the Holocene as the "paradisiacal Holocene" (Latour, 2017). When Crutzen highlighted that we were "no longer in the Holocene, but in the Anthropocene," he underscored the fact that Earth's conditions are changing so rapidly that the paradisiacal Holocene is disappearing. This shift—leading to an increasingly unpredictable Earth—is driven by the systematic crossing of six of the nine planetary boundaries (Rockström et al., 2023).[6] However, unlike previous Earth changes, these transgressions of planetary boundaries are now driven solely by human activity, not "blind" natural forces (Hamilton, 2018).

For this reason, Crutzen emphasised the central role of humanity in this new geological epoch, which he termed the Anthropocene. The word comes from the Ancient Greek ἄνθρωπος (ánthrōpos, "human") and καινός (kainós, "new" or "recent"), together meaning "the new age of humanity" (Crutzen, 2001)—an era in which humanity destabilises the Earth.

2.2 Face-To-Face with Gaia

Crutzen, drawing on extensive Earth System Science (ESS) research, characterises humanity in the Anthropocene as a "geological force" on par with natural phenomena such as volcanism, plate tectonics, hurricanes, and comets (Crutzen, 2002).[7]

In recent centuries, this growing human power has been collectively applied to build a modern "anthroposphere" (Vernadsky, 1928).[8] The anthroposphere

[6] The debate about the history of the Anthropocene is multifaceted, but broadly, an origin story can be synthesized that begins with a prelude during the Agricultural Revolution (approximately 12,000–7000 years ago). While this period saw localized forms of destabilisation, it was not yet a planetary phenomenon. The Anthropocene gained more distinct features during the early modern period (around 1500) with the emergence of global circulation networks and colonial expansion into the non-Western world. It became truly transboundary during the Industrial Revolution (from 1784) and ultimately revealed its full planetary scale with the onset of globalization (from 1945 onwards) (Ruddiman, 2003; Lewis & Maslin, 2015; Crutzen, 2002).

[7] This is not merely metaphorical. Beyond the aforementioned planetary boundary transgressions, studies have revealed that Earth has tilted by 80 cm in less than two decades due to human activities—specifically, the extraction of groundwater for irrigation and drinking purposes (Seo et al., 2023).

[8] An frequently used term that highlights the role of technique is 'technosphere', a sphere that functions "according to a quasi-autonomous dynamics" (Haff, 2014).

includes all human-made structures such as houses, factories, schools, mines, hospitals, farms, roads, airports, ports, and computers, and is regarded as crucial infrastructure that has undeniably enhanced human well-being (Ritchie, 2024).[9]

On the other hand, the Earth itself gains power in the Anthropocene. The Australian philosopher Clive Hamilton describes this Earth as "a sublime Earth, appearing to us as beautiful when it is calm but terrifying when it is enraged, a familiar yet strange planet—Mother and monster" (Hamilton, 2018, p. 65). In the Anthropocene, humanity increasingly faces Earth not as "Mother Earth" but as a "monster" in its "frightening" and "fury-driven" reality. Bruno Latour captures this terrifying experience as coming "face-to-face with Gaia"[10] (Latour, 2017).[11] Earth system scientists, once sober-minded, now speak of Gaia as an "awakened giant," "the irritable beast," or "Gaia who fights back" and seeks "revenge" (Hamilton, 2018, pp. 63, 64).

3 The Eco-philosophical Discourse and Its Polarisation Toward the Way Forward in the Anthropocene

3.1 Eco-philosophical Positioning

Based on the Earth System Scientific understanding of the Anthropocene, in which both the Anthropos and Gaia have gained power, an analysis of the current eco-philosophical discourse reveals two distinct typologies within the array of perspectives (Hamilton, 2018, p. 62).

The first typology is characterised by an affirmation of the new power of the Anthropos as a geological force, coupled with a denial of the increased power of the Earth. Ecophilosophical positions within this typology include "ecomodernists," "ecopragmatists," and "geoengineers".[12] These positions hold significant political and economic power today and largely align with "the system," or "market capitalism" (Hamilton, 2018, p. 102). Their eco-philosophical stance suggests that

[9] The anthroposphere, or technosphere, currently weighs a total of 1.1 trillion kilograms, surpassing the weight of the entire biosphere.

[10] The term Gaia was developed by James Lovelock and Lynn Margulis in the late 1960s and early 1970s to denote the Earth as a self-regulating, complex system. Lovelock chose this name, inspired by Greek mythology, on the suggestion of Lord of the Flies author William Golding. At the time, Golding was influenced by Steiner's Anthroposophy, in which the concept of Gaia had been circulating for some time, and he had assisted Lovelock in enrolling his intellectually disabled son in a Steiner school. In the book The Gaia Hypothesis: Science on a Pagan Planet, philosopher of science Michael Ruse argues: "Lovelock would deny it himself, but without Steiner, there would have been no Gaia hypothesis" (Ruse, 2013).

[11] He describes this first encounter as: "I could no longer evade it: I had to comprehend what was imposing itself upon me in the rather alarming form of a force that was simultaneously mythical, scientific, political, and likely religious" (Latour, 2017, p. 13).

[12] Thinkers belonging to this category include Michael Shellenberger, Stewart Brand, Mark Lynas, and David Keith.

they "continue to chase the modernist dream in the Anthropocene, namely that we can use technology to make nature bend to human will" (Hamilton, 2018, p. 102). This position is sustained in the Anthropocene largely because its proponents "overlook that the Earth has been jolted awake from its peaceful Holocene slumber and is now on an uncontrollable collision course" (Hamilton, 2018, p. 103).

In contrast, the second typology views the Earth—conceived as Gaia—as an active and powerful entity, while relativising the power of the Anthropos. Positions within this typology include "posthumanists," "ontological pluralists," "vital materialists," and "object-oriented ontologists" (OOO).[13] This category currently holds the most influence in the academic environmental humanities discourse and is widely embraced by climate activists. These positions seek "to undermine the modern cherished notion of a passive nature that can be controlled by humans" (Hamilton, 2018, p. 102). In addition to this reconceptualisation of nature, they engage with "the human after modernity," seeking to "dethrone humans, deny our uniqueness, and emphasize that we do not occupy a separate position in relation to inorganic and organic non-humans" (Hamilton, 2018, pp. 102, 114).

3.2 Navigating the Anthropocene

The two distinct typologies each envision a different way forward in the new geological epoch of the Anthropocene, shaped by their contrasting understandings of the Anthropocene reality. Typology 1 views the environmental damage resulting from modern progress as a necessary collateral effect.[14] This damage is perceived as a sacrifice made in the construction and current functioning of the anthroposphere, which increasingly sustains the livelihoods of ~8 billion people. From this perspective, what is now needed is to render the anthroposphere sustainable through technological innovation and more efficient resource management (Ritchie, 2024). The guiding principle here is 'eco-economic decoupling,' which posits that the anthroposphere can continue to grow while its impact on nature is, at first, 'relatively' decoupled and, in the medium to long term, 'absolutely decoupled' (Nordhaus & Shellenberger, 2015). In this view, a better—or at least more manageable—Anthropocene is on the horizon (Ellis, 2012).[15]

[13] Thinkers who fall within this category include Jane Bennet, Timothy Morton, Isabelle Stengers, Bruno Latour en Donna Haraway (Hamilton, 2018).

[14] As a result, life expectancy, individual prosperity, health, and mobility have all rapidly improved over the past centuries (Ritchie, 2024).

[15] Erle Ellis reinforces this optimism by stating, "A good, or at least a better, Anthropocene is within our reach. To create that future, we must move beyond the fear of exceeding natural limits and the nostalgic hope of returning to a pastoral or pristine era. Above all, we should not view the Anthropocene as a crisis but as the beginning of a new geological era filled with human-led opportunities" (Ellis, 2012, para 4).

In contrast, typology 2 fundamentally understands humanity as 'earthbound' (Latour, 2017). This conception emphasises that humans are intrinsically 'tied' to the Earth and its limits. Since the Earth is no longer a 'mother' as in the Holocene, but rather a 'monster' in the form of Gaia. And therefore the 'earthbound' anthroposphere is built on an inherently unstable foundation. This new reality is understood as the 'intrusion of Gaia' or the 'New Climate Regime' (Latour, 2017; Stengers, 2009). According to thinkers in typology 2, humanity will be compelled to 'think ecologically' in this new reality. Ecological thinking is conceived as "a practice and a process of becoming fully aware of how human beings are connected with other beings—whether animal, vegetable [...] the rocks, the air, and the water" (Morton, 2010, p. 7). However, in the new Climate Regime, this interconnectedness transcends our traditional understanding of connectedness (Morton, 2010, p. 7). This means that we must avoid conceptualising the nature of non-humans in the terms of the peaceful Holocene. In this new reality, the power of non-humans has increased, and that of Humans decreased.

3.3 A War of Worlds

Bruno Latour's title for his fifth Gifford Lecture Series, War of Worlds: Humans against Earthbound, aptly captures the fundamental differences between the two distinct ways of navigating the Anthropocene (Latour, 2013). For Latour, this polarisation extends so far that he characterises it as a 'war of worlds,' in which, in world 1, the 'Humans', or 'Moderns', of typology 1 live, while in world 2, the 'Earthbound' of typology 2 exist.

In the world of the 'Moderns,' the guiding principle of action is 'decoupling,' which seeks a complete or 'absolute decoupling' between the anthroposphere (culture) and the Earth (nature). This reflects a desire for absolute freedom and autonomy. In contrast, in the world of the 'Earthbound,' the aim is an 'absolute coupling' between the anthroposphere (culture) and Gaia (nature), demonstrating a longing—and fear—for embeddedness and connectivity. Latour's use of the term 'war of worlds' is intentional, as these principles are diametrically opposed: the pursuit of 'unbounded freedom' on one side and 'bounded connectedness' on the other.[16] If the 'alchemy of Anthroposophy' can rebalance current 'environmental commitments,' then constructive anthropocentrism must respond to this challenge by opening a pathway toward a 'shared world' that balances freedom and embeddedness, autonomy, and connectivity (McKanan, 2017).

[16] The authors Bruno Latour and Clive Hamilton are not the only ones theorising this polarisation. Outside of the eco-philosophical discourse, it is the author Charles Mann who identifies the fundamental opposition between, on the one hand, the 'wizards' (advocating for technological innovation) and, on the other hand, the 'prophets' (advocating for limits to growth) (Mann, 2018). Additionally, Arne Naess presents his classic distinction between 'shallow ecology' (Modern) and 'deep ecology' (Earthbound) (Naess, 1973).

4 In Search of the Roots of the Environmental Convictions of World 1 and 2

4.1 The Framework of the 'Evolution of Consciousness'

The foundational roots of World 1, inhabited by the 'Moderns' who yearn for unbounded freedom, and World 2, where the 'Earthbound' exist in bound connectivity are deeply embedded in cultural, philosophical, and even spiritual traditions. To understand these roots, which seem to permeate the intense rhetoric of both worlds, and to uncover potential avenues for rebalancing their fundamental convictions, it is necessary to shift from the political-philosophical level to the ontological level. This entails a move from questions of power and principles of action to a deeper investigation into the nature of being itself.

An interdisciplinary framework known as the 'evolution of human consciousness,' was initially developed by Steiner (1910, 2006). Since Steiner's death in 1925, the framework of 'the evolution of consciousness' has been continuously expanded and enriched. Owen Barfield (1898–1997) introduced concepts that solidified Steiner's theosophically oriented ideas, grounding them more firmly and supplementing them with new anthropological and linguistic insights (Fuccia, 2016; Vernon, 2020). Building on Barfield's work, American philosopher Richard Tarnas furthered this exploration by reducing its teleological and Eurocentric elements, while highlighting its pluralistic nature (Tarnas, 1991, 2002, 2006). The following investigation into the spiritual roots of the two 'worlds' will synthesise the contributions of Steiner, Barfield, and Tarnas. It aims to provide an ideal–typical overview of their key concepts and associated terminology.

4.2 The Key Notion of Participation

The foundational premise of their ontological research is that human understanding of being is not universal, but pluralistic in nature. This means that human convictions, and more fundamentally, their pre-cognitive experiences, vary across different cultural and historical epochs. From this pluralistic ontological perspective, being itself cannot be described by a universal, eternal set of traits. Instead, worlds develop and perish, and it is the framework of the evolution of consciousness that characterises the nature of these worlds. The understanding of this nature of worlds is based on the criteria of whether and how humans participate, or fail to participate, in society, nature, and the sacred (Fuccia, 2016).

From this standpoint, the central hypothesis of Steiner, Barfield, and Tarnas is that, particularly within the Western evolution of consciousness, there has been a gradual transformation of worlds. This shift is marked by a fundamental change in the way humans participate, moving from a state of 'original participation' through different phases of 'withdrawal of participation' to a current 'absence of participation' (Barfield, 1957; Vernon, 2020).

4.3 Original Participation

The state of 'original participation' is characterised by a human self with permeable boundaries, indicating an inherent interconnectedness with the surrounding world, which is imbued with non-human agency. Within this reality, conscious intelligence and inwardness are not confined to individuals but are perceived as permeating both the self and the external world.[17]

Steiner describes the human experience in this state as an "instinctive connection with the spiritual world" because "consciousness was spread out more widely," and humans felt themselves "as part of a sublime organism" (Steiner, 1910, pp. 180, 181). During this period, maintaining a constant effort through ritual participation was essential for survival, as it ensured a good relationship with this 'other side' of reality (Bellah, 2011).

This mode of being should not be merely idealised as a womb-like embeddedness; it was also a reality that could be dangerous and capricious in its divine-natural aspects, while at the same time offering protection and meaning.[18]

4.4 Withdrawal of Participation

The transition from 'original participation' to 'absence of participation' is marked by a gradual contraction and narrowing of consciousness. Initially, the human self was experienced as inherently interconnected with the world, its boundaries fluid and open to non-human agency. Over time, this state began to change, leading to a progressive differentiation between self and world.

As this transformation unfolded, the human self-gained a heightened sense of autonomy and agency, becoming more distinct from nature and the broader existential matrix. This shift, however, came at a cost—what was once perceived as an animated, meaning-filled world gradually lost its sense of inherent intelligence and sacred presence. Meaning, which had previously been distributed throughout the cosmos, became increasingly concentrated within the human self alone.

This process, often described as the disenchantment of the world, involved the retreat of the gods and the growing independence of human beings. It was not a sudden rupture but rather a slow and varied progression, unfolding at different rates across cultures. While the withdrawal of participation enabled greater individual

---

[17] The notion of 'the porous self' is also a central concept in the work of Taylor (1989, 2007).

[18] Barfield tries to explain the phenomenology of 'original participation' as follows: "Anyone who finds it difficult to form any conception of participation, that is, of self and no-self identified in the same moment of experience, should reflect on that while peculiar realm of semi-subjectivity which still leads a precarious existence under the name of 'instinct'—or on those 'irresistible' impulses, on which psychiatrists are inclined to dwell. Many of us know what panic feels like, and ordinary men are proud of their sexual vigour or ashamed of the lack of it, although the act is readily acknowledged in retrospect to be at least as much something that is done to, or with, them by an invisible force of nature (…)" (Barfield, 1957, p. 73).

freedom and rational understanding, it also led to an increasing sense of separation from the deeper, interconnected fabric of existence. As Tarnas explains:

> The history of the human mind's movement from a state of participation mystique to a more fully differentiated mode of awareness is in many respects the history of the human mind itself. Impelled by the powerful human drive to achieve ever-greater autonomy relative to the conditions of existence, virtually the entire evolution of human consciousness has served this psychological and epistemological impulse to distinguish the human self from the world, subject from object, part from whole. The Promethean project seems to be intrinsic to the human condition. Yet this project has been carried out most vigorously and brilliantly by the Western mind, above all by the modern mind, that avatar and apex of Promethean progress. (Tarnas, 2006, p. 22)

4.5 Absence of Participation

In the current stage of human consciousness, participation has largely withdrawn, leading to a distinct separation between the self and the world. The boundaries of the self have become rigid and well-defined, creating a clear division between inner experience and the external environment.[19]

Consciousness and inwardness, once perceived as diffused throughout both the self and the world, are now confined exclusively within the individual. This shift results in the external world being experienced as a collection of inert objects, devoid of agency or intrinsic meaning. The human self, now isolated, perceives itself as autonomous and empowered, acting according to its own will rather than in response to an interconnected cosmic order.

With this transformation, there is no longer a necessity to engage with externally imposed meanings or participate in structures of traditional cultural authority—whether religious, social, or political. The sense of embeddedness in a larger, living cosmos has faded, replaced by an increasing emphasis on individual agency and subjective interpretation of reality.

4.6 The Roots of the Environmental Convictions

After this brief outline of the evolution of human consciousness, we can now attempt to illuminate the cultural-spiritual roots of the 'environmental convictions' of the worlds of 'the Moderns' and 'the Earthbound'. Firstly, the desire of 'the Moderns' (typology/world 1) to attain unbridled freedom by achieving, through technical ingenuity, an 'absolute decoupling' between the anthroposphere and the Earth in the Anthropocene aligns with the modern mode of consciousness known as 'absence of participation', as it perpetuates humanity's Promethean quest for emancipation from the matrix of an external order (Tarnas, 2006). Martin Lewis

[19] Also, this notion returns as the 'buffered self' in the work of Taylor (1989, 2007).

articulates this Modern 'environmental conviction', he calls "Promethean environmentalism" (typology 1), that "will continue to burn Prometheus's flame" in a heroic maintainace of the "the wildly creative, if at times wildly destructive, course of human ascent." (Lewis, p. 15).[20]

Every form of restraint or reversal of this initiated further movement of 'ascent' has to be prevented at all costs, because "to create that future, we must move beyond the fear of exceeding natural limits and the nostalgic hope of returning to a pastoral or pristine era" (Ellis, 2012). Ecopragmatist Steward Brand gave this promethean project the motto: 'We are as gods and might as well get good at it' (Brand, 2009).

World 2 wherein the 'Earthbound' live, reaches a diametrically opposed conclusion, for them, the "absence of participation" is not liberation but rather a form of 'spiritual autism' as the eco-theologian Thomas Berry laments:

> We talk only to ourselves. We do not talk to rivers, we do not listen to the wind and the stars. We have broken the great conversation. By breaking that conversation, we have shattered the universe. All the disasters now happening are a consequence of this spiritual 'autism'. (Berry et al., 2016, p. 66)

Their environmental conviction of 'absolute coupling' lies in overcoming this 'original sins of modern anthropocentrism' (Hamilton, 2018). Therefore, they draw inspiration from the world of 'original participation,' which is reflected in the great interest in premodern ontologies—where humans and non-humans coexisted in an enchanted world—such as 'animism,' 'totemism,' and 'analogism' (Descola, 2014).

It is for this reason that in the (new) ontologies being developed, the agency is taken away from the modern subject to be redistributed among non-humans, who, just as in the world of 'original participation,' are once again attributed with agency, while the human loses a central position. This is also what the Dutch philosopher Jos de Mul notes when he writes: their aversion to the vertical ontology and human exceptionalism associated with anthropocentric humanism leads them to advocate for a horizontal ontology (flat ontology) in which all living beings, and even inanimate objects, are attributed with comparable agency (De Mul, 2023).

[20] Adding that "but we must learn to do so as responsible adults rather than as pyromaniacal adolescents (Lewis, p. 15).

5 Constructive Anthropocentrism and the Possibilities for a Rebalancing of the Eco-philosophical Discourse

5.1 Participative Turn

Already at the end of the eighteenth century, various thinkers such as Goethe, Schelling, Hegel, Coleridge, and Emerson—in line with the critique of the 'earthbounds' on the 'anthropocentric sin' resulting from the absence of participation—recognised the need to explore a new form of participation. Tarnas writes about this shared project: 'Each of these thinkers gave his own distinct emphasis to the developing perspective, but common to all was a fundamental conviction that the relation of the human mind to the world was ultimately not dualistic but participatory' (Tarnas, 1991, p. 466). It is Rudolf Steiner who, at the beginning of the twentieth century, takes up this tradition again.

However—and here this position differs from the contemporary attack of the 'earthbounds' on modern anthropocentrism—'this participatory epistemology, developed in different ways by Goethe, Hegel, Steiner, and others, can be understood not as a regression to naive participation mystique, but as the dialectical synthesis of the long evolution from the primordial undifferentiated consciousness through the dualistic alienation' (Tarnas, 1991, p. 467). Thus, these thinkers attempt to explore a third form of participation that seeks to unite the thesis of 'original participation' and the antithesis of 'absent participation' into a synthesis. This project can be characterised as a 'participative turn' (Ferrer & Sherman, 2008).

This new form of participation can be described as 'reciprocal participation' (Vernon, 2020, introduction). In this new form of participation, "the inner life of the individual is felt to belong to him—or herself—the gain of withdrawal of participation—but also reflects the inner life of nature, and the cosmos" (Vernon, 2020, introduction). Thus, "the individual has a soul that is not cut off; an interiority that can reflect and reflect on life and its meaning', but people also have a subjectivity that can forge purpose and intentions" (Vernon, 2020, introduction).

5.2 Maturation of Romanticism

Since the concept of reciprocal participation has its roots in Romanticism—a cultural period often associated with an idealised, subjective worldview rather than a realistic or practical one—Barfield's book title suggests that Steiner's anthroposophy represents "romanticism" that "comes of age" (Barfield, 1942).

Romanticism, with its focus on emotion, imagination, and the inner world, failed to fully develop a grounded and pragmatic alternative to the detached or passive forms of participation it critiqued. According to Barfield, this failure can be attributed to the immaturity of Romanticism, which placed excessive emphasis on idealisation, subjectivity, and fantasy. As a result, it was unable to engage with reality in a meaningful way, instead remaining entrenched in speculative visions rather than facing the truth or taking moral responsibility. However, Steiner's anthroposophy, while indebted to Romanticism and sometimes even close to its pitfalls, managed to transcend these limitations by developing a more realistic and grounded approach to reciprocal participation. Unlike Romanticism, which primarily left behind artistic or philosophical impressions, Steiner's work—along with the efforts of the farmers, economists, doctors, teachers, and medical professionals inspired by him—demonstrated tangible, real-world applications. "In my view," McKanan writes, "Steiner is worth studying because his students have built an interconnected global network of schools, farms, intentional communities, clinics, innovative businesses, and other "initiatives" that bring healing energies to both nature and society" (McKanan, p. 1). It is this 'global network' that succeeded in making the fruits of reciprocal participation visible in concrete results, shaping practical solutions for various societal and ecological needs.

5.3 Defining Constructive Anthropocentrism

In his analysis of this 'interconnected global network' McKanan conceptually defines the eco-philosophical stance behind these practices as "constructive anthropocentrism."[21]

The term "constructive" in this context refers to a "cosmic order" to which these practices are not only related but also embedded (Verdonk, 2015). This order is multifaceted, encompassing cultural, natural, and spiritual dimensions. As a result, it fosters active participation in local cultural traditions, such as annual festivals and rituals, the natural rhythms of the seasons and planetary movements, as well as spiritual and artistic contemplation.

The term "constructive" is paired with "anthropocentrism" because, within the practices McKanan examines, there is a deliberate encouragement of creativity, innovation, and the ability to create. This relationship to the external order is not mimetic; rather, it is co-creative. In this light, we can interpret the statement by German artist Joseph Beuys,[22] who asserted that:

> [o]nly art is capable of dismantling the repressive effects of a senile social system that continues to totter along the deathline: to dismantle in order to build 'A SOCIAL ORGANISM AS A WORK OF ART'… EVERY HUMAN BEING IS AN ARTIST who—from his state of freedom—the position of freedom that he experiences at first-hand—learns to determine the other positions of the TOTAL ART WORK OF THE FUTURE SOCIAL ORDER. (Tisdall, 1974, p. 48)

Beuys' words reflect the ethos of these practices, which emphasise human agency in shaping the world in accordance with the interconnected cosmic order.

[21] See Chap. 5 in this publication for many examples.

[22] Beuys was a member of the Anthroposophical Society, headquartered in Dornach, Switzerland, which has over 40,000 members spanning all continents, cultures, and religions.

5.4 Rebalancing of the Earthbound and the Moderns

Constructive anthropocentrism recognises the modern freedom and 'world-creating' potential of the "Moderns" (world 1), acknowledging humanity's capacity to shape and transform the world. However, it takes issue with the notion that such freedom is not inherently participatory. Instead, constructive anthropocentrism critiques the idea of unbounded freedom, arguing that this type of freedom is disconnected from the larger cosmic and ecological frameworks that sustain life (Saito, 2020). This unrestrained freedom leads to a state where humanity becomes increasingly detached from the cosmic order and the natural world, driven by an insatiable desire for domination and control. The metaphor of "the Promethean fire" captures this destructive urge, suggesting that, while human creativity is powerful, it often becomes distorted when it disregards its place within the broader ecological system.

In contrast, constructive anthropocentrism proposes a recognition of a cosmic order that transcends human subjectivity, aligning it more closely with the "Earthbounds" (world 2). Like the Earthbounds, it understands human subjectivity as part of a vast, interconnected network of active agents—elements, seasons, planets, and other cosmic forces—that must be taken into account. However, the divergence comes in the rejection of the notion that human agency should be subordinated or relativised within a non-hierarchical "Mesh" or "Gaia" system, as the Earthbounds suggest (Latour, 2013; Morton, 2010). While the Earthbounds argue for decentering the human being, constructive anthropocentrism maintains that human creativity and responsibility are essential. It insists that human agency must not be relegated to mere passivity or subordination but should actively engage with the larger cosmic order. In this view, the human role is one of co-creation emphasising the need for human creativity and responsibility in fostering a harmonious and sustainable relationship with both the Earth and the broader cosmos.

5.5 Symbiocene

The "Humans" and the "Earthbounds" represent two distinct approaches to navigating the Anthropocene, as our analysis of eco-philosophical discourse reveals. The 'Moderns,' on one hand, strive to achieve a 'good Anthropocene' through the concept of 'absolute decoupling,' wherein primarily anthropocentric interests are prioritised, often at the expense of ecological considerations (Parrique et al., 2019). In contrast, the 'Earthbounds' advocate for an 'exit Anthropocene,' proposing an 'absolute coupling' that emphasises ecocentric concerns and relativises human interests in relation to the larger ecological and cosmic system.

In response to these divergent paths, constructive anthropocentrism offers an alternative approach for navigating the Anthropocene, one that seeks a model of 'free coupling.' This concept envisions a mutually beneficial relationship in which both human and ecological interests are equally respected and served. Such a vision requires the skill of reciprocal participation between the Anthropos (human beings) and Gaia (the Earth). Constructive anthropocentrism posits that a balanced and cooperative relationship with the natural world is essential for fostering a sustainable future.

This future world, which Australian climate philosopher Glenn Albrecht terms the 'Symbiocene,' represents a potential path forward in which humans learn to freely bind themselves to an external order, aligning their actions with the rhythms and needs of the planet (Albrecht, 2019). Constructive anthropocentrism, in this context, offers a potential roadmap toward the Symbiocene, emphasising the importance of human responsibility in co-creating a mutually beneficial relationship with the Earth.

5.6 Greening the Desert

The exhausted landscape where John and Molly Chester started The Biggest Little Farms, can be seen as a metaphor for the Earth's increasingly degraded condition (100 million hectares of healthy land are lost every year[23]). But John and Molly—with the help of Adam—demonstrate that regenerating the land is indeed possible, and at the same time provide a context for human flourishing.[24] This example demonstrates the creative potential when people switch from destructive to a constructive anthropocentrism.

Also inspired by biodynamic agriculture, Ibrahim Abouleish from Egypt began a groundbreaking project in the 1970s to green the desert near Cairo. This initiative evolved into SEKEM, which was recently honoured with a Champion of the Earth award (December 2024). The award citation highlighted that "SEKEM demonstrates that it is possible to make food systems work for both people and the planet, which is essential for addressing environmental crises like desertification and restoring humanity's balance with nature" (United Nations Environment Programme, 2024).

[23] Figures in October 2023 of the United Nations, Convention to combat Desertification.

[24] A recurring question when watching the documentary is: who funds this project? In the case of The Biggest Little Farm, little information is available on this matter. However, on a broader scale, the Triodos Bank announced in late 2024 an investment of 500 million euros in Nature-Based Solutions. In alignment with constructive anthropocentrism, they state: "The world does not revolve solely around humans nor is it purely 'natural,' so we aim to invest in projects where 'the relationship between humans and nature is seen as one of mutual dependency and care,'" grounded in the fundamental conviction that "if we take good care of the planet, the planet will take care of us. This can be true on a local ecosystem level, up to a global level and everything in between" (Triodos Bank, 2024).

SEKEM's ambitious goals include helping 15,000 Egyptian farmers transition to biodynamic agriculture by 2024, increasing to 40,000 by 2025, and 250,000 by 2028. For decades, many farmers in Egypt have depended on chemical fertilisers and pesticides to boost yields, which has degraded soil health, caused land erosion, and accelerated desertification. This issue is especially urgent, as 96% of Egypt's land is desert (Moselhy, 2018). To date, SEKEM has regenerated 70,000 ha of land, and their plan is to expand this number significantly in the years ahead.

6 Conclusion

However hopeful biodynamic-inspired regenerative practices are, the prospects for collectively entering this symbiotic future are diminishing, as a 'destructive anthropocentrism' continues to exacerbate ecological instability. If no constructive anthropocentric turn is made soon, a Promethean conflict with a vengeful Gaia will be the daily reality of a growing amount of people. This would be an apocalyptic scenario. The choice is still ours: the continuation of a destructive but short-term profitable but destructive anthropocentrism, or a turn toward long-term prosperity by entering the Symbiocene through a shift to constructive anthropocentrism. All intellectual and practical resources are needed to make this shift. It is the seeds sown by Rudolf Steiner—and taken on by his students—that are one of these yet collectively unexplored resources. It is time to acknowledge their collective value as a polar star in the darkened night of the Anthropocene, pointing to a new, greener, and more human era of the Symbiocene.

References

Albrecht, G. (2019). *Earth emotions: New words for a new world*. Cornford Press.

Angus, I. (2018). *Facing the anthropocene: Fossil capitalism and the crisis of the earth system*. Monthly Review Press.

Apricot Lane Farms. (n.d.). *Our story*. Apricot Lane Farms. Retrieved November 13, 2024, from https://www.apricotlanefarms.com/our-story/

Barfield, O. (1942). *Romanticism comes of age*. Barfield Press.

Barfield, O. (1957). *Saving the appearances: A study in idolatry*. Wesleyan University Press.

Bellah, R. (2011). *Religion in human evolution: From the paleolithic to the axial age*. Harvard University Press.

Berry, T., & Tucker, M. (2011). *Dreamer of the earth: The spiritual ecology of the father of environmentalism*. San Francisco: Counterpoint.

Brand, S. (2009). *Whole earth discipline: An ecopragmatist manifesto*. New York: Viking.

Crutzen, P. J. (2001). The anthropocene. *Nature*. https://doi.org/10.1038/415023a

Crutzen, P. J. (2002). Geology of mankind. *Nature*. https://doi.org/10.1038/415023a

De Mul, J. (2023). *The place of humans in the Symbiocene: On the future of humans in a technologically driven world*. https://www.demul.nl/nl/publicaties/online-publicaties/6654-de-plaats-van-de-mens-in-het-symbioceen

Descola, P. (2014). *Beyond nature and culture*. University of Chicago Press.

Di Fuccia, M. V. (2016). *Owen barfield: Philosophy, poetry, and theology*. Cascade Books.

Ellis, E. (2012). *The planet of no return*. The Breakthrough Institute. https://thebreakthrough.org/journal/issue-2/the-planet-of-no-return
Ferrer, J. N., & Sherman, J. H. (Eds.). (2008). *The participatory turn: Spirituality, mysticism, religious studies*. State University of New York Press.
Haff, P. K. (2014). Humans and technology in the Anthropocene: Six rules. *The Anthropocene Review, 1*(2), 126–136. https://doi.org/10.1177/2053019614530575
Hamilton, C. (2018). *De provocerende aarde. Over het lot van de mens in het Antropoceen*. Klement.
Latour, B. (2013). *War of the worlds: Humans against earthbound [Lecture]. Gifford lecture series*. University of Edinburgh. https://www.ed.ac.uk/arts-humanities-soc-sci/news-events/lectures/gifford-lectures/archive/series-2012-2013/bruno-latour/lecture-five
Latour, B. (2017). *Oog in oog met Gaia. Acht lezingen over het Nieuwe Klimaatregime*. Octavo.
Lewis, S. L., & Maslin, M. A. (2015). Defining the Anthropocene. *Nature, 519*(7542), 171–180.
Mann, C. C. (2018). *The wizard and the prophet: Two remarkable scientists and their dueling visions to shape tomorrow's world*. Alfred A. Knopf.
McKanan, D. (2017). *Eco-alchemy: Anthroposophy and the history and future of environmentalism*. University of California Press.
Morton, T. (2010). *The ecological thought*. Harvard University Press.
Moselhy, N. (2018). *Final country report of the land degradation neutrality target setting programme*. https://www.unccd.int/sites/default/files/ldn_targets/2019-02/Egypt%20LDN%20TSP%20Country%20Report.pdf
Næss, A. (1973). The shallow and the deep, long-range ecology movement: A summary. *Inquiry: an Interdisciplinary Journal of Philosophy, 16*(1), 95–100.
Nigro, D. (2014). *Remembering a guru of green: Biodynamic consultant Alan York*. Wine Spectator. https://www.winespectator.com/articles/remembering-a-guru-of-green-biodynamic-consultant-alan-york-49613#:~:text=Alan%20inspired%20many%20others%20to,Biodynamics%20journal%20for%20a%20stint. Accessed 13 December 2024.
Nordhaus, T., & Shellenberger, M. (2015). *An ecomodernist manifesto*. The Breakthrough Institute. https://thebreakthrough.org
Parrique, T., Barth, J., Briens, F., Kerschner, C., Kraus-Polk, A., Kuokkanen, A., & Spangenberg, J. H. (2019). *Decoupling debunked: Evidence and arguments against green growth as a sole strategy for sustainability*. European Environmental Bureau.
Ritchie, H. (2024). *Not the end of the world. How we can been the first generation to build a sustainable planet*. Little, Brown Spark.
Rockström, J., Richardson, K., Crippa, P., et al. (2023). Earth beyond six of nine planetary boundaries. *Science Advances*. https://doi.org/10.1126/sciadv.ade2412
Ruddiman, W. F. (2003). The anthropogenic greenhouse era began thousands of years ago. *Climatic Change, 61*, 261–293.
Ruse, M. (2013). *The Gaia hypothesis: Science on a Pagan planet*. Chicago: University of Chicago Press.
Saito, K. (2020). *Capital in the Anthropocene*. Shueisha.
Seo, K.-W., Ryu, D., Eom, J., Jeon, T., Kim, J.-S., Youm, K., Chen, J., & Wilson, C. R. (2023). Drift of Earth's pole confirms groundwater depletion as a significant contributor to global sea level rise 1993–2010. *Geophysical Research Letters*. https://doi.org/10.1029/2023GL103509
Steiner, R. (1910). *Occult science: An outline*. Anthroposophic Press.
Steiner, R. (2006). *The evolution of consciousness: As revealed through initiation knowledge (G.A. 227)*. Rudolf Steiner Press.
Stengers, I. (2009). *In catastrophic times: Resisting the Coming barbarism*. University of Minnesota Press.
Tarnas, R. (1991). *The passion of the western mind: Understanding the ideas that have shaped our world view*. Ballantine Books.
Tarnas, R. (2002). Is the modern psyche undergoing a rite of passage? *ReVision, 24*(3), 48–60.
Tarnas, R. (2006). *Cosmos and psyche: Intimations of a new world view*. Viking Press.
Taylor, C. (1989). *Sources of the self: The making of the modern identity*. Harvard University Press.
Taylor, C. (2007). *A secular age*. Belknap Press of Harvard University Press.

Tisdall, C. (1974). *Art into society, society into art*. ICA.
Triodos Bank. (2024). *Financing the nature based solutions sector building a system that fosters prosperity while caring for a healthy earth*. https://www.triodos.com/en/press-releases/2024/triodos-bank-commits-to-financing-500-million-euro-in-nature-based-solutions
United Nations Environment Programme (UNEP). (2024). *Sekem: Champions of the earth 2024*. Retrieved December 21, 2024, from https://www.unep.org/championsofearth/laureates/2024/sekem
Verdonk, J. (2015). *De antroposofische orde: Religieus individualisme, gemeenschapsleven en maatschappelijke participatie in de antroposofische beweging van Nederland*. VU University Press.
Vernadsky, V. I. (1928). *The biosphere*. USSR Academy of Sciences Translated by D.B. Langmuir, Macmillan.
Vernon, M. (2020). *A secret history of Christianity: Jesus, the last Inkling, and the evolution of consciousness*. Christian Alternative.

Climate Psychology: From the Rediscovery of the Ecological Self to Atmospheric Consciousness

Stefan Ruf

Abstract

Against the backdrop of a theory of the evolution of consciousness, this article derives the obstacles and potentials for achieving a sustainable consciousness: (1) the current framework of perceiving the world, which strongly shapes our present identity, referred to as the Modern Self. It suggests a (false) experience of separation from nature and the cosmos. (2), soul qualities from earlier developmental stages, termed the Ecological Self, can help us become more sensitive to our connection with nature and its cycles. Strengthening this aspect of ourselves is essential for fostering sustainable action. (3), relying solely on this approach would lead to a regression to an older consciousness of local connectedness. Therefore, the inner conflict between the Ecological Self and the Modern Self must be acknowledged and held. From this dialectical tension, a transformation of consciousness can arise, leading to a socio-ecological global perception. This transformation enables us to experience both our connectedness and autonomy more fully, fostering sustainable global action. I refer to this consciousness as Atmospheric.

Keywords

Climate psychology • Atmospheric consciousness • Ecological self • Modern self • Evolution of consciousness • Dialectical awareness • Schema psychology • Sustainable consciousness • Environmental perception • Mind-behaviour gap • Troposphere/Macrosphere • Microsphere • Mesosphere • Psychological transformation • Global sustainability

S. Ruf (✉)
Maeander Youth Health Center, Berlin, Germany
e-mail: ruf@maeander-ggmbh.de

J. Kronenberg and E. T. Lammerts van Bueren (eds.), *On the Earth We Want to Live*, World Sustainability Series, https://doi.org/10.1007/978-3-031-98758-8_9

1 Introduction

This contribution to climate psychology aims to develop a psychology of transformation, drawing on theories from depth psychology, schema psychology, ego-state psychology, and transpersonal psychology, as well as models of the evolution of consciousness from Rudolf Steiner and Jean Gebser.[1] While theories of technical and social transformation have been established for some time, there is still a significant lack of psychologies addressing ecological transformation. The approach presented here includes both a "model of disorder"—a theory to explain our mind-behavior gap—and a "potential model," which outlines the capabilities we need to develop to navigate this extremely challenging transformation. The goal is a form of consciousness that perceives a different world-space: an "Atmospheric Consciousness."

Atmospheric Consciousness, as an attempt for a definition at the beginning of this essay, would lead to a different perception of the world and thus to a different relationship to both the outer and inner worlds: one experiences oneself in a different inner and outer space. These two spaces remain clearly distinct through an invisible boundary, allowing room for both individual autonomy and a more sensitive awareness of the dynamic local and global interconnectedness in which we all exist. This would, in turn, lead to more sustainable action.

2 Perception

We perceive the world in a distorted way. This is the only explanation for why we ultimately feel detached from its ecology, experiencing it as an "Um-Welt"[2] (environment) (Uexkuell, 2011) and perceiving ourselves as a small world that seems to exist independently from it.

If we were not suffering from this distorted perception, we would experience what we already know cognitively from climate science, ecology, and indirectly from the perspective of space travel: there is no environment. Instead, we are embedded in a web of relationships shared with all human and non-human beings that exist within this Earth's sphere. This web of relationships exists within the biosphere, and this vast yet finite biosphere is, from a climate science perspective,

[1] When discussing the evolution of consciousness, I will exclusively use the terminology of Gebser. This is because Gebser, who passed away in 1973, was able to incorporate the anthropological research findings of the 20th century to a much greater extent than Steiner, who passed away in 1925. In many areas, there are overlaps between the two theories. In other areas, the interpretations diverge or pertain to different levels. Fundamentally, the two theories differ significantly in that Steiner embeds his concept of the evolution of consciousness within a broader theory of the evolution of the entire cosmos (e.g., Steiner, 1954).

[2] In German, the term "Umwelt" carries a dual meaning: it refers both to the natural environment and, literally, to the "surrounding world," emphasising an externality that contrasts with the connectedness central to ecological thinking.

located in the troposphere, the lowest layer of the atmosphere. Modern humans know this—or are capable of knowing it—but they do not experience this atmospheric space. What we have collectively experienced over the past 500 years, triggered by a shift in consciousness during the Renaissance and amplified by 250 years of Enlightenment and industrialisation, can be summarised in the following premises:

1. We believe we have unlimited space at our disposal. Space in the vertical dimension: the universe, which begins above our heads, is boundless, so all emissions from the use of fossil fuels can simply be discharged into infinite outer space. Space in the horizontal dimension: until a few decades ago, this sense of vastness was also pervasive—the world itself seemed to offer immeasurable space. New continents were being discovered, appearing to provide endless room and unlimited resources. (von Weizsäcker and Wijkman, 2017)
2. The space between us is empty and lifeless, bridged only by electromagnetic waves, which our sensory organs receive and translate into sensory phenomena (Fuchs, 2000).
3. There is no religious "heaven"; "God is dead" (Nietzsche, 2022) and is "no longer needed as a hypothesis" (Laplace in Strick, 2009). The "spherical experience" of antiquity and the Middle Ages—that one lives in a space that provides enclosure and boundaries, serving as a gateway to other spheres formed by the planets and the fixed stars—has dissipated (Sloterdijk, 1998; Taylor, 2012).

This type of world experience, which evokes a perception of space as infinite and empty, omnipotent and isolating, is what I call the "Schema of Modernity", and I will describe it in more detail shortly. This worldview is being challenged by the ecological crisis, particularly the climate crisis. However, this "challenge" remains more intellectual than experiential. We do not perceive ourselves as part of the atmosphere. If we did, our world would look something like this:

> we are embedded in a vast, curved space—the troposphere—whose arched boundary extends over 40,000 km in all directions from the observer's perspective. These dimensions are unimaginable from a subjective point of view. And yet, this gigantic, curved space is finite: after 40,000 km, everything comes back to us—into our "rear space." The upper boundary is much closer: 12–16 km above our heads. These dimensions are easier to conceptualise, but the nature of this boundary presents a challenge: it is invisible. An invisible boundary arches above our heads, enabling the very existence of a living atmosphere (Wallace, 2006). This atmosphere is shaped by everything that exists on Earth: alpacas in the Andes, corals in the Great Barrier Reef, weathering limestone Alps, human beings, the forests of the taiga, and more. It is formed through the dynamic interactions between us, where gaseous, solid, liquid, and thermal elements all play a role. It weaves itself into the "in-between." This "in-between" is therefore not empty (Lovelock, 1992). Part of this atmosphere consists of climate gases, most notably CO_2, which allows the Sun's cosmic light to pass through but reflects some of its heat back, causing this vast, curved space to warm. This is how the climate crisis arises (Schellnhuber, 2015). We are thus intrinsically connected to and exist within the airy layer of this planet, which is part of its substance. In this sense, we do not live "on" the Earth but rather "in" the Earth.

If we were able to experience only a fraction of this interplay—this global interconnectedness—our lifestyle, which currently causes so much harm to others and ourselves, now and in the future, would be far more painful for us. If we had a sense for our actions, something akin to a sensory organ for the atmospheric, we would likely act quite differently as we currently do. To use a metaphor: if a child runs into the street in front of a driver, the driver will slam on the brakes, even if he is in a rush. We can assume this holds true even if the car is an SUV and the driver has an extremely high ecological footprint—someone who normally might not think much about the consequences of his actions on his "environment." Perception, which primarily functions through the visual system, can establish a relationship between the immediate environment and one's actions in fractions of a second. The driver sees himself in a relationship with the child and does not want to cause harm—so he brake (Schellnhuber, 2015).

We do not see this relationship in other contexts: we "do not see" it because it truly cannot be seen. Many boundaries and connections are invisible—at least not perceivable by the sense of sight. As long as we do not utilise other channels of perception for this, we are left to rely on intellect to guide our actions. But thinking is "airy"—it lacks the persistence of sensations and emotions and can easily be suppressed. Our perception, which was radically reshaped during the Renaissance by a shift in consciousness, became extremely focused on the sense of sight, on the visual (Gebser, 1992)—on perspective, the three-dimensional, and the face to face view of "here I am, and there you are." This focus misleads us and fosters a distorted worldview. As mentioned, this distorted perception is described here as the "Schema of Modernity."

3 The Schema of Modernity—The Modern Self

The concept of a schema originates from twentieth and twenty first-century psychology, first introduced by Piaget (Berk, 2005) and later developed by psychologists such as Grawe (Roediger, 2009) and Young et al. (2003). In psychotherapy, a schema describes an experiential schema. The relevant point here is that experience does not only mean thinking. An experiential schema consists of cognitions, emotions, bodily sensations, and behavioural impulses.[3]

[3] Young, the founder of schema therapy, defines a schema as "a broad pattern of memories, emotions, cognitions, and bodily sensations… that shape/influence our consciousness and guide our behaviour" (Young et al., 2003).

To illustrate, consider the schema of "inadequacy" (Young et al., 2003). This is a schema that most people have likely encountered in situations where they feel vulnerable—such as being poorly prepared for an exam. The unpleasant experience is not limited to thoughts but also includes feelings of fear or shame and bodily sensations like "feeling small." All of this alters our perception of the situation both externally and internally: the examiner may be perceived as powerful, while one's *self* is experienced as small and transparent. The more often such situations are encountered, the more deeply ingrained the schema becomes, and the more irrational the fear, shame, and physical sensations may become (Roediger, 2009). Exams are no longer experienced with a "normal" level of tension but as existential events, as if life and death were at stake. When caught in such an experience, a rational approach ("It's just an exam, and you're well-prepared") is almost entirely ineffective. Even if one intellectually knows this to be true, it "feels" entirely different. It feels different because the emotions say something else entirely, and, above all, the body—through its embodiment—sends sensations of "being small" and "being existentially threatened." The crucial point here is this: it is not merely thoughts that hold this power over us. In fact, upon closer inspection, it is the thoughts that hold the least power—emotions and sensations are far stronger in such situations (Lammers, 2007). Old emotions and sensations from the past rise into the present moment and distort our current perception. We perceive the present moment through a "lens" coloured by the past and remain unaware of it.

In Climate Psychology: Atmospheric Consciousness as a Path out of the Climate Crisis (Ruf, 2021), the concept of schemas is extended to collective or supra-individual problems. This is based on the hypothesis that the schema concept is more sociologically applicable than terms such as cognition, paradigm, or zeitgeist[4] because it acknowledges that the influence extends beyond thought. It reaches into feelings, bodily sensations, and thereby into collective perception. The aim here is to trace how, beginning in Europe over the past 500 years—particularly since the onset of the Enlightenment and industrialisation—a collective experiential schema has developed: a Schema of Modernity. This schema unconsciously shapes our modern world and self-experience as cognitive, emotional, and physical "background noise," much like how an exam situation shapes an individual's external and internal experience. Unlike an exam situation, however, this experiential pattern has continuously persisted on a collective level over centuries and into the present. For most people, this pattern is nearly always active. And because it continuously influences self- and world-perception, we identify with it so strongly that it can also be seen as a part of the self. Another term for this

[4] And also more holistic than the concept of emotion, which has only recently found its way into sociology (see, for example, Illouz, 2024: *Explosive Modernity*)—though this trend is fundamentally welcomed by the author.

"Modernity Schema" could therefore be: our Modern Self.[5] I will use this term as well moving forward.

If we attempt to translate this into an experience of the world, the Modernity Schema—or the Modern Self—might look something like this:

- **Thoughts**: Independence of God and nature, as one has outgrown them/"I am an island"/"reality is inanimate"
- **Emotion/Inner Attitude**: pride/loneliness/emptiness/restlessness
- **Bodily Sensation/Posture**: upright stance/being "in one's head"/feeling encapsulated
- **Impulse to Act**: actionism, drive for expansion, "growth".

This Modern Self within us has by no means only had negative effects on the evolution of human consciousness. Through individual freedom, an awareness of social issues and human dignity has emerged, leading to a global consciousness of such values as those described in the Universal Declaration of Human Rights and the 17 Sustainable Development Goals of the United Nations.[6]

And yet, this worldview and way of acting has led us into a profound existential crisis. If there is to be a social and technical transformation in the coming years that helps to prevent the worst, there must also be a psychological transformation. This means that this soul part must change. But how? Where is the energy, the impetus for this inner transformation, supposed to come from? Clearly, it cannot origin solely from cognitive dissonance in the literal sense of the term—realizing through thought that what one is doing does not align with what one ought to do—otherwise, the transformation would already have taken place, and the infamous mind-behaviour gap would not exist.

Neurobiologically, there is an explanation for this. Like any other schema, this experiential pattern is a neurogenic activation network that forms stable representations through repetition (neurological activation tendencies, NAT) (Roediger, 2009). It extends into embodiment through sensations and changes in posture: the

[5] The terminology in the schools of depth psychology varies depending on the school and is often contradictory. Terms such as "schema" and "part", "mode" (Young, 2023), "inner team" (Schulz von Thun, 1998), "ego-state" (Watkins, cited in Peichl, 2013), "complexes" (Jung, 1995a, 1995b), and "inner self-parts" (Peichl, 2013) are used. However, they are largely defined on a physiological level as cohesive neurobiological excitation patterns (e.g., Hüther, 2010) and are also described in very similar ways on a broad phenomenological level. For example, the definition of an ego-state by Watkins and Watkins: "An ego-state can be defined as an organised system of behaviour and experience, the elements of which are held together by a common principle and are separated from other ego-states by more or less permeable boundaries" (Peichl, 2013).

A more detailed differentiation, particularly regarding terminology in climate psychology, which is certainly important, goes beyond the scope of this article.

[6] This does not mean that such values did not exist in the past, locally, in European and especially non-European regions (see f.e. Graeber/Wengrow 2021: "The dawn of everything" (Anfänge))—what I am referring to here is a global consciousness.

embodiment (Storch et al., 2010). It can only be changed if a situation of dissonance arises that goes beyond the cognitive. One must experience that the inner experience does not match the external situation—by perceiving an inner aspect that experiences something different as the Modern Self.

This inner aspect must be able to perceive nature or ecology, to sense and feel it. From a psychoanalytic perspective, one might speak of object representation. From the theory of mirror neurons (Rizzolatti & Sinigaglia, 2007), one might argue that there must also be mirror neurons for the non-human environment. From the perspective of schema psychology, one would argue: there must be a part of the soul that senses ecology—and, by extension, sustainability. According to my hypothesis, we also carry this part within us: I call it the "Ecological Self". However, it is far less pronounced than the Modern Self, which is why most people barely identify with it.

4 The Ecological Self

From both a theory of the evolution of consciousness (see Gebser, 1992) and a conventional evolutionary perspective (Pääbo, 2014), we carry within us the evolutionary heritage of earlier stages: that is, we carry something of the Palaeolithic human within us—the hunter and gatherer. This human being was still deeply connected to the cyclical rhythms of the cosmos and nature through their migratory movements. They followed the buffalo, deer, and antelope herds they hunted and were intimately familiar with the rhythms of the rainy and dry seasons and the ripening of fruits they gathered. They had a deep, intuitive connection with the surrounding nature, which they experienced as alive and of which they felt themselves to be a part. This person was still an interconnected participant in the rhythmic processes of life. Gebser refers to this phase as the Magical Phase and its consciousness as a magical consciousness. According to Gebser and Steiner, we still carry something of this consciousness within us (Gebser, 1992; Steiner, 1954).

Similarly, we also carry something of the human being from the phase of settlement. Gebser calls this the Mythical Phase: The Neolithic human, who learned to read the seasons from the stars to guide their harvest, no longer moved with the rhythms of the surrounding nature. Instead, they chose a fixed location for their fields, homes, and temples. They no longer moved with time; rather, time—the cyclical course of the year—moved through them and their location. By this point, we can already detect a degree of separation from the rhythmic fabric of nature, a separation that has intensified over the millennia.

We can still experience something of these two states within and around us. We can still observe children playing as if magically enchanted by the underbrush. Adults, too, can experience moments of deep resonance with nature, whether in the mountains, by the sea, in the forest, or in a garden, where one can intensely observe the cycle of growth and decay. Psychologically speaking, there must be something like an object representation within us for the qualities and beings of nature.

We will call this side of ourselves our Ecological Schema or Ecological Self. This part of us can resonate with nature and experience it as a shared world. It has a connection to materiality and knows how precious it is. In this sense, it feels a sense of shame toward the concept of waste and rubbish (a typical invention of modernity, unknown in this form to our ancestors). It also has a connection to the cycles of nature. Both aspects have been deeply ingrained over millennia. For this reason, this side understands concepts like reuse and circular economy. This side feels sustainability at its core. It also feels profound pain, shame, and fear when it perceives the ecological state of the world around us. When we are able to feel this, something more is present than mere "cognitive dissonance." It is a cognitive/emotional/sensitive dissonance. This dissonance has the potential energy to act as a lever to transform the Modern Self, as will be further elaborated later, even though the Ecological Self is significantly less developed in most people compared to the Modern Self.

5 An Inner Conflict and Its Repression

We thus have two sides within us—the Modern Self and the Ecological Self—that perceive the world in fundamentally different ways. This means that modern human beings face an inner conflict. It is this inner conflict that is primarily responsible for the mind-behaviour gap in today's ecological challenges.

The Modern Self (MS) repeatedly recognises cognitively that the growth-based economic system, along with its associated technology, can lead to ecological catastrophe—but it neither senses nor feels this reality. Aside from that, it harbours a profound fear of losing control over the world should changes be made to its technology and economic system. It also fears, at least equally, the inconvenience and effort that a life closer to nature would inevitably bring.

The Ecological Self (ES), on the other hand, senses local environmental pollution and waste. However, in most people, it is so weakly developed that it can be easily repressed. Even where it is not repressed, it has two critical deficits: the ES, which originates from an older magical and mythical consciousness, is capable of understanding ecology only on a local level. It lacks the conceptual framework for the global, simply because in the world before the emergence of the great city-states and empires in Mesopotamia, Egypt, India, and similar regions, while networks surely existed, they were not at a global level nor included an understanding of a spherical planet.

Furthermore, the ES lacks a scientific understanding of non-linear complex systems, as provided by systems theory, chaos theory, or quantum theory. Its concept of development is also much more cyclical, tied to processes of growth and decay. The Modern Self, by contrast, has an ecologically (and depth-psychologically) distorted notion of development, characterised by "ever more" and "limitless growth." Behind this, however, lies a developmental impulse—a desire for growth that seeks to move beyond cyclical processes. This impulse may reflect a transpersonal developmental need that demands recognition (Figs. 1 and 2).

This inner conflict leads to unpleasant emotions such as uncertainty, fear, shame, guilt, anger, and disgust—feelings we do not want to experience. To avoid feeling these as much as possible, there are internal "defence mechanisms" (A. Freud, as cited in List, 2014) or coping modes (Young et al., 2003). From my perspective, I have identified seven particularly prominent ones:

- Apparent Consensus— > "Climate protection is very important to me, but..."
- Trivialising— > Downplaying the problem: "There have always been environmental issues."

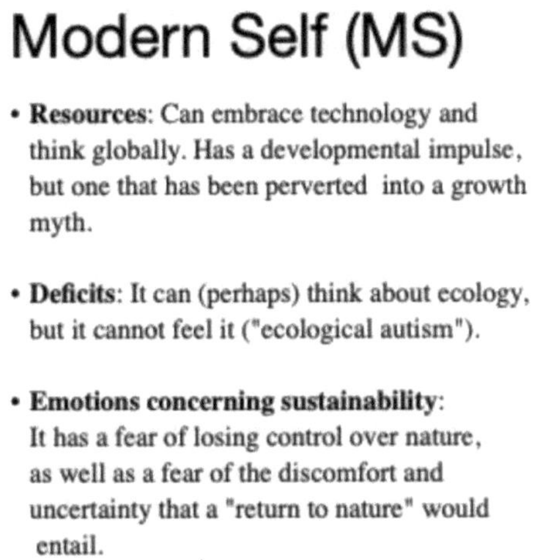

Fig. 1 Modern self (MS) (Ruf, 2021)

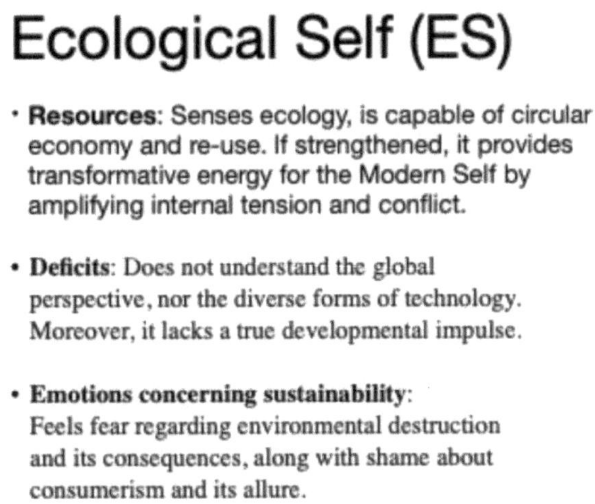

Fig. 2 Ecological self (ES) (Ruf, 2021)

- Denying— > Rejecting the problem: "There is no climate change"/"There is no man-made climate change."
- Projecting— > "The Chinese/Americans/Qataris are to blame and should act first…"
- A Mix of Denial and Projection— > Conspiracy theory.
- Cynical Acceptance— > "There's nothing we can do about it anymore…"
- Continuing in Schema-Conforming Ways— > "Let's finally implement technical solutions."

In order to move beyond defence mechanisms and paralysis, and to harness both the resources of the Ecological Self and the potential resources of the Modern Self, a different attitude is necessary. Since this involves a significant inner conflict, it seems logical to seek solutions in the field specifically dedicated to resolving inner conflicts: psychotherapy. Within psychotherapy, different approaches exist depending on the school of thought. The approach used here comes from DBT (Dialectical Behaviour Therapy, developed by Linehan (1993), for individuals with borderline personality disorder or complex trauma) and is called Dialectical Awareness.

6 Dialectical Awareness

In the therapy of individuals with borderline personality disorder, the goal is to bring together and integrate opposites that patients experience as intense and irreconcilable. For example, individuals with this condition often tend to idealise or demonise others, and this can shift over time regarding the same person—for instance, their therapist (Bohus & Wolf-Arehult, 2013). From a neurobiological and psychological perspective, there must be internal representations, or object representations, for both states—the "good therapist" and the "bad therapist"—within the patient, even though these representations do not align. The aim of therapy is to reconcile these opposites. It moves from an EITHER-OR to an AND. In this case, it means the therapist has both positive and negative aspects. Both are true. Patients must learn to hold this "AND" simultaneously within themselves. Therapists must also learn this: they must authentically maintain a stance that is, in itself, completely contradictory—acceptance AND challenge.

This means having an inner part that understands why the patient is behaving as they are—tense and aggressive—and sees that they likely cannot act differently. And, at the same time, challenging them to act differently. Many studies show that this tension between opposites, this tension of the "in-between," creates deeply transformative encounters (Bohus & Wolf-Arehult, 2013), which, in our terms, lead to a transformational experience by integrating polar opposites.

From a neurobiological and schema-therapy perspective, this can be explained as follows: both internal parts are neurological activation tendencies (NATs). When activated individually, the NAT is reinforced because it confirms what is seen, thought, felt, and sensed. The neural pathways and connectivity are strengthened. NATs can only change when they are activated simultaneously. This weakens the connectivity and releases energy for forming a new pattern. Change occurs when the simultaneous activation of both polar patterns creates cognitive dissonance, triggering a (volitional) impulse to establish a new neurological activation pattern (NAP) (Hüther, 2010; Roediger, 2009).

What does this mean in concrete terms for the inner conflict between the Ecological Self (ES) and the Modern Self (MS) described above? It means that the inner conflict caused by these two sides must not be repressed. The first step can involve smaller goal conflicts, which one must allow to confront them in their contradictions. These conflicts will almost always involve the fields of social, ecological, and individual needs. Here are three examples:

- Should I take the plane so I can still put the children to bed, or should I take the train, even though that means I won't see them anymore? (Social versus Ecological)
- I want my own children to be well-educated and financially secure. Should I invest my money in an ecological way or in a way that yields the maximum return? (Social versus Ecological)
- Should I finally go on the long-desired trip to Japan, even though it will severely impact my carbon footprint? (Individual versus Ecological).

These conflicts and the accompanying feelings of insecurity, irritation, shame, and fear must first be held—or better yet, endured. From this, a solution can arise in the specific situation that is the best possible one. "Best possible" does not necessarily mean ecologically sustainable in every case. One might consciously choose a socially harmonious solution at the expense of ecology. This decision must then be endured and taken responsibility for: this applies equally if one chooses the social or individual need.

However, the minor frustration caused by such decisions can create an energy for action—an effort to prioritise differently in other situations, gradually breaking the automatic pattern of consistently acting in an unsustainable way. More importantly, this approach can fundamentally change one's overall attitude: it leads to a stance beyond the two selves (attractors).[7]

C.G. Jung writes about such processes: "The back-and-forth of arguments and affects represents the transcendent function of opposites. The confrontation of positions creates a tension charged with energy, which gives rise to something living, a third... a new stage of being..."(Jung, 1995a, 1995b, Vol. 8). This is an attitude that acts from a consciousness no longer rooted exclusively in the Modern Self or the Ecological Self, but from something new. It is a stance that integrates ecological, social, and individual needs more effectively in real-life situations, gradually transforming one's life so that there are fewer conflicting situations

[7] There is an interesting analogy here between neurology and climate science: in both fields, there are attractors—vortices that exert a pull on their surroundings. In neurology, these are NATs; in climate science, hurricanes or cyclones. One occurs in our inner world, the other in the outer world (see Chaps. 8 and 9).

across these three fields. At some point, for example, one might no longer own a car, or it might no longer be a question whether to avoid investing money in a bank that is not ecologically responsible.

The more people adopt this path, the more society and its structures will be transformed—particularly in democracies. However, this process demands psychological abilities that, like those required in therapy for borderline or trauma patients, are not something "normal" individuals typically possess. One needs an attentive inner observer consciousness, as well as an inner resilience or centre, so that these processes can unfold without overwhelming the individual. How does one acquire these abilities?

7 Transformation Through Mindfulness and a Deeper Self

In recent decades, various approaches to mindfulness have had a significant influence on Western culture—primarily through meditation, yoga, and other body therapy methods such as Feldenkrais and Cantienica. In psychiatry and psychotherapy, there are now many well-researched methods that incorporate mindfulness (e.g., MBSR, MBCT, DBT).[8] Parameters for a mindful attitude include presence, focus, acceptance, and non-reactivity (e.g., Nhat Hanh, 1996). Numerous studies have demonstrated its effectiveness, with EEG and neuroanatomical changes documented (e.g., Piron, 2020; Siegel, 2007). With this mindful attitude, the focus of observation can be directed outward—for example, toward another person or a juniper tree, which I perceive with this open and understanding mindset. Or it can be directed inward, for instance, toward the inner conflict between the two parts of the self and the contradictory thoughts, feelings, bodily sensations, and impulses associated with it. These must be observed, named, and held—that is, the impulse to act must initially be resisted—before being brought into an exchange, an internal dialogue.

Mindfulness describes the power of perception, but what is the force that holds this inner conflict, brings it into dialogue, and transforms it into something new, a third, as C.G. Jung describes? Here, the therapeutic approaches mentioned above remain conceptually vague. In DBT, which draws inspiration from Zen Buddhist methods (Bohus & Wolf-Arehult, 2013), this is referred to as the "wise mind." In schema therapy, there is the loosely defined concept of the "Healthy Adult." C.G. Jung speaks of the "transcendent function" (Jung, 1995a, 1995b, Vol. 8). Gebser and Wilber refer to manifestations of the spiritual (Wilber, 2006). Rudolf Steiner described this earlier than both of them as a manifestation of the spiritual that comes into a more conscious connection with the soul. The interface where this

[8] MBSR = Mindfulness-Based Stress Reduction, aimed at reducing stress in cases of anxiety disorders and depression; MBCT, = Mindfulness-Based Cognitive Therapy, used for depression; DBT = Dialectical Behaviour Therapy, applied in cases of borderline personality disorder (see, for example, Heidenreich & Michalak, 2006, or Siegel, 2007).

occurs is what Steiner calls the "I". This "I" possesses a depth different from our everyday self (see e.g., Ewertowski, 2007; Steiner, 1904).

Subjectively, this results in an experience of a centre, a middle point, a deeper self (Storch, 2021), or the "I" (Steiner, 1904).[9] Strengthening this centre and perceiving it more clearly is partly a matter of practice, such as regular meditative exercises. However, it is also a phenomenon that many people who do not meditate regularly can distinctly experience. Most people sense this centre in the region of the heart, though some perceive it deeper or higher within the body.[10] Where it is experienced varies from person to person and is meant to be individual. What is relevant is that it can be experienced at all. This experience of the centre is essential for the present time as a sustaining and integrative force; otherwise, the conflict and shadow aspects cannot be borne.[11]

Steiner places significant emphasis on the perception of this centre in his instructions for a meditative path, as he considers it essential for developing the sustaining strength required when encountering external and internal dynamics that were previously unconscious. To this end, he provides exercises aimed at improving the perception of this centre, which he calls the "I" (Ich) and which should not be confused with the everyday ego (Kühlewind, 2002; Steiner, 1993). Additionally, he offers exercises designed to cultivate willpower and moral strength, which, in his view, are necessary for this process. Steiner refers to these as the "auxiliary exercises"—a somewhat misleading term, as he considers them essential for spiritual transformation processes (Steiner, 2019).

In this field of tension between mindful observation and the centre, a "Third Self" emerges from the sustained polarity between the Modern Self (MS) and the Ecological Self (ES). This Third Self is not identical to the centre (Steiners "I") but lies closer to it than the other two selves. It is urgently needed as the starting point for addressing the challenges of our time. This is the self in which an "Atmospheric Consciousness" (Ruf, 2021) can develop (see Fig. 3).[12]

[9] In the many different spiritual traditions, there is a similarly wide range of terms used to describe this centre: Atman, deep self, Buddha nature, Christ nature, or **"I"** (Jung, 1995a, 1995b, Vol. 9/2; Kühlewind, 2002; Piron, 2020). This diversity is comparable to the variety of terms used in psychotherapeutic schools to describe inner parts of the self (see FN No. 5). Of course, what is described here is not about profound spiritual enlightenment processes but rather the initial perception of an inner centre, a middle point capable of holding opposites in such a way that something new—a third—can emerge.

[10] Both the fact that it can be clearly felt by almost all people and the specific location where it is perceived are observations based on hundreds of patients and workshop participants. However, this would need to be objectively validated in studies.

[11] On this topic, Jung writes about this issue in typical therapeutic processes (1995a, 1995b, Vol. 8): "Even when there is enough intelligence to understand the question, there is often a lack of courage and self-confidence, or one is too mentally and morally sluggish or cowardly to make an effort..."

[12] It can only be briefly indicated here that a polar perspective and the integration of polarities into a third, a middle point, is a fundamental principle of anthroposophy. According to Steiner, for example, there are two cosmic forces that, in their one-sidedness, are destructive but can be

Fig. 3 Microspheric consciousness

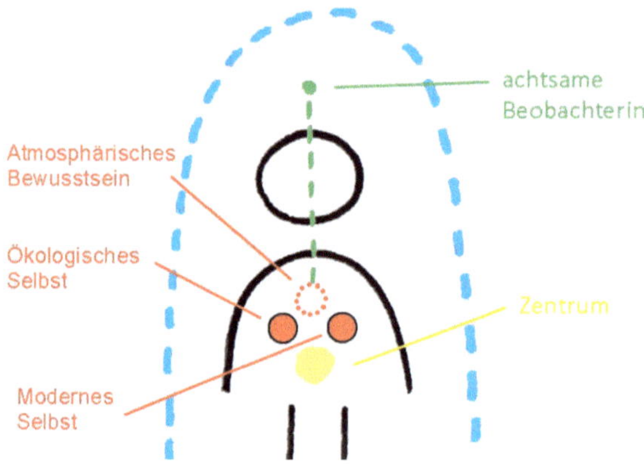

8 Atmospheric Consciousness

To achieve this, our ability to perceive globally must be cultivated. The inner transformation described above does not automatically enable us to perceive globally: it can help us recognise the one-sidedness of the Modern Self (MS). It can assist us in developing a deeper perception of the Ecological Self (ES), allowing us to better sense our local interconnectedness and mutual interaction with all things—an important starting point! However, because the ecological crisis is global, and everything is interconnected, perception must extend beyond the local. Visual perception alone cannot achieve this: it sees only visible boundaries and can only perceive as far as the horizon.

We therefore need perceptual channels that are not visual, as the MS has learned in its extreme one-sidedness, but rather have a more listening quality, like those available to the ES—channels that must, however, be extended to the global level. According to Gebser and other anthropological researchers, people of the ES epoch certainly had good eyesight; nonetheless, their world perception was not as exclusively fixated on the visual as it is in modern consciousness. Much of the intuitive connectedness they experienced with "the river," "the jaguar," or "the antelope" was not perceived through the visual system but rather had a quality resembling auditory perception—like antennas attuned to the local atmosphere (Kohn, 2023; Weber, 2018). Similarly, Rosa (2016) describes "resonance" as a form of perception.[13] The hypothesis in this treatise is that if we want to develop our capacity

highly conducive to development when brought into balance. He calls them Luciferic and Ahrimanic, and the two parts of the self can theoretically be associated with these forces: the Luciferic is more closely linked to the Ecological Self (ES), while the Ahrimanic is associated with the Modern Self (MS). The middle ground between these forces must be sought, and it is closely connected to this "I"-consciousness.

[13] Quote from Rosa (2016, p. 24): "When we love it, something like a vibrating wire is created between us and the world."

to perceive the global, then this sensorium—which must extend deeply into the global space—will have much to do with this "listening quality."

Something of this listening ability must connect with the consciousness of a global world and the centring force; only then can one enter an experience of atmospheric relationality. Ultimately, this atmospheric consciousness is a relational organ for the planetary: a relational organ of the subject to the Earth as a whole, and a relational organ of the subject to individual global actors. However, this way of forming relationships takes a completely different shape from what we are generally familiar with. Normally, when we enter into a relationship, it is a form of confrontation: from person to person, or from person to animal, plant, or landscape. Here, it is something entirely different. It is, in essence, a 360-degree relationship. What do we emit through our thinking, feeling, and acting in all directions into this vast atmospheric space, stretching 40,000 kms, until it "comes back to us from behind?" What do we send out into this space, which is curved and invisibly bounded by the troposphere's boundary approximately 12 kms above us?

To move forward here, the images we have previously created of this global space are not helpful. These images—embedded in the collective consciousness since NASA's Apollo 8 mission—portray the Earth as a beautiful blue planet in the dark expanse of space. However, these images convey the perspective of a normal form of relationship: that of confrontation. This type of relationship, in this case, is only possible through tremendous technical effort, which enables astronauts to inhabit a spacecraft with an artificially created atmosphere. Without such artificial means, a confrontational relationship with the Earth is not possible (Sloterdijk, 1998).

The natural relationship of human beings to the Earth would be one without confrontation—a 360-degree relationship within its curved space, where everything is interwoven. How can this be experienced? There is a way to perceive and practice this spatial awareness, even on a very small scale: in mindful encounters with ourselves.

9 Microsphere

Once again, a mindful observer consciousness is required, as in Chap. 7 described, where it was involved in perceiving the two parts within us. When this consciousness is directed toward the subjective inner space in a relaxed state with closed eyes, something extraordinary happens through the shift into mindful awareness: another space emerges—another inner space. When mindful awareness directs its perceptual focus inward, toward thoughts, feelings, and bodily sensations, nearly all people perceive a space that does not share the same boundaries as the physical body. In most cases, this space is significantly larger than the physical boundaries. The limits of this perceived space often vary across different areas, and most people place the boundary 10 to 30 cm in front of their chest.

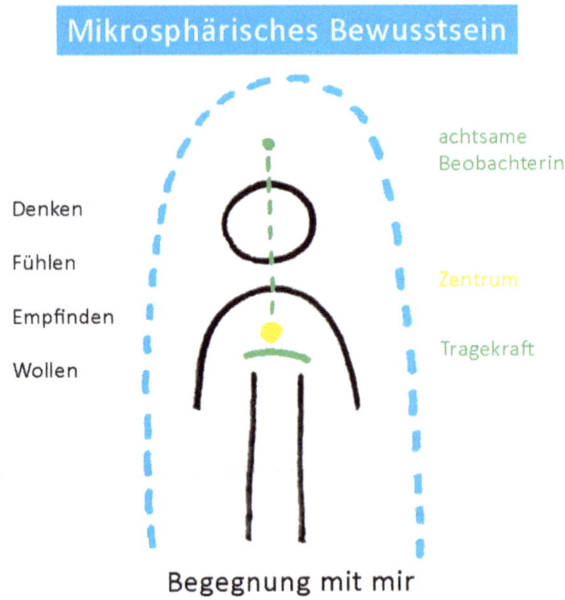

Fig. 4 The mindful observer, the dialectical tension between the ecological self (ES) and the modern self (MS), and the emergence of a new consciousness: the atmospheric consciousness. This process is supported by a centre

Five aspects of this perceptual exercise are particularly relevant here: (1) The body-soul space that suddenly becomes perceptible is larger than the physical body.[14] (2) Although invisible, the boundaries of this space are clearly definable for almost all people.[15] (3) A kind of domed, 360-degree space emerges around the perceiving subject, which is not perceived visually but evidently through a different sense. (4) Within this space, sensations, feelings, and thoughts interact—one might also say, different parts interact. (5) And yet, there is a centre that cannot be imagined without the rest: point and circumference (Fig. 4).

Thus, within the small scale of our microsphere, a similar atmospheric space can be perceived as in the larger scale of the macrosphere of the atmosphere. We experience a spherical, domed space surrounding our body, with a defined but invisible boundary. This boundary, depending on our inner state, may be closer to or farther from the body. Its extent fluctuates, much like the boundary of the

[14] See, for example, Fuchs (2000), who adopts Merleau-Ponty's distinction between Körper and Leib. Körper refers to the body as an object, something to be examined from the outside like a foreign body. Leib, on the other hand, is the body as subject, which we experience from an internal perspective. This corresponds to Buber's (1992) levels of relationship:

[15] There are experiential reports from hundreds of workshop participants and patients of the author. Of course, thorough scientific research in this area would be both necessary and desirable.

troposphere, which is not fixed but shiftable. Additionally, we experience a centre or deeper self within us.[16] We are thus surrounded by this small atmosphere, this microsphere, in which thoughts, feelings, and bodily sensations occur—just as we are surrounded by the large atmosphere, the macrosphere, in which many interactions also take place.

Exercises to become more sensitive to this space can be found in various meditation techniques, ranging from body scans in MBSR to spiritual embodiment practices (Storch, 2021), from Aikido to Qi Gong, to anthroposophical imagination meditations and therapeutic eurythmy exercises.[17]

This personal inner space can be opened by moving from the relational level of "I-to-Me" to the level of "I-Thou." To do this, one must cross the invisible boundary surrounding the microspheric space with one's mindful observer consciousness and open one's senses to the other. By doing so, one enters (or creates?!) a new space of encounter—the mesospheric space—which will only be outlined briefly here:

Mesosphere: This is the space where, in Buber's dialogical philosophy, the "I-Thou" relationship takes place, as opposed to the "I-It" relationship, in which the other is alienated as an object (Buber, 1992). It is the space described by Sloterdijk in his philosophy of spheres (Sloterdijk, 1998) and by Fuchs in his spherical anthropology (Fuchs, 2000): a space of encounter at the local level, a sphere of exchange between two or more actors, which can also include non-human beings. This space also has a spherical boundary, the size of which is variable. It is a space to which highly sensitive individuals are particularly perceptive (Aaron, 2005), especially regarding the moods, emotions, and energies exchanged within it. Hermann Schmitz explores this in detail in his philosophy of atmospheres (see, e.g., Schmitz, 2020).

Exercises to increase sensitivity to this space can be found in certain Eastern martial arts, notably Aikido, in Bohm's dialogue process, and in group eurythmy.

If this "local I-Thou encounter space" is opened again and one crosses the invisible threshold of the boundary of the mesospheric space, one enters the macrospheric space—the space of the Earth's atmosphere, the perception of the troposphere.

---

[16] Most participants locate it in the heart region. In many spiritual traditions from both East and West, this inner instance is described. Steiner and the Western esoteric tradition refer to this inner instance as the **higher self** (…). We will return to this quality in the final chapter.

[17] Eurythmy, as "visible singing" or "visible speech," seeks to make the effects of language and sound—effects that are inherently invisible in spatial terms because they are not visual—perceptible to the eye (e.g., Steiner, GA 278, 2001).

10 Macrosphere/Atmospheric Consciousness

Through these exercises in the realms of the microsphere and mesospheric space, contemporary human beings have the opportunity to at least begin developing a global perception and relationship to the world. This means "perceiving" the world-space of the troposphere in the way it has long been understood on a cognitive level: as limited, curved, filled, interwoven, shared, and vast—at least in its initial stages.

Exercises to further cultivate sensitivity to this space include, in addition to imaginative and perceptual exercises focused on the atmospheric/macrospheric realm that I offer in my workshops, meditative work with Steiner's Calendar of the Soul (Steiner, 1994).[18] Ultimately, however, all the aforementioned exercises and steps in inner development prepare individuals to enter this final but crucial field of perception.

11 Conclusion

This has been a brief outline of a possible path of psychological transformation toward a consciousness capable of translating global sustainability into action. Many people today have already undergone such a process and might describe it using different terms. Alternatively, if introspection and psychology are not their fields, they might not describe it at all but simply act. However, such action often becomes mere activism if it is not rooted in a deeper experience of the self in relation to the world.

If there is still a chance to ensure that the necessary transformative processes of the coming decades do not become excessively disruptive and catastrophic for a large number of human and non-human beings, a sufficient number of people must also understand the psychological aspect of transformation. First theoretically, but then practically—by creating inner and outer spaces for this transformation. These spaces should ideally be group spaces, as we should not experience ourselves as "islands," as is typical of the Modern Self. In this sense, this text is intended as an introduction to a concept that aims to initiate a shared practice and has already been applied in seminar contexts in recent years.

[18] In the *Calendar of the Soul*, Steiner provides a meditation for each week of the year, aiming to relate processes in the cosmos and nature (oriented to the climate of the Northern Hemisphere) to processes within the inner soul—thus creating a connection between the outer world and the inner world.

References

Aaron, E. N. (2005). *Sind Sie hochsensibel? Wie Sie Ihre Empfindsamkeit erkennen, verstehen und nutzen* [The highly sensitive person: how to recognize, understand, and make use of your sensitivity]. MGV Verlag.

Berk, L. E. (2005). *Entwicklungspsychologie* [Developmental psychology]. Pearson Education.

Bohus, M., & Wolf-Arehult, M. (2013). *Interaktives Skilltraining für Borderline-Patienten* [Interactive skill training for borderline patients] (2nd ed.). Schattauer Verlag.

Buber, M. (1992). *Das dialogische Prinzip* [The dialogical principle]. Schneider.

Ewertowski, J. (2007). *Die Entdeckung der Bewusstseinsseele* [The discovery of the consciousness soul]. Verlag Freies Geistesleben.

Fuchs, T. (2000). *Leib Raum Person: Entwurf einer phänomenologischen Anthropologie* [Body, space, person: Outline of a phenomenological anthropology]. Klett-Cotta.

Gebser, J. (1992). *Ursprung und Gegenwart. 1. Teil: Die Fundamente der aperspektivischen Welt; 2. Teil: Die Manifestationen der aperspektivischen Welt* [Origin and present. Part 1: The foundations of the aperspectival world; Part 2: The manifestations of the aperspectival world]. Deutscher Taschenbuch Verlag.

Graeber, D., & Wengrow, D. (2021). *Anfänge (The dawn of everything: A new history of humanity).* Stuttgart, Klett-Cotta

Heidenreich, T., & Michalak, J. (2006). *Achtsamkeit und Akzeptanz in der Psychotherapie* [Mindfulness and acceptance in psychotherapy]. DGVT Verlag.

Illouz, E. (2024). *Explosive moderne.* Berlin, Suhrkamp Verlag.

Jung, C. G. (1995a). Die transzendente Funktion [The transcendent function]. In *Die Dynamik des Unbewussten* (Gesamtausgabe Band 8). Walter-Verlag.

Jung, C. G. (1995b). *Aion* (Gesamtausgabe Band 9/2). Walter-Verlag.

Kohn, E. (2023). *Wie Wälder denken: Eine Anthropologie jenseits des Menschlichen* [How forests think: An anthropology beyond the human]. Matthes & Seitz.

Kühlewind, G. (2002). *Aufmerksamkeit und Hingabe* [Attention and devotion]. Verlag Freies Geistesleben.

Lammers, C. H. (2007). *Emotionsbezogene psychotherapie.* Stuttgart, Schattauer Verlag.

Linehan, M. (1993). *Skills training manual for treating borderline personality disorder.* The Guilford Press.

List, E. (2014). *Psychoanalyse* [Psychoanalysis]. Facultas Verlag.

Lovelock, J. (1992). *Gaia: Die Erde ist ein Lebewesen* [Gaia: A new look at life on earth]. Scherz Verlag.

Nhat Hanh, T. (1996). *Der Weg zu einem achtsamen Leben* [The miracle of mindfulness]. Herder Verlag.

Nietzsche, F. (2022). *Die fröhliche Wissenschaft* [The gay science] (8th ed.). Kröner Verlag.

Pääbo, S. (2014). *Die Neandertaler und wir: Meine Suche nach den Urzeit-Genen* [Neanderthal man: In search of lost genomes]. S. Fischer.

Peichl, J. (2013). *Die inneren Trauma-Landschaften* [The inner trauma landscapes]. Schattauer Verlag.

Piron, H. (2020). *Meditationstiefe: Grundlagen, Forschung, Training, Psychotherapie* [Depth of meditation: Basics, research, training, psychotherapy]. Springer Verlag.

Rizzolatti, G., & Sinigaglia, C. (2007). *Mirrors in the brain: How our minds share actions and emotions.* Oxford University Press.

Roediger, E. (2009). *Praxis der Schematherapie: Grundlagen - Anwendung - Perspektiven* [Practice of schema therapy: Fundamentals—Application—Perspectives]. Schattauer Verlag.

Rosa, H. (2016). *Resonanz: Eine Soziologie der Weltbeziehung* [Resonance: A sociology of world relations]. Suhrkamp Verlag.

Ruf, S. (2021). *Klimapsychologie. Atmosphärisches Bewusstsein als Weg aus der Klimakrise.* Frankfurt, Info3 Verlag

Schellnhuber, H. J. (2015). *Selbstverbrennung. Die fatale Dreiecksbeziehung zwischen Klima, Mensch und Kohlenstoff*. München, Bertelsmann Verlag.
Schmitz, H. (2020). *Atmosphären*. Freiburg/München, Verlag Karl Alber
Schulz von Thun, F. (1998). *Miteinander Reden Band 3: Das Innere Team und situationsgerechte Kommunikation*, Hamburg, Rowohlt Verlag
Siegel, D. (2007). *Das achtsame Gehirn*. New York, Arbor-Verlag.
Sloterdijk, P. (1998). Sphären-Trilogie (Sphären I-Mikrosphärologie, Blasen, Frankfurt am Main 1998, Sphären II-Makrosphärologie, Globen, Frankfurt am Main 1999, Sphären III-Plurale Sphärologie, Schäume, Frankfurt am Main, alle Suhrkamp).
Steiner, R. (1954). *Ders.: Anthroposophische Leitsätze*. GA 26, Dornach.
Steiner, R. (1994). *Anthroposophischer Seelenkalender*. Dornach, Rudolf Steiner Verlag
Steiner, R. (1904). *Ders.: Theosophie*. GA 9, Dornach 1987.
Steiner, R. (1993). *Ders.: Wie erlangt man Erkenntnisse der Höheren Welten*. GA 10, Rudolf Steiner Verlag.
Steiner, R. (2001). *Ders.: Eurythmie als sichtbarer Gesang*. GA 278, Dornach, Rudolf Steiner Verlag.
Steiner, R. (2019). *Ders.: Die Nebenübungen – Sechs Schritte zur Selbsterziehung*. Rudolf Steiner Verlag
Storch, M. (2021). *Spiritual Embodiment*. München, Arkana Verlag.
Storch, M., Cantieni, B., Hüther, G., & Tschacher, W. (2010). *Embodiment*. Bern, Verlag Hans Huber.
Strick, H. K. (2009). *Laplace, der Lehrer Napoleons*. Spektrum der Wissenschaft, Spektrum Verlag.
Taylor, C. (2012). *Ein säkulares Zeitalter*. Frankfurt, Suhrkamp Verlag.
Uexkuell, T. V. (2011). *Psychosomatische Medizin, 7*. Auflage, München, Elsevier.
Wallace, J., & Hobbs, P. (2006). *Atmospheric Science*. Burlington, Academic Press.
Weber, A. (2018). *Indigenialität*. Berlin, Nicolai Publishing & Intelligence.
Weizsäcker, E. U. V., & Wijkman, A. (2017). *Wir sind dran. Was wir ändern müssen, wenn wir bleiben wollen – Eine neue Aufklärung für eine volle Welt*. Gütersloh/München, Gütersloher Verlagshaus.
Wilber, K. (2006). *Integrale Psychologie. Geist, Bewusstsein, Psychologie, Therapie*. Freiamt, Arbor Verlag.
Wolf, C. (1982). *Kindheitsmuster*. Berlin und Weimar, Aufbau-Verlag.
Young, J. E., Klosko, J. S., & Weishaar, M. E. (2003). *Schematherapie. Ein praxisorientiertes Handbuch*. Paderborn, Junfermann Verlag

A New World—The Cosmopolitan Human Being on a Common Earth

Ha Vinh Tho

Abstract

The world faces a profound socio-ecological crisis rooted in fragmentation: disconnection from self, others, and nature. A reductionist worldview prioritizing economic growth over well-being has caused burnout, isolation, and environmental exploitation. Transforming this requires a shift toward interconnectedness and holistic values like compassion and sustainability. Drawing on personal experiences, the author explores frameworks like Anthroposophy and Gross National Happiness (GNH). GNH prioritises well-being, equity, and ecological harmony, challenging GDP's limits. Its nine domains—psychological well-being, health, education, and more—model balanced development. Initiatives like Vietnam's Peaceful Bamboo Family and Happy Schools adapt holistic principles to diverse contexts, while GNH in business redefines success through sustainability and happiness, exemplified by Bitis' transformation. This vision integrates spiritual, cultural, and ecological dimensions, calling for systemic change to build a resilient, inclusive, and harmonious future.

Keywords

Cosmopolitan human being • Gross National Happiness (GNH) • Anthroposophy • Holistic development • Environmental sustainability • Social cohesion • Interconnectedness • Well-being • Ethical leadership • Transformative education • Mindfulness • Systemic change • Sustainable business • Cultural preservation • Inclusive governance

H. V. Tho (✉)
Eurasia Learning Institute for Happiness and Wellbeing, Palézieux, Switzerland
e-mail: havinhtho@gmail.com

Gross National Happiness Center Bhutan, Thimphu, Bhutan

Traning International Committee of the Red Cross, Geneva, Switzerland

J. Kronenberg and E. T. Lammerts van Bueren (eds.), *On the Earth We Want to Live*, World Sustainability Series, https://doi.org/10.1007/978-3-031-98758-8_10

1 Introduction: The Quest for Compassion Amid Global Crisis

The world is currently facing a severe socio-ecological crisis characterised by environmental degradation (United Nations Environment Programme [UNEP], 2024), social inequality (World Inequality Database [WID], n.d.), and a growing disconnection between humans and nature. One comprehensive study from the German Centre for Integrative Biodiversity Research (iDiv) shows that the average distance between human populations and natural areas has increased (Cazalis et al., 2022). This crisis arises from a fragmented worldview that values economic growth over well-being, competition over cooperation, and narrow individualism over community placing immense strains on social structures and ecosystems. Tackling this urgent challenge calls for transformative solutions that reconnect us with ourselves, each other, and the natural world.

At the heart of this crisis lies a profound disconnection manifesting in three interlinked forms from: self, others, and nature (Scharmer, 2018).

Disconnection from self is seen in rising levels of stress, burnout, and loss of meaning, fueled by a culture that glorifies productivity over personal well-being. This disconnect erodes mental health and the ability to find joy, trapping individuals in cycles of anxiety and exhaustion (Petitta & Ghezzi, 2023).

Disconnection from others weakens social bonds, increases isolation, and fuels conflict. As communities fracture, polarization, and intolerance grow, fostering a climate of distrust and division. This breakdown in empathy not only affects individual well-being but also undermines societal resilience, contributing to violence, discrimination, and unrest.

The disconnection from nature is perhaps the most visible, with environmental exploitation, biodiversity loss, and climate change highlighting a mindset that views nature as a commodity rather than a living system of which we are a part. This worldview disrupts the balance that sustains life, endangering ecosystems and threatening our own survival.

The systemic disconnects we face stem from a reductionist worldview that prioritises economic growth over the well-being of people and the planet.[1] This mindset fuels exploitation, valuing material wealth above compassion, solidarity and sustainability, misaligning societal values with the true needs of humans and the environment.

Addressing this crisis requires a radical shift to a holistic perspective that recognises the interdependence of all life. We need integrative approaches that prioritise well-being, empathy and ecological balance, fostering a culture of connection and respect for all living beings. This transformation demands not only policy changes but also a deep cultural and spiritual shift towards inclusivity and sustainability.[2]

[1] A 2021 report from the Intergovernmental Panel on Climate Change (IPCC) found that economic growth and industrial activities are the largest contributors to climate change.
[2] The "planetary boundaries" framework identifies critical limits that should not be exceeded to maintain the Earth's systems. The framework, published by Rockström et al. (2009), underscores

Society stands at a crossroads: one path driven by nationalism, unchecked economic growth, and isolationism, neglecting the long-term impact on people and the planet; the other guided by compassion, inclusion and ecological responsibility, promoting interconnectedness and stewardship. Our choices will shape the future—will we continue on a fragmented path, or embrace a holistic vision that redefines our relationships with each other and the Earth?

This perspective is rooted in Indigenous cultures, spiritual traditions and philosophies that recognise life's interconnected nature offering timeless guidance for today's challenges.

I will draw on my personal experiences to make the call for a holistic worldview more relatable and impactful.

2 From Revolution to Reflection: A Personal Journey Through Political Ideals, Conflict and the Search for a Holistic Worldview

As a young person, I participated in the student revolution in France in May 1968. My father was Vietnamese, and my mother was French, so the war in Vietnam directly impacted my family. Even abroad, I felt deeply concerned and engaged in political action to contribute to societal change. Many from our generation believed that post-World War II society, with its materialistic outlook and singular focus on economic development, failed to address our deeper aspirations (Suri, 2007). The hypocritical stance of Western powers—preaching human rights and democracy while exploiting the rest of the world through neocolonial practices—drove many young people in Europe and the US toward communism and socialism as alternatives. Figures like Ho Chi Minh, Fidel Castro and Che Guevara became romantic heroes to our young, somewhat naïve minds.

However, when Russian tanks invaded Prague in the summer of 1968, my hope that communism could be a viable alternative to capitalism was shattered. That same summer, my family was returning to Vietnam, but the war was raging more than ever. As a Vietnamese citizen, I would have been required to join the army, but as someone committed to non-violence, I was not willing to risk being conscripted. This led me to travel through Asia and reconnect with my Buddhist roots, discovering Buddhism, meditation and the spiritual path. Although I considered becoming a Buddhist monk, my commitment to social and political engagement remained strong, and I found myself grappling with how to balance inner transformation with societal change.

---

the interconnectedness of the Earth's systems—climate, biodiversity, ocean chemistry, etc.—and how unsustainable human activity disrupts this balance, threatening long-term sustainability.

This journey led me to discover Rudolf Steiner and anthroposophy, prompting me to study Eurythmy[3] at the Goetheanum in Dornach. After completing my studies, I worked as a teacher at a Waldorf School[4] in Germany and later joined a Camphill[5] community in Switzerland.

Later, I had the opportunity to return to Vietnam shortly after the war. Moved by the suffering of many children affected by the conflict, my wife, colleagues, and I founded the Eurasia Foundation[6] to support curative education and social therapy in Vietnam. We initially focused on projects for children and young adults with disabilities and later expanded into fields such as ecology, social renewal, education, and Gross National Happiness applied to business.

I mention these aspects of my life because a worldview is not just a theoretical or philosophical perspective—it is deeply rooted in our lived experiences. To truly understand someone's worldview, it is important to know the personal experiences that have shaped their ideas.

Later, I left Camphill and became Director of Learning and Development at the International Committee of the Red Cross (ICRC), working in some of the most challenging environments worldwide. The ICRC is a neutral, independent and impartial organisation with a mission rooted in international humanitarian law, specifically the Geneva Conventions.[7] It primarily operates during armed conflicts and other situations of violence, providing protection and assistance to victims without discrimination.

This position exposed me to war zones and revealed that the systemic and structural dimensions of conflict often drive much of the destruction. I realised that systems are not natural laws; they reflect our thoughts, emotions, actions and interactions. To change systems sustainably, we must also change the mindsets that created them. Similarly, to shift consciousness, we must transform the cultural, economic and political systems that shape us.

While I valued my work at the ICRC, I felt dissatisfied because humanitarian action, though essential, does not address the root causes of wars. In Darfur, I saw that the conflict was not just about ethnic tensions or local power struggles

---

[3] Eurythmy is an expressive movement art developed by Rudolf Steiner that combines dance, music and speech to make visible the sounds and rhythms of language and music through harmonious gestures and movements.

[4] Waldorf school is an educational institution based on the principles of Rudolf Steiner that emphasises holistic, developmentally appropriate and arts-integrated learning to nurture students' intellectual, artistic, and practical skills in a supportive and creative environment.

[5] Camphill was founded in 1939 by Dr. Karl König and a group of his colleagues and friends in Scotland. The community was established as a place where individuals with developmental disabilities could live and work together in a supportive, inclusive environment based on the principles of anthroposophy.

[6] https://eurasia-foundation.org.

[7] The Geneva Conventions are a set of international treaties that establish the standards of humanitarian treatment during war, protecting non-combatants, prisoners of war and the wounded.

but was deeply connected to systemic issues like climate change and global power dynamics. The conflict, often labeled as ethnic or resource-based, is intertwined with environmental degradation, prolonged droughts and diminishing resources, which have escalated competition and tensions between groups.

Furthermore, international actors with economic interests in Sudan's resources have played a contradictory role, offering aid while pursuing exploitative practices. Their involvement complicates the conflict, often prioritising access to resources over local stability, perpetuating violence and poverty. These complex dynamics highlight a gap in traditional humanitarian responses: they alleviate immediate suffering but rarely address the underlying causes, such as environmental degradation, systemic inequalities and external exploitation.

My experience in Darfur highlighted the need for a holistic approach that addresses the socio-economic systems and mindsets perpetuating conflict. It was a pivotal moment in my search for alternative development models, ultimately leading me to Bhutan's concept of Gross National Happiness (GNH). Before exploring GNH, it's important to reflect on how worldviews shape societies.

3 The Power of Worldview in Shaping Society

Until recently, worldviews were primarily shaped by local cultures, religions and traditions, creating diverse identities but also divisions and conflicts (Geertz, 1973). Today's global challenges, like climate change and social inequality, demand a search for universally human values that unite us while respecting local differences.

Worldviews influence how individuals and societies perceive reality, shaping values, decisions and priorities. Spiritual perspectives, especially those emphasising interconnectedness, challenge materialistic views, fostering a sense of unity and purpose.

In Bhutan, pre-Buddhist shamanic traditions see nature as inhabited by spirits, leading to a deep reverence for the environment. This animistic view influences daily practices, fostering sustainable and mindful interactions with nature.

The Bhutanese approach teaches that environmental conservation is both an ethical and spiritual duty, deeply embedded in cultural values. This respect for nature extends to community living, where decisions balance human needs with environmental harmony, offering valuable lessons for global sustainability. Bhutan is the only country that is not only carbon neutral but even carbon negative. This means that it absorbs more carbon dioxide than it emits.

Today, holistic worldviews are resurging within movements for social justice and sustainability, promoting inclusive and compassionate approaches to global challenges. Balancing both global and local perspectives is key, as it embraces the wisdom of diverse traditions while fostering a shared sense of humanity. For example, movements like Indigenous Land Rights, the Buen Vivir in Latin America and the Transition Towns exemplify this dynamic.

This interplay shapes identities in our interconnected world, encouraging openness and solidarity while honoring cultural roots.

Having grown up in a diplomatic family and having experienced various cultures, I have seen the value in diverse worldviews. Recognising the legitimacy of different perspectives helps decolonise our systems and supports the quest for universally human values that promote a harmonious, inclusive world.

4 My Experiences with Anthroposophy

As mentioned, my journey to balance inner spiritual development with social transformation led me to explore Rudolf Steiner's work and anthroposophy. I was struck by how seamlessly Steiner integrated spiritual insights with practical applications across various fields, rooted in a spiritual worldview that offers a broader understanding of the human being and the cosmos (Steiner, 1910).

Steiner outlined a clear spiritual path that individuals could independently follow to gain deeper insights, coupled with practical methods for engaging with society (Steiner, 1904).

Developed in the early twentieth century, anthroposophy offers a unique perspective on integrating spiritual and material aspects to address social challenges. It posits that humans are not merely physical beings but also possess spiritual dimensions that connect them to the wider cosmos. This worldview emphasises that personal spiritual development is not just a private pursuit but a pathway to understanding and transforming the world.

According to Rudolf Steiner, humans are understood as a threefold being consisting of the body, soul and spirit. The physical body connects us to the material world; the soul mediates our inner experiences, emotions and thoughts; and the spirit connects us to higher, transcendent realities, providing a deeper purpose and insight into our place within the cosmos. Steiner's concept of the threefold human being highlights the interdependence of these aspects and suggests that true social change requires nurturing all dimensions of human existence.

For example, Waldorf education[8] emphasises holistic child development by nurturing intellectual, emotional and spiritual growth, creating learning environments that respect the individual's spiritual journey. Anthroposophy challenges us to address social issues by recognising the threefold nature of the human being—body, soul and spirit—emphasising that true social harmony arises when all dimensions are acknowledged and nurtured. It suggests that social structures and solutions must not only meet material needs but also support the soul's emotional and intellectual well-being and the spirit's search for purpose and connection, integrating these elements to foster a balanced and harmonious society.

[8] Waldorf education is a holistic approach to learning that emphasises the intellectual, artistic and practical development of students, integrating academics with creativity and hands-on experiences, and is based on the educational philosophy of Rudolf Steiner.

This approach contrasts sharply with purely materialistic or technocratic solutions, offering a more balanced, humane way of engaging with the complex challenges of contemporary society.

Rudolf Steiner's contributions to sustainable development and a holistic, spiritual worldview are both profound and far-reaching, yet his recognition in academia and mainstream culture remains surprisingly limited. Anthroposophy offers innovative approaches in fields like education, agriculture and social renewal, all of which resonate with current global challenges. For instance, Steiner's development of biodynamic agriculture (see Chap. 4.6)—a holistic method that predates and informs today's regenerative agriculture—emphasises ecological sustainability, soil health and cosmic rhythms. These principles align closely with contemporary sustainability goals, yet they are rarely highlighted in major forums like COP[9] or in discussions around the Sustainable Development Goals.[10]

This lack of recognition can be partly attributed to the insular nature of the anthroposophical movement, which often remains confined within its own circles and hesitates to engage in open dialogue with the broader public. To truly fulfil its potential in addressing social challenges, anthroposophy must open itself to dialogues with other cultures, spiritual traditions and like-minded groups. By embracing diverse perspectives and engaging in mutual learning, anthroposophy can enrich its own practices and contribute more effectively to the global conversation on creating a more harmonious and spiritually integrated society. As a result, many of Steiner's forward-thinking ideas are underrepresented in international policy discussions and environmental platforms, missing critical opportunities to contribute to global conversations on issues like climate change and food security.

In contrast, Steiner's legacy could gain greater appreciation if anthroposophical institutions actively engaged with wider society, presenting their work in a way that resonates with contemporary values and challenges. To fully unlock the potential of anthroposophy, it is essential to critically engage with Steiner's legacy, appreciating his spiritual teachings while acknowledging the context-specific biases of his time. For example, the notion of European cultural superiority was widely accepted during the era of Western colonial dominance. Today, a more critical perspective on colonialism has emerged, along with greater recognition of the value of non-European cultures.

By separating his enduring wisdom from outdated cultural views, we can enrich our understanding and application of his ideas in a way that is relevant today. This critical approach does not diminish Steiner's contributions but rather enhances their impact, allowing anthroposophy to emerge as a dynamic, evolving framework that

[9] Conference of the Parties (COP) is the annual global climate summit where representatives from countries gather to negotiate and assess progress in addressing climate change under the United Nations Framework Convention on Climate Change (UNFCCC).

[10] The Sustainable Development Goals (SDGs) are a set of 17 global goals established by the United Nations in 2015 to address urgent challenges such as poverty, inequality, climate change and environmental degradation, aiming for a more sustainable and equitable world by 2030.

speaks to the needs of our time and contributes to building a more holistic and compassionate world.

At the same time, it is noteworthy that Academia can often be insular. When I began my PhD, I aimed to assess Steiner's educational impact with scientific rigor, but finding a willing supervisor was a challenge. Despite the influence of Waldorf education, Steiner's work is frequently dismissed as too esoteric for mainstream research, reflecting academia's tendency to reject unconventional ideas that don't fit established paradigms.

This mutual isolation restricts opportunities for meaningful dialogue and collaboration, missing opportunities to contribute to broader discussions on sustainability and education, and social renewal while academia's reluctance to engage with alternative worldviews stifles innovative research.

Examples like Otto Scharmer's Theory U,[11] Ibrahim Abouleish's Sekem initiative[12] or Arthur Zajonc's work with the Mind and Life Institute[13] demonstrate how anthroposophical ideas can gain traction when presented transparently and supported by evidence. Scharmer connects anthroposophy with contemporary leadership and management, making it accessible to leaders in global change, SEKEM has brought biodynamic agriculture on the world stage, while Zajonc weaves Steiner's insights into dialogues between science and spirituality. These approaches show that engaging with broader contexts can elevate Steiner's contributions, fostering a more inclusive and transformative dialogue on global challenge.

5 Addressing Social Challenges Through Gross National Happiness

The Social Framework of Gross National Happiness (GNH)

My experiences with the International Committee of the Red Cross (ICRC) profoundly influenced my search for a different development model—one that addresses the complex needs of individuals and communities beyond mere economic growth.

In 2011, the Royal Government of Bhutan submitted a resolution to the United Nations titled "Happiness: Towards a holistic approach to development" (United Nations General Assembly, 2011); it calls for a holistic approach to development

[11] Otto Scharmer's Theory U is a transformative leadership framework that guides individuals and organisations through a process of deep listening, co-sensing and co-creating, enabling them to let go of old patterns and access emerging future possibilities to drive profound innovation and change.

[12] Ibrahim Abouleish was the founder of SEKEM, an innovative Egyptian initiative that combines biodynamic agriculture, social development, and cultural renewal, demonstrating a holistic approach to sustainable development and bridging traditional values with modern technologies for community transformation. Also see Chap. 5.22 SEKEM in this publication.

[13] Arthur Zajonc, a physicist and contemplative scholar, played a pivotal role in the Mind and Life Institute, an organisation that fosters dialogue and collaboration between science, contemplative wisdom and education to deepen our understanding of the mind and promote human flourishing.

that prioritises happiness, well-being and sustainability over purely economic indicators like GDP, aiming to inspire global policy changes towards more balanced and humane progress. This vision deeply resonated with me. The resolution was adopted by consensus, and in 2012, UN Secretary-General Ban Ki-moon convened a high-level meeting in New York to explore ways to operationalise these ideas. Inspired by this bold vision, Bhutan's Prime Minister J.Y. Thinley established the GNH Centre to further investigate these questions and I was selected as its director. This role led me to live in Bhutan for 7 years, where I worked to translate the concept of Gross National Happiness (GNH) into practical, actionable programmes (Ha Vinh, 2022).

Conventional development models like GDP often mistake economic growth for the ultimate goal, prioritising the production of goods and services over broader societal objectives such as environmental sustainability, social equity and well-being. This focus on economic expansion can result in policies that treat people merely as resources, ignoring the true purpose of development: enhancing individual and collective well-being.

In contrast, the Gross National Happiness (GNH) framework views economic growth as a means to achieve broader objectives, including health, education, environmental conservation and overall happiness. GNH emphasises that when economic growth is treated as the end goal, it can lead to environmental degradation, increased social inequality and neglect of mental health and community cohesion.

By prioritising well-being over economic output, GNH challenges conventional approaches, advocating for a development model that aligns economic policies with the real goal: improving quality of life sustainably and meaningfully. This holistic perspective aims to ensure that economic activity supports human flourishing rather than defining it.

The GNH Framework Is Structured Around Four Main Pillars

1. Sustainable and equitable socio-economic development emphasises economic growth that benefits everyone, ensuring that resources and opportunities are distributed fairly to reduce inequality and improve the overall quality of life.
2. Preservation and promotion of culture aims to maintain the rich cultural heritage, values and traditions that shape identities and provide a sense of belonging, fostering a resilient and cohesive society.
3. Good governance ensures transparent, accountable and inclusive decision-making processes that prioritise the well-being of citizens, creating trust and enabling effective implementation of policies.
4. Conservation of the environment focuses on protecting natural ecosystems, recognising that environmental health is crucial for human well-being and that sustainable use of resources is essential for future generations.

The balance between these four pillars is essential for healthy development, as it ensures that economic progress does not come at the expense of environmental sustainability, cultural integrity or social equity, thereby fostering a holistic and harmonious society.

The GNH framework and Rudolf Steiner's concept of the threefold social order (Chap. 4.10; Steiner, 1999) share three core dimensions—economic, cultural and political—each reflecting essential aspects of societal well-being. Steiner uniquely applied the ideals of the French Revolution—Liberty, Equality and Fraternity—to these spheres, illustrating how liberty should guide cultural life, equality should govern political and legal structures, and fraternity should underpin economic activity. This alignment offers a profound way to understand the distinct roles and interconnections of these societal dimensions, highlighting their importance in fostering a balanced and harmonious social order.

The GNH framework further extends this by adding a crucial fourth dimension: ecology.

1. **Cultural Dimension: GNH Preservation and Promotion of Culture|Steiner Cultural/Spiritual Sphere—Liberty**

In the GNH framework, the preservation and promotion of culture emphasise maintaining cultural heritage, values and traditions that foster a sense of identity and belonging. This aligns with Steiner's cultural or spiritual sphere, which advocates for the principle of *Liberty*—freedom in education, arts, religion and spiritual life. Both frameworks recognise that cultural flourishing and individual freedom to explore and express creativity are essential for a resilient and cohesive society. Liberty in this context means enabling individuals to grow and contribute to culture without undue interference, ensuring that cultural development is driven by inner values and inspirations.

2. **Political Dimension: GNH: Good Governance/Steiner: Political/Legal Sphere—Equality**

Good governance in the GNH framework involves transparent, accountable and inclusive decision-making processes that prioritise the well-being of all citizens, fostering trust and effective policy implementation. This mirrors Steiner's political or legal sphere, which is guided by the principle of *Equality*—where every individual has equal rights, opportunities and a voice in the governance processes. Both approaches stress that governance must serve the collective good, ensuring justice, protecting individual rights, and promoting participatory democracy. Equality in governance is crucial for maintaining social harmony and addressing the needs of the community equitably.

3. **Economic Dimension: GNH Sustainable and Equitable Socio-Economic Development/Steiner: Economic Sphere—Fraternity**

The economic pillar of GNH focuses on sustainable and equitable socio-economic development, advocating for economic growth that benefits everyone and reduces inequality. This aligns with Steiner's economic sphere, governed by the principle of *Fraternity*—cooperation and mutual support in economic life. In Steiner's view, economic activities should be driven by the spirit of brotherhood/sisterhood, where the needs of all are considered, and resources are distributed fairly. Both frameworks emphasise that economic processes should not exploit but instead contribute to the well-being of society, reflecting a commitment to mutual care and shared prosperity.

4. **Ecology GNH-Specific: Conservation of the Environment**

The GNH framework uniquely includes the conservation of the environment, highlighting the importance of ecological sustainability as a foundation for overall well-being. This dimension emphasises the protection of natural ecosystems and the sustainable use of resources, recognising that environmental health is intimately connected to human health and societal flourishing. The principle underlying this approach is *Interdependence.* While not explicitly mentioned in the Threefold Social Order, Rudolf Steiner's ecological vision is rooted in his holistic worldview, emphasising the interconnectedness of all life. He pioneered biodynamic agriculture in 1924, advocating sustainable farming that treats the Earth as a living organism. Steiner believed modern industrial farming depleted the Earth's life forces, leading to environmental harm, while biodynamic methods rejuvenate the soil and promote ecological balance. His spiritual ecology stressed humanity's responsibility to care for the planet, viewing ecological crises as both material and spiritual. Steiner's ideas resonate with modern sustainability movements, emphasising the need for harmony between human activity and nature's intrinsic value.

6 The Balance of Dimensions and Principles

Both the GNH framework and Steiner's threefold social order highlight the importance of maintaining a balance between these dimensions. In GNH, the integration of sustainable development, cultural preservation, good governance and ecological conservation ensures that no single aspect dominates to the detriment of others. Steiner's concept similarly stresses that the spheres of Liberty, Equality and Fraternity must work independently but harmoniously, ensuring that cultural freedom, social justice and economic cooperation guide societal development.

The Nine Domains of GNH

During the UN High-Level Meeting on Happiness and Wellbeing, the central theme was to explore new approaches to development that go beyond traditional economic indicators like Gross Domestic Product (GDP). However, a critical question arose during the discussions: "How can one replace a robust statistical measure like GDP with a vision or ideal like happiness?" This question highlighted the need for a concrete, measurable framework to assess progress.

Bhutan developed the GNH Index, a multidimensional measure that uses nine domains to quantify well-being comprehensively (Ura et al., 2011). These nine domains capture the diverse aspects of human life and environmental health, offering a more holistic measure of progress than GDP, which focuses solely on economic output. The nine domains of the GNH Index are:

1. Psychological Well-being: Assesses the mental and emotional health of individuals, including life satisfaction, optimism, spirituality and prevalence of negative emotions like stress and anxiety.
2. Health: Measures physical health, nutrition and access to healthcare services, highlighting the importance of a healthy lifestyle and the availability of quality medical care.
3. Education: Evaluates educational attainment, knowledge, skills and values, focusing on lifelong learning, critical thinking and the alignment of education with cultural and social values.
4. Time Use: Analyses how people spend their time, balancing work, leisure and sleep, emphasising the importance of rest, recreation and time spent with family and community.
5. Cultural Diversity and Resilience: Assesses participation in cultural activities, preservation of traditions and the sense of cultural pride, ensuring the continuity and vitality of cultural heritage.
6. Good Governance: Evaluates citizens' perceptions of government performance, political participation, trust in institutions and the transparency and accountability of public officials.
7. Community Vitality: Measures the strength of social relationships, trust among community members and the sense of belonging, highlighting the importance of social support and community cohesion.
8. Ecological Diversity and Resilience: Monitors the environmental health, sustainable use of resources and individuals' attitudes and behaviors towards conservation, recognising the interdependence between humans and nature.
9. Living Standards: Assesses income, material well-being, housing quality and financial security, ensuring that people's basic needs are met and that economic development benefits all.

The nine domains of the GNH Index offer a holistic framework that goes beyond GDP to assess the development of countries, organisations and institutions. By integrating economic, social, environmental and cultural factors, GNH captures the full spectrum of human well-being, reflecting the diverse needs of individuals and communities. Unlike GDP, which focuses solely on economic output, GNH measures quality of life through indicators that prompt deeper questions: Are our communities thriving? Are people happy and fulfilled? Is our environment sustainable?

For organisations, GNH encourages evaluating their impact on employees, communities and the environment, promoting policies that enhance overall well-being rather than just financial success. Its adaptability allows GNH to guide not only national development but also the growth of companies and institutions, steering them toward sustainable and meaningful progress. By prioritising holistic measures, GNH fosters balanced and thriving societies, recognising that true progress lies in enhancing the economic, spiritual, emotional and social well-being of all.

7 Insights from My Work and Practice

During the time in Bhutan, my mission was to translate the profound principles of Gross National Happiness into practical applications. This work centered on developing programmes that embodied GNH's core values, such as fostering compassionate societies, promoting holistic education and driving social innovation that prioritises well-being over profit. The experience of implementing GNH-inspired projects demonstrated that integrating spiritual values into development practices could lead to more sustainable and compassionate outcomes.

Creating a Camphill Inspired Community in Vietnam: The Peaceful Bamboo Family
My family and I spent many years in a Camphill Community in Switzerland called Perceval. The Camphill movement is a global network of intentional communities that provide support and meaningful work for people with developmental disabilities, mental health needs and other social vulnerabilities. Founded on the principles of anthroposophy by Karl König (1902–1966) in 1939, Camphill communities emphasise a holistic, inclusive approach to care, fostering equality, respect and the development of each individual's potential through shared living, biodynamic agriculture and social integration. The movement's core values include community life, mutual respect and the belief in the spiritual dignity and worth of every person.

The work of Eurasia in Vietnam in the field of special education led, after a decade of development, to the creation of a Camphill-inspired community: "The Peaceful Bamboo Family" (Eurasia Foundation, n.d.). Our intention was to adapt the fundamental principles of Camphill to the Vietnamese context, avoiding the imposition of foreign practices and methods while preserving the universally human values embedded in this approach in a way that respects Vietnamese culture, traditions, values and socioeconomic context.

Blending Camphill principles with mindfulness inspired by Thich Nhat Hanh's (1926–2022) Socially Engaged Buddhist approach,[14] along with the concept of Gross National Happiness—though not Vietnamese in origin, it comes from a Buddhist country with common cultural roots—gradually shaped the unique character

[14] Thich Nhat Hanh's socially engaged Buddhism is a practice that combines mindfulness, compassionate action, and nonviolent principles to address social, environmental, and political issues, encouraging practitioners to actively apply Buddhist teachings to create positive change in the world.

of our community. The leadership and coworkers of the community are all Vietnamese, although many friends from around the world often visit to support us. Our community is fully integrated into the local context and strongly supported by the local authorities.

This experience illustrates how anthroposophical insights and practices can be thoughtfully adapted to local cultures and contexts without losing their core values, while avoiding the dangers of a neocolonial approach that imposes a foreign culture. In our work, we consciously avoided a direct transplant of Western practices. Instead, we focused on retaining the essence of Camphill's core values—such as the emphasis on community, inclusivity and respect for the dignity of every individual—and integrating them with local cultural elements that resonate deeply with the Vietnamese people.

One example is the importance of celebrating festivals, which are central to the rhythm of life in both Camphill communities and Vietnamese culture. We honor local traditions by celebrating the full moon and new moon, reflecting moments of reflection and renewal, and incorporating major Vietnamese festivals such as Tết (Lunar New Year), which brings the community together to welcome the new year with gratitude and hope. We also celebrate the Mid-Autumn Festival, emphasising family, unity and appreciation for nature's bounty, connecting deeply with the values of togetherness and community. These celebrations not only enhance the sense of belonging and shared joy but also bridge Camphill's universal principles with the rich cultural heritage of Vietnam, creating a truly integrated and respectful community life. By grounding the community's leadership in local hands and fostering a true intercultural exchange, we ensured that the Peaceful Bamboo Family was not an external imposition but a locally rooted, culturally sensitive initiative. This approach allowed us to honor both the universal values of anthroposophy and the rich traditions of Vietnam, creating a unique and authentic community that reflects both worlds without compromising its foundational principles.

Transforming Education

While my pedagogical background is rooted in Waldorf education, Eurasia's mission was to develop an approach that could be applied to public schools in diverse cultural and socio-political contexts worldwide. Inspired by Steiner's educational philosophy (see, Steiner, 1919), we reflected on how to adapt these ideas for broader acceptance. Transforming education requires addressing both inner development and systemic change, as exemplified by the Happy Schools Project, which we developed in Vietnam and is now expanding to many other countries.

The Happy Schools programme (Eurasia Learning Institute, n.d.), initiated by the Eurasia Learning Institute for Happiness and Wellbeing, aims to create a school environment that prioritises the well-being of students, teachers and the wider community. It focuses on three core areas: self-care through mindfulness and emotional intelligence, fostering empathy and compassionate relationships, and reconnecting with nature to promote environmental stewardship. The programme integrates social and emotional learning (SEL), mindfulness and education for sustainable development, aiming to enhance mental health, social cohesion and academic performance. By emphasising these elements, Happy Schools foster positive, inclusive and supportive learning spaces.

This shift moves education from a narrow focus on cognitive skills and standardised testing to a broader perspective that values emotional, social learning, creativity and mental well-being, fostering compassionate and self-aware individuals. However, inner transformation alone is insufficient without systemic change, which involves advocating for educational policies that prioritise well-being and holistic development.

This led Eurasia to collaborate with the OECD/PISA[15] survey to ensure that educational systems are not only measured by academic results, but also include the overall well-being of students. This resulted into to the recent creation of the PISA Happy Life Dashboard,[16] which uses nine dimensions—inspired by the nine domains of GNH—to measure student happiness, belonging and life satisfaction, encouraging policymakers to redefine success in education. Together, these approaches create a feedback loop where mindset transformation inspires systemic change, and progressive policies further cultivate healthier mindsets, ultimately building an education system that nurtures happiness, compassion and holistic development, empowering young people to both succeed and positively transform society.

GNH in Business

In a world facing urgent environmental and social challenges, sustainability is essential for our planet and society's future. Two powerful leverage points for transforming society are education, which shapes future citizens and leaders, and business, which has a central influence today.

Gross National Happiness (GNH) offers a transformative approach to business, shifting the focus from profit maximisation to holistic well-being. This paradigm emphasises that a successful organisation measures not just financial performance but its contribution to the happiness of employees, communities and the environment. By integrating GNH principles, businesses can foster a culture that values sustainability and happiness alongside material success.

Anthroposophy also reimagines business by emphasising human dignity, ethical responsibility and sustainable practices. Examples include Weleda, behind Dr. Hauschka products (see Chap. 5.27), which integrates biodynamic farming. They are a leader in natural cosmetics committed to holistic health and ecology, and GLS Bank,[17] which applies anthroposophical principles to finance, focusing on

[15] The PISA survey (Programme for International Student Assessment) is a global study conducted by the OECD every 3 years to evaluate the education systems of participating countries by measuring 15-year-old students' abilities in reading, mathematics and science.

[16] The PISA Happy Life Dashboard is a tool developed by the OECD to measure and compare the well-being of students worldwide, focusing on factors such as life satisfaction, school climate and social-emotional skills alongside academic performance.

[17] Also see Chap. 5.8 Freie Gemeinschaftsbank in this publication for a similar initiative.

transparent and sustainable banking that supports social and environmental initiatives. These companies demonstrate how anthroposophy can guide businesses toward practices prioritising people and the planet, showing that business can be a force for positive transformation when grounded in values beyond profit.

To implement GNH in business, we use Scharmer's (2007) Theory U, inspired by anthroposophy, to drive change within companies. Theory U facilitates transformative change through deep listening, sensing and co-creating from the emerging future. This approach helps businesses connect with their deeper purpose, align operations with holistic well-being, and foster innovative solutions that benefit society and the environment. By applying Theory U, businesses can create pathways to profound change that integrate GNH values, encouraging a shift towards more conscious, sustainable and happiness-oriented practices.

Adopting GNH in business requires a shift in leadership and management. Leaders prioritise the well-being of employees, recognising that a happy and fulfilled workforce is more productive, innovative and dedicated. This approach embeds happiness into core values and operational strategies, transforming organisations into living entities with essential dimensions.

Implementing GNH Involves Three Key Aspects

1. Vision, Mission and Values: Businesses redefine their goals to include happiness, community well-being and sustainability, aligning long-term success with positive societal impact.
2. Developing Happiness Skills: Training in mindfulness, empathic communication and emotional intelligence fosters a supportive and collaborative workplace.
3. Measuring What Matters: The GNH Index, with its nine domains, offers a comprehensive framework for assessing organisational well-being, guiding businesses toward balanced, sustainable practices aligned with stakeholder wellbeing.

Real-World Applications of GNH in Business (Ha Vinh, 2022)

The application of GNH principles in business is not just theoretical; it has been successfully implemented in various organisations around the world. For instance, companies like Mainetti and Biti's in Vietnam, or B. Grimm in Thailand have embraced the GNH framework, transforming their corporate cultures and operational strategies to prioritise the happiness of their employees and the well-being of their communities.

8 Conclusion

The socio-ecological crises of today call for a profound shift from fragmented, economically driven models to holistic approaches that prioritise human well-being, ecological health and social cohesion. The dominant focus on economic growth often overlooks the interconnectedness of people, nature and society, leading to unsustainable practices and growing inequalities. To address these challenges, we need a transformative vision that integrates spiritual, ethical and cultural dimensions into social, economic and educational systems, embracing values such as empathy, inclusivity and ecological stewardship.

Frameworks like Gross National Happiness (GNH), anthroposophy and various cultural and spiritual traditions provide pathways to reimagine education, business and governance. These holistic worldviews challenge conventional paradigms, redefining progress beyond economic metrics to focus on happiness, resilience and ecological integrity. By advocating for policies that honor life's interconnectedness, we can build systems that support balanced and thriving societies.

Incorporating spiritual and holistic perspectives into global policies can redefine sustainable development as an ethical endeavor, aligning economic activities with human and ecological well-being. This vision prioritises long-term resilience, inclusivity and harmony with nature, fostering a global movement that balances cosmopolitan ideals with local wisdom. By embracing these holistic perspectives, we can create a future where true success is measured by the well-being of all living beings and a harmonious planet.

References

Cazalis, V., Loreau, M., & Barragan-Jason, G. (2022). A global synthesis of trends in human experience of nature. *Frontiers in Ecology and the Environment*. https://doi.org/10.1002/fee.2540

Eurasia Foundation. (n.d.). *The peaceful bamboo family*. Retrieved February 23, 2025, from https://eurasia-foundation.org/peaceful-bamboo-family/

Eurasia Learning Institute. (n.d.). *Happy schools project in Vietnam*. Retrieved February 23, 2025, from https://www.elihw.org/happy-schools

Geertz, C. (1973). *The interpretation of cultures: Selected essays*. Basic Books.

Ha Vinh, T. (2022). *A culture of happiness: How to scale up happiness from people to organizations*. Parallax Press.

Intergovernmental Panel on Climate Change. (2021). Climate change 2021: The physical science basis. In V. Masson-Delmotte, et al. (Eds.), *Contribution of working group I to the sixth assessment report of the intergovernmental panel on climate change*. Cambridge University Press. https://www.ipcc.ch/report/ar6/wg1/

Petitta, L., & Ghezzi, V. (2023). Remote, disconnected, or detached? Examining the effects of psychological disconnectedness and cynicism on employee performance, wellbeing, and work–family interface. *International Journal of Environmental Research and Public Health, 20*(13), 6318. https://doi.org/10.3390/ijerph20136318

Rockström, J., Steffen, W., Noone, K., Persson, Å., Chapin, F. S., Lambin, E., Lenton, T. M., Scheffer, M., Folke, C., Schellnhuber, H. J., Nykvist, B., de Wit, C. A., Hughes, T., et al. (2009). Planetary boundaries: Exploring the safe operating space for humanity. *Ecology and Society, 14*(2), 32.

Scharmer, C. O. (2007). *Theory U: Leading from the future as it emerges*. Society for Organizational Learning.

Scharmer, C. O. (2018). *The essentials of Theory U: Core principles and applications*. Berrett-Koehler Publishers.

Steiner, R. (1904). *Knowledge of the higher worlds and its attainment*. Anthroposophic Press.

Steiner, R. (1910). *An outline of occult science.* Anthroposophic Press.
Steiner, R. (1919). *Allgemeine Menschenkunde als Grundlage der Pädagogik (GA 293).* Rudolf Steiner Verlag.
Steiner, R. (1999). *Towards social renewal: Rethinking the basis of society* (M. Barton, Trans.). Rudolf Steiner Press.
Suri, J. (2007). *The global revolutions of 1968: A Norton casebook in history.* W. W. Norton & Company.
United Nations General Assembly. (2011). *Resolution adopted by the general assembly on 19 July 2011: Happiness—towards a holistic approach to development (A/RES/65/309).* United Nations. https://digitallibrary.un.org/record/715187
United Nations Environment Programme. (2024). *Global resources outlook 2024.* International Resource Panel. https://www.unep.org/resources/Global-Resource-Outlook-2024
Ura, K., Alkire, S., & Zangmo, T. (Eds.). (2011). *Gross national happiness: Development with values.* The Centre for Bhutan Studies.
World Inequality Database. (n.d.). *World inequality database.* https://wid.world

Biodynamic Agriculture—And the Future of Agriculture

Nadia El-Hage, Helmy Abouleish, and Ueli Hurter

Abstract

Agriculture, derived from "agri" (field) and "culture" (cultivation), embodies humanity's evolving interaction with Nature. From the Neolithic Revolution's settlement farming to the Industrial Revolution's mechanisation and the genomic and digital revolutions, agriculture has shaped civilisations while often harming biodiversity and ecosystems. Now at a crossroads, the future of agriculture demands balancing environmental stewardship and co-creative approaches. Biodynamic farming, rooted in Rudolf Steiner's 1924 Agricultural Course, offers a sustainable approach by treating farms as self-contained living organisms interconnecting soil, plants, animals, human and cosmic activity. Practices such as humus renewal, cosmic rhythms, and biodynamic preparations enhance soil vitality, reduce emissions, and foster biodiversity. Inspired by pioneers like Dr. Ibrahim Abouleish, biodynamics transforms landscapes and livelihoods, demonstrating economic and ecological viability through true-cost accounting in Egypt. With transparency, regenerative methods, and a foundation of love for Nature and humanity, agriculture can transition to a sustainable, equitable, and culturally rich future.

N. El-Hage (✉)
Swette Center for Sustainable Food Systems, Arizona State University, Tempe, AZ, USA
e-mail: Nadia.El-Hage@asu.edu

H. Abouleish
SEKEM, Cairo, Egypt
e-mail: helmy.abouleish@sekem.com

U. Hurter
Section for Agriculture at the Goetheanum, Dornach, Switzerland
e-mail: ueli.hurter@goetheanum.ch

Co-leader Section for Agriculture at the Goetheanum - Freie Hochschule für Geisteswissenschaft & Executive Council of the General Anthroposophical Society, Dornach, Switzerland

J. Kronenberg and E. T. Lammerts van Bueren (eds.), *On the Earth We Want to Live*, World Sustainability Series, https://doi.org/10.1007/978-3-031-98758-8_11

Keywords

Biodynamic agriculture • Rudolf Steiner • Sustainable farming • Regenerative agriculture • Holistic agriculture • Farm organism • Soil vitality • True-cost accounting • Agroecology • Climate stewardship • Associative economy • Agricultural individuality • Cosmic rhythms • Food system transformation • Economy of love • SEKEM

1 Agriculture Through the Ages

Nadia El Hage

1.1 Introduction

In Latin, 'agri' means 'field' and 'culture' means 'cultivation'. However, the Latin word 'culture' also encompasses 'culture,' defined as 'the integrated pattern of human knowledge, belief and behaviour' (The New Encyclopedia Britanica, 1986, p. 784).

Thus, agriculture, in its various forms across different eras, is the reflection of the evolution of human consciousness. In evolutionary terms, evidence of self-awareness can be taken as evidence for consciousness, or whatever our intellect can make of our sensible observations of the physical world. Considering that humans have not only a material life but also a spiritual life, anthroposophy as foundation of Biodynamic Agriculture goes beyond in describing human consciousness as a quality that can grow clearer and clearer through imagination, inspiration, and intuition - all faculties that are given by a non-sensorial spiritual environment (Steiner, 1969 [1910]).

1.2 Agriculture 1.0: Neolitic Revolution (The Walking Development Phase)

1.2.1 Activating Nature's Vital Forces: Domestication

For nearly 200,000 years, humanity strived instinctively in Nature, interlinked with creative forces acting in the Earth and the cosmos. Ancient people looked up to the stars with a sense of sacredness, recognising prey, tracks, and weapons among the constellations or observing stars high in the sky during birth or migration seasons of animals. The seasonal regeneration of the elements—soil, water, air, and fire—sustained a nomadic lifestyle centred on gathering wild food and hunting. As the Earth evolved through the last great ice sheets retreat, forming fertile soils that hosted a rich world of plants and animals, humans began their journey of developing contemplative consciousness, as documented in the Vedic texts of ancient India (see e.g., Chand, 1984; Easwaran, 2007; Griffith, 1896; Whitney, 1905).

Although the cultivation of food was practiced sporadically 12,000 years ago, it became widely established only in the past 7000 years. The shift from palaeolithic diets to cultivating crops and raising animals for food has occurred independently in several parts of the world by capitalising on the hunter–gatherers' vast knowledge and use of wild plants; archaeobiology research identifies 24 different origin sites around the planet (Fuller & Denham, 2021). Food species developed under the guidance of priests who sought to transform the vital element by activating the forces that lead to fruiting and ripening of all plant organs (root, stem, gem, leave, flower, seed) and to transform animal souls. People also began to alter their environments as they cultivated, vivifying soils through irrigation and drainage, scratching land with plough and hoe, and transforming wild biotopes into cultivated landscapes and gardens.

Neolithic agriculture marked the transition from nomadism to sedentary life, transforming how people interacted with the environment and themselves. It suggested that natural limits could be overcome, bringing about a fundamental shift in the structural processes of human history and the nature of cognition (Bocquet-Appel, 2011).

1.2.2 Building Fences: Reverence Farming

In 529 AC, Benedict of Nursia, founder of the Benedict Order, sought to transform the Earth with his 'pray and work' motto for monks living communally, as they practiced all the arts of human labour, including growing their own food and raising their own animals. By a kind of (super)natural process, monasteries emerged as societal leaders, excelling not only in prayer and liturgy but also in agriculture, animal husbandry, medical care, architecture, music, and other aspects of a healthy and integrated human life. With the church at the centre of the communal territory, villages grew around it, reflecting the principle of the point and circle, analogous to the shaping principle of all organisms. From the seventh to the fifteenth century, this Western-Christian impulse introduced crop cultivation and animal husbandry within villages and parishes, creating an integrated and closed agricultural organism based on synergistic associations and crop rotations (Klett, 2021). However, the natural separation between agriculture and nomadic pastoralism persisted in vast grasslands, as seen in Eurasian steppe or the Great Rift Valley herding.

Roman law, which spanned over a thousand years of jurisprudence in Europe, Latin America, and parts of Africa and was continued by Byzantine law in the Near East, introduced the concept of 'rights' over physical objects. Roman Law introduced the doctrine of ownership and rights over 'things,' thus the possibility of personally disposing of the land, turning it into private property that is inherited or sold on the market. Romans did not consider the human body a thing, but they considered animals as things, in as much as animals could be owned and have a cash value. The geopolitical imperialistic Roman vision extended to agriculture, triggering the development of agricultural estates and food storage facilities, while slavery of the agricultural labour force was common (Buck, 1983).

In non-colonised territories, spirituality continued to guide agricultural practices, as witnessed by Indigenous Peoples reverence to Nature' spirit across continents. Indigenous Peoples manage the land through the lens of life, recognising the spirit and vitality of water, soil, plants, and animals, resulting in interactions that perpetuate natural systems, resilience, and continuity (FAO, 2021).

While rational thinking started with the Greco-Latins, Roman Law marked the transition from priestly wisdom to relationships regulated by human intelligence. By the fifteenth century, the world came to be conceived as external. In the sixteenth century in Europe, science introduced the concept of particles and the need to become objective observers, leaving aside the qualitative sensations of the natural world. Between the seventeenth and nineteenth centuries, famines and wars fuelled peasant migration to cities, undermining Central European agrarian culture, while mechanistic science gradually eroded spirituality in Western agriculture.

1.3 Agriculture 2.0: Industrial Revolution (The Speaking Development Phase)

1.3.1 Controlling Nature: Chem-Tech

The industrial revolution that started in 1760 coincides with what is dubbed the Anthropocene, or the Earth era marked by human activities (see e.g., Crutzen & Stoermer, 2000; Zalasiewicz et al., 2011). With the advent of fossil fuels, nitrogen was converted into chemical fertilisers via the Haber-Bosch process, field mechanisation and irrigation expanded rapidly, while thousands of toxic molecules were introduced into the environment through pesticides and herbicides, and high-yielding monocultures replaced integrated plant and animal systems. Initially, this development positively impacted the human population that grew five times faster than before. Today, the Green Revolution is recognised as a primary driver of ecosystem degradation, biodiversity loss, and climate change, while industrially processed foods are linked to the pandemic of non-communicable diseases (Daisy & Giridhara, 2021; Glaeser, 2009; Saha, 2023). *Industrial agriculture* marked human emancipation from Nature, as physical forces and substances were managed by technology. Atomistic science and the objective observation of Nature suffocated human experience and hence, the moral self-perception of peasants.

1.3.2 Creating Life: Gene-Tech

The turn of the millennium was marked by the advent of genomic agriculture that promised wonders. By introducing exogeneous genes or tweaking DNA, plants and animals with new traits are believed to accelerate adaptation to climate change, drought, or other natural stress, while also responding to new quests of novel foods. Over a quarter of a century has passed since genetically modified crops were introduced on less than 7% of agricultural lands by 2019 failing to deliver on their promises (AgbioInvestor, 2024; Potapov et al., 2022). Meanwhile, herbicide use has increased due to resistance build-up (Klein & Benbrook, 2024).

New genomic techniques, currently in the making, follow the same failing narrative but with increased risks of non-targeted mutations in plants and their progenies, let alone unknown human health hazards due to new toxins and allergens (Koller & Cieslak, 2023).

Most importantly, enhanced seeds are controlled by just seven firms, which dominate 71% of the global seed market, disenfranchising farmers, stripping consumers of their right to choose, and accelerating the alarming loss of biodiversity (ETC Group, 2015). Three-quarters of plant varieties have become extinct in the past century. The current one-dimensional search for single-factor solutions has replaced the agricultural organism concept, even in so-called 'integrated' management systems. *Genomic agriculture* goes beyond longing for free self-determination, as researchers experiment creating transgenic life. This further emancipation from Nature laws is yielding self-oblivion.

1.4 Agriculture 3.0: Digital Revolution (The Thinking Development Phase)

1.4.1 Farm-Less Culture: Digitalisation and Artificial Intelligence

With the pace of change accelerating, digital agriculture is rapidly developing, promising to replace farmers with self-driven tractors and artificial intelligence robotics designed to optimise resource use efficiency, while claiming to mitigate the negative externalities brought about by the Green Revolution. Dubbed 'precision agriculture,' this technological army crosses the entire supply chain, from smart machinery and Apps on the field, through GPS and drones over the landscape, cloud computing, food market blockchains, robotic delivery, and e-learning services (Ding et al., 2023). This technological arsenal utilising genetically modified plants and animals, artificial proteins, insects cultured in easily controlled tanks, and 3D-printed items clearly favours medium to large estates (Ceccarelli et al., 2022).

Rather than relying on traditional human knowledge passed down through generations, scientism and technology promise to free humans from work, while a few corporations detaining Big Data make profits by capitalising on the world population knowledge and emotions to create regenerative AI (Liu, 2024). By 2050, digital agriculture is expected to lead to the abandonment of 300 million farms, the forced migration of one billion people, and a radical transformation of peasant food systems, which currently feed 70% of humanity (IPES-Food & ETC Group, 2021). *Digital agriculture* replaces the division of labour of industrial agriculture with the emancipation of food production from farming and human labour, as laboratories and automation seek to produce 'foods.'

1.4.2 The Quest for Quality: Resurrection of Agriculture

While agriculture used to be the cultural foundation of civilisations, its future role in the development of people and nature is at a crossroad. Mainstream reductionist science, coupled with globalisation, has propagated the mechanistic metaphor—as opposed to the organism metaphor—within modern knowledge. This metaphor promotes agricultural and other systems that prioritise quantitative over qualitative processes, with constituent elements (practices and ideas) held together mechanically, often at the expense of openness and freedom. Clearly, agriculture is in a state of transition and all institutions today are talking about 'food system transformation,' though no agreement exists on what type of transformation is to be pursued (El-Hage Scialabba et al., 2023).

On the other hand, the failures of the last centuries to control Nature and the current attempts to control the global population thoughts and deeds is prompting the acceleration of individual consciousness, everywhere. Already in the nineteenth century, a group of farmers questioned Rudolf Steiner about agricultural clues to heal the land, giving birth to *biodynamic agriculture* (Pfeiffer, 1958) and the many forms of today's scattered pockets of agroecological agriculture around the globe.

Although atomistic science is still mainstream, the slow but steady development of quantum physics supports the demand for quality, not just quantity. In fact, several quantum phenomena, such as non-locality and entanglement, support the biodynamic concept of the farm organism, based on inter-connectedness and synchrony (as explained in the following section). Emerging insights into environmental and human microbiomes highlight nutritional prerequisites such as diverse, seasonal, and vital foods—concepts described by Rudolf Steiner (1924). These foods are produced in harmony with cosmic rhythms by farmers aware of Earth's evolutionary trajectory and their role in enriching soil and creating harmonious landscapes.

For true self-determination, the agri-cultural vocation and its many associative forms carries the seeds to reconfigure social life and give an impulse to renewal, for a future based on human dignity and the conservation of the natural means of production. This is where Biodynamic Agriculture has been contributing to this vision of sustainable farming for over 100 years, as elaborated upon in the sections below.

2 The Biodynamic Impulse

Ueli Hurter

2.1 Introduction

For 100 years, biodynamic farmers and gardeners, processors and traders, researchers and breeders have been among the pioneers of sustainable development (McKanan, 2017). They contribute globally through concepts such as the "farm organism," soil-based animal husbandry, innovative practices in plant breeding, networking within value chains, new forms of land ownership, vibrant food workshops, fertilisation using the preparations and an awareness of the stars, the collectively managed 'Demeter' brand, and unconventional research methods (Klett, 2021).

What is the source of this innovative strength, which inspires so many people and encompasses such diverse areas? Its origin lies in Rudolf Steiner's *Agricultural Course* (Steiner, 2022). Delivered in 1924 as a series of eight lectures to an audience of 130 people on the Koberwitz estate (now Kobierzyce) near Wrocław, this course can be regarded as the birthplace of a new form of agriculture (McKanan, 2017; Paull, 2011, 2019). The biodynamic movement as we know it today emerged from this foundation. For many participants, the course was a life-changing experience, leading them to dedicate themselves to this new 'art of farming'. A sense of enthusiasm carried these individuals, enabling them to pass on this impulse to subsequent generations, while deepening their understanding and broadening their experience over time. From this commitment to biodynamic agricultural culture, a century-long history has unfolded, spreading across the globe (Bind & Hurter, 2025).

The *Agricultural Course* had a profound influence not only on organic and ecological agriculture but also on the emerging science and practice of sustainability (Brocket al., 2019a, 2019b). It can be considered a significant source of inspiration for the engaged civil society of the twentieth century. One notable example is Rachel Carson, whose landmark book *Silent Spring* (1962), regarded as a milestone in the development of public ecological consciousness, was influenced through her acquaintance with biodynamic methods and Rudolf Steiner's course. This connection was made via her friend Marjorie Spock, an environmental activist and biodynamic gardener (McKanan, 2017).

2.2 Core Themes of Biodynamics

2.2.1 The Agricultural Wholeness

Understanding and shaping agriculture as a living whole is one of the fundamental principles of the biodynamic impulse. Rudolf Steiner speaks of the "farm organism" and even of agriculture as an individuality. Biodynamic farming views the farm organism as a productive interplay of soil, plants, and animals, shaped by the human being's work of cultivation (Steiner, 2022).

This foundational principle can be understood as a renaissance of pre-scientific and pre-industrial agriculture. Alternatively, it can be seen as the beginning of a rational eco-management system and a prototype for circular economy practices. The reality is that the traditional farm organism dissolved over the course of the twentieth century. When assessed according to business management criteria, it breaks down into its individual components. Only in specialised operations—such as dairy farms, arable farms, orchards, or vegetable farms—can the factors of labour, capital, and expertise be employed in a relatively profitable manner.

All the internal services that a mixed farm provides for its individual branches through mutual exchange—for example, farmyard manure from livestock farming, straw from arable farming, or health potential from a diverse landscape—disappear and must be purchased externally as operational inputs. Mineral fertilisers, chemical pesticides, herbicides, and industrial feed make their way in. Thus, the industrialisation of agriculture is carried out, and in many countries during the twentieth century, this became a widespread reality touted as "progress." This progress, however, is simultaneously the acute problem of industrial agriculture, with its ecological and social deficiencies. At this juncture, where the complexity of the agricultural sector is examined from ecological, social, and economic perspectives, the seemingly unprofitable mixed farm re-emerges as having numerous advantages (Klett, 2021).

The biodynamic movement has always understood itself as a space where the principle of agricultural wholeness is upheld and promoted in diverse ways, despite the real-world economic challenges. This includes the effort of each generation to re-engage with and deepen their understanding of this principle.

2.2.2 The Farm Organism

> If agriculture is understood as an organism—whether as a farm, a village, or a valley—it refers to a productive ecological system cultivated from the underlying nature. The model or prototype for this can be seen in natural organisms, particularly in mammals, where individual organs serve the whole, and the organism is the organising principle that transcends the individual organ systems. Similarly, in the agricultural organism, individual farm branches become the organs of the farm organism. This perspective opens a new understanding of each part, now seen as an organ that serves the whole and derives a significant part of its purpose from this whole. (Steiner, 2022)

For example, the dairy herd is not merely a source of income through milk sales; rather, the herd—and ruminants in general, which thrive on roughage—becomes the digestive system of the entire organism. The internal contribution of manure as a driver of sustainable soil fertility, the shaping of the landscape through grazing, or the structuring of the daily rhythm for the farm community through twice-daily milking becomes just as important as the income from milk production. Accordingly, breeding, husbandry, and feeding practices are developed and organised to integrate the dairy herd effectively as an organ within the organism.

The farmer or farming community primarily designs the farm organism as a whole, which is their central entrepreneurial task and achievement. Managing and cultivating the individual organs is secondary and must always follow the principles of the overarching organism. However, this orientation toward the whole cannot be taken as an absolute standard. The farm must also remain financially viable and adaptable to current market conditions.

The principle of the organism is self-contained; it maintains a certain degree of autonomy from its environment. This is achieved, first, through a high degree of internal diversity: biodiversity through landscape design, genetic diversity through adaptable plant varieties and animal breeds, and economic diversity through varied production, processing, and marketing. Second, it is achieved through a closed material cycle involving manure, soil, and feed. In biodynamic farming, it is self-evident that this is not an eternal cycle of inert matter but that substances are imbued with the characteristics of the beings from the natural kingdoms to which they belong. There are "quantum leaps" for substances as they move from the mineral soil to the living plant to the ensouled animal. This is particularly evident in the case of nitrogen, which has a completely different quality when it is released from an internal material cycle than when it is introduced externally in the form of highly soluble mineral fertilisers (Bind & Hurter, 2025).

For biodynamic practitioners, every transition of a substance from one natural kingdom to another is of special interest. This is especially true for the renewal process through humus in the soil. According to Steiner, the fertilisation question is the "most interesting question of all". At this point, he expands the scientific foundation with the concept of the "agricultural individuality" (Paull, 2011).

2.2.3 Individuality as a Concept of Agricultural Culture

Rudolf Steiner introduced the concept of "agricultural individuality" as a cultural idea in agriculture, breaking away from the confines of classical agronomy. In cultural studies, "individuality" refers to the potential of each human being to develop beyond collective affiliations and to live autonomously. A person with great individuality can influence their social context and time far beyond their immediate circumstances.

The human being as individuality now becomes a model for agricultural wholeness, elevating it beyond the concept of the organism. Each farm, each farming community, possesses a unique, multi-layered character, specific site qualities, and its own history, shaped by the particular influences of the current situation, location, surrounding environment, seasons, and interactions. This imbues biodynamic agriculture with a richness and diversity of individual expressions and practices.

A place developed in the sense of wholeness and nurtured over years—whether a farm, garden, park, or valley—recreates within itself all the elements that the encompassing natural world has produced. It is within this tension between the specific and the universal that the identity of a farm is rooted. The identity strength of the farm, which is becoming increasingly important even for its economic success, is an expression of its agricultural individuality.

Biodynamic agriculture is not a standardised system with rigidly prescribed methods or fixed guidelines set in stone. Rather, it is an invitation to engage in creative cultural work with and upon the Earth. In this, agriculture inherently seeks a productive engagement; it aims to and can serve as a producer of food and textile raw materials, of ecosystem services, and of socially enriching living spaces (Bind & Hurter, 2025; Klett, 2021).

2.2.4 Opening Moments for Future Forces

The stream of substances circulating within the farm reaches the animals, primarily the cows or other ruminants. These animals subject this feed to a "cosmic-qualitative analysis" during the process of ruminating (Steiner, 2022). This means that the feed is comprehensively perceived as it is eaten and digested by the ruminating cow. The result of this "analysis"—that is, what the animal experiences from the feed—is imparted back into the substance as it transitions in digestion from feed to manure.

The manure produced by the cow is, on the one hand, the same substance as the feed, but it is now fully imbued with the forces the cow has impregnated into it. When this manure is applied to the fields as fertiliser, it brings not only nutrients but also formative forces for the plants. In this sense, over the years, a process of self-encounter occurs, which signifies individualisation.

Emerging microbiome research shows that this process creates a farm-specific soil biome, often referred to as the "homefield advantage". This is critical for ensuring sustainable yields, the health of the farm, and its resilience.

Wherever a life cycle completes itself and a new one begins, there is an opening in the Earthly causal chain for future forces. In particular, the human act of closing the material cycle through the carefully managed farmyard manure represents such a moment of opening. The closed organism is not an end in itself; rather, it forms the foundation for an opening that allows future forces to flow in.

2.2.5 The Biodynamic Preparations

As a technique to introduce future forces into the natural causal chain, Rudolf Steiner introduced the preparations in his Agricultural Course (Steiner, 2022). These preparations act as bio-stimulators, stimulating and "dynamising" life processes. There are two spray preparations: the horn manure preparation and the horn silica preparation. They are complementary.

The horn manure preparation is sprayed onto the soil at sowing time. It supports the connection between the plant and the soil, particularly in root formation, and stimulates what winemakers refer to as "terroir." The horn silica preparation is sprayed onto the developing cultivated plant, aiding in leaf, flower, or fruit formation through an enhanced connection with light and warmth.

In addition, there are six compost preparations:

1. **Yarrow preparation** (*Achillea millefolium*), made from harvested blossoms prepared within the bladder of a stag.
2. **Chamomile preparation** (*Chamomilla officinalis*), made from blossoms prepared in the intestines of a cow.
3. **Nettle preparation** (*Urtica dioica*), prepared without an animal casing and buried in the soil for one year.
4. **Oak bark preparation** (*Quercus robur*), made from oak bark prepared in the skull of a domesticated animal.
5. **Dandelion preparation** (*Taraxacum officinale*), made from blossoms prepared in the mesentery of a cow.
6. **Valerian preparation** (*Valeriana officinalis*), made as a juice from pressed blossoms.

These compost preparations are added in minute quantities to farmyard compost, manure, and liquid manure. The treated fertiliser exhibits a harmonious decomposition process, and when the matured fertiliser is applied, it strengthens the cultivated plants in a comprehensive way. For instance, the oak bark preparation has a salutogenic effect, protecting plants from fungal diseases.

All the preparations can be made by farmers themselves and are applied independently on the farm. The knowledge of how to create and use these preparations is managed as open-source technology within the global biodynamic movement. The preparations are an essential tool for the independence of farmers, as they allow the fertility and resilience of the farm and its crops to develop year after year. This, in turn, enhances the quantitative and qualitative reliability of yields, which is the foundation of a farm's success (Brock et al., 2019a, 2019b).

2.3 Social Future Designs in the Biodynamic Movement

2.3.1 Three Generations of Farms Since the 1950s

After the Second World War, work in Germany had to be rebuilt from the ground up. In many other countries, national organisations were developed from the initial beginnings. Many farms underwent conversion, a process that, in each individual case, represented a comprehensive existential challenge for all those involved. Beyond the technical and agricultural demands, this was often a social challenge as well. Any deviation from the conservative, traditional conditions in villages or the rapidly advancing agricultural science and industry was considered heretical (Hurter, 2014).

In a second wave, during the 1970s and 1980s, many farms were established by young people from urban areas. The farms that could be taken over were often in challenging locations. In these cases, good agricultural practices often had to be learned through numerous mistakes. However, the social constraints of traditional agriculture were not present on these farms, which allowed diverse new social forms to emerge both within the farm communities and around the farms.

In a third wave, it is primarily highly specialised farms—mostly viticulture, fruit cultivation, and vegetable farming enterprises—that have integrated biodynamic methods into their practices. In many respects, for instance, viticulture farms are the opposite of the idealistically founded "dropout farms." They are monocultures with high capital investments, and the wine market is a highly specialised sector. Correspondingly, the people involved and their cultural and social contexts are also different. It was a significant challenge for the movement to integrate the new with the old, but this effort has been successful (Hurter, 2014).

Not all countries and regions have seen the development of these three generations of farms; this applies mainly to Northwestern European countries. In other areas, such as the Romance-speaking countries of Europe, the first generation is largely absent. In many countries in South America or in India, there is only one generation of biodynamic farms. Historically, the movement has spread radially across the world from a central point. However, from the perspective of engagement, every place where the biodynamic impulse is existentially practised becomes its own centre.

2.4 Scientific, Legal, and Economic Initiatives

In the second half of the twentieth century, organisations representing biodynamic agriculture in various countries made it a priority to scientifically document and "prove" biodynamic methods. This led to the establishment of dedicated research institutes, such as the Research Institute for Biodynamic Agriculture in Darmstadt, Germany (1950); the Institute of the Nordic Research Ring in Järna, Sweden (1956); the FiBL (Research Institute of Organic Agriculture) in Switzerland (1973); the Louis Bolk Institute in the Netherlands (1976); and the Michael Fields Agricultural Research Institute in the USA (1984). Collaboration with academic institutions also took place, and starting in 1973, the University of Giessen under Eduard von Boguslawski (1905–1999) saw the first dissertations on biodynamic topics.

A notable project is the DOK trial in Switzerland, initiated through political efforts and implemented as a long-term study by state research institutions in collaboration with FiBL. Since 1978, it has compared three farming systems: biodynamic (*D*), organic (*O*), and conventional (*K*) (Maeder et al., 2002; Rodas-Gaitan et al., 2022). This academically rigorous exploration and substantiation of biodynamics has been crucial for societal acceptance and the ability to engage in dialogue within today's knowledge-driven society. While it primarily serves as a foundation for societal recognition, it also plays a supporting role in the development and refinement of biodynamic methods.

In addition to formal research institutes, farmers organised into regional groups in many countries have always seen themselves as a research community. Together with individuals from various professional backgrounds, they have engaged intellectually and practically with advancing the biodynamic impulse. In some regions, such as Northern Germany, this approach has been particularly prominent, giving rise to well-organised and content-rich learning communities.

Starting in the late 1960s, many biodynamic farms became laboratories for new forms of social organisation. The issue of land ownership was addressed, and diverse models were developed to transfer farms into nonprofit stewardship. Long-term, transgenerational legal and financial responsibility for the land and buildings was often assigned to a legal entity committed to the biodynamic impulse. The transfer of ownership from the original family to the new steward was usually enacted as a gift. These stewards could be both the farm operators and members of broader society who viewed themselves as representatives of non-agricultural communities. Removing farms from the traditional inheritance system allowed farmers to establish freer, more collegial forms of collaboration. This led to the creation of numerous community farms and farms directly linked to other economic activities (Hurter, 2014).

During the same decades, a growing market for Demeter products emerged. Depending on the country, the promotion and protection of the Demeter brand were managed either by an independent Demeter association or by a Demeter commission within the biodynamic association. Regional partnerships with processing companies and consumer organisations were established, integrating questions of nutritional value into their work. The Demeter symbol, founded in 1928, is globally recognised as a representation of the universal biodynamic impulse, facilitating international trade flows.

In 1997, Demeter International was founded to coordinate branding policies among certifying Demeter associations worldwide. Today, there are around 300 export-oriented Demeter projects in over a dozen countries, mainly in the Global South. These projects are certified by the International Certification Office (ICO) and often serve as focal points for promoting regional and socially sustainable development.

In 2020, the Biodynamic Federation Demeter International (BFDI) was established through the merger of Demeter International and the International Biodynamic Association (IBDA). This unification brought together the certifying Demeter associations and biodynamic organisations into one global federation, creating strong foundations for growth. New associations have since emerged in Latin America, Africa, and Asia, with the biodynamic movement now active in over 50 countries worldwide, spanning all continents where agriculture is possible.

The Section for Agriculture at the Goetheanum has supported this global movement for over 100 years through research, education, and networking. Its annual Agricultural Conference, now attended by 800 participants from approximately 40 countries, is the longest-running specialised conference on sustainable agricultural culture.

In the 1980s, shops and wholesale businesses were established in many regions, often inspired by Steiner's ideas on social threefolding. Collaboration with individuals and institutions from the broader organic movement quickly emerged. Today, the organic retail market is structured differently across countries and is currently undergoing intense adaptation in the post-COVID era.

Over the past decade, Demeter has successfully entered the supermarket sector in its strongest markets. This expansion has significantly increased the trading volumes of certain products, enabling corresponding growth in production. There are also noteworthy partnerships for research and development projects, such as breeding and marketing dual-purpose chicken breeds. However, the movement aims to avoid submitting to the market power of large actors and instead seeks economically transparent and fair solutions for distributing costs and revenues across the value chain.

A new type of "marketing," which originated in biodynamic circles and is now spreading across many countries, is Community Supported Agriculture (CSA). In this model, a farm's annual budget is covered collectively by a community of consumers, who in return receive the farm's products.

2.5 The Biodynamic Movement and the Development of Organic Agriculture

Biodynamic agriculture has evolved both as a part of the anthroposophical movement and as a branch of the organic movement. The organic movement has various roots, one of which is the biodynamic impulse. In this sense, there were no alternative farming methods or corresponding movements in 1924 when biodynamics was founded (Paull, 2011). The book *An Agricultural Testament* by Sir Albert Howard (1873–1947) was not published until 1940, and Lady Eve Balfour (1898–1990) began her work in the 1930s. *Nature et Progrès* was only founded in 1964. Rudolf Steiner and his collaborators were, therefore, among the very first to explore new approaches to agriculture.

In Switzerland, the organic-biological movement emerged in the 1940s, clearly differentiating itself from biodynamic agriculture. Hans and Maria Müller (1891–1988 and 1894–1969), who led this movement, were well aware of the biodynamic approach but could not engage with its "dynamic" aspect—or, one might say, with anthroposophy. Consistently, they built their framework entirely on biology, providing it with a scientific foundation through Hans Peter Rusch (1906–1977) and his book *The Cycle of Living Substances (1980)* (Der Kreislauf der lebendigen Substanz). Reverence for life was also an essential inner motivation for these groups and a reason to reject the use of chemicals.

After periods of tension between the two groups, collaboration in Switzerland eventually materialised, particularly in the founding of FiBL (1973) and Bio Suisse (1981). As such, Steiner's insights and the contributions of representatives of the biodynamic impulse and the Demeter label have significantly shaped the development of today's organic movement.

Even now, the biodynamic movement continues to pioneer essential questions that later become critical across the entire organic world, such as plant and animal breeding, education and research, and socio-economic issues.

3 An Agriculture of the Future

Helmy Abouleish

Dr. Ibrahim Abouleish (1937–2017) transformed the Egyptian desert into a blooming oasis, contributing to the potential unfolding of thousands of people. When thinking about the future of agriculture, the development of people and their interconnection with Nature cannot be but at the core of considerations.

As discussed in the previous sections, taking the term 'agriculture' literally leads us to an understanding that extends far beyond the structure of most agricultural practices today. An agriculture of the future is about more than treating soils or protecting the environment. Agriculture can only be sustainable if it represents a meaningful co-creation between people and Nature—a true cultural act—and if it incorporates the entire farm organism and the full value chain of farming activities.

What kind of farming enables us to interconnect with Nature, rather than merely interacting with it? What kind of challenges does agriculture need to handle? And how can a true agriculture transition from a niche practice to a mainstream method?

3.1 From a Climate Polluter to a Climate Steward

Over the past 200 years, industrial agriculture, land use change, and forestry have largely contributed to climate change. Global emissions from the agrifood system represented 31% of anthropogenic emissions in 2020 (Food and Agriculture Organisation of the United Nations, 2022). Besides, conventional farming is significantly contributing to biodiversity loss due to its intensive use of chemical fertilisers and pesticides. It does deplete our soils and hence, contributes to food insecurity on the long run. In fact, it guarantees higher quantities, rather than sustainable practices, often giving the impression that mass production results in huge yields. Farmers who directly interact with soils, and stand at the very beginning of the value chain, receive the smallest part of the added value. They are ever more under pressure to produce with the easiest and cheapest conditions to survive (Czyżewski et al., 2018; Sustain, 2022). Similarly, processors focus on working with the smallest possible investment. Nowadays, the agrifood system is far from functioning as a value chain that interacts with Nature and forms the foundation of our existence.

However, these current challenges can be transformed into opportunities if we turn around and think about what kind of solutions regenerative agriculture offers and especially, how sustainable agricultural practices can become mainstream. The 100-year-old biodynamic philosophy and practical approaches certainly still represent a comprehensive guideline on how to interact with Nature and produce food sustainably. Rudolf Steiner's (2029 [1924]) Agricultural Course teaches us that the farm is not merely a factory for food production but a living organism that interconnects with humans in the same way as with animals and all natural elements. It recognises the interconnectedness of all life and the necessity of working with natural rhythms, rather than against them. Biodynamic practices, such as the use of specific preparations and the timing of planting according to lunar cycles, are designed to enhance the vitality of the Earth, ensuring that agricultural productivity is achieved without depleting the land's inherent life forces. By avoiding chemical fertilisers and pesticides, building healthy soils, planting trees, and utilising renewable energy, greenhouse gas emissions are reduced, water is conserved, biodiversity is promoted, and healthy food is produced for human wellbeing.

The fact that biodynamic agriculture currently offers the most productive and efficient solutions for current and future challenges is proven by scientific evidence (see e.g., Brock et al., 2019a, 2019b; Santoni et al., 2022). The most well-known study is the DOK trial, conducted under the supervision of the Swiss Research Institute of Organic Agriculture (FiBL). Established in 1978, the DOK trial compares the cultivation of different crops under conventional, organic and biodynamic conditions (Research Institute of Organic Agriculture, n.d.). As a long-term comparative study, the DOK trial demonstrates that biodynamic farming, which uses significantly less fertiliser and energy, has the lowest impact on climate change while maintaining the highest soil vitality (Research Institute of Organic Agriculture, n.d.). This one example clearly shows how biodynamic agriculture provides the best approach for an agriculture that corresponds to future requirements, such as being climate-friendly.

The Intergovernmental Panel on Climate Change (2019) estimates that global croplands and grasslands could sequester between 0.4 and 8.6 gigatons of CO_2-equivalent per year if managed sustainably. Planting trees on the agricultural land in practices such as Agroforestry offers further potential and is required for a healthy biodynamic farm. Trees sequester additional carbon, contributing to global CO_2 savings alongside other ecosystem services provided by forests. Furthermore, the elimination of artificial fertilisers and pesticides in biodynamic agriculture contributes to mitigating climate change, as the production of nitrogen fertiliser, along with N_2O emissions on farm contribute to over 1% of global greenhouse gas emissions (or 10% of agricultural emissions). Nitrous oxide emissions are up to 300 times more potent and more damaging to the climate than CO_2, causing severe damage to groundwater and air (Greenhouse Gas Protocol, 2024).

Cumulative savings from enhanced soil carbon sequestration, on-farm tree planting, and the elimination of fossil fuel-based inputs such as fertilisers and pesticides could, through a global shift to biodynamic agriculture, result in climate-neutral farming and prevent nearly half of global greenhouse gas emissions (Smith, 2016; Nair & Nair, 2014; Gattinger et al., 2012; Regenerative Agriculture Initiative, 2020; Climate Action Tracker, 2020). In the face of this huge potential, it is highlighted that currently, only 2% of the worlds' agricultural lands are certified organic, a small pocket that includes biodynamic farming (Willer et al., 2021). How can we overcome this discrepancy?

3.2 The True Costs of Agriculture

Integrating biodynamic management into mainstream agriculture requires highlighting its contribution to cheapest and best quality food production. This can be unveiled by applying true-cost accounting methods, as applied in Egypt to compare organic/biodynamic and conventional crops. The study results clearly show that production costs for biodynamic crops are significantly lower than those of conventional crops (Seada et al., 2016). If externalised costs—currently borne by society or future generations—were reflected in market prices, their impact would become evident to everyone. The research concludes that organic/biodynamic agriculture enables a reduction of the environmental and health damage costs, and therefore, results in better cost effectiveness and profitability in the long-term (Seada et al., 2016). These findings align with the 2023 Food and Agriculture Organization of the United Nations (FAO) (2023) report, which estimated the hidden costs of the global agrifood system at 12.7 trillion USD. Agrifood systems considers the journey of food from the growing and harvesting via processing, packaging, distributing until consuming and disposing. According to the FAO report, over 70% of hidden costs are health-related, while environmental costs—primarily associated with greenhouse gas and nitrogen emissions—account for 20% and are nearly equivalent to one-third of agricultural value added.

Such data clearly reflects that a holistic farming system is needed to consider not only the short-term repletion of people's stomachs by calories but to consider the human as a holistic being, along with the surrounding environment and source of life. How can these positive impacts be translated into reality, empowering people to act accordingly?

3.3 Heading Towards the Mainstream

As previously mentioned, biodynamic soils sequester significant amounts of carbon, offering a tangible way to address climate change, biodiversity loss, and soil erosion. Accordingly, one concrete step that could be undertaken at the moment is to motivate farmers, as much as possible, to apply biodynamic standards in order to change their role from polluters to stewards and co-creators. This could be achieved by providing incentives to farmers for their efforts in reducing greenhouse gas emissions, recognising their contributions as an essential environmental service. With such additional income, producers can access the market with lower prices and be competitive with conventional products. In doing so, biodynamic farmers would be empowered with higher incomes, improved livelihoods, while providing healthy food for all.

In Egypt, more than 13,000 farmers have already been reached thanks to the Economy of Love Carbon Credits (see Chap. 5.5). However, it is important to note that this is only one stepping stone towards future agriculture. Eventually, holistic agricultural systems will be established, eliminating the need for additional payments through carbon credits. This shift will occur when most consumers recognise agriculture not only as a source of food but also as a system that sustains the environment and fosters human potential. If the true costs of agricultural products were reflected in agricultural policies and producers were paid fair prices, sustainable agriculture would naturally become the most cost-effective and efficient method of food production (Food & Agriculture Organization of the United Nations, 2018; Institute for Agriculture & Trade Policy, 2017).

3.4 100% Transparency Along the Whole Value Chain

While in the past centuries, the agrifood system involved a relatively small circle of people, the contemporary agrifood value chain is wide and comprehensive. Hence, the challenge is to include everyone with the same interconnectedness: humans to humans, and humans to nature. As human beings, we have the chance to be aware of Nature as an evolving organism with which we co-evolve.

Farmers, who stand at the beginning of value creation, receive the smallest share of production value and often face pressure to prioritise high yields and efficiency. Similarly, other actors down the food value chain lose awareness of their connection to Earth. At last, the end consumer is even further removed from feeling connected to Nature or being part of a holistic organism. To address this, transparency helps to reconnect people to Earth again. To this end, Rudolf Steiner proposed an associative economy as a means for producers and consumers to interact and foster mutual awareness.

All those involved in the food value chain need to be taken into account, starting from the farmer but going as far as the trader who supplies the raw and packaging materials for a product, as well as to the forestry worker who cuts down the tree for the paper from which a package is made, or the driver who transports the paper. Transparency from start to finish needs to be ensured, not least because only then people can make an informed purchasing decision.

In a nutshell, this means that an agriculture of the future needs to enable that four basic questions can be answered: (i) what is the impact of the product and its production on the social environment? (ii) what is the impact of the product and its production on the natural environment? (iii) what is the impact of the product and its production on the development and potential of people? (iv) what is the true price of the product?

These four questions can be answered satisfactorily only when enough transparency will be created. Consequently, care could also be taken to ensure that all activities along the food value chain do not harm the environment and people, but instead, help them.

When looking at current standards or efforts to improve farmers or processors conditions, it becomes obvious that there are some attempts to consider environmental or social aspects. However, the important the cultural dimension is not taken into account, when it is precisely cultural development of both humans and Nature that is crucial for a sustainable future. In addition to a healthy social system, individuals require cultural inputs to support their personal development. This is essential not only for farmers but for everyone involved in the food value chain. The opportunity for lifelong learning must be at the core of all productive activities.

3.5 Interconnected by Love

How can we implement and ensure such cultural development opportunities? The answer may seem cliché or overused, but it is simple: it's all about love. That was one reason for naming the new Egyptian standard *Economy of Love* (see Chap. 5.5). If we, Earth inhabitants, understand ourselves as part of a holistic organism, a feeling of love to everyone and everything within this organism guides us in all our dealings.

What does this mean with regard to agriculture? Besides treating the soils, plants and animals with love, handling further parts of the agrifood system requires love as well. For instance, the processer needs to receive a fair part of the added value, as well as the trader. And they need to have the same chance to learn and develop their individual potential. Finally, love for the consumer is required, which translates to offering a high-quality product that nourishes not only the body, but also the soul and spirit.

With such round the clock consciousness, agriculture becomes an act of co-creation, rather than mere production management. Processing and marketing become means for people to support one another in nourishing their minds and souls. Consuming nutritious and vital foods empowers body and mind for potential unfolding.

3.6 Conclusion

Human civilisation has experienced a rapid evolution in its relationship with Earth and Nature. From being embedded in Nature, to hunting and gathering, to cultivated agriculture, humanity now faces the challenge of industrial large-scale monocultures, which undermine long-term food security, climate stability, and the prosperity of both humanity and Earth.

The organic agriculture movement, though only about 100 years old, has significant potential as a solution to many interlinked challenges in social, ecological, and economic sustainability. Biodynamic agriculture goes a step beyond organic farming, not only by emphasising the avoidance of chemical inputs and monoculture practices but also by embracing the philosophy that human beings are an inherent part of Nature, with a responsibility to co-create Earth's evolution. Thus, biodynamic agriculture advocates not only for regenerating but for renewing practices, moving beyond protection to stewardship and the co-creation of Earth as a living organism. Human, Earth, and cosmic rhythms are aligned, and the act of culture is emphasised in the practice of agriculture as a guiding vision for a sustainable world.

References

Part 1 (Dr. Nadia El-Hage)

AgbioInvestor. (2024, February). *Global GM crop area review*. AgbioInvestor.

Bocquet-Appel, J.-P. (2011). When the world's population took off: The springboard of the Neolithic demographic transition. *Science, 333*(6042), 560–561. https://doi.org/10.1126/science.1208880

Buck, R. (1983). Agriculture and agricultural practice in Roman Law. In *Historia – Einzelschriften* (Vol. 45). Franz Steiner Verlag. ISBN 878-3-515-04040-2.

Ceccarelli, T., Chauhan, A., Rambaldi, G., Kumar, I., Cappello, C., Janssen, S., & McCampbell, M. (2022). Leveraging automation and digitalization for precision agriculture: Evidence from the case studies. Background paper for *The State of Food and Agriculture 2022*. FAO Agricultural Development Economics Technical Study No. 24. Food and Agriculture Organisation of the United Nations. https://doi.org/10.4060/cc2912en

Chand, R. R. (Trans.). (1984). *Samaveda: Sanskrit text with English translation.* Bharatiya Vidya Prakashan.

Crutzen, P. J., & Stoermer, E. F. (2000). The Anthropocene. *Global Change Newsletter, 41*, 17–18. https://www.igbp.net

Daisy, J., & Giridhara, B. (2021). Lessons from the aftermaths of Green Revolution on food system and health. *Frontiers in Sustainable Food Systems, 5*, 644559. https://doi.org/10.3389/fsufs.2021.644559

Ding, H., Tian, J., Yu, W., Wilson, D. I., Young, B. R., Cui, X., Xin, X., Wang, Z., & Li, W. (2023). The application of artificial intelligence and big data in the food industry. *Foods, 12*(24), 4511. https://doi.org/10.3390/foods12244511

Easwaran, E. (Trans.). (2007). *The Bhagavad Gita* (2nd ed.). Nilgiri Press.

El-Hage Scialabba, N., Zitterbarth, S., & Shah, T. M. (2023). *State of the debate on agri-food systems transformation.* In *FORESEE (4C)—The transformation of agri-food systems in times of multiple crises (4Cs: Climate, COVID-19, conflict, cost of externalities)* (Report No. 2). TMG—Think Tank for Sustainability. https://doi.org/10.48565/foresee.2023.2

ETC Group. (2015, December). Breaking bad: Big Ag mega-mergers in play. *ETC Group Communiqué*, 115. https://www.etcgroup.org/content/breaking-bad-big-ag-mega-mergers-play

FAO. (2021). The White/Wiphala paper on indigenous peoples' food systems. *Rome.* https://doi.org/10.4060/cb4932en

Fuller, D. Q., & Denham, T. (2021). Coevolution in the arable battlefield: Pathways to crop domestication, cultural practices, and parasitic domesticoids. In T. R. Schultz, R. Gawne, & P. N. Peregrine (Eds.), *The convergent evolution of agriculture in humans and insects* (pp. 177–208). MIT Press.

Glaeser, B. (2009). The Green Revolution revisited: Critique and alternatives. *Routledge.* https://doi.org/10.4324/9780203840443

Griffith, R. T. H. (Trans.). (1896). *The Rig Veda: An anthology of one hundred and eight hymns.* E.J. Lazarus.

IPES-Food & ETC Group. (2021). *The long food movement: Transforming food systems by 2045.* International Panel of Experts on Sustainable Food Systems.

Klein, K., & Benbrook, C. (2024, March 13). *Comments submitted to the USMCA genetically engineered corn dispute resolution tribunal.* Friends of the Earth. https://foe.org/wp-content/uploads/2024/03/Friends-of-the-Earth-USMCA-Mexico-Corn-comments.pdf

Klett, M. (2021). *Von der Agrartechnologie zur Landbaukunst: Wesenszüge des biologisch-dynamischen Landbaus.* Verlag am Goetheanum.

Koller, F., & Cieslak, M. (2023). A perspective from the EU: Unintended genetic changes in plants caused by NGT—their relevance for a comprehensive molecular characterisation and risk assessment. *Frontiers in Bioengineering and Biotechnology, 11*, 1276226. https://doi.org/10.3389/fbioe.2023.1276226

Liu, F. (2024). Artificial intelligence in emotion quantification: A prospective overview. *CAAI Artificial Intelligence Research, 3*, 9150040. https://doi.org/10.26599/AIR.2024.9150040

The New Encyclopædia Britannica. (1986). *Micropædia: Ready reference* (15th ed., Vol. 3). Encyclopædia Britannica, Inc.

Pfeiffer, E. (1958). Preface. In R. Steiner (Ed.), *Agriculture course: The birth of the biodynamic method* (G. Adams, Trans.). Rudolf Steiner Press (Original work published 1924).

Potapov, P., Turubanova, S., Hansen, M. C., Tyukavina, A., Zalles, V., Song, X.-P., Khan, A., Pickens, A., Shen, Q., & Cortez, J. (2022). Global maps of cropland extent and change show accelerated cropland expansion in the twenty-first century. *Nature Food, 3*(1), 19–28. https://doi.org/10.1038/s43016-021-00429-z

Saha, M. (2023). Left behind by the Green Revolution. *Issues in Science and Technology, 39*(4), 90–92. https://doi.org/10.58875/ESUJ6182

Steiner, R. (1924). *Nutrition and health* (GA 354). Rudolf Steiner Archive. https://rsarchive.org/Lectures/GA354/English/AP1987/NutHlt_index.html

Steiner, R. (1969). *Occult science: An outline* (G. Adams & M. Adams, Trans.). Rudolf Steiner Press (Original work published 1910).

Whitney, W. D. (Trans.). (1905). *Atharva Veda Samhita.* Harvard University Press.

Zalasiewicz, J., Williams, M., Haywood, A., & Ellis, M. (2011). The Anthropocene: A new epoch of geological time? *Philosophical Transactions of the Royal Society A: Mathematical, Physical and Engineering Sciences, 369*(1938), 835–841. https://doi.org/10.1098/rsta.2010.0339

Part 2 (Ueli Hurter)

Bind, R., & Hurter, U. (2025). *From the agriculture course to sustainable farming: 100 years of the biodynamic movement*. Floris Books.

Brock, C., Geier, U., Greiner, R., Olbrich-Majer, M., & Fritz, J. (2019a). Research in biodynamic food and farming—A review. *Open Agriculture, 4*(1), 743–757.

Carson, R. (1962). *Silent spring.* Penguin Books.

Hurter, U. (Ed.). (2014). *Agriculture for the future: Biodynamic agriculture today. 90 years since Koberwitz.* Verlag am Goetheanum

Klett, M. (2021). *Von der Agrartechnologie zur Landbaukunst: Wesenszüge des biologisch-dynamischen Landbaus.* Verlag am Goetheanum.

Mäder, P., Fliessbach, A., Dubois, D., Gunst, L., Fried, P., & Niggli, U. (2002). *Soil fertility and biodiversity in organic farming.*

McKanan, D. (2017). *Eco-alchemy: Anthroposophy and the history of nature.* University of California Press.

Paull, J. (2011). *From anthroposophic farming to organic agriculture (1924–1940).* https://considera.org/wpwp/wp-content/uploads/2021/02/The_Pioneers_of_Biodynamics_in_Great_Britain-from-AS-to-organic-ag-1924-to-1940.pdf

Paull, J. (2019). *The pioneers of biodynamics in the USA: The early milestones of organic agriculture in the United States.* https://orgprints.org/35567/

Rodas-Gaitan, H., Fritz, J., Dahn, C., Köpke, U., & Joergensen, R. G. (2022). Biodynamic compost effects on soil parameters in a 27-year long-term field experiment. *Chemical and Biological Technologies in Agriculture, 9*, 74. https://doi.org/10.1186/s40538-022-00344-w

Steiner, R. (2022). Landwirtschaftlicher Kurs: Geisteswissenschaftliche Grundlagen zum Gedeihen der Landwirtschaft (9. vollständig überarbeitete und erweiterte Neuauflage). Rudolf Steiner Verlag.

Part 3 (Helmy Abouleish)

Brock, C., Geier, U., Greiner, R., Olbrich-Majer, M., & Fritz, J. (2019b). Research in biodynamic food and farming—A review. *Open Agriculture, 4*(1), 743–757. https://doi.org/10.1515/opag-2019-0064

Climate Action Tracker. (2020). *The role of agriculture in climate solutions.*

Czyżewski, B., Matuszczak, A., & Miśkiewicz, R. (2018). Public goods versus the farm price-cost squeeze: Shaping the sustainability of the EU's Common Agricultural Policy. *Technological and Economic Development of Economy, 25*(1), 82–102. https://doi.org/10.3846/tede.2019.7449

Food and Agriculture Organization of the United Nations. (2018). The state of food and agriculture 2018: Migration, agriculture and rural development. *FAO.* https://doi.org/10.4060/I9549EN

Food and Agriculture Organization of the United Nations. (2022). *Greenhouse gas emissions from agrifood systems: Global, regional, and country trends, 2000–2020* (FAOSTAT Analytical Brief 50). FAO.

Food and Agriculture Organization of the United Nations. (2023). The state of food and agriculture 2023—Revealing the true cost of food to transform agrifood systems. *FAO.* https://doi.org/10.4060/cc7724en

Gattinger, A., Jawtusch, J., Muller, A., & Mäder, P. (2012). No-till agriculture—A climate smart solution? *Nature Climate Change, 2*(1), 10–12.

Greenhouse Gas Protocol. (2024). *IPCC global warming potential values.* https://ghgprotocol.org/sites/default/files/2024-08/Global-Warming-Potential-Values%20%28August%202024%29.pdf

Institute for Agriculture and Trade Policy. (2017). *The true cost of food: How the current food system undermines our health and environment.* IATP. https://www.iatp.org/documents/true-cost-food

Nair, P. K. R., & Nair, V. D. (2014). Carbon sequestration in agroforestry systems. *Environmental Science & Policy, 38*, 204–211.

Regenerative Agriculture Initiative. (2020). *Biodynamic agriculture: A pathway to climate neutrality.*

Research Institute of Organic Agriculture. (n.d.). *DOK experiment: Long-term study on biodynamic, bio-organic and conventional farming systems.* FiBL. Retrieved January 11, 2025, from https://www.fibl.org/en/themes/projectdatabase/projectitem/project/404

Santoni, M., Ferretti, L., Migliorini, P., Vazzana, C., & Pacini, G. C. (2022). A review of scientific research on biodynamic agriculture. *Organic Agriculture, 12*, 373–396. https://doi.org/10.1007/s13165-022-00394-2

Seada, T., Mohamed, R., Fletscher, T., Abouleish, H., & Abouleish-Boes, M. (2016). *The future of agriculture in Egypt: Comparative study of organic and conventional food production systems in Egypt* (Version 1.0). Carbon Footprint Center, Heliopolis University for Sustainable Development.

Smith, P. (2016). Soil carbon sequestration and biochar as negative emission technologies. *Global Change Biology, 22*(3), 1315–1324.

Sustain. (2022). *Unpicking food prices: Where does your food pound go, and why do farmers get so little?* University of Portsmouth. https://www.sustainweb.org/reports/dec22-unpicking-food-prices/

Willer, H., Trávníček, J., Meier, C., & Schlatter, B. (Eds.). (2021). *The world of organic agriculture: Statistics and emerging trends 2021.* Research Institute of Organic Agriculture (FiBL) and IFOAM—Organics International.

The Food Organism—Perspectives on a Healthy Food System

Jasmin Peschke

Abstract

Sustainable development encompasses various aspects of nutrition, from global food security to individual diets, as well as the strengthening of willpower through nutrition. Additionally, a plant-based and diverse diet can reduce diet-related diseases and significantly lower greenhouse gas emissions. The foundation for this lies in healthy foods that grow in healthy soil, are processed and traded in a healthy environment, and possess a rich microbiome. This microbiome stimulates the microbiome of consumers, fostering resilience. Achieving the necessary transformation of the food system requires value-based action from all stakeholders and the cultivation of relationships. This begins with the individual, who hones their capabilities accordingly. In this way, the food system evolves into a food organism, where its actors function as independent organs, collaborating as equals to achieve the shared goal of ensuring a sustainable future for humanity and the Earth.

Keywords

Food organism • Sustainable nutrition • Biodynamic agriculture • Food systems • Holistic nutrition • Regenerative food • Gut microbiome • Sustainable diet • True cost accounting • Planetary health diet • Social sustainability • Food sovereignty • Mindful eating • Nutritional competence • Interconnectedness

J. Peschke (✉)
Section for Agriculture at the Goetheanum, Dornach, Switzerland
e-mail: jasmin.peschke@goetheanum.ch

J. Kronenberg and E. T. Lammerts van Bueren (eds.), *On the Earth We Want to Live*, World Sustainability Series, https://doi.org/10.1007/978-3-031-98758-8_12

1 Introduction

Until a few years ago, nutritional science confined itself to the consideration of nutrients, adopting a reductionist perspective that disregarded the food system as a whole. However, there has been a shift towards a systemic approach that increasingly examines the interconnections between food production, nutrition and health. A diverse diet is intrinsically linked to diversity in production and is already a building block for a varied, health-promoting diet and Sustainable Development. Nutrition and Sustainable Development are tightly interwoven. In fact, transitioning the diets of the wealthiest 20% of the global population to plant-based, sustainable diets could reduce more than 50% of climate and environmental damage (Tian et al., 2024). Nutrition, in combination with agriculture, can therefore be considered one of the most significant levers for Sustainable Development.

Unsustainable agricultural practices, which have caused soil depletion and declining product quality, were the impetus for Rudolf Steiner's lecture series on establishing the principles of biodynamic agriculture in 1924. In these lectures, Steiner explored the interconnections between cultivation, food quality and nutrition. He argued that food should not merely deposit nutrients somewhere in the body but should instead serve to invigorate it. From this foundation, he developed a new understanding of plant, animal and human nutrition (Steiner, 2022). Biodynamic agriculture and nutrition can thus be seen as a cohesive form of Sustainable Development. Here, development is aligned with ecological, social and economic dimensions, while placing the individual human being at the centre as the primary agent of change.

When people think of sustainability, few associate it with the retention of employees in a company. However, when employee tenure is long-lasting, this reflects a sustainable organisational culture. This is referred to as social sustainability. Social sustainability also encompasses the relationship between nutrition and health, which cannot be separated from ecological and economic sustainability. Exploring the concept of sustainable nutrition reveals a wide array of aspects shaped by the focus of the analysis and the underlying worldview. Human lifestyles, accelerated by global crises, are leading to increasingly poor societal health, which is inherently unsustainable. Moreover, the unequal global distribution of food means that many communities are unable to maintain a sustainable lifestyle or preserve their health.

A defining feature of these crises is that they are human-made, yet accountability is seldom assumed. This reflects a lack of connection and relationships. Active engagement with nature or the Earth is rare (Eisenstein, 2021; Homans, 2022). Similarly, there is little reflection on the consequences of one's decisions. For example, when eggs are consumed daily, people often fail to question their source, how the hens are raised and fed, or what happens to male chicks. Such considerations are largely overlooked, often due to convenience. Furthermore, many people rely on technological means to maintain interpersonal relationships, which remain distant as a result. Awareness of one's own genuine needs is also limited. This disconnection leads to a disregard for the consequences of personal actions, resulting in the exploitation of resources and unhealthy conditions at all levels.

The following sections will discuss various aspects of nutrition that are connected to Sustainable Development. Additionally, a concept will be introduced for reimagining nutrition and the food system in a way that ensures sustainability for all involved, including the Earth, plants and animals. Through such transformation, it is possible to preserve the resources needed for a healthy life, contributing to a sustainable future for all living beings on Earth.

2 Nutrition and Health—A Question of Relationship

Nutrition and health are directly interconnected. A balanced and varied diet made up of fresh, organic or biodynamic foods, freshly prepared and attractively presented, provides the best foundation for staying healthy. Recent dietary recommendations by the German Nutrition Society (DGE) go beyond mere nutrient recommendations (DGE, 2024). As early as 2015, the Brazilian Ministry of Health, in its Dietary Guidelines for the Brazilian Population, advocated for fresh foods and communal eating. This already points to a sustainable form of nutrition that extends beyond nutrient supply.

In reality, however, malnutrition is steadily on the rise. Malnutrition refers, on the one hand, to insufficient food availability, leading to hunger. Currently, up to 733 million people (FAO, 2024) suffer from a lack of adequate food through no fault of their own, and this number increases annually. The causes include prolonged conflicts, recurring extreme weather events or economic difficulties. Economically disadvantaged and vulnerable populations are particularly affected, many of whom live in small-scale farming systems. Additionally, over 2.33 billion people worldwide cannot afford a healthy diet (FAO, n.d.). This has a direct impact on social sustainability, encompassing health and the right to sufficient and nutritious food.

On the other hand, overconsumption is prevalent, with diet-related illnesses such as cardiovascular diseases, Type 2 diabetes and obesity continuing to rise. These so-called "diseases of civilisation" highlight how unhealthy society's way of life has become. According to the World Obesity Atlas (2024), by 2035, 54% of the global adult population will be overweight (BMI 25–30) or even obese (BMI $\geq$ 30), increasing from 0.81 billion in 2020 to 1.53 billion in 2035. The Atlas also identifies a clear link between the greenhouse gas emissions associated with a country's development and rising obesity rates—both in the Global South and the Global North—underscoring the interconnectedness of environmental factors and obesity. In this sense, current dietary practices are unsustainable for health preservation. The more affluent a society becomes, the more it distances itself from the origins of food production, consuming increasingly processed and ultra-processed products. This, in turn, leads to a higher prevalence of obesity, overweight and other illnesses (Kliemann et al., 2023).

Despite the wealth of knowledge about nutrition, this knowledge is poorly implemented, as evidenced by the inability to halt or reverse these negative trends, let alone bring about behavioural change. The underlying issue lies in the lack of relationship. The growing divide in malnutrition results in increasingly unhealthy conditions and diseases. This situation, characterised by both hunger and overnutrition, is unsustainable for health and the overall development of all life.

The specific nutritional situation in Central Europe reflects shifting food trends, which in themselves are unsustainable due to their lack of consistency. A few years ago, coconut oil was the "superfood," added to everything, but this trend has since been replaced multiple times by other products. Currently, the focus is on protein. A wide range of enriched products, from protein cereals and bread to protein shakes, is marketed. Yet healthy consumers in this region, who generally do not suffer from protein deficiencies, derive no benefit from these additions. On the contrary, excessive protein consumption can burden health (DGE, 2021). The primary beneficiary of this trend is the food industry, which, at least temporarily, creates a new product segment.

A more established trend is veganism. This dietary choice eliminates all animal-derived products, such as milk, cheese, eggs, and meat, from the diet. Strict vegans also avoid wearing leather shoes and replace wool sweaters with cotton or microfibre clothing. Even natural cosmetic products containing beeswax as a consistency agent are avoided (The Vegan Society, 2024).

Many vegan products fall into the category of ultra-processed foods manufactured on an industrial scale, consuming energy, raw materials and transport resources. All ultra-processed foods require scrutiny regarding their sustainability. On the one hand, it is essential to evaluate whether they conserve resources and contribute positively to the future of life on Earth. On the other, their health effects are equally significant and directly impact social sustainability. Recent statistics indicate that in Germany, 50% of daily caloric intake comes from ultra-processed foods (DGE, 2023), including soft drinks, ready meals, sweets and snacks. Such products are unsustainable as they fail to contribute to personal health maintenance. Increased consumption of ultra-processed foods correlates with a rise in diet-related illnesses (DGE, 2023; Kliemann et al., 2023). As early as 2010, Wallinga warned that subsidising ultra-processed foods has harmful health effects.

Hanni Rützler (2024a, 2024b), a food trend expert, describes a clear trend towards more sustainable nutrition. Increasing emphasis is placed on traceable enjoyment with a clear conscience, meaning climate-friendliness and sustainability are shaping dietary behaviour. This includes plant-based diets, "cultured meat" products, and reducing food waste through recycling leftover food. Terms like "regenerative food" are being introduced to highlight the importance of producing food through regenerative agriculture. The origin of food thus contributes to its quality and enjoyment, bridging the disconnection described by Homans (2022).

Building and maintaining relationships is key to better health and greater sustainability in nutrition. Relationships with others and the environment can only thrive if one has a relationship with oneself. This foundation can be cultivated and developed, including the development of personal nutritional competence. Such competence fosters independence from recipes, diet apps and influencers. It enables individuals to align with their own values and live them.

Mindful eating, as described by Chozen Bays (2017), intuitive eating according to Ehrlichmann (2017), and the "three questions" presented by Peschke (2021) are simple methods to establish a relationship with oneself. These can be easily integrated into daily life. By starting with the sensory perception of what is on the plate and how it tastes, individuals can nurture their relationship with themselves and better understand their own needs. Something particularly delicious may spark curiosity about its origin. For example, eggs from a neighbour's hens taste better than those from a supermarket shelf.

3 The Food and Nutrition Industry—Who Pays the Price?

In the food economy, including food processing and retail, the topic of sustainability is highly prominent, as evidenced by the "Sustainable Natives" (see Chap. 5). Key considerations include resource conservation, energy efficiency and the carbon footprint of food transport, as well as soil and humus regeneration and toxin-free production. Regional and seasonal raw materials and products make a positive contribution, as does avoiding palm oil and other ingredients from monocultures, which is viewed favourably in terms of sustainability and promoting biodiversity in the producing countries. The type of agriculture from which ingredients are sourced, whether organic or even biodynamic, significantly influences sustainability. Sect. 4 of this volume explains and illustrates through research findings that biodynamic agriculture is the most sustainable form of farming.

The question of whether the treatment of employees within companies is socially sustainable is often raised. High employee turnover is an indicator of dissatisfaction among staff. Each replacement costs the company money and leads to the loss of expertise that must then be rebuilt. Measures to foster a pleasant, appreciative working environment and a partnership-based leadership model can save money while contributing to employee satisfaction. Corporate social responsibility is a key topic for international raw material sourcing and trade in general. Fair trade labels and standards regulate socially responsible value chains. Supply chain laws focus on social, environmentally friendly and nature-protective components.

In Germany, the conventional food trade is dominated by a few major players, such as Aldi, Lidl, Edeka and Rewe, who dictate the range of products available. The shelf space allocated to a product group depends on what manufacturers pay or the price they are willing to accept. The driving motivation is to achieve ever-greater financial profit and larger market shares. In contrast, Swiss consumers value quality and "Swissness" over the lowest possible price. This is why Aldi Suisse focuses on Swiss quality. "Swissness" is positively connoted and secures a positive image and market share for the company. When Swiss Toblerone chocolate, iconic for its depiction of the Matterhorn, shifted production abroad, it was forced to forgo its "Swissness." The Matterhorn image could no longer be used, weakening the emotional connection of consumers to their Swiss chocolate. It remains to be seen how the new strategy of the Swiss retailer Migros, which is now targeting discount pricing, will unfold. It is still unclear who ultimately bears the cost of these low-priced offers: agriculture, the environment or others?

In the conventional food economy, everything revolves around profit maximisation. The focus is on market shares, shelf space in retail and, consequently, sales and profits. Sustainable production often plays a secondary or even negligible role. For example, in the summer of 2024, the CEO of Nestlé, who had prioritised sustainability, modernity and trends, was unexpectedly replaced by a new CEO tasked with expanding market share (Neue Zürcher Zeitung, 2024). Nevertheless, companies that publish sustainability reports or embrace corporate social responsibility (CSR) or corporate sustainability reporting directives (CSRD) strive to maintain a positive image. Other companies, such as those with an anthroposophical background or within the organic food sector, incorporate forward-looking business practices into their core philosophy and brand identity. These businesses prioritise circular economy principles, "true costs," and resource conservation.

In the food industry, a distinction can be made between purely materialistic thinking aimed at profit maximisation—where input and output are balanced without accounting for true external costs—and sustainable practices that operate systemically and within cycles. Sustainable practices aim to minimise profit at the expense of others or the environment. The goal is not exploitation but shared development. Chapter 5 provides a comprehensive overview of such examples.

4 Food Production—Dominating or Collaborative

Conventional agriculture is marked by monocultures, soil degradation, air, water and environmental pollution, biodiversity loss, and significant contributions to the climate crisis (Utopia, 2024). The goal is to achieve the highest possible gross margin in the short term with the optimal use of operating resources. This means producing as much food as possible at the lowest prices. Soil is viewed as a transformer of nutrients rather than a living entity. Everything is controlled, and disruptions are eliminated. The focus is on mastering the natural growth of plants and animals. However, this approach shows that conventional agricultural food production is unsustainable—not only are resources consumed rather than preserved, but the environment is also polluted (Hülsbergen et al., 2023), and diseases are fostered. Furthermore, agricultural subsidies in the Global North destabilise local markets in the Global South.

Many solutions for sustainable production are sought in technology. For food crops, this mindset leads to highly controlled cultivation. Weeds are combated, as are diseases, with chemical-synthetic pesticides. Lettuce, vegetables and other foods are produced indoors, often in urban areas close to consumers. Producing food near the point of consumption saves transport costs and reduces the carbon footprint. For vertical farming, several advantages are cited: drinking water is not polluted by fertilisers, can be recycled and reused within the system, and pesticides are unnecessary since plants are not affected by pests. Additionally, production is independent of seasons and weather influences. Precision planning ensures targeted production, with technical personnel or even robots replacing farmers (ETH Zurich, 2020).

Livestock is often housed in high-density closed barns, where the goal is to automate as many processes as possible. Vaccines and the preventive use of antibiotics are employed to protect animals—mainly pigs, poultry and cattle—from diseases and to promote growth (WHO, 2017). Due to the risk of infection, these barns can no longer be entered.

Nature, in essence, is deemed unnecessary because it is seen as a source of disruption. Unpredictable weather, exacerbated by climate change, poses significant challenges, leading to the desire to shield farming and livestock operations from external influences. However, this approach has negative consequences. Mass-produced, agro-industrial products still need to be transported to consumers, resulting in a carbon footprint that undermines sustainability. Other adverse effects include the consumption of imported animal feed—often at the expense of deforested rainforests—greenhouse gas emissions and the disposal problem of manure.

On average, almost 100kg of meat is consumed per person annually in the United States, approximately 60kg in Europe and 15kg in Africa. Consumption is rising, driven by population growth and increasing economic prosperity, which leads to higher meat consumption (OECD/FAO, 2021). In Western countries, meat consumption is declining but still significantly exceeds the German Nutrition Society's recommendation of 300g per week, which equates to roughly 16kg per year (DGE, 2024). This trend is mirrored in other Western countries.

Globally, one-third of grain produced is used as animal feed in conventional livestock farming. Agricultural land suitable for food production is instead used to produce feed, creating competition between human and animal nutrition (Notz, 2019). This is unsustainable—not only for the animals, which could thrive on alternative feed, but also for humans, as high meat consumption is unhealthy.

For those unwilling to give up meat yet opposed to industrial livestock farming, lab-grown meat offers a future alternative. This "clean meat" or slaughter-free meat is described as a sustainable option. It promises technological solutions for sustainability, especially climate friendliness. Production does not require farms but involves tissue-specific stem cells taken from live animals and cultured in suitable nutrient media. These media often contain growth factors derived from the umbilical cord blood of cattle embryos. Due to ethical concerns, alternatives are now being sourced from genetically modified microorganisms, plants or algae (Willinger, 2024). While some issues remain unresolved—such as energy consumption in labs and the production of nutrient solutions—research is heavily supported, with alternative solutions actively developed. In Germany, the government is investing millions in the so-called innovation space NewFoodSystems, where research and industry collaborate to develop new foods without agriculture.

This technological approach to sustainability creates a detachment from nature, leading to "disconnectedness." The principle is to control and standardise everything, eliminating disruptions to ensure a hygienic and "clean" process. However, this results in significant dependence on the agro-food industry and corporations. Nature becomes an industrial landscape, and food transforms into an industrial product. The vital quality of food, formed through agricultural production, suffers as a result. Whether our food comes from a laboratory in the future is not just a question of food production but also of our fundamental attitude towards nature, plants, animals and ourselves. Do we want to move further and further away, or do we see ourselves as part of a whole to which we are connected and with which we wish to remain connected? (Peschke & Sprügel, 2024).

The separation from real life is evident in the sterile environment of lab-grown meat production, where there is no microbiome to stimulate the gut microbiome. Following the principle of supplementation, solutions are being developed to inoculate products with microorganisms. However, this replaces the naturally diverse and variable microbiome with a uniform solution.

5 Collaborative Food Production

Sustainable food production, as practiced in biodynamic agriculture, follows a fundamentally different approach. This approach embraces complementarity and strengthens systems. The agricultural enterprise is understood as an organism in which diverse relationships and interdependencies are consciously cultivated.

Disruptions are not avoided but are managed, leading to greater robustness and vitality. The farmer nurtures the soil and creates the conditions for plants to thrive and for the well-being of animals. Thus, growth processes are indirectly promoted, and production occurs in harmony with the surrounding environment. This includes fostering soil life and fertility through composted manure fertilisation, implementing extensive crop rotation and intercropping practices. Other measures include sowing flowering buffer strips at field edges to attract specific insect species for natural aphid control.

Livestock is kept in conditions that allow natural behaviours, such as lower stocking densities, provision of roughage for cows and minimisation of chronic stress. Through species-appropriate and well-adapted feeding practices, the animals maintain optimal gut health, recover more quickly from infections or avoid illness altogether. Some farms even slaughter animals on-site to avoid the stress associated with slaughterhouses.

This form of sustainable agriculture is characterised by an attitude of care and respect towards all living beings, including soil, plants and animals. Practices and measures are adapted to the specific situation, avoiding standardised processes.

The foundation of this approach lies in trust in the intrinsic nature of a plant or animal and fostering their capacity for self-regeneration.

In this system, relationships, diversity and vitality are nurtured. An open-minded attitude, paired with measures adapted to specific circumstances, forms the basis. This can be described as a collaborative attitude in contrast to the dominating attitude prevalent in technological systems (Napel et al., 2006).

6 The Sustainable Diet—Less Meat and More Diversity

Against the backdrop of the urgent need for sustainable development for Earth and its living beings, achievable only if all sectors rethink and develop new concepts, the EAT-Lancet Commission introduced the Planetary Health Diet in 2019 (EAT Forum, 2024). This diet advocates for a plant-based, mixed diet with reduced meat consumption and describes a way of eating that is environmentally sustainable, socially responsible and healthy for both the planet and people. Since excessive meat consumption is neither healthy nor sustainable, the dietary guidelines of both the German and Swiss Nutrition Societies have included recommendations since 2024 to reduce meat consumption and to increase the intake of plant-based foods (DGE, 2024; SGE, 2024).

A varied daily diet contributes to health in multiple ways. Different ingredients stimulate the senses through their aromas, colours and shapes, enhancing enjoyment and sensory perception. This conscious sensory engagement with meals also improves their digestibility. The diversity of foods consumed promotes a diverse microbiome, which positively impacts gut health (Wicaksono et al., 2023).

Globally, there is limited diversity in the foods people consume. Although individual countries may offer a broader array of edible crops to their consumers, on a global scale, a reduction in dietary diversity is evident. Between 1961 and 2009, food supply homogeneity increased by 16.7% (Khoury et al., 2014). As urbanisation progresses and diets increasingly resemble the Western Diet, the global menu is becoming more uniform (Khoury & Jarvis, 2014). At the same time, diet-related diseases are on the rise, with nutrition-related cardiovascular diseases remaining the leading cause of death worldwide (WHO, 2024). Globally, just three crops—rice, maize and wheat—account for 50% of dietary calories. This monotony harms not only the climate and human health but also the resilience of food systems, which are becoming increasingly homogenised (Khoury et al., 2016). Reduced agricultural diversity leads to declining genetic resources, expanding monocultures, soil depletion and diminished resilience (WHO, 2015).

Dietary diversity is directly linked to agricultural diversity. A varied meal on the plate not only enhances enjoyment but also supports a diverse gut microbiome, which in turn fosters better health. The potential for diversity in food is vast—out of 20,000 to 50,000 edible plants, only around 150 to 200 species are regularly consumed (Federal Office for Agriculture & Food, 2024). Utilising this potential requires openness, flexibility and a willingness to innovate, whether in farming practices or culinary preparation. This calls for boldness in exploring new possibilities—a necessary transition that can bring change at all levels. Dwivedi et al. (2017) even argue for collaboration among scientists, the food industry, breeders and farmers to develop new plant varieties and livestock breeds that can address the global polycrisis and lay the foundation for healthy, sustainable living.

7 The Food System—Everything Is Connected

Everything is interconnected, and focusing solely on the cultivation of vegetables or grains is insufficient without also considering processing, trade and resource use. Sustainable Development must be based on the contributions of all involved. If one part fails, the success of Sustainable Development is jeopardised. This has led to a systemic approach that examines not just individual areas but all stakeholders and their interconnections.

The term "food system" refers to everything that serves human nutrition: from food production to processing, trade and consumption. In some cases, inputs (e.g., energy and other resources) and outputs (e.g., waste) are also considered. By definition, a food system encompasses all areas of a society relevant to food supply (Spektrum der Wissenschaft, 2024). For the FAO, a food system consists of elements (environment, people, inputs, processes, infrastructures, institutions, etc.) and activities related to food production, processing, distribution, preparation and consumption (FAO, 2014). Food systems essentially involve interactions, inputs and outputs, and commodity flows. When sustainability is a criterion for food systems, the United Nations' Sustainable Development Goals (SDGs) provide benchmarks and standards for development in 17 areas.

The systemic approach relies on collaboration among various players. Communities are formed to pursue shared goals, giving purpose to each food system. All elements of a food system are interconnected, forming a whole with interactive relationships. This means that healthy food can only arise under healthy conditions in production, processing and trade. Systems, such as ecosystems, are self-contained in that they can reach a new equilibrium after an external disruption. The aim of a system is to re-establish equilibrium and maintain its integrity, even when exchanging with other systems.

This applies to the concept of food systems, which are largely self-contained and maintain their integrity, adapting but not necessarily evolving. This perspective does not align with our reality. For example, in a biodynamic farm, the natural foundation—soil, landscape and climatic conditions—plays a significant role in agricultural practices and activities, which are fundamentally about development. Biodynamic farming cannot exist without its surrounding environment, which even includes the influence of the cosmos.

It may appear that humans are missing from the food system. While human nutrition is its purpose, humans themselves are often not explicitly recognised as an element. However, it is humans who establish and sustain the food system, shaping the relationships between its components. This idea is embraced by the Conscious Food Systems Alliance (CoFSA, n.d.), initiated under the United Nations Development Programme (UNDP). The Alliance asserts that the transition to sustainable living begins with humans, their attitudes and capabilities. It explicitly highlights the human component, advocating for a new mindset as a prerequisite for transformation. This Alliance is a movement of professionals from the fields of food, agriculture and consciousness-building. Their shared goal is to support individuals across all areas of food and agricultural systems in cultivating inner capacities for systemic change and regeneration, echoing Zen Master Thich Nhat Hanh's principle, "The way out is in" (Nhat Hanh, 2015).

Humans themselves must embody the change they wish to see. This involves developing inner capacities, a focus of the Inner Development Goals (IDG, n.d.). Systemic or structural change cannot be achieved without a shift in consciousness and the development of requisite skills. Einstein famously noted that the problems we face cannot be solved with the same thinking that created them. Therefore, the prevailing worldview of exploitation, profit maximisation and reliance on technical or material solutions must be overcome.

The IDGs offer a pathway, describing five key dimensions. First, everything begins with the individual, starting with their connection to themselves (Being). Then, perception and thinking are needed to abstract and understand complexity and meaning. The next goal is building relationships, caring for others and the environment, rooted in mindfulness, appreciation and empathy. Social skills must be developed for collaboration and communication, with trust as the foundation. The fifth dimension is action, driven by courage, creativity, optimism and perseverance. Transformation of the food system can only succeed when individuals and communities work on their inner development.

This system-based perspective is a vital foundation for our concept of the "food organism." However, it differs in that we focus on mutual development rather than merely achieving balance within the whole. Advancing the systemic concept means envisioning the system as open and dynamic, reflecting actual conditions.

8 The Food Organism—Mindfulness and Joint Development

Living beings are organisms. By definition, an organism (from the Greek organon, meaning "tool") is, in general terms, a structured whole that develops dynamically. One can imagine it as an entity in which all parts not only relate to each other in form and function but are also fundamentally shaped by the whole. In an organism, the parts are not merely externally juxtaposed like components of a lifeless mechanical structure but are internally interconnected with the whole, forming a unified process (AnthroWiki, 2024). An organism, therefore, is not merely an aggregation of matter. Aristotle argued that since an organism has a purpose and is thus goal-oriented, it is more than the sum of its parts. Life shapes the organism, which is inherently alive. Each part contributes to the whole by fulfilling its function, thereby also being alive (Meyer, 2012).

An organism absorbs impulses and transforms them. Its connection to its surroundings—the correspondence with its environment—is intrinsic to the concept of an organism. Despite external influences, it can form a boundary and maintain its integrity. This is possible because all organs comprising the organism are interlinked, and each contributes uniquely to the overall identity.

Rudolf Steiner emphasised, "Here, the external appearance is governed by an internal principle; the whole operates within each organ. The external appearance, the spatial juxtaposition of parts, can still be observed after the destruction of life, for it persists for some time. But what we have before us in a dead organism is, in truth, no longer an organism. The principle that permeates all the individual parts has disappeared" (Steiner, 2022).

An organism can only function if its organs serve the whole, perceive each other, and adapt to one another. Using this as a metaphor, we can advance the concept of the food system into a food organism. In doing so, the human being takes centre stage. With a mindful attitude towards the well-being and development of all participants (organs)—whether farmer, processor or chef—we lay the foundation for healthy nutrition and food culture. The human being is potentially the ordering principle, carrying the idea and the development of the whole. This requires connectedness—a capability unique to humans.

Nature becomes culture when something new emerges that does not exist in nature alone and would not exist without humans. This transformation is evident in agriculture, which thus becomes agri-culture. It is especially true in cooking and table culture.

All consumers eat food and prepare meals. A diverse diet of high-quality food forms the foundation for a healthy life, fostering joy in work and productivity. Looking beyond one's own plate, a sustainable food culture that relies on healthy soils and conditions becomes a matter of interest. How is the farmer, whose dairy cows had to stay in the barn during last summer's heat, coping? Can the gardener grow enough cucumbers to meet the demands of the market? Everything is interconnected.

9 The Food Organism—Based on Biodynamics

In the biodynamic food organism (Peschke, 2024), food originates from biodynamic agriculture, which requires the breeding of seeds and livestock breeds. Harvested products are partially processed and, in any case, traded—whether through organic markets, supermarkets or farm shops. Meals are created through preparation and cooking. Cooking is a uniquely human, creative activity, as humans are the only beings on Earth who cook, thereby bringing something new into existence. Education and training are necessary in all areas.

Everything is interconnected, forming a circle that remains open to its surroundings, much like a typical organism that interacts with both external and internal environments.

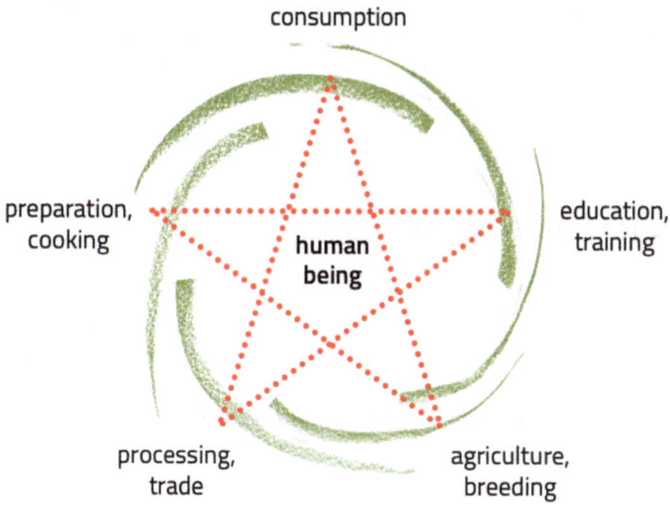

Fig. 1 The Biodynamic Food Organism with its components and the human being at its centre, connecting the components. All components are interrelated, and the whole receives and emits impulses from the external environment (Peschke, 2024)

The components operate independently while serving the whole. The driving force, overarching principle, idea and purpose of a food organism is the human being. Humans are its creators. This is symbolised by the five-pointed star at the centre of the circle (see Fig. 1). Within the circle, neighbouring components are directly connected to one another. Through the human being, represented by the five-pointed star, all elements are interconnected. Meal preparation requires guests to eat, as well as trade and processing to obtain ingredients. Knowledge and skills (education) are also essential to create an appealing meal. The foundation lies in the fruits of the field from agriculture.

Thus, all elements play a common melody, and the conductor is the human being. The conductor understands every instrument. Each component is autonomous, develops in alignment with the whole and contributes to the whole's development. All serve one idea: the idea of human nourishment as the basis for physical, mental and spiritual development on Earth. The human being is made the foundation. The biodynamic food organism exists through humans and for humans. What matters is the attitude with which the human acts within the whole, not just the high-quality end product.

The open circle receives and emits impulses. A biodynamic food organism positively impacts its surroundings and thus the future of both humans and Earth. One could say it promotes the health of all living beings. This does not differentiate between the health of humans, plants, animals or soil. It is about one health, one life, and shared development with Earth.

A food organism is shaped by the interactions, relationships and connections between its components and can therefore also be considered a community. In general, crises and challenges can only be resolved together with others, through mutual support, pooling of strengths, encouragement and the pursuit of a shared goal.

10 Encounter—The Key Skill

An encounter is a process of building a relationship—turning towards the other with interest and empathy and perceiving them. It is like a heart-to-heart conversation. Charles Eisenstein (2021) observes that humanity has broken its dialogue with life both outside and within itself, leading to the crises humanity now faces. He argues that people have severed their connection with all of creation. Eisenstein sees the solution in cultivating the ability to love, which is nothing other than a deeply interested form of relationship-building, mindfulness and appreciation. In short, reconnection can only occur through the human heart.

Thus, relationship-building is the foundation for a healthy, sustainable life on Earth. This requires the development and nurturing of distinctively human capacities such as interest, appreciation and mindfulness—towards oneself and all living beings. Through our cultural and creative forces, we can become co-creators of harmony in the world. Earth can then develop sustainably, providing humanity with the foundations for life and opportunities for growth. A fundamental skill in this process is the ability to encounter. Life revolves around encounters: selecting food with awareness, eating with enjoyment or engaging in conversations with others and oneself. The farmer also encounters his animals, observing how the crops grow. Through this perception, they cultivate a relationship from which their actions arise.

Collaboration within a community requires social competence from its members. This includes being at peace with oneself to engage fruitfully with others. This is where the Inner Development Goals (IDGs) and the Conscious Food Systems Alliance (CoFSA) come into play, supporting the development of personal skills. Building relationships and nurturing encounters are also integral parts of the food organism and essential prerequisites for its positive development. A brochure describes exercises focusing on nutrition, food quality and the mindful experience of the annual cycle (Peschke, 2024).

11 Conclusion

Ehrenfried Pfeiffer, an anthroposophical natural scientist and leading advocate of biodynamic agriculture, lived in Dornach, Switzerland, during Rudolf Steiner's time. One day in 1924, Pfeiffer asked Steiner why it was so difficult to translate spiritual insights into action. Steiner replied that it was a question of nutrition. He explained that the general state of nutrition no longer provided the strength needed for the spiritual to manifest in the physical (as cited in Meyer, 2014, p. 149). This raises an important question about the quality of food. The quality of food must be such that it strengthens the will to act. Through its sensory profile and authenticity in digestion, food offers something intangible to the human being—something that must be overcome, strengthening him in the process. The forces contained within food stimulate forces within the human organism (Steiner, 2022), thus promoting vitality.

Healthy foods, such as those grown on healthy, fertile soil in organic or biodynamic agriculture, are essential for human health. They contain more secondary plant compounds (Baranski et al., 2014), offering greater colour, flavour and aroma. This engages the senses and enhances enjoyment, fostering mindfulness while eating and strengthening one's relationship with oneself. Cultivating this relationship, beginning with food and extending to social interaction, fosters vitality and resilience. Additionally, organic and biodynamic foods stimulate the human gut microbiome with their own diverse microbiomes (Wicaksono et al., 2023), providing the foundation for health and mental-spiritual performance.

Thus, the circle is complete: with healthy biodynamic foods, the will to act can be strengthened. The question of willpower is more relevant than ever. Many people wish to eat healthily, but this desire does not always translate into consistent purchasing and eating habits. Many problems in sustainable development stem from the inability to align personal values with consistent actions. Overcoming this "mind-behaviour gap" and contributing positively to the Earth's future begins with mindful eating, guided by the "three questions" (Peschke, 2021).

The "three questions" provide a simple method to strengthen one's relationship with oneself and with food. They are asked before, during and after a meal:

Before eating: What is on my plate?

While eating: How does it taste?

After eating: How do I digest and tolerate what I have eaten?

Conscious awareness of food leads to curiosity about its origins and "biography." This fosters connection and prevents one from being a mere pawn of advertising and the food industry. Taking responsibility in this way contributes to a sustainable world.

References

AnthroWiki. (2024). *Organism*. Retrieved December 10, 2024, from https://anthrowiki.at/Organismus

Baranski, M., Srednicka-Tober, D., Volakakis, N., Seal, C., Sanderson, R., Stewart, G. B., Benbrook, C., Biavati, B., Markelova, H., Gromadzka-Ostrowska, J., Rembialkowska, E., Skwarlo-Sonta, K., Tahvonen, R., Janovska, D., Niggli, U., Nicot, P., Leifert, C., et al. (2014). Higher antioxidant and lower cadmium concentrations and lower incidence of pesticide residues in organically grown crops: A systematic literature review and meta-analyses. *British Journal of Nutrition, 112*(5), 794–811. https://doi.org/10.1017/S0007114514001366

Bays, J. C. (2017). *Mindful eating: A guide to rediscovering a healthy and joyful relationship with food* (Revised ed.). Shambhala. ISBN 978-1611804652

Bröder, J., Tauer, J., Liaskos, M., & Hieronimus, B. (2023). Verzehr stark verarbeiteter Lebensmittel und ernährungsmitbedingte Erkrankungen: Eine systematische Übersichtsarbeit. In Deutsche Gesellschaft für Ernährung (Ed.), *15. DGE-Ernährungsbericht* (pp. V2–V40). Deutsche Gesellschaft für Ernährung. https://www.dge.de/fileadmin/dok/wissenschaft/ernaehrungsberichte/15eb/15-DGE-EB-Vorveroeffentlichung-Kapitel9.pdf

Conscious Food Systems Alliance (CoFSA). (n.d.). *Conscious food systems*. Retrieved December 10, 2024, from https://consciousfoodsystems.org/

Deutsche Gesellschaft für Ernährung (DGE). (2021). *Ausgewählte Fragen und Antworten zu Protein und unentbehrlichen Aminosäuren*. Retrieved December 9, 2024, from https://www.dge.de/gesunde-ernaehrung/faq/ausgewaehlte-fragen-und-antworten-zu-protein-und-unentbehrlichen-aminosaeuren/

Deutsche Gesellschaft für Ernährung (DGE). (2023, December 13). *Wie wirken stark verarbeitete Lebensmittel auf die Gesundheit?* Retrieved December 9, 2024, from https://www.dge.de/presse/meldungen/2023/wie-wirken-stark-verarbeitete-lebensmittel-auf-die-gesundheit/

Deutsche Gesellschaft für Ernährung (DGE). (2024). *Gut essen und trinken – die DGE-Empfehlungen*. Retrieved December 9, 2024, from https://www.dge.de/gesunde-ernaehrung/gut-essen-und-trinken/dge-empfehlungen/

Dwivedi, S. L., Lammerts van Bueren, E., Ceccarelli, S., Grando, S., Upadhyaya, H. D., & Ortiz, R. (2017). Diversifying food systems in the pursuit of sustainable food production and healthy diets. *Trends in Plant Science, 22*(10), 842–853. https://doi.org/10.1016/j.tplants.2017.06.011

EAT Forum. (2024). *The planetary health diet and you*. Retrieved December 11, 2024, from https://eatforum.org/eat-lancet-commission/the-planetary-health-diet-and-you/

Ehrlichmann, M. (2017). *Einfach ehrlich essen*. Hirzel Verlag. ISBN 978-3777626628

Eisenstein, C. (2021). *Beitrag Landwirtschaftliche Tagung Goetheanum 2021*. Sektion für Landwirtschaft. Retrieved December 12, 2024, from https://www.sektion-landwirtschaft.org/digitale-tagung/videos and https://charleseisenstein.org

ETH Zurich. (2020). *Feeding the world with integrated vertical farming*. Retrieved December 10, 2024, from https://ethz.ch/en/industry/industry/news/data/2020/03/feeding-the-world-with-integrated-vertical-farming.html

FAO/HLPE. (2014). *Food losses and waste in the context of sustainable food systems: A report by the High Level Panel of Experts on Food Security and Nutrition of the Committee on World Food Security* (p. 29). Food and Agriculture Organization of the United Nations (FAO). Retrieved from https://www.fao.org/3/i3901e/i3901e.pdf

Federal Office for Agriculture and Food. (2024). National programme on plant genetic resources. Retrieved December 11, 2024, from https://www.ble.de/DE/Themen/Landwirtschaft/Biologische-Vielfalt/Nationales-Fachprogramm-Pflanzen/nationales-fachprogramm-pflanzen_node.html

Food and Agriculture Organization of the United Nations (FAO). (2024, July 24). *Hunger numbers stubbornly high for three consecutive years as global crises deepen: UN report*. Retrieved December 9, 2024, from https://www.fao.org/newsroom/detail/hunger-numbers-stubbornly-high-for-three-consecutive-years-as-global-crises-deepen--un-report/en

Food and Agriculture Organization of the United Nations (FAO). (n.d.). *Understanding food insecurity*. Retrieved December 9, 2024, from https://www.fao.org/interactive/state-of-food-security-nutrition/en/

Homans, F. (2022). *Addressing the disconnect between farm & fork. Impact* (pp. 3–4) College of Food, Agricultural and Natural Resource Sciences.

Hülsbergen, K.J., Schmid, H., Chmelikova, L., Rahmann, G., Paulsen, H.M., Köpke, U. (2023). *Umwelt- und Klimawirkungen des ökologischen Landbaus. Weihenstephaner Schriften* (Vol. 16). Verlag Dr. Köster. ISBN 978-3968310350

Inner Development Goals (IDG). (n.d.). *Inner development goals*. Retrieved December 12, 2024, from https://innerdevelopmentgoals.org

Khoury, C. K., Achicanoy, H. A., Bjorkman, A. D., Navarro-Racines, C., Guarino, L., Flores-Palacios, X., Engels, J. M. M., Wiersema, J. H., Dempewolf, H., Sotelo, S., Ramírez-Villegas, J., Castañeda-Álvarez, N. P., Fowler, C., Jarvis, A., Rieseberg, L. H., & Struik, P. C. (2016). Origins of food crops connect countries worldwide. *Proceedings of the Royal Society B: Biological Sciences, 283*(1832), 20160792. https://doi.org/10.1098/rspb.2016.0792

Khoury, C. K., & Jarvis, A. (2014). *The changing composition of the global diet: Implications for CGIAR research (CIAT Policy Brief No. 18)*. Centro Internacional de Agricultura Tropical.

Khoury, C. K., Bjorkman, A. D., Dempewolf, H., Ramirez-Villegas, J., Guarino, L., Jarvis, A., Rieseberg, L. H., & Struik, P. C. (2014). Increasing homogeneity in global food supplies and the implications for food security. *Proceedings of the National Academy of Sciences, 111*(11), 4001–4006. https://doi.org/10.1073/pnas.1313490111

Kliemann, N., Rauber, F., Bertazzi Levy, R., Viallon, V., Vamos, E. P., Cordova, R., et al. (2023). Food processing and cancer risk in Europe: results from the prospective EPIC cohort study. *The Lancet Planetary Health, 7*(3), 219–232. https://doi.org/10.1016/S2542-5196(23)00021-9

Meyer, F. M. (2012). Aristoteles über die Psyche als Prinzip und Ursache des Lebens. *Peitho/ examina Antiqua, 1*(3), 115–142.

Meyer, T. (Ed.). (2014). Pfeiffer: Ein Leben für den Geist (1899–1961) (4th ed., p. 149). Perseus Verlag. ISBN 978–3907564318

Ministry of Health of Brazil. (2015). *Dietary guidelines for the Brazilian population*. Retrieved from http://www.b4fn.org/resources/publications/publication-item/dietary-guidelines-for-the-brazilian-population/

Napel, J., ten, Bianchi, F., & Bestman, M. W. P. (2006). Utilising intrinsic robustness in agricultural production systems. *Invention for a sustainable development of agriculture* (pp. 32–54). TransForum.

Nhat Hanh, T. (2015). *The way out is in* (1st ed., p. 2). Thames & Hudson. ISBN 978-0500110287

Neue Zürcher Zeitung. (2024). *Nestlé appoints a new CEO*. Retrieved December 9, 2024, from https://www.nzz.ch/wirtschaft/ld.1845212

Notz, C. (2019). Gegen den Welthunger hilft Gras für das Rindvieh. *Bioaktuell, 2*, 13.

OECD/FAO. (2021). *OECD-FAO agricultural outlook 2021–2030*. OECD Publishing. https://doi.org/10.1787/19428846-en

Peschke, J. (2021). *Vom Acker auf den Teller: Was Lebensmittel wirklich gesund macht*. AT Verlag. ISBN 978-3039021116

Peschke, J. (2024). *From food system to food organism: Steps to a succesful transformation*. Goetheanum. Retrieved December 12, 2024, from https://www.goetheanum.org/fileadmin/landwirtschaft/Ernährung/Peschke_Ernaehrungsorganismus_EN.pdf

Peschke, J., & Sprügel, L. (2024). *Will animal products be produced in the lab in future?* Section for Agriculture. Retrieved December 12, 2024, from https://www.sektion-landwirtschaft.org/en/sv/will-animal-products-be-produced-in-the-lab-in-future

Rützler, H. (2024). *Future food studio*. Retrieved November 18, 2024, from https://www.futurefoodstudio.at/

Rützler, H. (2024). *Food trends 2025: Was und wie wir in Zukunft essen werden*. Zukunftsinstitut. Retrieved December 9, 2024, from https://www.zukunftsinstitut.de/zukunftsthemen/food-trends-hanni-ruetzler

Schweizerische Gesellschaft für Ernährung (SGE). (2024). *Schweizer Lebensmittelpyramide*. Retrieved December 10, 2024, from https://www.sge-ssn.ch/ich-und-du/essen-und-trinken/ausgewogen/schweizer-lebensmittelpyramide/

Spektrum der Wissenschaft. (2024). *Food system: Definition*. Retrieved December 10, 2024, from https://www.spektrum.de/lexikon/ernaehrung/ernaehrungssystem/2684

Steiner, R. (2022). *Einleitungen zu Goethes naturwissenschaftlichen Schriften* (5th ed.). Rudolf Steiner Verlag. ISBN 978-3-7274-0013-1

Steiner, R. (2022). *Landwirtschaftlicher Kurs* (7th ed., p. 74). Rudolf Steiner Verlag. ISBN 978-3727400588

Tian, P., Zhong, H., Chen, X., Feng, K., Sun, L., Zhang, N., Shao, X., Liu, Y., & Hubacek, K. (2024). Keeping the global consumption within the planetary boundaries. *Nature, 635*(625), 625–630. https://doi.org/10.1038/s41586-024-08154-w

The Vegan Society. (2024). *Vegan trademark standards.* Retrieved December 9, 2024, from https://www.vegansociety.com/vegan-trademark/vegan-trademark-standards

Utopia. (2024). *Conventional agriculture: Characteristics and disadvantages.* Retrieved December 10, 2024, from https://utopia.de/ratgeber/konventionelle-landwirtschaft-kennzeichen-und-nachteile_162674/

Wallinga, D. (2010). Agricultural policy and childhood obesity: A food systems and public health commentary. *Health Affairs, 29*(3), 405–410. https://doi.org/10.1377/hlthaff.2010.0102

Wicaksono, W. A., Cernava, T., Wassermann, B., Abdelfattah, A., Soto-Giron, M. J., Toledo, G. V., Virtanen, S. M., Knip, M., Hyöty, H., & Berg, G. (2023). The edible plant microbiome: evidence for the occurrence of fruit and vegetable bacteria in the human gut. *Gut Microbes, 15*(2), 2258565. https://doi.org/10.1080/19490976.2023.2258565

Willinger, G. (2024). Fleisch aus der Retorte. *Spektrum der Wissenschaft, 4*, 44–49.

World Health Organization, & Secretariat of the Convention on Biological Diversity. (2015). *Connecting global priorities: Biodiversity and human health: A state of knowledge review* (p. 29). World Health Organization. ISBN 978-9241508537

World Health Organization (WHO). (2024). *The top 10 causes of death.* https://www.who.int/newsroom/fact-sheets/detail/the-top-10-causes-of-death

World Health Organization (WHO). (2017). *Guidelines on use of medically important antimicrobials in food-producing animals.* World Health Organization.

World Obesity Federation. (2024). *World Obesity Atlas 2024.* World Obesity Federation.

Socio-ecological Medicine—An Anthroposophical Approach

David Martin and Georg Soldner

Abstract

Socio-ecological medicine, exemplified by anthroposophic practices, integrates physical, emotional, and spiritual health with sustainability and social justice. This approach, rooted in Rudolf Steiner's vision, emphasises self-regulation, autonomy, and meaningful care, aligning with e.g. Positive Health models. Anthroposophic medicine promotes therapies like eurythmy, external applications, and biodynamic nutrition, reducing reliance on pharmaceuticals and fostering ecological harmony. Its emphasis on patient empowerment, ethical responsibility, and collaboration addresses modern challenges like overmedicalisation, antimicrobial resistance, and environmental degradation. This holistic model fits seamlessly with socio-ecological agriculture, economics, and education as it aims to enhance resilience and dignity, offering a sustainable path forward for individuals, communities, and the planet.

Keywords

Socio-ecological medicine • Anthroposophic medicine • Integrative health • Positive health • Self-regulation • Holistic healthcare • Eurythmy therapy • Biodynamic nutrition • External applications • Ethical medicine • Sustainable healthcare • Patient empowerment • Environmental health • Antimicrobial resistance

D. Martin
Medical Theory, Integrative and Anthroposophic Medicine, Institute of Integrative Medicine, University of Witten/Herdecke, Witten, Germany
e-mail: david.martin@uni-wh.de

G. Soldner (✉)
Medical Section, Goetheanum - Freie Hochschule für Geisteswissenschaft, Dornach, Switzerland
e-mail: georg.soldner@medsektion-goetheanum.ch

J. Kronenberg and E. T. Lammerts van Bueren (eds.), *On the Earth We Want to Live*, World Sustainability Series, https://doi.org/10.1007/978-3-031-98758-8_13

1 Introduction

Health, and therefore medicine, cannot be fully understood without a holistic, socio-ecological approach. This is the reason why definitions of health, such as it being "a state of complete physical, mental and social well-being and not merely the absence of disease or infirmity" (World Health Organization, 1946, para. 1), have been developed into more encompassing and dynamic ones. For example, Machteld Huber, redefined health as "the ability to adapt and self-manage in the face of social, physical, and emotional challenges" (Huber as cited in Thomas, 2023, para. 2). Her concept, known as "Positive Health," shifts the focus from a state to the capacity for resilience and autonomy. Huber's model includes six dimensions: Body functions, mental functions, spiritual dimension, quality of life, social participation, and daily functioning (Valentijn, 2021). Indeed, reducing health to the absence of pain or malaise would not only ignore many conditions that do not cause discomfort, but would also incur the risk of interpreting strenuous, uncomfortable challenges, development processes, and transitions of body, soul, and spirit as diseases.

Similarly, all definitions of health that rely purely on norms, be it of weight or blood pressure, run the risk of generalisation and of missing what health is really about. These are reasons why there can be no final, globally accepted definition of health, and therefore also of medicine. Defining health, and therefore medicine, will always remain a contextualised, socio-ecological act by those involved in a given situation. Therefore, health, including medicine, are socio-ecological realities to their very core. Anthroposophic endeavours in health and medicine were pioneering socio-ecological perspectives and activities from the very start contributing to a socio-ecological just future of humankind and Earth.

In 1924, Rudolf Steiner (1924/2004) saw human health as inseparable form sustainable agriculture that is designed for biodiversity, sustainability and soil development[1] and thus serving healthy nutrition of all life. Steiner had already pointed out malnutrition (e.g. too much protein for infants) and anthroposophic paediatricians were and are leading in promoting breastfeeding for everyone. Anthroposophic hospitals were among the first breastfeeding-friendly hospitals in Germany—promoting what is available to every woman, regardless of social status, and reducing the rate of avoidable deaths and serious infections. In this context, the principles of anthroposophic agriculture and medicine align closely with the contemporary One Health approach (Matthes et al., 2024).

[1] See https://www.fibl.org/en/themes/projectdatabase/projectitem/project/404 for a large number of publications comparing biodynamic and conventional farming, including a Science publication on the DOK Experiment: Long-term study on bio-dynamic, bio-organic and conventional farming systems – the world's longest running trial.

2 Anthroposophic Medicine

In anthroposophic medicine, the leading health question is the question of self-regulation, on a vital, emotional and spiritual level (meaningful, conscious lifestyle, low use of animal food, critical relationship to alcohol, etc.) (Girke, 2016). The anthroposophic lifestyle, which also includes spiritual aspects, was explicitly tested against other forms of lifestyle regarding its effects on the development of early childhood stress and allergies (Alm et al., 1999; Stenius et al., 2010; Swartz et al., 2015). The latter seminal Lancet study by Alm et al. (1999) led to many other trials including the trials on milk consumption and allergy (Waser et al., 2007), which have also been reflected in studies on the microbiome in recent years, continuing to show that the anthroposophic lifestyle promotes biodiversity in nature *and* high level of microbial life in humans (Muhie, 2023; Santoni et al., 2022).

Pharmaceutically, studies such as the Parsifal study (Alfvén et al., 2006), but also the anthroposophic medicine outcomes study (Hamre et al., 2014), and the IIPCOS study (Hamre et al., 2017) demonstrate that anthroposophic doctors use fewer pharmaceutical drugs, including antibiotics. Modern pharmaceutical practices are associated with high environmental pollution, often in poorer countries, but also worldwide—evident in the widespread contamination of water with metformin (Zheng et al., 2024), linked to the exponential rise in type 2 diabetes. It has been proven that more and more people are losing the ability to self-regulate their blood sugar, blood pressure, body weight, etc. Polypharmacy has been proven to endanger old people with an increased risk of falls (Zia et al., 2015), and it pollutes wastewater and groundwater to a high degree, for example through diclofenac (Sathishkumar et al., 2020), in complete contrast to the phytotherapeutic use of *Arnica montana*—whereby helenalin in arnica has a stronger anti-inflammatory effect than diclophenac (Greco et al., 2022).

An excellent paradigm for dealing with the self-regulation of the human organism is that of dealing with fever. The development of FeverApp[2] and FeverFriend[3] supports reducing unfounded fears in a socially barrier-free manner. They also support the reduction in the use of unnecessary medicines such as paracetamol, ibuprofen and antibiotics in early childhood and strengthen confidence in the ability to self-regulate. The background here is the conviction that the warmth regulation of humans, which differs significantly from animals (naked skin, high need for thermoregulation, especially of the brain) (Jablonski, 2010), is not a purely physical matter, but has to do with the fact that the soul and spiritual aspect, and thus ultimately the question of self-efficacy, is directly related to the "embodiment" of healthy self-regulation. This pertains especially to the release of energy in the self-aware, self-organising (I-organisation) human organism with its "physiology of freedom" (Steiner & Wegman, 1924).

[2] https://www.feverapp.de/.

[3] https://feverfriend.eu/.

Artistic therapies and eurythmy therapy are also suitable for strengthening self-efficacy and ameliorating many chronic conditions (Hamre et al., 2007, 2009), including patients with cancer (children and adults) and anorexia nervosa (Bar-Sela et al., 2007; Simmance, 2015).

Anthroposophic clinics and hospitals are available to all patients in Germany and regularly receive a very high satisfaction score in patient evaluations: Many patients feel more comprehensively seen and empowered by anthroposophic medical professionals (Baars et al., 2017, 2018). Biographical work is important in anthroposophic medicine and in all areas of anthroposophic One Health concepts (Baars, 2005).

A core enquiry is about how human consciousness is categorically different from all physical-molecular processes. Brains may assimilate, associate, and correlate, but brains do not think, people do. Anthroposophic medicine aims to not confuse functions with their physical prerequisites. Thinking and consciousness are enabled by the brain but not caused by the brain (Fuchs, 2018, 2021). They are the result of the activity of whole, embodied, living entities. Anthroposophic medicine addresses patients and their relatives at their various levels of consciousness and in their values. This is central, for example, in end-of-life care (spiritual care) (Heusser et al., 2006), but also e.g. in obstetrics. In terms of its frequency and its chronic consequences, the caesarean section is without a doubt a much too frequently performed procedure today (Angolile et al., 2023). Anthroposophic Hospitals such as the Filderklinik and the Gemeinschaftskrankenhaus Herdecke prove that anthroposophic medicine can contribute to a reduction in the caesarean section rate in the direction of 15%, while in many developed countries the rate is significantly higher (Betrán et al., 2016). Data from 2010 to 2018, spanning 154 countries and accounting for 94.5% of global live births, reveal that 21.1% of women gave birth via caesarean section, exceeding the World Health Organization's (WHO) recommended range of 10–15% as an acceptable standard (Betrán et al., 2021).

Anthroposophic medicine promotes, as with caesarean section, fever, balanced nutrition, the active skills of patients in a comprehensive way (Berger et al., 2024; Mascher et al., 2023; Michalsen et al., 2021), in terms of what characterises the "blue zones" (describe as the regions of the world where people live longer and healthier lives than average) (Buettner & Skemp, 2016), as well as the Positive Health model.

Anthroposophic medicine assumes that a person's actions in relation to their environment do not remain without consequences for their future fate, and thus, promotes reverence for life. Human freedom, human responsibility—which encompasses the responsibility for preserving our own and each other's human dignity, including the human dignity of the disabled—are core values of anthroposophy and anthroposophic medicine. The human being is respected as an "embodied self" with its individual biography and values. In contrast, trends in neurobiology today threaten to declare humans themselves as an illusion of a "body machine" whose self is in principle a brain projection (Fuchs, 2020). This neurobiological paradigm has profound consequences e.g. for the professional relationships to patients in medicine. As a consequence, such positions amount to *ethical nihilism*, which ultimately denounces all ethical norms as arbitrary statements of a consciousness "invented" and dominated by matter.

In contrast, anthroposophy is based on the evidence of the human self and the fact that human thinking in no way represents the mere projections of physical structures and processes (Fuchs, 2018, 2021), but follows its own laws, for example in mathematics. Likewise, the question of *what is good* for people—as well as for all living beings and the Earth as a whole—has a very independent and central meaning in anthroposophy beyond any materialistic reductionism (Steiner, 1894a, 1894b).

Anthroposophy, without fundamentally rejecting animal experiments, emphasises that humans and especially higher animals have an emotional and spiritual experience of their own bodies and environment. However, medicine and cosmetics that are based on millions of often painful, torturous and fatal animal experiments are fundamentally problematic from an anthroposophic point of view. Mainstream science has long denied animals what they undoubtedly have: The ability to suffer is in a way made all the more profound in that the animal lacks the conscious perspective of meaning that can make some temporary suffering seem bearable for humans. Therefore, people bear a moral responsibility when they cause suffering to animals in factory farming and animal testing.

3 Socio-ecological Aspects of Anthroposophic Medicine

Like health, the terms "social" and "ecological" cannot be confined to any abstract or finite definition. Recognising this carries significant implications: ultimately, what is deemed social or ecological can only be determined by those directly involved in a specific context—whether at a local or global scale. This challenge is compounded by the fact that what may seem to represent a lack of health, sociability, or ecology at one level might, from a broader perspective, be viewed as a healthy, sociable, and ecological solution.

This raises an important question: If the determination of what is social and ecological is best made by those directly involved in a particular context, can artificial intelligence serve as an assessor of what is healthy, social, and ecological? With its rapidly advancing capabilities, including modes of data collection that increasingly resemble human senses, artificial intelligence is becoming more adept at identifying correlations between various factors and markers of health, sociability, and ecology. Such markers might include longevity, peace, and stability. However, there may be circumstances where longevity, peace, and stability are not the ultimate indicators of health, sociability, or ecology. Indeed, there may even be instances where health, sociability, and ecology themselves are not the ultimate objectives.

This train of thought brings us directly to what Rudolf Steiner (1894a, 1894b) called "ethical individualism" in his work "Philosophy of Freedom". Steiner proposed "Philosophy of Spiritual Activity" as title for this book, which is very fitting given that the choice of goal(s) to be striven for in a particular situation is ultimately a matter beyond any definition, norm or habit. It can be a free, highly contextualised act that—though it may defy all hitherto known parameters—may in hindsight be judged as healthy, sociable and ecological. Human history is full of such more or less free and contextualised acts. Taking examples from the field of medicine one could say that it is neither healthy, nor ecological nor social to be using chemotherapy in a highly isolated cell being prepared for autologous stem cell transplantation. Yet many people, especially children, have been enabled long-term survival by such means. Oncologists and their patients constantly face the question as to what is the next best step to take. Different patients will choose different paths, and we have time-and-again witnessed how decisions which seemed irrational, asocial, unecological and doomed to fail have in some way advanced humanity, if only in terms of gaining helpful knowledge for the next generation of patients.

Hence, health and medicine include a very individual dimension of meaning. Even on their deathbeds, people want their lives to have been—and still be—meaningful for others (Büssing, 2024). Interestingly, dimensions of spirituality play an important role in all the blue zone areas of the world. People who live a long time tend to live socially, ecologically and spiritually.

The capacity for autonomy, i.e. for individuals and communities to take care of their own health is a very important aspect of a socio-ecological medicine. Today, 70% of death globally are related to noncommunicable chronic diseases (NCDs) and Western lifestyle (World Health Organization, 2024). Lifestyle changes are the most effective prevention of the predominating NCDS (cardiovascular diseases, diabetes, chronic obstructive lung diseases, cancer), of dependency on medicines and procedures that are a burden on the environment and that lead to dependency and social isolation or strain. The spiritual activity kindled by anthroposophy or other paths of spiritual awakening can be strong contributors to self-development and health maintenance by engendering life-long learning and ensouled mental and physical exercise, such as eurythmy and other arts. It is also a strong antidote against damaging consumerism and induces a shift from profit-oriented to socio-ecologically oriented lifestyles and businesses.

This book presents examples of how anthroposophy is contributing to social and ecological health on Earth. The field of health and medicine also offers many examples worldwide, ranging from highly concentrated[4] to highly potentised remedies,[5] from individual biography work and art therapies to group eurythmy therapy sessions, from neurosurgery[6] to external applications[7] to ecotherapy in therapeutic gardens, from small therapy centres to 600 + bed hospitals (Martin, 2020), and from health applications[8] to nationwide trials.[9]

A social, ecological medicine is one that helps the patient develop treatments themselves in an autonomous, ecological manner that promotes self-efficacy. Anthroposophic medicine attempts to do this in many ways. The Vademecum: External Applications in Anthroposophic Nursing (Pflege Vademecum)[10] offers patient empowering and highly ecological external applications that can help promote well-being, vitality and function. Given that these external applications are often applied by another person, this side of anthroposophic medicine has an inherently social aspect to it. All these external applications involve substances that are readily taken directly from nature and seemingly seamlessly brought back into the cycle of nature. The social aspect of medicine is enhanced throughout in the way therapists, nurses and doctors all work together to produce as holistic a picture and treatment of the human being as possible.

It would have been very advantageous to anthroposophic medicine if it had engaged in rigorous research earlier and more intensively. But at least now research intensity is picking up (Kienle & Kiene, 2010).

Anthroposophic doctors were notoriously early in recognising non-ecological and ultimately non-social and unhealthy aspects (Soldner & Stellmann, 2014), such as the detrimental effect of television (Glöckler et al., 2024), and screen-time on children (Schwarz et al., 2024a, 2024b; Zink, 2024), of an over-intellectualised education, of an overuse of painkillers, antidepressants, and sleeping pills and of an overuse of antibiotics and the danger of antimicrobial resistance.

4 Conclusion

In summary, anthroposophic medicine is in its very core intended to be a socio-ecological medicine and contributes in many ways to infuse such qualities into mainstream medicine. Anthroposophic medicine provides a holistic, socio-ecological approach that integrates physical, emotional, and spiritual health. Rooted in Rudolf Steiner's vision, it emphasises self-regulation, autonomy, and

[4] Vademecum anthroposophische Arzneimittel, www.anthromedics.org www.misteltherapie.org.
[5] https://www.wisscom.ch.
[6] https://www.uni-wh.de/ansprechpartner/wolfram-scharbrodt.
[7] www.pflege-vademecum.de.
[8] www.warmuptofever.org, www.feverapp.de, www.feverfriend.com.
[9] www.entaier.de, www.bildschirmfrei-bis-3.de, www.feverapp.de, and many others.
[10] https://www.pflege-vademecum.de/?locale=en.

meaningful interactions, promoting therapies that empower individuals while respecting the interconnectedness of humans, nature, and society. By addressing modern challenges like overmedicalisation, environmental sustainability, and ethical care, anthroposophic medicine aligns with global movements like the Positive Health model and One Health. Its inclusive therapies—such as external applications, eurythmy, and integrative treatment plans—foster long-term well-being while respecting ecological and social principles. This text highlights how anthroposophic medicine bridges conventional and alternative paradigms, advocating for patient empowerment, ethical responsibility, and sustainable health practices. It offers a pathway to a more just and sustainable future for humanity and the planet.

References

Alfvén, T., Braun-Fahrländer, C., Brunekreef, B., Von Mutius, E., Riedler, J., Scheynius, A., Van Hage, M., Wickman, M., Benz, M. R., Budde, J., & PARSIFAL Study Group. (2006). Allergic diseases and atopic sensitization in children related to farming and anthroposophic lifestyle–the PARSIFAL study. *Allergy, 61*(4), 414–421.https://doi.org/10.1111/j.1398-9995.2005.01024.x

Alm, J. S., Swartz, J., Lilja, G., Scheynius, A., & Pershagen, G. (1999). Atopy in children of families with an anthroposophic lifestyle. *The Lancet, 353*(9163), 1485–1488. https://doi.org/10.1016/S0140-6736(98)09344-1

Angolile, C. M., Max, B. L., Mushemba, J., & Mashauri, H. L. (2023). Global increased cesarean section rates and public health implications: A call to action. *Health Science Reports, 6*(5), e1274. https://doi.org/10.1002/hsr2.1274

Baars, T. (2005). How biographical experiences affect a research and training programme in biodynamic agriculture at Kassel University. In B. Haverkort, K. van 't Hooft, & W. Hiemstra (Eds.), *Moving worldviews: Reshaping sciences, policies and practices for endogenous sustainable development* (pp. 364–380). ETC/Compas.

Baars, E. W., Koster, E. B., & Verhoef, J. (2017). The contribution of anthroposophic medicine to self-management: An exploration of concepts, evidence, and patient perspectives. *Complementary Medicine Research, 24*(4), 225–231. https://doi.org/10.1159/000478981

Baars, E. W., Kiene, H., Kienle, G. S., Heusser, P., & Hamre, H. J. (2018). An assessment of the scientific status of anthroposophic medicine, applying criteria from the philosophy of science. *Complementary Therapies in Medicine, 40*, 145–150. https://doi.org/10.1016/j.ctim.2017.10.006

Bar-Sela, G., Atid, L., Danos, S., Gabay, N., & Epelbaum, R. (2007). Art therapy improved depression and influenced fatigue levels in cancer patients on chemotherapy. *Psycho-Oncology, 16*(11), 980–984. https://doi.org/10.1002/pon.1175

Berger, B., Scheffer, C., Jenetzky, E., & Martin, D. D. (2024). Emotional regulation competencies for people with type 1 diabetes mellitus (T1D)—A multimodal training program (current state). *The Mind*, 3. https://doi.org/10.61936/themind/2024121227

Betrán, A. P., Ye, J., Moller, A. B., Zhang, J., Gülmezoglu, A. M., & Torloni, M. R. (2016). The increasing trend in caesarean section rates: Global, regional and national estimates: 1990–2014. *PLoS ONE, 11*(2), e0148343. https://doi.org/10.1371/journal.pone.0148343

Betrán, A. P., Ye, J., Moller, A.-B., Souza, J. P., & Zhang, J. (2021). Trends and projections of caesarean section rates: Global and regional estimates. *BMJ Global Health, 6*(6), e005671. https://doi.org/10.1136/bmjgh-2021-005671

Buettner, D., & Skemp, S. (2016). Blue Zones: Lessons from the world's longest lived. *American Journal of Lifestyle Medicine, 10*(5), 318–321. https://doi.org/10.1177/1559827616637066

Büssing, A. (2024). Spiritual Care—Standardisierte Erfassung von existenziellen und spirituellen Bedürfnissen und ihre Implementierung in Versorgungseinrichtungen. In A. Büssing, A.

Giebel, & T. Roser (Eds.), *Spiritual Care & Existential Care interprofessionell: Handbuch spiritueller und existentieller Begleitung in der Gesundheitsversorgung* (pp. 359–369). Springer Berlin Heidelberg. https://doi.org/10.1007/978-3-662-67742-1_35

Fuchs, T. (2018). *The ecology of the brain: The phenomenology and biology of the embodied mind.* Oxford University Press.

Fuchs, T. (2020). Embodiment and personal identity in dementia. *Medicine, Health Care and Philosophy, 23*(4), 665–676. https://doi.org/10.1007/s11019-020-09973-0

Fuchs, T. (2021). *Das Gehirn – ein Beziehungsorgan: Eine phänomenologisch-ökologische Konzeption* (6. Aufl.). Kohlhammer Verlag.

Girke, M. (2016). *Internal medicine: Foundations and therapeutic concepts of anthroposophic medicine.* Salumed-Verlag.

Glöckler, M., Goebel, W., & Michael, K. (2024). *Kindersprechstunde: Ein medizinisch-pädagogischer Ratgeber.*Urachhaus.

Greco, A., Mennet-von Eiff, M., Christoph, M., Keusgen, M., Edelhäuser, F., & Peifer, C. (2022). Wie kann die Pharmazie zur Nachhaltigkeit beitragen? *Pharmazeutische Zeitung, 47*, 48–52.

Hamre, H. J., Witt, C. M., Glockmann, A., Ziegler, R., Willich, S. N., & Kiene, H. (2007). Anthroposophic art therapy in chronic disease: A four-year prospective cohort study. *Explore, 3*(4), 365–371. https://doi.org/10.1016/j.explore.2007.04.008

Hamre, H. J., Witt, C. M., Kienle, G. S., Meinecke, C., Glockmann, A., Willich, S. N., & Kiene, H. (2009). Anthroposophic therapy for children with chronic diseases: A two-year prospective cohort study in everyday practice. *BMC Pediatrics, 9*, Article 39. https://doi.org/10.1186/1471-2431-9-39

Hamre, H. J., Kiene, H., Ziegler, R., Tröger, W., Meinecke, C., Schnürer, C., Vögler, H., Glockmann, A., & Kienle, G. S. (2014). Overview of the publications from the anthroposophic medicine outcomes study (AMOS): A whole system evaluation study. *Global Advances in Health and Medicine, 3*(1), 54–70.https://doi.org/10.7453/gahmj.2013.097

Hamre, H. J., Glockmann, A., Heckenbach, K., & Matthes, H. (2017). Use and safety of anthroposophic medicinal products: An analysis of 44,662 patients from the EvaMed pharmacovigilance network. *Drugs - Real World Outcomes, 4*(4), 199–213. https://doi.org/10.1007/s40801-017-0118-5

Heusser, P., Braun, S. B., Bertschy, M., Burkhard, R., Ziegler, R., Helwig, S., Wegberg, B.V., & Cerny, T. (2006). Palliative in-patient cancer treatment in an anthroposophic hospital: II. Quality of life during and after stationary treatment, and subjective treatment benefits. *Forschende Komplementärmedizin/Research in Complementary Medicine, 13*(3), 156–166. https://doi.org/10.1159/000093489

Jablonski, N. G. (2010). The naked truth: Why humans have no fur. *Scientific American, 302*(2). https://www.scientificamerican.com/article/the-naked-truth-why-humans-have-no-fur1/

Kienle, G. S., & Kiene, H. (2010). Review article: Influence of *Viscum album L.* (European mistletoe) extracts on quality of life in cancer patients: A systematic review of controlled clinical studies. *Integrative Cancer Therapies, 9*(2), 142–157. https://doi.org/10.1177/1534735410369673

Martin, D. (2020). 100-Year Anniversary of Anthroposophic Medicine as an Integrative Medical System. *Complementary Medicine Research, 27*(6), 375–378.

Mascher, A., Pelzer, F., Duncan, L. J., Martin, D. D., Baumgartner, S., & Berger, B. (2023). The introspective patient experience of mistletoe therapy in cancer: A qualitative study. *Integrative Cancer Therapies*, 22.https://doi.org/10.1177/15347354231198474

Matthes, H., Baars, E. W., Brinkhaus, B., Christoph, M., Edelhäuser, F., Grah, C., Gründemann, C., Keßler, C., Martin, D., Michalsen, A., & Willich, S. N. (2024). The Earth as a living organism: Contribution of integrative medicine to the healing of our planet (One Health). *Complementary Medicine Research, 31*(5), 477–483.https://doi.org/10.1159/000540226

Michalsen, A., Jeitler, M., Kessler, C. S., Steckhan, N., Robens, S., Ostermann, T., Kandil, F. I., Stankewitz, J., Berger, B., Jung, S., Kröz, M., & Büssing, A. (2021). Yoga, eurythmy therapy and standard physiotherapy (YES-Trial) for patients with chronic non-specific low back pain:

A three-armed randomized controlled trial. *The Journal of Pain, 22*(10), 1233–1245. https://doi.org/10.1016/j.jpain.2021.03.154

Muhie, S. H. (2023). Concepts, principles, and application of biodynamic farming: A review. *Circular Economy and Sustainability, 3*(2), 291–304. https://doi.org/10.1007/s43615-022-00184-8

Santoni, M., Ferretti, L., Migliorini, P., Vazzana, C., & Pacini, G. C. (2022). A review of scientific research on biodynamic agriculture. *Organic Agriculture, 12*(2), 373–396. https://doi.org/10.1007/s13165-022-00394-2

Sathishkumar, P., Meena, R. A. A., Palanisami, T., Ashokkumar, V., Palvannan, T., & Gu, F. L. (2020). Occurrence, interactive effects and ecological risk of diclofenac in environmental compartments and biota. *Science of The Total Environment, 698*, Article 134057. https://doi.org/10.1016/j.scitotenv.2019.134057

Simmance, C. M. (2015). *Psychophysiologischer Wirksamkeitsnachweis eines multimodalen Therapiekonzepts bei Jugendlichen mit Anorexia nervosa* (Doctoral dissertation, Universität Tübingen). https://publikationen.uni-tuebingen.de/xmlui/bitstream/handle/10900/69278/Dissertation_Cara_Simmance.pdf

Steiner, R., & Wegman, I. (1924). *Fundamentals of therapy: An extension of the art of healing through spiritual knowledge.* https://ivaa.info/wp-content/uploads/2021/09/Fundamentals_of_Therapy.pdf

Soldner, G., & Stellmann, H. M. (2014). *Individual paediatrics: Physical, emotional and spiritual aspects of diagnosis and counselling.* MedPharm Scientific Publishers.

Steiner, R. (1894). *Die Philosophie der Freiheit: Grundzüge einer modernen Weltanschauung. Seelische Beobachtungsresultate nach naturwissenschaftlicher Methode.* Rudolf Steiner Online Archiv. Retrieved January 22, 2025, from http://anthroposophie.byu.edu/schriften/004.pdf

Steiner, R. (1894). *The philosophy of freedom: Basics of a modern worldview. Psychological observation results according to the scientific method.* Rudolf Steiner Online Archive. Retrieved January 22, 2025, from http://anthroposophie.byu.edu/schriften/004.pdf

Steiner, R. (1924/2004). *Agriculture course: The birth of the biodynamic method* (Revised ed.). Rudolf Steiner Press.

Stenius, F., Swartz, J., Lindblad, F., Pershagen, G., Scheynius, A., Alm, J., & Theorell, T. (2010). Low salivary cortisol levels in infants of families with an anthroposophic lifestyle. *Psychoneuroendocrinology, 35*(10), 1431–1437. https://doi.org/10.1016/j.psyneuen.2010.05.010

Swartz, J., Lindblad, F., Arinell, H., Theorell, T., & Alm, J. (2015). Anthroposophic lifestyle and salivary cortisol are associated with a lower risk of sensitization during childhood. *Pediatric Allergy and Immunology, 26*(2), 153–160.

Schwarz, S., Büssing, A., Krafft, H., Streit, B., & Boehm, K. (2024a). Parents and siblings as role models in dealing with digital screen media: Findings from a media fasting intervention. *Journal of Woman's Reproductive Health, 3*(2), 14–25. https://doi.org/10.14302/issn.2381-862X.jwrh-24-5099

Schwarz, S., Krafft, H., Maurer, T., Lange, S., Schemmer, J., Fischbach, T., Emgenbroich, A., Monks, S., Hubmann, M., Martin, D., & Martin, D. (2024b). Screen time, nature, and development: Baseline of the randomized controlled study 'Screen-free till 3'. *Developmental Science, e13578.* https://doi.org/10.1111/desc.13578

Thomas, R. (2023, June). A positive approach to health. *WALA Heilmittel GmbH.* https://www.wala.world/en/health/a-positive-approach-to-health.html

Valentijn, P. (2021, April 20). From a new definition of health towards person-centred care. *Essenburgh.*https://www.essenburgh.com/en/from-a-new-definition-of-health-towards-person-centred-care/

Waser, M., Michels, K. B., Bieli, C., Flöistrup, H., Pershagen, G., von Mutius, E., Mutius, E., Ege, M., Riedler, J., Schram-Bijkerk, D., & Braun-Fahrländer, C., PARSIFAL Study Team. (2007). Inverse association of farm milk consumption with asthma and allergy in rural and suburban populations across Europe. *Clinical & Experimental Allergy, 37*(5), 661–670.https://doi.org/10.1111/j.1365-2222.2006.02640.x

World Health Organization. (1946). *Constitution of the World Health Organization.*https://www.who.int/about/governance/constitution

World Health Organization. (2024, December 23). *Noncommunicable diseases.* https://www.who.int/news-room/fact-sheets/detail/noncommunicable-diseases

Zheng, Y., Shao, Y., Zhang, Y., Liu, Z., Zhao, Z., Xu, R., Ding, J., Li, W., & Zhang, H. (2024). Metformin as an emerging pollutant in the aquatic environment: Occurrence, analysis, and toxicity. *Toxics, 12*(7), Article 483. https://doi.org/10.3390/toxics12070483

Zia, A., Kamaruzzaman, S. B., & Tan, M. P. (2015). Polypharmacy and falls in older people: Balancing evidence-based medicine against falls risk. *Postgraduate Medicine, 127*(3), 330–337. https://doi.org/10.1080/00325481.2014.996112

Zink, N. (2024). „Je weniger Bildschirmzeit, desto besser": Das Wichtigste zur neuen Leitlinie zum Bildschirmmediengebrauch. *Pädiatrie, 36*(1), 55–57. https://doi.org/10.1007/s15014-024-5369-5

Earthbound Pedagogy—Waldorf Pedagogy

Albrecht Schad and Margareta Leber

Abstract

Adolescents and young adults are finding it increasingly challenging to connect with the world. What are Waldorf education's responses to this issue? During the early years of life, up until the beginning of formal schooling, it is crucial that children are presented with imaginative and vivid individual stories that depict the world pictorially, allowing them to experience themselves as active participants within their environment. This fosters a connection with their surroundings. From around the ages of 9–10, children begin to perceive themselves as separate from the world for the first time. At this stage, natural science education is introduced, with a methodological approach that develops the individual from the whole and re-establishes connection to the world through feelings. In upper school, scientific inquiry takes precedence, aiming to cultivate an interest in and engagement with the world.

Keywords

Waldorf education • Earthbound pedagogy • Anthroposophical education • Sustainability • Environmental education • Child development • Ecological awareness • Holistic learning • Experiential learning • Natural science education • Goethean science • Steiner pedagogy

A. Schad (✉)
Methodology and Didactics of Upper Secondary Education, Biology, Geography, Freie Hochschule Stuttgart, Stuttgart, Germany
e-mail: albrecht.schad@t-online.de

M. Leber
Pedagogy, Anthropology, Methodology and Didactics of the Class Teacher, Freie Hochschule Stuttgart, Stuttgart, Germany

J. Kronenberg and E. T. Lammerts van Bueren (eds.), *On the Earth We Want to Live*, World Sustainability Series, https://doi.org/10.1007/978-3-031-98758-8_14

1 Introduction

Historically, it has been observed that at the beginning of adulthood, there is a period of happiness, characterised by anticipation and excitement about the future. Adolescents and young adults tend to seek out new experiences and are often willing to take risks and embrace adventure. Since the mid-1950s, research has suggested that life satisfaction follows a U-shaped curve: it begins with a state of happiness that characterises the period of growing up through to young adulthood. From around the mid-20s, life satisfaction begins to decline, progressively decreasing as a result of the reality of life. It is not until the age of 60 and beyond that happiness levels begin to rise again. However, this well-known U-shape of life satisfaction no longer holds. Young adults are no longer as happy as older individuals; they are not even as happy as those in midlife, as a new study from the United Kingdom concludes (Blanchflower & Bryson, 2024) (see also Braun, 2024).

Today, this stage of life is often characterised by anxiety, uncertainty, and a lack of confidence in the future. The period of young adulthood was long regarded in research as the happiest time of life. This, however, has changed. Possible reasons for this shift include global crises (such as wars and climate change), the erosion of values, and the threat to democracy, but above all, the influence of social media. The authors emphasise that this finding is not influenced by the constraints of the COVID-19 pandemic, as this trend has been measurable since 2011 (Blanchflower & Bryson, 2024). One might also consider the potential impact of the launch of the first LTE (4G) smartphone in 2010, particularly in the Global North. The underlying issue appears to be that, unlike in the past, young people today have less trust in and positive feelings towards the world even before they have experienced it. The study mentions that adolescents and young adults have not yet been able to build a "happiness buffer." There is a sense that young people are hesitant to engage with the world, with the planet itself. This is also reflected in the increasing difficulty they experience in forming relationships with others in real life. This represents a significant problem. It is proposed here that the causes of this phenomenon are not solely found in this age group but are rooted much earlier. This will be further explored below.

So, what needs to be done if we want to provide young people with the opportunity to connect with the Earth and support them in doing so? An important concept comes into play here: sustainability. What is possible in order for us, as humans and as a species, to relearn sustainable development and sustainability on all three levels?

- **Ecological sustainability**: nature should be preserved for future generations, ensuring the maintenance of ecological balance.
- **Economic sustainability**: a durable, iable foundation for earning and prosperity through the economy. This involves a moderate approach to one's own needs.

- **Social sustainability**: all members of society should globally benefit from the achievements of the democratic community of values and from justice (see also Schieren, 2010).

As early as 1986, a study was published on the significance of environmental awareness (Langeheine & Lehmann, 1986). The result is sobering. Well-intentioned environmental education increases knowledge and awareness of ecological problems to a significant extent. However, no behavioural changes can be observed. For concrete behaviour and the relationship with the Earth, even the best-intended environmental education is irrelevant. The study even suggests that it is counterproductive to constantly tell children and adolescents how we are destroying the Earth. They then want nothing more to do with the topic. This is a natural psychological defence mechanism. The same can be observed globally. At the peak of our knowledge about global and ecological interconnections, humanity is destroying the Earth to an unknown and now existentially threatening extent (Storch, 1988).

How can this significant discrepancy between knowledge and action be resolved? In the 1970s, the concept of environmental education was introduced. However, as early as the beginning of the twentieth century, concepts under the name "nature and homeland education" can be found. There are even earlier precursors.

2 The Early Ecologists

We may rightly regard Rudolf Steiner (1861–1925) as an early ecologist, even though he could not have possessed the natural scientific knowledge available to us today. Steiner's holistic, ecological perspective is evident from the quotations provided below. However, even before Steiner, there were individuals who viewed the world and the Earth in a holistic manner. As early as 1792, Goethe wrote:

> In living nature, nothing occurs that is not in relation to the whole; if our observations appear isolated, or if we regard experiments as isolated facts, this does not mean they are so. The only question is: how do we find the connection between these phenomena, these occurrences? (Goethe, 1792, p. 33)

As an astute observer of life, Goethe understood that, in living systems, everything is always interconnected. "This fact has been elaborated by modern natural science with great thoroughness over the past 100 years in an admirable manner. Noteworthy examples include the ecology of the Earth and human physiology."

As early as 1819, Goethe, in a sense, spoke as the first ecologist about the purification of the elements—Earth, water, and air—in his *West-östlicher Divan:*

> When earth and water are purified,
>
> The sun will gladly shine through the air,

Where, worthy of it, life will thrive,

Bringing healing and virtue to all.

(Goethe, 1819)

Goethe (1749–1832) here presents a central question for our future: either we follow the legacy of the Parsis, who have just spoken, and purify and preserve the Earth, water, and air, or we choose the path of destruction—self-destruction and world-destruction. (Schad, 2023a, 2023b).

Alexander von Humboldt (1769–1859) also warrants brief mention. In 1845, in the first volume of his monumental work Kosmos, he speaks of the "mingling of the atmosphere with more or less harmful gaseous exhalations" (von Humboldt, 1845, p. 166). Humboldt thus already recognised the onset of environmental degradation caused by the burning of fossil fuels. Although the role of carbon dioxide released through combustion in climate change was not identified until 1896 by Svante Arrhenius, the downright "clairvoyant" and realistic assessments of the world by these two great minds, Goethe and Humboldt, are all the more admirable. At this point, one can also mention other early ecologists, such as Henry David Thoreau, George Perkins Marsh, John Muir, and Emerson.

Rudolf Steiner understood himself as someone, who builds upon the work of Goethe. He likely does not mention A. Humboldt in his own writings, but in terms of content and intent, he certainly aligns with the ideas of Alexander von Humboldt. Goethe and Alexander von Humboldt had a close personal and scientifically motivated relationship, which was highly fruitful for both. Steiner always starts from a holistic view of both the human being and the Earth, portraying the Earth as a living organism:

> If you put single plants in front of the child and demonstrate different things from them, you are doing something, which is not reality. A plant by itself is not a reality. If you pull out a hair and examine it as though it were a thing by itself, that would not be a reality either. […] Therefore you must not teach Geography and Geology by themselves, and then Botany separately. That would be absurd. Geography must be taught together with a description of the country and observation of the plants, for the earth is an organism and the plants are like the hair of this organism. The child must be able to see that the earth and the plants belong together, and that each portion of soil bears those plants which belong to it. [...] Thus the only right way is to speak of the plants in connection with the earth, and to give the child a clear feeling that the earth is a living being that has hair growing on it. The plants are the hair of the earth. [...] In order to understand how the earth is really a part of plant life you must find out what kind of soil each plant belongs to; the art of manuring can only be arrived at by considering earth and plant world as a unity, and by looking upon the earth as an organism and the plant as something that grows with this organism. (Steiner, 1924/1989, p. 44)

> [...] it will have become clear to you that our entire earth, which we as whole humanity inhabit, is a kind of great living being, and that we ourselves are included as members within this great living being. In various lectures I have spoken about particular living phenomena of this being, our earth. (Steiner, 1918, pp. 145–146)

> As physical being, Man is not a reality in himself, but only when regarded as one with the whole Earth, just as a hand severed from the human organism cannot be regarded as in any true sense a reality: it dies, it is only thinkable in connection with the organism. A rose, when plucked, dies, and as a reality it is only conceivable in connection with the rose-tree which is rooted in the Earth; so too, to estimate Man in his entirety, in his totality, one cannot regard him as simply enclosed in the boundaries of his skin. (Steiner, 1920, p. 194)

3 The Founding of the Waldorf School

In the course of the 19th and early twentieth centuries, various reform pedagogical movements emerged. However, there was almost no exchange between these movements and the newly founded Waldorf School. The Wandervogel movement advocated for hiking in nature, experiencing community, and seeking meaning there. Susanne Rappe-Weber writes in the Historical Dictionary of Bavaria that this movement had an impact on reform pedagogy and also on the associated residential schools: "With the founding of the 'Freie Schulgemeinde Wackersdorf' (1906) in Thuringia, impulses from the Wandervogel were transferred to daily school life and tested, in the connection of community and individuality, nature experience and intellectual education…" (Rappe-Weber, 2017).

Steiner does not incorporate these efforts in any way. "The educational institutions for rural youth fared not much better under Steiner, who acknowledged their good intentions but at the same time criticised them, stating that they were 'a complete misunderstanding of what needs to happen for the culture of the future'" (Frielingsdorf, 2019, p. 77; Steiner, 1919/1932, 1992, p. 74). The representatives of the rural boarding schools had urgently advised Steiner to place the Waldorf school in the unspoiled natural environment. Steiner's response to this (paraphrased):

> Where the need is greatest, in the cities, that is where the new schools should be. And anyone who believes that one must go to the countryside as a city-dweller to find true life is mistaken, for they take their urban identity with them. In this case, only the external shell has changed, but nothing has been resolved. (Schad, 1994, p. 117; Schad et al., 2024)

Rather, the focus is on understanding the development of young people and being in inner contact with the child.

> We do not seek out special places, do not go to the forest or heath to show consideration, because we believe that what must truly be achieved for education can be attained when it is done from within our social structure, but where one truly knows what is at stake with the human being, how the human develops. (Steiner, 1924/1979, p. 161)

It is interesting to note that Rudolf Steiner does not engage with the demands and activities of his contemporaries. Instead, he establishes Waldorf pedagogy as a completely independent impulse, as if he were already foreseeing what Lehmann and Langeheine would discover more than 60 years later.

4 Relationship Between Humans and Earth

Before we turn to Waldorf education, it should be mentioned that human beings are by no means always destined to destroy nature. The view that humans can only destroy nature is still widespread. This idea traces back to Konrad Lorenz and his ethnological school. It is claimed that humans have a Paleolithic genetic makeup selected during the Pleistocene and that, under changed environmental conditions, they are incapable of doing anything other than destroying nature (Lorenz, 1984; Mohr, 1987; von Dithfurth, 1985). Ultimately, it is argued that nature will recover if humans disappear or are kept away from it.

Important facts are completely overlooked in this discussion. During the long developmental phase of humans as hunters and gatherers, we largely lived in harmony with nature. The cradle of humanity was in Africa. From there, humans migrated to Eurasia, then to Australia, and only much later, via the Bering Strait, to America (see also Storch, 2013, pp. 499–501). It is striking that, for example, after the appearance of humans in America, many large mammal species went extinct. The causes are still debated (Storch, 1988, p. 14). Some biologists propose active extermination by humans, while others attribute the disappearance of the megafauna to climate changes. This issue cannot be settled here. In Africa, humans lived alongside large animal herds until the twentieth century, which can be interpreted as a joint evolution of humans and large mammals (Storch, 1988, p. 14).

Humans have long lived in harmony with the surrounding nature. A significant development was the settlement of humans at least 12,000 years ago, which led to the domestication of plants and animals for the household. Indeed, in order to create a traditional rural cultural landscape, humans in Europe first had to clear, or rather destroy, the monotonous and species-poor forests. The wilderness was pushed back. "The subsequent creation of richly structured mosaics of orchards, hedgerows, field woodlands, forest parcels, etc., however, meant a tremendous enrichment of habitats and thus of the species spectrum, compared to the previous species-poor mature forests" (Ellenberg, 1978; Feldmann, 1987; Kremer, 1988; Schad & Suchantke, 1990), (see also Suchantke, 1993; Vahle, 1991, 2007, 2020). This cultural landscape, which one might also call a cultural savannah, is purely a human creation. There is hardly a single square metre in Europe that is not shaped by humans. The cultural landscape is an event created by humanity.

Through human intervention, plants and animals from other landscapes were able to establish themselves:

> from the high mountains, from the coasts, from the eastern steppes, and from the Mediterranean region. Thus, a landscape form emerged that aligns with the natural savannah because it achieves, more than any other landscape form, the rich variety of species and life forms of plants and animals, with humans integrated into it. The similarity between the cultural and the natural savannah is so great that botanists who travelled through East Africa about a hundred years ago referred to it as 'orchard steppe.' (Schad, 2023a, 2023b, p. 165)

The cultural landscape is only viable through the nurturing hand of humans. The creation of cultivated plants is also an example of the collaboration and further development of nature by humans. Nature can become more diverse through human involvement, and humans can continue to develop it. This constitutes a spectacular evolution-advancing act (see Suchantke, 1993, p. 111).

> A look at the cultivated plants we grow reveals this. Spelt, emmer, wheat, barley, oats, rye, peas, lentils, broad beans, turnip and radish species, rapeseed, plums, and quinces all originate from the Near East and the Mediterranean. Added to this are carrots, celery, fennel, and asparagus. South and Central America contribute significantly to our diet with maize, potatoes, tomatoes, peppers (paprika), bush and pole beans, sunflowers, Jerusalem artichokes, and pumpkins. Not least, tobacco also belongs to this group. From India, we have cucumbers and eggplants, while peaches and apricots come from China. Many of these species have become so familiar to us that we assume they originated here. Central Europe contributes almost nothing. It remains uncertain whether apples, pears, and cherries were cultivated in Central Europe. (Schad, 2023a, 2023b, pp. 168–169)

Humankind created new ecosystems in collaboration with nature, in which it was an entirely integrated part of the whole. These ecosystems could not exist without humanity; furthermore, humans were "the generating, shaping, and sustaining factor, naturally in accordance with the natural conditions of the soil, climate and so on" (Suchantke, 1993, p. 111).

It can be studied dramatically in the Alps what it means when humankind withdraws. Alpine pastures that are no longer cultivated, and thus fall into neglect, show signs of bush encroachment, species loss, and severe erosion. The avalanche risk increases significantly because snow holds better on mown meadows than on unmown ones. The observations made here can be demonstrated for all cultural landscapes (Suchantke, 1993).

The following illustration shows the increase in biodiversity of (wild) plants along the transition from natural to cultural landscapes in Europe. The figure also indicates that with the Industrial Revolution, the increase in species levels off and begins to decline more drastically. Farmers had, in their own interest, learned to handle nature gently and adapt to the respective conditions. This gradually ceased and was replaced by a focus on yield maximisation and the exploitation of nature (Fig. 1).

It is not about protecting nature by excluding humans. Instead, we should cooperate with nature based on knowledge and consideration. Both sides depend on this relationship. As we have seen, however, knowledge alone is not enough; it must be grounded in an emotional connection to the world and through the experience of being able to collaborate with nature. How can we experience ourselves as a unity with the world?

This brief digression was necessary to make it clear that humans can not only live together with nature but can also take it along on a shared evolution from which both can benefit.

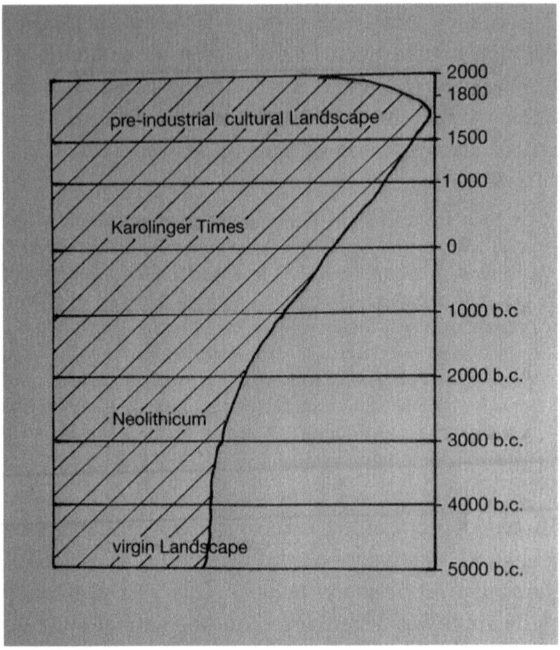

Fig. 1 Increase in the variety of (wild) plants on the way from natural to cultural landscape and their drastic decline in recent times. According to Sukopp (1978) from Riedl (1991)

5 Fundamentals of Waldorf Education

So, what is to be done? How does Waldorf education approach the issue of sustainable development? What can Waldorf education contribute to help children and young people, and ultimately adults, connect with the Earth?

Before founding the Waldorf School, Rudolf Steiner researched the question of how the physical body is related to the soul. Solving this question was one of the prerequisites for founding Waldorf education. In 1917, he first published his findings on this in a few pages (Steiner, 1917). In the realm of the soul, we can distinguish three activities: thinking, feeling, and willing.

> Thinking has the nervous system as its bodily foundation. This is widely accepted today. Feeling has the bodily foundation of the organs of the middle system, which work rhythmically throughout life, namely the respiratory and cardiovascular systems. And willing has the bodily foundation in the metabolic organs of the abdominal area and the limbs. In thinking, we can be fully awake, in feeling, we move more in a dreamy state of consciousness, and willing as an activity is not consciously experienced, but only 'in that very dull degree, which is present in sleep'. (Steiner, 1917, p. 153)." (Schad et al., 2024)

In the over 100 years that Waldorf education has existed, many colleagues worldwide have further developed this pedagogy. To elaborate on this in detail would exceed the scope of this contribution. Therefore, three recommended publications are provided for those interested: *Die Waldorfschulen und ihre Menschen* (Waldorf Schools and Their People) by Nana Göbel (2019), *Waldorfpädagogik in der Erziehungswissenschaft* (Waldorf Education in Educational Science) by Volker Frielingsdorf (Frielingsdorf, 2012), and *Handbuch Waldorfpädagogik und Erziehungswissenschaft* (Handbook of Waldorf Education and Educational Science), edited by Jost Schieren (Schieren, 2016).

Anyone who wishes to accompany children and young people in their development must embark on a path of personal development. This certainly includes thoroughly engaging with the general development of children and adolescents. The foundational work for Waldorf teachers is Foundations of Human Experience (Allgemeine Menschenkunde, GA 293) by Rudolf Steiner, together with Practical Advice to Teachers (Methodisch-Didaktisches, GA 294) and Discussions with Teachers (Seminarbesprechungen und Lehrplanvorträge, GA 295). These texts continue to enable teachers to gain a deeper understanding of the human essence, which is further corrected and developed through lived practice in the school. Moreover, the content of each subject requires constant and thorough professional development and engagement with relevant material. Teachers should strive continuously to cultivate an active relationship with the world, with the living Earth, and with humanity. This appears to be a prerequisite for guiding children on their path to connecting with the Earth and practising social and cultural sustainability. In a concluding lecture from 6 September in his Courses on the Foundations of the Art of Education (Kurse zur Begründung der Erziehungskunst, Steiner, 2019a, 2019b, GA 295, pp. 185–186), Rudolf Steiner encouraged practising the following qualities: (1) the teacher should be a person of initiative, (2) have an interest in all worldly and human existence, (3) never compromise with the untruthful, (4) remain fresh and not become embittered, (5) permeate oneself with imaginative capacity, and (6) strengthen one's sense of spiritual responsibility.

The developmental task of each child is to integrate their spiritual-soul nature with their living physical body. This intricate process extends over approximately two decades and may encounter obstacles due to health or emotional crises. The Waldorf school curriculum is designed to align with the child's developmental stages, adjusting annually to correspond with specific age groups. Consequently, all children advance to the next grade together, as the curriculum is tailored to support their development at each stage. This approach will be elaborated upon in the following discussion. Children facing developmental challenges receive individualised support through remedial teachers, the school physician, therapeutic eurythmy, and other interventions. Successfully integrating the child's spiritual-soul nature with their living physical body facilitates a harmonious connection with the Earth and earthly conditions.

6 Development of the Child and Adolescent and Their Relationship to the World

1. **Early Years**

In their early years, the child perceives themselves as an imitative being, still entirely at one with the world. Jean Piaget highlighted this as early as 1926:

> If the child does not distinguish between the psychic and the physical world, if, at the beginning of their development, they draw no precise boundaries between their self and the external world, one must be prepared for them to regard numerous entities, which for us adults are lifeless, as living and endowed with consciousness. (Piaget, 1926)

Up until approximately the age of nine, the child does not differentiate between subject and object. Teaching intellectual content at this stage disrupts the child's natural connection to the world and effectively impairs the foundation for ecological experience. Ecological experience means feeling one with the world through active engagement. At this age, the child does not merely live emotionally in their surroundings; the objects of their environment also live within them.

At Waldorf schools, the phenomena and beings of nature are presented to children of this age through vivid storytelling, where animals and plants feel and act like human beings. These "meaningful stories," as Steiner called them, are not natural science lessons but feature characters akin to fables, embodying distinct moral and ethical qualities. Children love these stories and usually want to hear them repeatedly. They develop affection and admiration for the positively portrayed characters and aversion to negative and harmful ones. This is a moral education that does not rely on submission to rules but allows children to remain inwardly free. They are given the space to develop sympathy for good and antipathy for evil themselves. This provides a much firmer foundation for future self-determined action than any behaviour adopted through coercion (Schad & Suchantke, 1990, pp. 276–277). At this stage, the child primarily lives through action.

As early as 1917, two years before the founding of the first Waldorf school, Rudolf Steiner emphasised the importance of not approaching growing children with factual explanations but instead presenting the lives of animals, plants, and so forth in a vivid and illustrative way. "Whether or not these things happen exactly as described is entirely unimportant; what matters is that they are meaningful" (Steiner, 1917/2013, p. 203). From this, the literary genre term "meaningful story" originates.

Particularly in the first three school years (ages 7–9), teachers are encouraged to present the entire world to children through imaginative, individual stories. "It is important, especially in the very first school year, to awaken the child to their surroundings—a spiritual awakening, so they learn to truly connect themselves with their environment" (Steiner, 1919/2019, p. 196). This approach fosters both an awakening to the surrounding world and an emotional connection with it.

2. From the Age of 9/10: The Beginning of Science Education

The child's relationship with the world undergoes a fundamental change around the age of 9 or 10. For the first time, we can truly speak of an "environment" in the child's experience, as they begin to perceive the world as separate from themselves. A gap opens between the world and the child. Through this experiential distancing, the child can come to understand the world, as recognition requires a degree of separation from the object of knowledge.

This marks the appropriate time to begin science education. Rudolf Steiner emphasised two key principles in this context: Firstly, all science teaching should start with the human being.

> It is of great importance to know that one thoroughly spoils what should be achieved in natural history lessons if one does not begin with a discussion of the human being. You may rightly say: 'When the child is nine years old, we can tell them little about the natural history of the human being.' However little it may be, whatever can be taught to the child about the human being should be given as a preparation for all other natural history lessons. You must be aware, in doing this, that the human being is, in a certain sense, a synthesis, a summary of all three kingdoms of nature; that the other kingdoms are united in the human being on a higher level. You need not tell this to the child, but through the course of your teaching, you must evoke in the child a feeling that the human being is such a summary of all other kingdoms of nature. You will achieve this if you give the discussion of the human being the necessary emphasis, if, in the way you treat the human being, you evoke in the child the impression of the importance of the human being within the whole order of the world. (Steiner, 1975, p. 97)

Steiner aims to ensure that through this approach, the child can feel at one with the world, and indeed, with the entire cosmos. Secondly, he emphasised that in lessons, for example in botany, teaching should not start with isolated or dissected plants. Instead, he insisted, "we should never present a plant to children of a young age other than in connection with the Earth and in connection with the forces of the Sun" (Steiner, 1986, p. 161). This represents a truly ecological approach. Such a method fosters an understanding of interconnections and relates to the child's experiential world.

As a practical example, a brief excerpt from a main lesson book on botany from Class 5 is presented here. The class teacher, Norbert Dolderer, dictated these short texts to his pupils:

> 19 April: After the cold winter days, as the sun climbs higher each day and its light streams more powerfully into the earthly realm, as the air and soil gradually warm, and as spring rain encourages the seeds in the soil to germinate and grow, there is almost no patch of earth where a little plant does not emerge from the dark ground. Everywhere, growth, budding, and blooming can be seen!
>
> Where the forces of the earth, water, air, light, and warmth work together, living plants can develop their roots, stems, leaves, flowers, and fruits.
>
> 20 April: Like a golden-yellow carpet, it can be seen on meadows. Yet, it also sits in cracks in walls and even pushes through asphalt and concrete with great force.

For generations, children have woven crowns and chains from its hollow stems, while mothers have worried about the "milk stains." As a green, juicy plant, it is a favoured food for many animals, though farmers value it less as hay.

In early spring, we find a small rosette of leaves nestled close to the ground, appearing right after the snow melts.

21 April: In this rosette sit the firm green buds, which later will be carried upward by the stems. There, the green petals opens and reveals its golden-yellow splendour. What looks like one large flower is an illusion! It consists of well over 100 small blossoms! They are densely packed in a "basket," all sitting on the same receptacle and enclosed by green petals.

In the warm sunlight, the green petals open. In the evening and on cool, rainy days, it closes again.

22 April: One day, the sepals remain closed even in the morning. At the top opening, we can see the tiny blossoms inside gradually wither and are pushed outward. And then, one morning, the sepals bend sharply downward, revealing a delicate, white sphere: the dandelion clock. As soon as the wind blows, the seeds detach from the white cushion and are carried away by their parachutes. Wherever they land, they anchor themselves with tiny barbs, and the following year, a new plant grows.

(Norbert Dolderer, Class Teacher)

This fundamental idea of the methodological approach—starting from the whole and moving to the parts—can be applied not only to natural science subjects but across the entire breadth of the curriculum. Let us consider this in relation to mathematics. Typically, teaching progresses additively, beginning with the smallest number, one, and moving to increasingly larger numbers. This approach, however, reinforces the notion that the world is composed of separate elements. An alternative proposal would be to approach mathematics differently: to present.

> The one as an expression of essential unity, as large as the entire world, and to introduce the emergence of numbers as a process of dividing, even breaking apart into multiplicity, using vivid storytelling. Numbers should be introduced not additively but divisively. This approach avoids the confusion of numbers with quantities and, in an age-appropriate way, fosters thinking in terms of groups or sets. Steiner's (1997, p. 13) proposal is clear: not starting with the elements and then the connections, but the reverse—the whole as the origin, with the parts forming out of it. This method begins influencing the ecological dimension from the very first arithmetic lesson. Schad and Suchantke (1990, p. 275).

In Class 1, all four basic arithmetic operations are introduced relatively quickly, without being sharply separated from each other. Does this not constitute an overwhelming demand on the children? It would indeed be overwhelming if the expectation were to solve abstract calculations such as $12 \div 4 = 3$ or $20 \div 4 = 5$. However, Rudolf Steiner's intention in recommending the almost simultaneous introduction of the four operations can only be understood in the context of his methodological approach. He begins not with individual elements but with the whole of a given quantity, which can then be divided in various ways. In other words: the starting point is not the problem but the result, and the task is to find as many suitable problems as possible. This method fosters freedom, rooted in initiative and inner self-activity.

Through this methodological approach, we support the children in forming a connection with the Earth. Steiner foresaw this with great clarity even then:

> In this way, we are able to introduce the child to life so that it fits in, grasps wholes, and does not always proceed from the lesser to the greater. This exerts an extraordinarily strong influence on the child's soul life. When the child becomes accustomed to adding, it develops precisely that moral disposition that primarily cultivates a tendency toward the desirable. However, when one moves from the whole to the parts, the child develops an inclination not to overdevelop desire but instead to develop what can be called, in the noblest sense of the word, composure, moderation. And there is an intimate connection between what pleases and displeases a person morally and the way one has learned to deal with numbers. (Steiner, 1922, pp. 109–110)

The methodological approach, therefore, is to develop the individual elements from the whole. As already mentioned, this approach can also be applied to many other subjects at this age.

When introducing letters, there are several methods. One option is to write the word as a whole, e.g., "BATH", [German: 'Bad'] on the board and then, through speaking, extract the individual letters. Another approach involves drawing:

> Consider the following. Take the word 'mouth' ['Mund' in German]. (…) If you ask the child to draw a mouth, but to do so as a painting, letting them add splashes of red colour, and then have the child pronounce the word and say: Now don't say the whole word, just start it—M—and gradually turn the upper lip into this M, then you get the M from the painted mouth we began with. (Steiner, 1924/1989, p. 35)

A widely used method involves telling stories and creating beautiful illustrations on the blackboard, from which the letter is "extracted." This tends to yield the individual letters, which are then assembled (a synthetic approach).

When painting, the classic method involves starting with abstract colour fields, which are then brought into relation with each other and eventually form, for example, a fish.

In geology, one might begin with a description of mountain ranges, such as the limestone Alps or the central crystalline Alps. Only then might one show a piece of limestone, granite, or a crystal such as calcite, quartz, or tourmaline.

It is essential for children aged approximately 7–14 that teaching begins with a vivid, experience-rich depiction, allowing the child to emotionally connect with the essence of the animal, plant, or landscape. Only after this connection is made does the lesson move into nature to encounter the actual animal, plant, or stone. Such outings could be walks or more extensive excursions. This creates a sense of re-encountering an old friend. Without this preparatory inner connection, the subject remains alien and without relation to the child.

Mareike Schmitz provides a compelling account of how effective this can be. At the start of the school year, she took her Class 1 pupils for a walk in a nearby park to let them move around after a long bus journey. She was shocked to observe how hostile the children were toward the plants and animals: they chased them away, tore them apart, destroyed them, and trampled on them. She then sought to foster an inner connection between the children and the living beings:

> Here I recalled Rudolf Steiner's advice that one must develop an 'artistic sense' for children of this age, "to truly bring everything to life. The teacher must animate; the teacher must let the plants speak, let the animals act morally; the teacher must be able to transform everything into fairy tales, fables, and legends..." (Steiner, 1921/22). So, in the classroom, I became the raindrop or the earthworm, the bird or the butterfly, the bud on the tree, or the blade of grass trembling in the wind. And when we were outside again, the children pointed out all these things to me! I never referred to our shared outdoor experiences while telling stories, but I noticed a gentle awakening in some children to the phenomena of nature. (...) Soon, it became our path, our stream, our tree, and park; they became our squirrels and our blackbirds. When the stream dried up in summer, the children were sad and upset ('Where has our water gone?'), and they insisted we must give the stream its water back. This was accompanied by an inner urge to discover the cause of the drought—and they succeeded in finding it. (Schmitz, 1989, p. 93).

There are also interesting approaches in the current educational discourse concerning earth-connected pedagogy (see also Rademacher & Roller, 2024).

7 Upper School

In the upper school, scientific understanding moves to the forefront. Now, previous experiences and new perspectives are consciously and thoughtfully examined. The lessons take on a scientific character, focusing on cultivating an appropriate mode of thinking in all subjects. Steiner advises, particularly for this age, counteracting the danger of excessive introspection by fostering an interest in the world. This is a truly ecological approach: to perceive, explore, and recognise what one owes to the world. One might also call it ecological thinking.

The ecological dimension of Waldorf education even extends to the humanities, as can be demonstrated by two examples from history lessons. A classic example is the history curriculum in Class 10, where the second cycle of historical study begins. Walter Johannes Stein, a German and history teacher at the first Waldorf school, suggested in 1923 that the history lessons on antiquity in Class 10 should start with an exploration of the geographical and climatic conditions of Greece. This would provide a context for vividly illustrating cultural development (cf. Steiner, 2019a, 2019b, p. 41; Lindenberg, 2008; Esterl, 2005). By beginning with the natural conditions of the external world, independent of human influence, the aim is to direct the students' focus—away from the inner emotional challenges of puberty—towards objective connections. This enables them to experience the world as a meaningful whole (cf. Bartoniczek, 2014).

For Class 9, the influence of the ecological approach in history lessons can be seen in Albert Schmelzer's narrative about Louis XIV. Schmelzer begins his account with a description of the palace complex of Versailles. He leads the students through the gardens and architecture into the interior of the palace, drawing attention to the furnishings and the light streaming through the windows. He then imagines an unusual procession entering the stage on its way to the king's bedroom. What follows is a detailed depiction of the Lever du Roi, the French king's morning awakening ceremony (cf. Schmelzer, 2000, p. 10). Through his narrative, Schmelzer directs the students' attention from the grandeur of the palace to its focal point—the figure of Louis XIV lying in bed.

Schmelzer continues this method the next morning, deepening the previous day's content during the main lesson by introducing the concept of absolutism. In line with Steiner's understanding of concepts, Schmelzer aims to develop an open, dynamic concept. He suggests, for example, comparing French absolutism with governance in Russia or the German states. Alternatively, one could explore connections with philosophy, architecture, and baroque garden design. Another possibility is to compare it with the concept of rule in ancient Egypt under the pharaohs or with 20th-century dictatorships (cf. Schmelzer, 2000, pp. 17–18). While the first day moves from the grand to the detailed, the second day broadens the perspective, embedding the topic into a larger context.

8 Curriculum of the Waldorf School Exemplified by Biology Lessons

Class 4:

The block begins with the study of human biology. A central theme is the form and structure of the human being, expressed in the fundamental division into head, chest, and abdomen-limbs. From this foundation, the observation of animals such as the octopus, mouse, deer and horse is developed.

Class 5:

The study of zoology is continued, with an added focus on functional aspects (e.g. eagle, lion, cow). A first block on botany is introduced, addressing the evolutionary and chronological appearance of plants. The plants are considered in relation to their connection with and colonisation of the Earth.

Class 6:

The study of botany is further expanded. For instance, plants significant to human beings, such as food crops and medicinal plants, may be explored. In the geology block, the rocks of the Earth are studied, beginning with mountain formations. A new subject is also introduced: gardening. Pupils gain hands-on experience working with the soil. They discover the labour-intensive process of producing food, both for human beings and for animals under human care.

Class 7:

The teaching of nutrition and health should take place in the final years of middle school, though it does not necessarily require a separate block. According to Steiner:

> "In natural history, in physics lessons, even in the extended geography lessons, or even in history lessons, the teaching of nutrition and health can be integrated. From this, you will see that it is not necessary to include it as a separate subject in the curriculum, but that it must live within the teaching in such a way that it is interwoven into the lessons."
>
> (Steiner, 1919/2019, p. 202)

Classes 7–10:

During these four school years, a period of significant physical and emotional transformation, pupils are accompanied in their biology lessons by the study of human biology. The aim is to help pupils develop a positive relationship with their own bodies.

Class 11:

The inner instability characteristic of puberty begins to find equilibrium. The curriculum retraces the path taken in Classes 4–6 in reverse order. Main topics include unicellular organisms, cell theory, lower plants and human embryology, which also raises the question of the origin of life and the boundary between the living and the non-living.

Class 12:

The focus shifts to higher plants. The curriculum is rounded off with a survey of the animal kingdom and the evolution of humanity, completing the biology education of the Waldorf school years (Fig. 2).

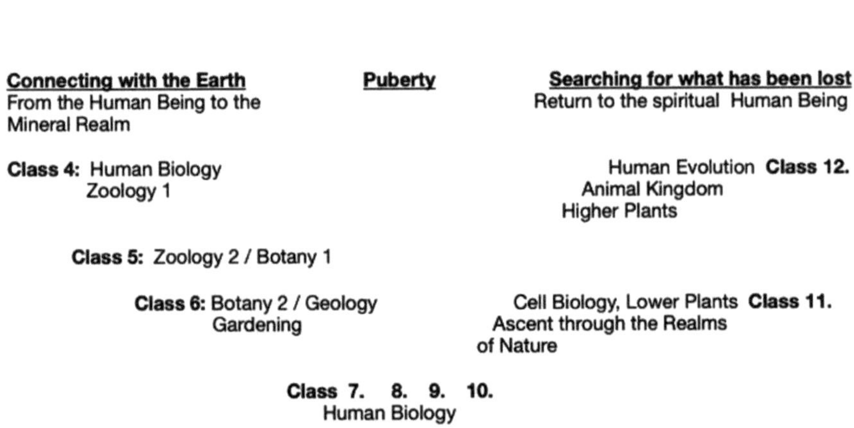

Fig. 2 Waldorf Curriculum (Schad)

> We are presented with a curriculum that does not explain the world from the perspective of the lifeless but instead approaches it multiperspectivally from the reality of the spiritual, the soul, and the living. The other realms of the world are derived in such a way that the lifeless is the last to emerge. During the class teacher years, up to Class 6, the children are guided on their path of connecting with the Earth. By Class 6, they have fully arrived in the earthly realm. At this stage, geology and gardening appear on the curriculum. This marks the descending arc of the curriculum, where children are gradually supported in forming a connection with the Earth. However, this is accompanied by a loss of the connection to the cosmic realm from which the journey began.
>
> Over the following four school years, we guide pupils through the years of puberty. Childhood comes to an end, and they become fully grounded in the earthly realm. Engaging with earthly conditions transforms the pupils. The maturity of the Earth begins, and with it, emotionally challenging years ensue. The discovery of their inner soul life, the birth of the astral body—as Steiner refers to it—extends over many years. The biology curriculum accompanies this phase with human biology. These years are also marked by a minimum incidence of cancer and a peak in suicides.
>
> The ascending arc of the curriculum in Classes 11 and 12 has an entirely different character. The blocks now follow the reverse path. The young people are supported in their search for what they lost during the descending path. The search for their spiritual home begins anew. (Schad, 2014, p. 20)

One can observe that the descending arc of the curriculum represents the decisive years in which the child connects with the Earth. This is to be achieved through the sequence of topics and the method employed.

At the Waldorf school, however, it is by no means limited to merely observing nature. A great deal of practical engagement is also planned:

How wonderful it is when the kindergarten has a small garden where children can play and take part in caring for plants. It would be ideal if the maintenance of garden beds were carried out by those working there, with the children helping in this process.

In Class 3, there is an agriculture block, which could take place in the school's own garden or on the farm of a friendly farmer. What matters is the experience of interconnections and development over time. This includes learning that one must first care for plants and soil before earning the right to harvest.

It is particularly noteworthy that this agriculture block already occurs in Class 3. The focus is primarily on action. Botany does not come until Class 5, and the sequence is critical. Excursions into nature, school trips and hiking also play an essential role.

Nature study begins in Class 4, and from Class 6, a new subject is introduced: gardening, which can continue through to Classes 9 or 10.

It must be said, however, that gardening is not an invention of Waldorf pedagogy. In Germany's public school system, gardening had already developed as a tradition rooted in the ideas of Pestalozzi and Kerschensteiner as part of the work school movement. Steiner was evidently aware of this and adopted the practice, but his rationale for including gardening is entirely different. In the early 1920s, Steiner stated:

> People who have gone through such lessons in school will be able to make decisions about whether a method or a measure in agriculture is right or wrong, not because they have learned it but out of the certainty of their feelings. Moral forces are also trained through such teaching. The impact of such lessons will only show later in the social attitudes of adults. (Oral communication to Lilly Kolisko, from the notebook of Mrs Michels, published by Rudolf Krause, 1984, p. 13)

Steiner envisioned gardening lessons as having a positive effect on social attitudes later in life. This is a crucial point because social attitudes are not only important toward other human beings but also toward the living being of the Earth.

Steiner suggested that gardening lessons, conducted over many years, should culminate in the grafting of fruit trees. This proposal highlights the core intention: humanity's ability to advance nature, something that was long customary in the past. How can nature be advanced? Through breeding and refinement. This creates new forms of plants and animals as well as human-designed ecosystems. The school garden is the best example of this. This is why it was emphasised earlier that human beings have a culture-creating capacity, the ability to work with nature and to carry it forward in a shared development and evolution. Steiner already had this vision in mind—a grand vision indeed.

There are also various practicals, such as agricultural and forestry internships, where pupils work on a farm or in a forestry operation.

All these activities aim to foster a nurtured and tangible relationship with nature and the Earth. Through this, a sense of shared responsibility can develop.

At the Waldorf School, there is no ecology class in the strict sense, apart from in the upper grades. Even there, it is embedded in broader contexts, such as geography or biological studies. The rationale for this lies in what has been explained thus far. Simply lecturing intellectually about how we are destroying the world achieves little to nothing. Similarly, providing extensive knowledge about ecological relationships at an early age is of little practical benefit.

What matters is ensuring that during the first seven to eight years of life, children are primarily engaged in action. From Class 4, when nature studies begin, everything should be linked to feelings, building an emotional connection to the world. Only in the upper grades, once the child has connected with the Earth, is what has been experienced through action and emotionally bonded to, then intellectually understood. After the holistic, image-oriented approach of the class teacher years, the upper school introduces analytical methods. Yet even here, non-metric scientific methods must not be neglected:

- Projective geometry in mathematics
- Incorporating psychosomatics into human biology
- Considering the environment in cell theory
- Including co-evolutionary aspects
- Viewing the Earth as a whole, as a living being (Class 10), which forms the basis for the globalisation block in Class 12.

Essentially, all subjects can contribute to helping children and young people develop a positive relationship with the Earth.

9 Conclusion

We will not be able to solve the environmental crisis if we fail to understand that the Earth is a living being (see A. Schad's contribution: The Earth as a Living Being) and if we do not internalise this reality as part of our lived experience. Moreover, we will not be able to address these challenges unless the methodical approach outlined here fosters in children and young people a connection with the Earth, a sense of reverence for the beauty of the world, and a willingness to help both other human beings and the Earth itself.

Anthroposophy contributes to an Earth-connected pedagogy that integrates not only ecological, social and economic aspects but also the fundamental questions of the existence and future of the Earth and humanity in a positive way. For this, it is crucial to understand and take into account the developmental principles governing children and young people. Waldorf education is a pedagogy that, so to speak, works from the ground up: first, it prioritises supporting the activities of children. Then, their actions are emotionally connected to the world. Finally, in the upper school, everything can and must be processed intellectually. This sequence is essential to enable children, in the course of their development, to form a connection with the Earth and with humanity. The order matters!

Every morning, the pupils recite, as a whole class, the so-called Morning Verse at the beginning of the main lesson. We would like to conclude this contribution with this verse, which is spoken daily from Class 5 until the end of upper school. Through this practice, the connection between the human being and the minerals, plants, animals, other human beings, the Earth and the cosmos can shine forth in the classroom every day (Fig. 3).

I do behold the world,

Wherein there shines the sun,

Wherein there gleam the stars,

Wherein there lie the stones.

The plants, they live and grow,

The beasts, they feel and live,

And we to spirit give

A dwelling in our souls.

I do behold the soul

That living dwells in me.

God's Spirit lives and weaves

In light of sun and soul,

In heights of world without,

In depths of soul within.

To Thee, O Spirit of God,
I seeking turn myself,
That strength and grace and skill
For learning and for work
in me may live and grow.
—Rudolf Steiner, Sept. 1919 (p. 351)[1]

Fig. 3 Wild Carrot (Daucus carota), 9. Juli 2024, Tornuv Mlyn, Tschechien (Schad)

[1] The translation is based on various translations found in US Waldorf schools, modified by Douglas Kennedy, Stuttgart.

References

Bartoniczek, A., (2014). *Die Zukunft entdecken: Grundlagen des Geschichtsunterrichts* [*Discovering the future: Fundamentals of history teaching*] (p. 634). Stuttgart, Germany: Verlag Freies Geistesleben.

Blanchflower, D., & Bryson, A. (2024, June 6). The global loss of the U-shaped curve of happiness. Global Interdependence Center.

Braun, A. (2024, February 20). Junge Erwachsene sind immer unglücklicher [Young adults are increasingly unhappy]. Tagesschau.

Ellenberg, H. (1978). *Vegetation Mitteleuropas mit den Alpen in ökologischer Sicht* [*Vegetation of Central Europe with the alps from an ecological perspective*]. Stuttgart, Germany: Ulmer Verlag.

Esterl, D. (2005). *Was geschieht in Geschichte? Zum Geschichtsunterricht an der Waldorfschule* [*What happens in history? On history teaching at Waldorf schools*] (p. 71). Stuttgart, Germany: Verlag Freies Geistesleben.

Feldmann, R. (1987). *Industriebedingte sekundäre Lebensräume als sicherheitswissenschaftliches Problem* [*Industry-induced secondary habitats as a safety science problem*] (Habilitation thesis). Wuppertal, Germany: Bergische Universität-Gesamthochschule Wuppertal.

Frielingsdorf, V. (2012). *Waldorfpädagogik in der Erziehungswissenschaft* [*Waldorf education in educational science*]. Weinheim, Germany: Beltz Juventa.

Frielingsdorf, V. (2019). *Geschichte der Waldorfpädagogik* [History of Waldorf education] (p. 77). Weinheim, Germany: Beltz Verlag.

Göbel, N. (2019). *Die Waldorfschulen und ihre Menschen* [*The Waldorf schools and their people*] (Vol. 3). Stuttgart, Germany: Verlag Freies Geistesleben.

Krause, R. (1984). *Zum Gartenbau-Unterricht an den Waldorfschulen* [*On horticulture teaching at Waldorf schools*] (p. 13). Wanne-Eickel, Germany: Hiberniaschule.

Kremer, B. P. (1988). Lebensraum Dorf—Chancen für die Natur im ländlichen Siedlungsbereich [Habitat village—Opportunities for nature in rural settlements]. *Natur und Museum, 118*, 225–239.

Langeheine, R., & Lehmann, J. (1986). *Die Bedeutung der Erziehung für das Umweltbewusstsein: Ergebnisse pädagogisch-empirischer Forschung zum ökologischen Wissen und Handeln* [*The importance of education for environmental awareness: Results of pedagogical-empirical research on ecological knowledge and action*]. Kiel, Germany: IPN.

Lindenberg, C. (2008). *Geschichte lehren: Thematische Anregung zum Lehrplan* [*Teaching history: Thematic suggestions for the curriculum*] (3rd ed., pp. 220–224). Stuttgart, Germany: Verlag Freies Geistesleben.

Lorenz, K. (1984). *Das sogenannte Böse: Zur Naturgeschichte der Aggression* [*On Aggression*] (Original work published 1963). Munich, Germany: Piper Verlag.

Mohr, H. (1987). *Natur und Moral: Ethik in der Biologie* [*Nature and morality: Ethics in biology*]. Darmstadt, Germany: Wissenschaftliche Buchgesellschaft.

Piaget, J. (1926). *La représentation du monde chez l'enfant* [*The child's conception of the world*]. Paris, France: Alcan. (German translation: *Das Weltbild des Kindes*. Frankfurt, Germany: Ullstein, 1980).

Rademacher, M., & Roller, C. (2024). Die Zukunft der schulischen Bildung im Fokus [Focusing on the future of school education]. *Biologie in unserer Zeit, 54*(1), 11–13.

Schad, W., & Suchantke, A. (1990). Ökologische Krise und Waldorfpädagogik: Die Umweltproblematik als Denkfolge und Erziehungsaufgabe [Ecological crisis and Waldorf education: Environmental issues as a consequence of thinking and educational task]. In F. Bohnsack & E. -M. Kranich (Eds.), *Erziehungswissenschaft und Waldorfpädagogik* [*Educational science and Waldorf education*] (pp. 276–277). Weinheim, Germany: Beltz Verlag.

Schad, W. (1994). *Erziehung ist Kunst* [*Education is art*] (p. 17). Stuttgart, Germany: Verlag Freies Geistesleben.

Schad, A. (2023a). *Vom Leben unserer Erde* [*On the life of our Earth*] (p. 165). Stuttgart, Germany: Verlag Freies Geistesleben.
Schad, A. (2023b, June). Wir leben auf dem Planeten der Hühner [We live on the planet of chickens]. *Newsletter der Pädagogischen Sektion Dornach.*
Schad, A., Wiechert, C., & Kullak-Ublick, H. (2024). Besinnung auf die Waldorfpädagogik [Reflection on Waldorf education]. *Newsletter der Pädagogischen Sektion Dornach.*
Schad, A. (2014). Vom Konzept des Waldorflehrplans am Beispiel des Faches Biologie [On the concept of the Waldorf curriculum using the example of biology]. *Rundbrief der Pädagogischen Sektion am Goetheanum, Michael, 2014*(52), 12–21.
Schieren, J. (2010). Bewusstseinsgrundlagen der Nachhaltigkeit [Foundations of sustainability consciousness]. *Anthroposophie: Vierteljahresschrift zur anthroposophischen Arbeit in Deutschland, 64*(1), (Ostern) (Nr. 251).
Schieren, J. (Ed.). (2016). *Handbuch Waldorfpädagogik und Erziehungswissenschaften* [*Handbook of Waldorf education and educational science*]. Weinheim, Germany: Beltz Juventa.
Schmelzer, A. (2000). *Wer Revolution machen will…: Zum Geschichtsunterricht der 9. Klasse an Waldorfschulen* [*Whoever wants to make a revolution…: On history teaching in 9th grade at Waldorf schools*] (p. 10). Stuttgart, Germany: Verlag Freies Geistesleben.
Schmitz, M. (1989). Die Naturferne der Stadtkinder [The estrangement of urban children from nature]. *Erziehungskunst, 53*(2), 91–96.
Steiner, R. (1917). *Von Seelenrätseln* [*On mysteries of the soul*] (p. 150). Dornach, Switzerland: Rudolf Steiner Verlag.
Steiner, R. (1917/2013). *Die spirituellen Hintergründe der äußeren Welt* [*The spiritual backgrounds of the outer world*] (GA 177, p. 203). Dornach, Switzerland: Rudolf Steiner Verlag.
Steiner, R. (1918). *Vortrag St. Gallen, 30. März 1918, GA 181, p. 145 ff.* [*Lecture at St. Gallen*]. Retrieved from https://rsarchive.org/Lectures/GA181/English/FB1986/19180330p01.html
Steiner, R. (1919/2019). *Erziehungskunst. Seminarbesprechungen und Lehrplanvorträge* [*The art of education. Seminar discussions and curriculum lectures*] (GA 295). Dornach, Switzerland: Rudolf Steiner Verlag.
Steiner, R. (1919/2019). *Erziehungskunst. Methodisch-Didaktisches* [*The art of education. Methodical-didactic*] (GA 294). Dornach, Switzerland: Rudolf Steiner Verlag.
Steiner, R. (1919). *Wahrspruchworte* [*Words of prophecy*] (GA 40, p. 351). Dornach, Switzerland: Rudolf Steiner Verlag.
Steiner, R. (1919/1932, 1992). *Allgemeine Menschenkunde als Grundlage der Pädagogik. Erster Teil* [*General anthropology as the basis of education. Part one*] (9th rev. ed., GA 293). Dornach, Switzerland: Rudolf Steiner Verlag.
Steiner, R. (1920). *Vortrag Dornach, 9. Mai 1920, GA 201, p. 194* [*Lecture at Dornach*]. Retrieved from https://rsarchive.org/Lectures/GA201/English/RSP1972/19200509p01.html
Steiner, R. (1922). *Die geistig-seelischen Grundkräfte der Erziehungskunst* [*The fundamental spiritual forces of the art of education*]. In Lecture delivered on August 21, 1922 (GA 305, p. 112). Dornach, Switzerland: Rudolf Steiner Verlag. (Note: Publication year "1979" appears in some citations; here, the lecture's delivery year is primary.)
Steiner, R. (1924/1989). *Die Kunst des Erziehens aus dem Erfassen der Menschenwesenheit* [*The art of educating from the understanding of human nature*] (GA 311, p. 35 ff). Dornach, Switzerland: Rudolf Steiner Verlag. Retrieved from https://rsarchive.org/Lectures/GA311/English/AP1982/19240814w01.html. (Note: Includes lecture from Torquay, 14 August 1924.)
Steiner, R. (1924/1979). *Anthroposophische Menschenkunde und Pädagogik* [*Anthroposophical anthropology and pedagogy*] (GA 304a). Dornach, Switzerland: Rudolf Steiner Verlag. (Duplicate entry noted; this appears earlier.)
Steiner, R. (1924/1979). *Anthroposophische Menschenkunde und Pädagogik* [*Anthroposophical anthropology and pedagogy*] (GA 304a, p. 161). Dornach, Switzerland: Rudolf Steiner Verlag.
Steiner, R. (1975/1997). *Methodisch Didaktisches. GA 294, Teil zwei, 7. Vortrag* [*Methodical-didactic.* GA 294, Part Two, Lecture 7, p. 97]. Dornach, Switzerland: Rudolf Steiner Verlag.
Steiner, R. (1978). *Die gesunde Entwickelung des Menschenwesens: Eine Einführung in die anthroposophische Pädagogik und Didaktik* [*The healthy development of the human being: An*

introduction to anthroposophical pedagogy and didactics] (GA 303, 10th lecture). Dornach, Switzerland: Rudolf Steiner Verlag.

Steiner, R. (1986). *Gegenwärtiges Geistesleben und Erziehung* [*Contemporary spiritual life and education*]. In Lecture delivered on August 13, 1923 (GA 307, p. 161). (Ambiguity: Original lecture date vs. publication date; clarification needed.)

Steiner, R. (2005). *Wahrspruchworte* [*Words of prophecy*] (p. 351, GA 40). Dornach, Switzerland: Rudolf Steiner Verlag. (Ambiguity: Reprint date vs. original publication; verify edition.)

Steiner, R. (2019a). *Konferenzen mit den Lehrern der Freien Waldorfschule in Stuttgart 1919 bis 1924* [Conferences with the teachers of the Freie Waldorfschule in Stuttgart 1919–1924] (Bd. III, 5th revised and expanded ed., GA 300c). Basel, Switzerland: Rudolf Steiner Verlag.

Steiner, R. (2019b). *Allgemeine Menschenkunde—Methodisch-Didaktisches Seminar. Studienausgabe* [*General anthropology—Methodical-didactic seminar*] (GA 293, GA 294, GA 295). Dornach, Switzerland: Rudolf Steiner Verlag.

Storch, V. (1988). Prozesse der Umweltzerstörung [Processes of environmental destruction]. In F. Cube & V. Storch (Eds.), *Umweltpädagogik* [*Environmental education*] (p. 18). (Ambiguity: Publisher details missing.)

Storch, V. (2013). *Evolutionsbiologie* [*Evolutionary biology*] (p. 499 ff.). Berlin, Germany: Springer Spektrum.

Suchantke, A. (1993). *Partnerschaft mit der Natur—Entscheidung für das kommende Jahrtausend* [*Partnership with nature—A decision for the coming millennium*]. (Ambiguity: Publisher and location details are not provided.)

Sukopp, H. (1978). Veränderungen von Flora und Vegetation durch den Menschen [Changes in flora and vegetation due to humans]. In G. Olschowy (Ed.), *Natur und Umweltschutz in der Bundesrepublik Deutschland* [*Nature and environmental protection in the Federal Republic of Germany*]. (Ambiguity: Page numbers and publisher details are missing.)

Vahle, H. -C. (2007). *Die Pflanzendecke unserer Landschaft: Eine Vegetationskunde* [*The plant cover of our landscape: A study in vegetation science*]. (Ambiguity: Publisher details are not provided.)

Vahle, H. -C. (2020). Die Kulturlandschaft als Organismus und ihr Entwicklungspotential [The cultural landscape as an organism and its development potential]. In B. Rosslenbroich (Ed.), *Perspektiven einer Biologie der Freiheit: Autonomieentwicklung in Natur, Kultur und Landschaft* [*Perspectives on a biology of freedom: Autonomy development in nature, culture, and landscape*] (pp. 450–489). (Ambiguity: Publisher details are missing.)

Vahle, H.-C. (1991). Die Idee der Kulturlandschaft am Beispiel Nordwestdeutschlands [The idea of cultural landscapes exemplified in northwestern Germany]. *Die Drei, 61*(7/8), 581–612.

von Goethe, J. W. (1792). Der Versuch als Vermittler von Objekt und Subjekt [The experiment as mediator of object and subject]. In *Faust* (Vol. 25, p. 33).

von Goethe, J. W. (1819). *West-östlicher Divan, Buch des Parsen, Vermächtnis alt persischen Glaubens* [*West-Eastern divan, book of the parsees, legacy of ancient Persian faith*].

von Humboldt, A. (1845). *Kosmos* [*Cosmos*] (p. 166). Stuttgart, Germany: Cotta'scher Verlag.

von Dietfurth, H. (1985). *So laßt uns denn ein Apfelbäumchen pflanzen – es ist so weit* [*So let's plant an apple tree—it's time*]. Munich, Germany: C. Bertelsmann Verlag.

Weber, S. (2017). Wandervogel [Wandering birds]. In *Historisches Lexikon Bayerns* [*Historical Dictionary of Bavaria*]. Retrieved from https://www.historisches-lexikon-bayerns.de/Lexikon/Wandervogel

Earthbound Economics

Nathaniel Williams

Abstract

This section introduces Rudolf Steiner's ideas on associative economics (one facet of his vision for a threefold social ordering of society) with particular attention to their intersection with salient issues in the contemporary discourse on sustainable development. The article shows how ideas of an associative form of economics from a century ago, and large scale projects they inspired, focused on central questions dealt with in the contemporary discourse on sustainable development: developing economic theory and practices that include ecological and scientific awareness of natural boundaries and systems; a unique critique of economic theory that focuses entirely on growth; ways of framing, and working together, while managing common pool resources and associated dilemmas of the commons; and the inherent requirement of global thinking and action in economic and ecological matters. Steiner's suggestion that an imaginative and pictorial method in economic science is also presented, which he described as akin to certain forms of thinking in the life sciences and botany, and necessary for revealing certain corrections of pseudo-economic judgments. Besides revealing the commodification of land, labour and capital as such pseudo-judgements, this variety of thinking leads to a proper estimation of the role of value destruction through the widespread funding of education and culture, and an implicit shift in focus from quantity and consumption to quality and meaning.

N. Williams (✉)
Youth Section at the Goetheanum - Freie Hochschule für Geisteswissenschaft, Dornach, Switzerland
e-mail: nathaniel.williams@goetheanum.ch

J. Kronenberg and E. T. Lammerts van Bueren (eds.), *On the Earth We Want to Live*, World Sustainability Series, https://doi.org/10.1007/978-3-031-98758-8_15

Keywords

Associative economics • Social threefolding • Sustainable development • Ecological economics • Rudolf Steiner • Planetary boundaries • True price • Ethical economics • Commons governance • Natural capital • Economic justice • Goethean science • Economic democracy • Education and economy

1 Introduction

This article contributes to the exploration of the intersection between Rudolf Steiner's ideas on associative economics and significant trends in contemporary discourse on sustainable development. It examines how ideas and select initiatives from the first two decades of the movement for the Threefold Social Order (Steiner, 1972) relate to essential dimensions of this discourse, such as: the inclusion of ecological conditions and natural capital in economic measurements; a unique interpretation of the reasons for destructive and anti-social tendencies associated with an overemphasis on economic growth; the inherent necessity of thinking on a planetary scale; organisational forms required for action on this scale; the necessity of a more differentiated concept of natural scientific practice and judgement; and how this judgement, combined with associative economic culture and organisation, points towards decentralised responses and solutions to the existential ecological, social, and economic challenges characteristic of the Anthropocene.

2 Summary of Social Threefolding

The idea of the Threefold Social Order was first articulated by Rudolf Steiner in *Towards Social Renewal* (Die Kernpunkte der sozialen Frage) in 1919. Steiner had published ideas on social and political questions earlier in his career but not in the same breadth and detail. The only other significant publication dedicated to social or political questions from his hand was *The Philosophy of Freedom*, first published in 1894, where he articulated a vision of freedom and ethical individualism.

Until the turn of the century, Steiner was most widely known as a Goethe scholar, having edited the natural scientific writings of Goethe as well as written and spoken widely on the great German writer and researcher. He was also active for five years as a teacher of history, public speaking, literature, and science at a school for workers in Berlin. From the turn of the century until his death, he was most widely known as a spiritual philosopher associated with the theosophical movement and the founder of anthroposophy. Steiner's political and social activity rose to an unprecedented level during the year of 1919 in the aftermath of WWI, when he stood as a public figure and social entrepreneur at the head of the movement for "social threefolding" (Schmelzer, 2017).

The evolution of his social ideas related to his own growth, especially in a variety of holistic, living thinking. Inspired by one of his teachers, Franz Brentano, he developed this new thinking while exploring the interrelationship between the mental faculties of thinking, feeling, and willing, and the human body—respectively, the nervous system, the rhythmic system, and the metabolic system (Steiner, 1996). Steiner relates how this thinking, capable of apprehending a differentiated yet interconnected living organism, when turned toward society, reveals an emergent threefold sociology. These emergent sociologies can be widely understood by a general public, which was important as Steiner saw their effectiveness as dependent on widespread legitimacy, suggesting they would need to be understood as three basic social sensibilities, accessible even to school children, and compared them with the basic skills of arithmetic and maths (Steiner, 1992, p. 57). He saw these sensibilities as connected with the slogan of the French Revolution, the ideals of liberty, equality, and fraternity.

Steiner characterises these three interacting sociologies as inherently animated by different values and demands. These were also dependent on one another; each social activity could reveal the justification through the partiality and deficiency of the other. Through stimulating the three activities and bringing them into relationship, an emergent, living social order was possible (Steiner, 1992, pp. 21, 110). The three spheres are the economy, the democratic rights state, and an independent cultural life, animated by the social ideals of fraternity, equality, and liberty, respectively.

Steiner blamed an inability to consciously differentiate these three tendencies in modern society, and to understand and guide their interrelations, for the social suffering and chaos leading to WWI and the chaotic aftermath of 1919. Steiner saw the decadence of middle European culture and the abstract social ideas of those in power as incapable of coming to terms with the radical transformations of modern society. Reliance on faded cultural inheritances of Europe left education, research, and the arts vulnerable to corruption by commercial and political interests. Steiner suggested the receptivity of the working class to the image of culture as ideology, as an illusion, was due to the fact that they were inoculated with modern, materialistically oriented science, while suffering the indecencies and injustices implicit in modern capitalism in a society shaped by an uncomprehending ruling class. Steiner saw in this the reason for the appeal of Marxian materialism (which he vehemently opposed), and while he advocated for energetic initiative in three directions, he saw the greatest significance in the renewal inherent in a cultural arena not tethered to the state or economic interests, that could address the challenges of modernity and science without falling into a crude materialism. The launching of the first Waldorf school in 1919 was one contribution in this direction.

In 1922 Steiner gave a course to university students where he developed his vision of associative economics (Steiner, 1972). Alongside *Towards Social Renewal*, these lectures are the most important sources for those who want to understand Steiner's economic ideas. Steiner engages the radical shift that began with the industrial revolution and the related, essential questions in political economy: intensification of the division of labour and specialisation, the significance of modern finance, the commodification of land, labour, and capital, the role of nature and the state, and monetary theory. He insists that one of the most important aspects of this transformation is the implicit cooperative spirit at work in the modern economy. Steiner saw great importance in the fact that with the intensification of the division of labour, and global trade, it was increasingly impossible to work for oneself. The division of labour steers each individual in society toward "conditions prevalent in the overall organism; economically, it precludes egoism" (Steiner, 1972, pp. 36–47; see also Steiner, 1992, p. 121).

Steiner humorously criticises all moralising about the social evils of capitalism as meaningless, as if one were to suggest: "If my mother-in-law had four wheels and a handle in front, she would be a bus" (Steiner, 1972, p. 133). Steiner did emphasise the importance of framing economic activity as an initiative of cooperation, mutual aid, and interdependence, but instead of moralising he set out to discover emergent potentials for selflessness within the economic processes of production, distribution, and consumption, and how these might be accommodated and activated through social organisation.

For Steiner, insistence on the central role of competition, self-interest, and national interest was at odds with the inherent, cooperative tendency in modern economics and its global, trans-national character. While the modern economy emerged through a focus on mass production, the time had come to create associations of producers and consumers that allowed for free agreements of cooperation and investment of capital in line with actual needs (Steiner, 1992, p. 112). Steiner characterises the union movement, and the form through which workers and management engaged in negotiations, as political and not economic in character. He suggested if each individual affected by an enterprise (again, including consumers as well as producers and distributors) is enabled to bring their unique economic insight, interests, and abilities into associations, a proper awareness of the limitations of regional and geographic realities and the needs of those creating a commodity could emerge.[1]

Steiner saw the economic process as unfolding in civil society, as it was understood in the traditional liberal interpretations of political economy, yet he considered faith in the invisible hand of Adam Smith as inadequate. What was needed was a framing of the economy as a cooperative undertaking for the mutual benefit of all, coupled with an associative method and organisational style. The associative method and organisational style activated a social and sober capacity for judgement in the distribution of goods, capital management, and decisions relating to production. Like Smith, Steiner saw the value judgement of the individual as a lynchpin in every economic process, but association and a cooperative

[1] It is worth noting that on a smaller, grassroots scale, this has proven a powerful way to finance agricultural initiatives (see Groh & McFadden, 1998; Bloom, 2016) and Mondragon's method for planning, developing, and financing diverse enterprises (Dresner, 2012).

framework allowed for a form of collective judgement by involving the insights and needs of each person concerned. True price, the amount of income required by a producer that enables them to meet their needs, and the needs of their dependents, until they can produce another commodity, was dependent on this dynamic process of association (Steiner, 1992, p. 72). This process presupposes a thinking that can follow the living fluctuations at the moments of exchange instead of static definitions and formulas (Steiner, 1972, pp. 33–36). The market itself is presented as a large cooperative venture wherein everyone who is able to work contributes according to ability and receives rights to withdraw from this same venture. These rights of withdrawal are issued in the form of currency: "Money, in a healthy social organism, can be nothing other than a draft on commodities produced by others, which the holder may claim from the overall social organism because he has himself produced and delivered commodities to this sector" (Steiner, 1992, p. 117).[2]

Steiner saw this social orientation as limited to the economy, and the economy as naturally bound by the conditions of nature, the biosphere, and geography on the one hand, and the democratic rights state on the other. His understanding of a rights state was an institution of agreements and legislation that limited itself to the creation of laws of a general character through democratic processes. All members of society of legal age would have a say in the creation of these laws. Many of Steiner's examples refer to labour laws that could protect a human life of decency (Steiner, 1992, p. 72). Steiner suggested that if the laws were created through fair democratic processes, and the state itself was not involved in economic enterprises but solely focused on questions of rights and duties, which would exclude lobbies and special interest groups (Steiner, 1992, p. 69), a widespread legitimacy was possible where these agreements would be acknowledged as a natural limitation on economic life.

Steiner's conception of the economy is limited to the production, distribution, and exchange of commodities and services. The economic organisations he called associations were to be exclusively concerned with their production and interchange. Commodities are goods that acquire value through immediate human needs, and which can be consumed (Steiner, 1992, p. 62). Steiner describes two dynamics of value creation in economics: the transformation of nature through labour (this can be as simple as harvesting food or extracting natural resources) and the exercise of intelligent management, innovation, and discovery, which lead to new efficiencies, organisation, and technologies (Steiner, 1972, lecture two).

For Steiner, modern capitalism represents a state of overreach. It not only trades in commodities but also in land, labour, and capital. Steiner suggests these are not actually commodities, and their commodification is a source of social grievances, inflation, and unjustifiable inequality that predictably lead to unrest. He does not

[2] Steiner acknowledged the necessity of providing for those unable to work, the elderly and children. Laws guaranteeing their provision were questions of the state, provision itself of the economy.

simply argue for the exclusion of land, labour, and capital from the market on ethical grounds but suggests that the anti-social effects connected with their trade point toward a confused form of economic judgement (Steiner, 1992, pp. 72–73). For Steiner, the status of economics as a science depends on a mobile and pictorial capacity of thinking that can reveal trading in land, labour, and capital as instances of misjudgement. Steiner frames the field of inquiry by focusing on the repetitive and cyclical processes of value creation and destruction and encourages the exercise of thoughts that can qualitatively accompany these. The result is that one begins to recognise when productive processes are not accompanied by a reciprocal consumption, resulting in a mobile form of equilibrium; then tendencies toward illness emerge, as with a living organism (Steiner, 1972, pp. 64–66; Steiner, 1973, pp. 62–63). The excessive focus on value creation and growth appears as a one-sided and socially destructive economic orientation. When the guidance of created values is not informed by the necessity of their immanent perishable constitution, but rather an intention to increase value creation through land or capital speculation (two activities that are not economic processes), an imbalance emerges comparable to illness in a living organism.[3]

Capital and land are not capable of effecting the actual destruction of value typical of commodity consumption. The purest form consumption processes take are cultural activities, for example, non-instrumental research, education, and art. While the teacher is a pure consumer who does not produce commodities, supporting teachers economically exercises the greatest possible impact on future economic production, more so than spending money on goods or making productive, interest-bearing loans (Steiner, 1972, lecture six). Implicit in this orientation is a shift in focus from economic production to cultural production (which in the short term is economic consumption), from the rivalrous to the non-rivalrous dimensions of experience, where not only is infinite growth possible, but a more profound dimension of the human experience.[4]

Association involves raising specific natural and social conditions in the economy to awareness to discover sound, and social, collective judgements and agreements in the cycles of production and consumption. Rudolf Steiner explicitly connects this type of thinking to organic processes and understanding, indicating its connection with the variety of judgement that Goethe developed in his phenomenological botanical studies (Steiner, 1922, seminar, pp. 38–39). In associations, the intention to cooperate is not enough; a certain style of communication that avoids hard definitions and abstract formulas is essential. This was associating

[3] Steiner's ideas about monetary design, which incorporate depreciation (the longer you hold it, the less it is worth), were also informed by this perspective, intending to bring money and the real, provisional nature of the economy of commodities closer together (Steiner, 1972, lecture 12).

[4] E. F. Schumacher suggested that modern economics is in need of wisdom, which reveals the "...hollowness and fundamental unsatisfactoriness of a life devoted primarily to the pursuit of material ends, to the neglect of the spiritual. Such a life necessarily sets man against man and nation against nation, because man's needs are infinite and infinitude can be achieved only in the spiritual realm, never in the material" (Schumacher, year, p. xx).

as a verb, in Steiner's sense, and was a sociology of the formation of collective judgements demanded by the modern economy.[5]

Taking this view, the institution of buying land appears as a pseudo-economic judgement. It is actually giving commodities for rights of use (Steiner, 1972, lecture five; see also Steiner, 1992, pp. 66–67). A landowner can make demands of people who use the land or live on it without themselves adding anything of value to the economy. Moreover, land that can be resold for profit without improvements leads to unjustified rights to commodities and services. Land use should be granted to individuals or enterprises based on socially beneficial use and not considered private property. Steiner suggested a similar approach for capital or the means of production, pointing out the important fact that these do not meet immediate needs as commodities even while they facilitate the creation of commodities that meet these needs (Steiner, 1985, pp. 38–52).

Another central dilemma for Steiner was trading money for labour, which this perspective reveals as trading goods for rights of command. This results in layers of confusion. It encourages workers to view money as an end instead of the meaningful cooperative enterprise itself, obscuring the implicit social impulse at work in the division of labour that precludes egoism in the economy. Instead of trading money, or goods, for rights of command, Steiner suggests determining income as means for mutual service (Steiner, 1992, p. 79). This would allow the inherent social tendency of the modern economy to break through the notion that one is working for one's wage (and oneself) and incorporate the insights and participation of the worker in the overall project as an intrinsically motivated individual. Labour markets encourage a hierarchical command culture where the employer expects rights of use that exclude intrinsic motivation of the worker. Further, there is no inherent connection between the needs for a decent existence and income. This is a major obstacle to the idea of associative judgement formation, which is dependent on the exchange of information and value judgements from diverse individuals in order to reach true pricing.

Steiner suggests that in order for economics to achieve the status of a science it will require thinking in imaginative pictures. Steiner contrasts conventional theoretical and abstract approaches to defining value with the value judgement of an experienced farmer who considers buying a plough:

> the picture that is composed before his mind—the picture of a certain sum of money and a plough—calls forth in him the immediate feeling that he can still afford to give so and so much more money, or else that he cannot. He has it directly out of his feeling-experience. Now even in the most complicated economic process, this feeling-experience is not to be eliminated. That is thinking in pictures. (Steiner, 1972, p. 130)

[5] During the early propaganda period, he emphasised that this was to inform all public speaking about economic issues (Steiner, 1971, pp. 84–85).

Steiner points out that these pictorial, feeling judgements, while essential, are inherently partial. It belongs to modern economic life that only through bringing perspectives together can sound economic, social (and ecological) judgements emerge. In this variety of socially oriented bargaining, each person can offer the value judgements their position allows, which are then supplemented by others to reach a true, while provisional, value judgement (Steiner, 1992, lecture ten).

How does this connect with the widespread social imaginary of classical liberal economics? (Taylor, 2004, pp. 69–82). This social imaginary casts people as basically self-interested on the whole. In an open society, with free markets, this naturally leads producers to compete by offering goods and services for lower prices, and consumers to look for deals. If a producer charges more than necessary, someone can step into their place by undercutting them. This dynamic, coupled with the law of supply and demand, offers the best possible solution for supplying society with sustenance, commodities, and basic needs at the most efficient rates. Steiner suggests judging egoistically in economics is both natural and necessary up to a point and that without associative economic culture or institutions, it will play a determining role. Through an overall cooperative framing, coupled with associative practices and institutions, this changes. The cooperative framing is a necessity; association will not do on its own.[6] Steiner suggests it is meaningless to expect an individual in a simple market situation to voluntarily offer to pay more than necessary for a commodity, but that through introducing a cooperative framing and associative style of communication, this changes. Association leads to a greater economic perspective that includes the rightful interests of others as well as ecological limitations (Steiner, 1972, lecture 10). The consumer's willingness to pay more, or perhaps consume less, arises out of a recognition of actual economic value judgements, not moral imperatives.

There is no preoccupation with growth in Steiner's economic ideas. The implications of his ideas suggest that production and consumption cycles would change significantly, as well as waste factors and inefficiencies, through association(s). Steiner also suggests the one-sided focus on production expresses a tendency toward illness that should be complemented by more capital flowing into the sphere of pure consumption: independent and accessible fostering of education, research, and the arts (instead of pseudo-commodities of land and capital).

While Steiner argued a century ago for the urgency of the liberation of cultural life, today the limitations imposed from the side of the planet have become existential, demanding the attention of people all around the world. What is the significance of the understanding in associative economics of nature as a co-constituting element, and inherent barrier, of the economy?

[6] As has been shown by recent research in experimental economics, the model of the rational, self-interested economic actor is not universal, and framing has been proven an important factor in how a person will act, namely competitively or cooperatively (Benkler, 2011).

3 Economics and the Limits of Nature: Dunlop, Stein and the World Economic Conference

The movement for the Threefold Social Order was a public and widely perceived movement in the early years after the First World War, even receiving attention in the United States (Schmelzer, 2017). In the years that followed, its influence was perhaps greater while its visibility receded. Through Dan McKanan's impressive study (2017), it is clear that anthroposophical initiatives and individual anthroposophists connected with these ideas played a central role during the emergence of the ecological movement in the United States. He reflects that: "...without the yeasty contributions of anthroposophical initiatives, there might have been no organic agriculture, no campaign to ban DDT, no community supported agriculture, and no environmental banks" (McKanan, 2017, p. 175).

McKanan's focus is on the environmental movement, but his study includes significant treatment of associative economics, banks, and finance. It is an important reference for anyone interested in the discourse on sustainable development, associative economics, and anthroposophy. The conclusion that anthroposophy provided essential "yeasty contributions" for the ecological movement might be understood to imply these ideas and projects take the form of powerful yet small innovations and contributions that only offer global visions of sustainable economic development for those who dare reckless extrapolation from such small-scale social imaginations. An example that counters this view can be found in the series of far-sighted projects pursued by Daniel Dunlop and Walter Johannes Stein, two of Rudolf Steiner's collaborators in the first two decades of the threefolding movement. Steiner's initial presentation of the Threefold Social Order, while indicating the global, cooperative, and particularly nature-bound character of the modern economy, can be understood in its far-reaching significance and detail through reference to their work and ideas.

Daniel Dunlop (1868–1935) was one of Rudolf Steiner's closest collaborators in Great Britain at the end of his life. Dunlop, an active member in the emerging field of the electric industry, led the establishment of the British Electrical and Allied Manufacturers Association (BEAMA) in 1911, inspired by his interest in the power of voluntary cooperation (Meyer, 2014). His ideas about cooperation were published in 1916 in the booklet *British Destiny: The Principles of Progress.* This work promoted the idea that the post-war responsibility of the British involved increasingly turning the Empire into associations dedicated to voluntary cooperation for the common good. He suggests an unprecedented context of the coming centuries will be trans-national globalism and interdependence, where exactly this orientation will be required. He sees the common juxtaposition between individuality and cooperation as the cardinal challenge, writing:

> In natural organisms Individuality and Co-operation are as exactly adjusted as the most delicate balance, but in most organizations of men one is always threatening the other, because it is not recognized that co-operation is necessary to give value to individual efforts. (Dunlop, 1916, p. 45)

On the one hand, an unfettered competition unleashes a race toward "ruinous reduction of prices" and the deterioration of quality connected with this. On the other hand, excessive coordination through centralised control blocks the fruitful contributions that individual efforts might bring into society. The inability to see the connection between these two tendencies, in a living and harmonious way, is the central defect of social and industrial experiments. The conclusion of the book proposes a national industrial federation arranged in the spirit of voluntary cooperation, which would bring together trade associations, employers' associations, trade unions and labour organisations, scientific and learned societies, agricultural cooperatives, and financial associations.

Dunlop went on to propose and inaugurate the World Power Conference in this spirit (known today as the World Energy Council). The idea, which was also discussed with Rudolf Steiner, was realised in 1924 (Villeneuve 2009, p. 958). In an article published before the conference, Dunlop calls for a new industrial revolution that moves beyond the ruthless desire for wealth and natural resources. A new industrial revolution that preserves the value of efficiency, "but conceived on different lines – conservation to the maximum degree of all sources of wealth left to us by Nature, consistent with improvement in scientific and industrial processes to make this conservation good" (Dunlop, 1924, p. 6). The conference was to deal with broader social and economic issues, not just technical ones. Dunlop focuses on the diminishing significance of national economic approaches and the emerging challenges of globalism, suggesting that tendencies of national egotism and competition need to be counterbalanced by international cooperation. He was convinced the best point of entry was through convening scientists, engineers, technicians, and business leaders from the energy sector, while not excluding statesmen and administrators. The first World Power Conference saw 1,700 delegates come together in July of 1924, from 40 countries. They met in the Conference Halls at the Palace of Engineering within the context of the Empire Exhibition (Wright et al., 2013, p. 13), with an opening address by the Prince of Wales.

This was a prelude for Dunlop, who harboured the intention to establish a World Economic Conference of a more general nature, dedicated to other areas of production while also embracing Steiner's suggestion that the associations for voluntary cooperation should include consumers. After Steiner's death in 1925, Dunlop continued his work within the World Power Conference and began an independent publication under its auspices called the World Survey (Dunlop and Stein, 1935). The first issue appeared in April of 1935, and its purpose was to gather statistics related to natural resources, geography, and population to encourage awareness of the planetary commons, which all people and nations share. Through this initiative, he hoped to contribute a counter-movement to the prevalent tendencies of nationalism and competition. The survey was envisioned as a global ledger that could show the interdependence and complementarity of the different regions on the planet and their corresponding economic advantages and disadvantages. It was also intended to model a culture of the free sharing of scientific and technical knowledge for the benefit of global industry and to promote the sense of interdependence in the modern, global economy. In 1933, Walter Johannes Stein (1891–1957) was invited to collaborate on the preparation of this survey and the establishment of a World Economic Conference. Stein had been a student and collaborator of Rudolf Steiner's in the fields of education and the movement for the Threefold Social Order and had written his dissertation on Steiner's interpretation of Goethe's approach to natural science (Stein, 1921).

In 1935, before the economic conference could be realised, Dunlop died. In 1937, a publication appeared by Stein under the title "The Earth as the Basis of World Economy," which helps to understand the vision he and Dunlop were developing for a World Economic Conference. Stein, reflecting on the fact that Dunlop's greater ideas were being forgotten, explains his motive to publish the ideas they were working on with the hope that: "...future promoters of a World Economic Conference may make use of the foundations which we [...] worked out together" (Stein, 1937, p. 8).

4 Economics and Geographical Internationalism

One of the intriguing sentences from Dunlop's hand in 1916 reads: "It is [...] evident that international Philosophy and Science, and even the study of Comparative Religions, have not been carried far enough yet to secure a working basis for physical or geographical internationalism" (Dunlop, 1916, p. 31). This thought takes on specificity in Stein's The Earth as the Basis of World Economy (1937), which argues that the chaos of the world economy can only begin to be harmonised when awareness of its natural foundations permeates general economic culture, understanding, and practice. The chaos, he suggests, arises largely from ignorance of the economy's connection to the greater natural world. The purpose of the book is to demonstrate that "[...] the earth as a whole is the basis of world economics." While the economy can be moulded by human beings, it remains dependent on the laws of nature. Acting within this field of natural forces without awareness can lead to unexpected consequences, potentially undermining the very conditions of life. Emphasising this point, Stein writes that:

> economic system is not the result of human activity alone, it is a part of the earth itself. It is the present-day tendency to be concerned only with individual nations and groups, and to regard the earth only as the stage on which the action takes place, as the mere provider of raw material. If we conceive economic life as forming part of the life of the whole earth, the complexion of the matter changes entirely, the inexorability of the economic forces largely disappears, and man is thus enabled to build up his own sphere. (Stein, 1937, p. 12).

Stein characterises the consciousness offered by natural science as containing a moral dimension that transcends group egotism and awakens a sense of wholeness and responsibility. Moreover, this sense of wholeness is necessarily global due to the very nature of the environmental sciences. Stein emphasises the impartiality that arises when scientists and technicians share their knowledge about the world, showing how it naturally leads to an awareness of ecological and economic interdependence that transcends special interests and group egotism. The moral dimension within the sensibility of the limited and interdependent nature of the Earth and its resources is central to Stein's message:

> The conception of the earth as a whole engenders a form of moral impulse which unlike that of the Orient as exemplified in its philosophy and its religion, is not moral through any conscious act of the will, but is based on an objective comprehension of the equilibrium of all surrounding forces. (Stein, 1937, p. 12)

The book includes sketches of relevant considerations, such as geology, the cosmological origins of metals, the oceans, the air and the Earth's temperature system, the circulation of chemicals, and the connection between the sun and planets, weather, and climate in relation to harvests. This scientific vision is presented as a source of inspiration for social life as well as ecological stewardship: "The living earth, regarded as the foundation of economic and social life, presents quite a different aspect in the form of understanding and solving problems, from the earth when looked upon—as it usually is—as a mere object of exploitation" (Stein, 1937, pp. 12–13).

5 Associative Economics, Geographical Internationalism, and Sustainable Development

How do current ideas on sustainable development and the heightened tension between ecological awareness and economics relate to associative economics as presented by Steiner, Dunlop, and Stein? Before sustainable development began to take on international significance through The Brundtland Report, the publication of The Limits of Growth, or the adoption of the Sustainable Development Goals by UN member states, we find early voices calling for the need of increased attention to the relationship between the economy, population growth, and the finitude of planetary resources.

On the most superficial level, the discourse of sustainable development takes its point of departure from this amplified awareness of the growth of human populations and economies and their connection to the limitations of the planet, its resources, and the ultimate instability of its biosphere. It is the culmination, or Great Acceleration, of globalisation, in its broadest sense, a part of the scientific and cultural development that led to the circumnavigation of the globe, communication technologies, the exploration of space as well as the continuation of the industrial revolution and the division of labour (Steffen et al., 2015). The resulting heightened awareness of the dependence of human economies on planetary boundaries motivated a re-evaluation of economic theories of development. In 1966 the economist Kenneth Boulding suggested humanity was in the middle of this shift in awareness from sensing the environment as characteristically open, with horizons and possibilities in all directions, to one that is closed, reflecting that: "economists in particular [...] have failed to come to grips with the ultimate consequences of the transition from the open to the closed earth" (Boulding, 2013, p. 3).

One problem with classical economic methodologies and accounting is that they traditionally exclude the costs and effects on the natural environment and the quality of life in their measures and ledgers and lead to distortions. Theories of development, which have relied heavily on Gross National Product, overlook existential information regarding the depletion of nature or destructive side effects of development. Herman E. Daly, a pioneer of ecological or environmental economics which conceives of the economy as a subset of the larger whole of the environment, saw a central thickness among economists in that: "Production is treated as a function of capital and labor alone – resources are omitted entirely" (Daly, 2007, p. 68). Drawing on the notion of John Stuart Mill's stationary state, he promoted theories of development without economic growth and ways of including externalities relating to the earth and the biosphere in economic measurements. This need for a new "policy framework for sustainable development," that addresses the distortions that concerned Daly in measures like the GDP, has influenced the work of the Organisation for Economic Co-operation and Development (OECD, 2004), the World Bank, and other international agencies.

As we have seen, the view of nature as co-constituting the economy, and not an externality, is a fundamental starting point in associative economics, with the original intentions of the World Power Conference, the World Survey, and the World Economic Conference as efforts in this direction. They were informed by the idea that national economic perspectives, when isolated or in competition, distort general understanding of the Earth as foundation of economics and its transnational character.

An important and challenging dimension of Dunlop and Stein's approach to voluntary, cooperative stewardship of planetary resources, inspired by the intrinsic wholeness these resources are embedded in, is that they do not easily fit with conventional conservative or progressive ideas. Their approach to the challenge of global resource stewardship aligns most clearly with ideas and case studies presented by Elinor Ostrom in Governing the Commons. Ostrom presents cooperative endeavours and forms of common-pool resource management that elude conventional economic and political ideas. In different ways, they demonstrate cooperative sharing of natural resources in situations that are neither regulated by government policy nor private ownership: "communities of individuals have relied on institutions resembling neither the state nor the market to govern some resource systems with reasonable degrees of success over long periods of time" (Ostrom, 2011, p. 1).

Ostrom discusses Garrett Hardin's "tragedy of the commons," a parable that suggests common resources will necessarily be ruined without government intervention and nationalisation or clear private ownership. Ostrom points out that while dominant political ideologies differ in their approach to governing the commons, they both typically see the state as the central arbiter: "accept as a central tenet that institutional change must come from outside and be imposed on the individuals effected" (Ostrom, 2011, p. 14). Ostrom also criticises theorists for their abstract treatment of social situations which are "institution-free", for instance, a rationalisation for nationalisation might leave out the most important specific issues about how officials are to come to significant technical understanding or be able to fund or regulate the policies they develop (Ostrom, 2011, p. 22).

Ostrom limits her focus to examples of CPR management impacting no more than 15,000 people. We need only consider that: "it is estimated that 90% of the world's fishermen and over half the fish consumed each year are captured in small-scale, inshore fisheries…" (Ostrom, 2011, p. 27),[7] to sense how cases such as the successful cooperative example of the inshore fishery of Alanya could attain global significance. Pooling their intimate insight and existential interest in finding a sustainable solution, the fishermen created an informed management solution without needing to fund a new division of the government, invest in the education of government workers, or devise a plan of oversight. Ostrom points out that acknowledging a general predisposition of goodwill and intelligence is important to rightly value such successes: "Instead of presuming that some individuals are incompetent, evil or irrational, and others are omniscient, I presume that individuals have very similar limited capabilities to reason and figure out the structure of complex environments" (Ostrom, 2011, p. 25). Her work demonstrates how good intentions are dependent on communication, sharing information, and cooperation involving those with intimate knowledge of and interest in the resources under consideration.

These observations align with the fundamental orientations of associative economics, which advocate for a cooperative framing of economic enterprise and the necessity of processes of concrete, specific sharing of information to come to collective judgements that can lead to true price in harmony with environmental conditions. We can situate Dunlop and Stein's efforts as an example of a vision for the cooperative, international stewardship and regulation of CPRs on the largest scale outside of the dominant social imaginaries of the central authority and Hardin's famous dilemma of the "tragedy of the commons."

6 Economics and the Living World

Economics concerns itself with understanding cycles of production and consumption, capital accumulation and finance, business cycles and stimulus measures, markets, and employment in so far as they relate to the management of basic needs of sustenance, commodities, and services. Economic projections are oriented toward rhythms of consumption and production in their connection with rhythms of reproduction and population. Thomas Malthus's Essay on the Principle of Population (1798) focused on these basic facets of biological life. The fact that Darwin's theory of evolution as a process of random mutation, natural pressure, and reproduction was in part inspired by Malthus's essay on the "Principle of Population" is meaningful in so far as it suggests an intimate connection of

[7] This is a figure that has not changed dramatically over the last 30 years according to the food and agriculture organisation of the UN.

the basic life processes of growth and decay, of birth and death, with questions of economics. John Maynard Keynes writes that if we investigate the past:

> we find that the economic problem, the struggle for subsistence, always has been hitherto the primary most pressing problem of the human race, not only of the human race, but of the whole of the biological kingdom from the beginnings of life in its most primitive form. (Keynes, 1930, pp. 366)

The idea of the dependence of economic growth on a finite ecological base, articulated by the likes of Boulding and Daly, can be characterised as static when compared to this cyclical field of phenomena and visions of interdependence and various states of dynamic equilibrium. This quality of awareness has not come predominantly from economists but from ecologists. Famous examples include Rachel Carson's Silent Spring as well as Aldo Leopold's A Sand County Almanac.

While Steiner, Dunlop, and Stein all advocate for the essential role that the sciences play in economics and resource management, they also emphasise the existence of different varieties of scientific thought. Steiner saw the quality of thinking needed in economics as intrinsically tied to the type of judgement that could understand dynamic life processes and living organisms. A facet of the discourse on sustainable development emphasises the importance of a differentiated understanding of types of scientific method and judgement. Donald Worster presented the intertwined history of economic and ecological awareness, situating Gilbert White's Natural History of Selborne as the beginning of what would grow into field ecology of the "Arcadian" variety (Worster contrasts this with an Imperial variety that he traces back to Francis Bacon). These ecologists have argued for specific orientations and methods that anticipate holism and life. Worster contrasts them with scientific approaches that seek to isolate phenomena, assign them functions that can be tested in highly simplified and controlled laboratory settings, or classified in fixed taxonomies (Worster, 1994, pp. 22–23). In associative economics, we find the "Arcadian" natural scientific orientation of Goethe explicitly positioned as an example of the pictorial form of insight related to a proper ability to understand the modern social organisation of the global economy.

Positioning the economy between the rights state and nature presupposes two socio-cultural activities. Each informs and regulates economic judgement and action on an individual level. One presupposes the legitimacy of legislation that emerges through democratic participation; the other, widespread ecological education and literacy. McKanan has pointed out how Goethe worked for a sensitivity in method that regarded:

> the experimental focus on "isolated facts" to be misleading because "in nature nothing happens that is not in connection with a whole" as an important influence during the emergence of the environmental movement through the levin of anthroposophy and the work of anthroposophist's. He urged scientists to conduct an open-ended series of experiences that would ultimately grow into "one single experiment, one experience present from manifold perspectives" allowing the experimenter a higher level of awareness. (McKanan, 2017, p. 181)

Steiner's intimate familiarity with Goethe's method was the subject of further research by the likes of Walter Johannes Stein, who explicitly suggested their central significance for both the future of environmental stewardship and economic development, and Steiner's student and collaborator Ehrenfried Pfeiffer, an expert witness in the trial that informed Rachel Carson's Silent Spring. Not unlike the influence that anthroposophy played in the emergence of the environmental movement, so too the remarkable community of researchers that have developed this orientation in the natural sciences is largely absent in the academic discourse on sustainability and phenomenology, which focuses more on thinkers like Husserl, Heidegger, and Merleau-Ponty (see Kockelmans & Kiesel, 1970; Heinemann, 1970).[8]

The implication is that certain orientations in the natural sciences can support success in the cooperative organisation of the economy in harmony with the limits of nature. This touches on an important dimension of political theory having to do with education. Within the greater effort of rethinking education in accordance with current ecological and social challenges—which span the spectrum from fresh approaches to economics driven by the protest of university students (Raworth, 2018) and the fourth Sustainable Development Goal—this is a particular inquiry into the varieties of thinking that can intensify both awareness of living ecology and the interconnected web of life, and strengthen sound judgement and sensitivity to processes of value creation and decay, production, and consumption in the economy.

Arguments in this direction have been articulated from different perspectives. Hannah Arendt suggested that the dominant "extra-terrestrial" reason, typical of modern science schooled by astronomy, is an insufficient instrument when facing social, political, and ecological problems of terrestrial, earthly life (Arendt, 1981, 1984). The famous biologist Adolf Portmann framed an expanded understanding of natural science as a necessity for addressing the ecological crisis, characterising it as a form of aesthetic education (Portmann, 1968). More recently, Jane Bennett has advocated for a focus on the education of ecological sensibilities that can catalyse widespread action and individual responsibility (Bennett, 2010, 2016). This perspective positions the interdependence of sustainable development goals and individual capacities of judgement and action on a local level.[9]

[8] For an introduction to this community of researchers, see the volumes of *Elemente der Naturwissenschaft* published between 1964 and 2024, contributions in Suchantke's *Ökologie*, and resources from Craig Holdrege and colleagues at the Nature Institute. The differentiation between reason and understanding in the teaching of natural sciences was recognised as of utmost importance from the first courses on science instruction for teachers that Rudolf Steiner conducted. In the book *Anthropologische Grundlagen der Waldorfpädagogik*, by an influential member of the worldwide Waldorf school movement, Ernst-Michael Kranich describes the development of judgement in the natural sciences (pp. 203–223).

[9] Recent efforts to compliment sustainable development goals with inner development goals can be understood from this perspective as well: https://innerdevelopmentgoals.org/.

One intriguing aspect of the picture that emerges from Steiner, Dunlop, and Stein is that it is possible to lift today's ecological externalities into economic value judgements in a decentralised, widespread, participatory manner. This contrasts strongly with the suggestion that the ecological crisis will necessitate a move from liberty toward authority (Eaton & Kostka, 2024), whether in the form of an authoritarian solution of a Hobbesian character or a benevolent eco-authoritarianism in line with the Aristotelian tradition. The decentralised positioning of individual value judgements integral to economic life and associations also contrasts with the emphasis on "political will," a term that often expresses the rational choice of the individual to dedicate themselves to a moral motive regardless of economic interests or value judgements. One example of this orientation can be found in the work of John Rawls's Theory of Justice, which regularly informs discourses on ecological and social justice (see Dresner, 2012, Chapter 8).

Associative approaches to economic life imply a widespread ecological literacy of the phenomenological variety. The latter facilitates a planetary awareness that informs economic association, judgement, and initiative. This, coupled with democratic legislation that enjoys widespread legitimacy, constitutes an essential boundary of economic life.

7 Conclusion

In the ideas of the threefold social order, and associative economics, we find a unique approach to addressing some of the central questions of sustainable development. With a backdrop that encourages a separation of work from income and the de-commodification of land and capital, this approach frames the preoccupation with economic growth of much conventional economic theory as misguided due to its lack of appreciation for the necessary co-constitution of the economy by planetary boundaries and a sense for the connection between equilibrium of production and consumption processes of commodities. Associative economics presents the form of planetary consciousness that arises through the environmental sciences and globalisation as an ethical counter-movement to group egotism and nationalism. The framing of economic enterprises as cooperative and social endeavours, coupled with associative forms of communication and organisation, facilitates true prices informed by awareness of social and ecological conditions.

The vision of the World Economic Conference is informed by the conviction that associations dedicated to this framing and approach, including producers, distributors, consumers, and individuals with pertinent scientific and technical knowledge, are in the best position to address the challenges of resource stewardship and ecological challenges, as opposed to international bodies consisting of representatives from political bodies and nation states. At the same time, it recognises the limits of forms of economic judgement in association and the rightful place of democratic regulation of economic processes. This orientation eludes conventional political and economic orientations typical in discourse on the "tragedy of the commons."

By recognising economic processes as primarily unfolding in the activities of production, distribution, and exchange of perishable commodities, it excludes land, labour, and capital from the market. The current tendency to seek perpetual economic growth, which sees excesses invested in land and capital, it sees as a misguided trend that fails to recognise the reciprocal necessity of value destruction through the non-rivalrous activities of (and most powerful form of investment) culture, in general, and education in particular. This involves an overall shift in the vision of development from one characterised by material consumption and quantity to non-rivalrous dimensions of meaning and quality.

Understanding associative economics involves differentiating between modes of understanding and judgement. Of most importance is a form of mobile, pictorial judgement that is connected with both economic processes and the life sciences. Given the global, widely participatory character of the economy, there is an implication that cultivating this form of judgement in general education is a necessity. This would facilitate awareness in individuals of both social and ecological dimensions of the value judgements through their participation in a global, associative economy.

Widespread, democratic education in aesthetic judgement in the natural sciences is a pedagogical task on which collective action in harmony with the ecological and social challenges of the "Anthropocene" depends, as much as associative economic practices. This addresses and clarifies an aspect of the dynamics that can lead to eco-authoritarian regimes that emerge through various states of exception and emergency as well as the shortcomings of strategies that promote moral imperatives and political will as primary drivers toward sustainable development.

References

Arendt, H. (1981). *The life of the mind: The groundbreaking investigation on how we think.* Houghlin, Mifflin, Harcourt.

Benkler, Y. (2011). *The Penguin and the Leviathan: How cooperation triumphs over self-interest.* Crown.

Bennett, J. (2010). *Vibrant matter: A political ecology of things.* Duke University Press.

Bennett, J. (2016). *The enchantment of modern life: Attachments, crossings, and ethics.* Princeton University Press.

Bloom, J. (2016). *Inhabiting interdependence: Being in the next economy.* Steiner Books.

Boulding, K. E. (2013). The economics of the coming spaceship Earth. In H. Jarrett (Ed.), *Environmental quality in a growing economy* (pp. 3–14). John Hopkins University Press.

Daly, Herman E. (2007) Ecological Economics and Sustainable Development. Edward Elgar Publishing.

Dresner, S. (2012). *The principles of sustainability.* Routledge.

Dunlop, D. N. (1916). *British destiny.* Path Publishing Company.

Dunlop, D. N. (1924). *Power Development: First World Conference.* The Daily Telegraph, 30 June.

Dunlop, D. N., & Stein, W. J. (Eds.) (1935). *World survey* (Vol. 1. p. 1). World Power Conference.

Eaton, S., & Kostka, G. (2024). Environmental authoritarianism: a critical review. In A. Wolf (Ed.), *The Oxford handbook of authoritarian politics.* Oxford University Press.

Groh, T., & McFadden, S. (1998). *Farms of tomorrow revisited.* Steiner Books.

Kockelmans, J. J., & Kisiel, T. J. (1970). *Phenomenology and the natural sciences: Essays and translations*. Northwestern University Press.
Kranich, E. M. (1999) *Anthropologische Grundlagen der Waldorfpädagogik*. Verlag Freies Geistesleben.
Leopold, A. (1970). *A sand county Almanac: With essays on conservation from round river*. Sierra Club.
McKanan, D. (2017). *Eco-alchemy: Anthroposophy and the history and future of environmentalism*. University of California Press
Meyer, T. H., & Dunlop, D. N. (2014). *A man of our time: a biography*. Temple Lodge Publishing.
Ostrom, E. (2011). *Governing the commons*. Cambridge University Press.
Portmann, A. (1968). Biologie als Aesthetische Erziehung. In *Biologie und Geist*. Suhrkamp.
Publishing, OECD (2004) *Measuring sustainable development: integrated economic, environmental and social frameworks*. Organisation for Economic Co-operation and Development, Paris.
Raworth, K. (2018). *Doughnut economics: Seven ways to think like a 21st century economist*. Chelsea Green Publishing
Schmelzer, A. (2017). The threefolding movement, 1919: A history. In *Rudolf Steiner's campaign for a self-governing, self-managing, self-educating society* . Rudolf Steiner Press
Stein, W. J. (1921) Die moderne naturwissenschaftliche Vorstellungsart und die Weltanschauung Goethes, wie sie Rudolf Steiner vertritt. Der kommende Tag
Stein, W. J. (1937). *The Earth as a basis of world economy*. The Present Age 2.
Steiner, R. (1971). *Anthroposophie, soziale Dreigliederung und Redekunst* (Vol. GA 339). Rudolf Steiner Verlag.
Steiner, R. (1972). In A.O. Barfield, T. Gordon-Jones (Trans.), *World-economy: The formation of a science of world-economics*. Rudolf Steiner Press
Steiner, R. (1973). *Nationalökonomisches Seminar*. Rudolf Steiner Verlag.
Steiner, R. (1985). *The renewal of the social organism* (CW 24). Steiner Books
Steiner, R. (1992). In F. T. Smith (Eds.), *Towards social renewal: Rethinking the basis of society*. Rudolf Steiner Press
Steiner, R. (1996). *Riddles of the soul*. Mercury Press.
Steffen, W., Broadgate, W., Deutsch, L., Gaffney, O., & Ludwig, C. (2015). *The trajectory of the anthropocene: The great acceleration. The anthropocene review 2* (pp. 81–98). SAGE Publications
Suchantke, A. (Ed.) (1998). Ökologie. Verlag Freies Geistesleben
Taylor, C. (2004). *Modern social imaginaries*. Duke University Press
Villeneuve, C. (2009). *Rudolf Steiner in Britain: A documentation of his ten visits, 1902–25* (Vol. 1–2). Temple Lodge Publishing.
Worster, D. (1994). *Nature's economy: A history of ecological ideas*. Cambridge University Press.
Wright, R., Shin, H., & Trentmann, F. (2013). *From World Power Conference to World Energy Council: 90 years of energy cooperation, 1923–2013*. World Energy Council.

Goetheanism as a Contribution to Sustainable Development

Vesna Forštnerič Lesjak, Matthias Rang, and Meinhard Simon

Abstract

The Goetheanistic method represents a holistic scientific approach that fosters new forms of thinking for an appropriate-for-life and contextually integrated conscious engagement with all organisms—be it a plant, a landscape, microorganisms, or the Earth as an organism, taking into account its atmospheric layers. These forms of thinking originate from the observation of natural phenomena, whether living beings or life processes, and serve as the foundation for future actions. In following this path, the researcher undergoes a profound inner, productive transformation to acquire new abilities. This article presents the methodological steps of a Goetheanistic approach to the aforementioned areas of the natural sciences.

Sections 1–4 were written by Vesna Forštnerič Lesjak, Sect. 5 by Matthias Rang, and Sect. 6 by Meinhard Simon. The manuscript that resulted was then further revised collaboratively by the authors.

V. Forštnerič Lesjak (✉) · M. Rang
Co-Leader, Natural Science Section at the Goetheanum, Dornach, Switzerland
e-mail: vesna.forstneric@goetheanum.ch

M. Rang
e-mail: matthias.rang@goetheanum.ch

M. Simon
Emeritus Professor Aquatic Microbial Ecology, Carl von Ossietzky University Oldenburg, Oldenburg, Germany
e-mail: meinhard.simon@uni-oldenburg.de

J. Kronenberg and E. T. Lammerts van Bueren (eds.), *On the Earth We Want to Live*, World Sustainability Series, https://doi.org/10.1007/978-3-031-98758-8_16

Keywords

Goetheanism • Holistic science • Sustainable development • Ecological consciousness • Scientific methodology • Phenomenology • Rudolf Steiner • Johann Wolfgang von Goethe • Nature observation • Organic thinking • Metamorphosis • Landscape ecology • Microbiome • Prokaryotic life • Biosphere • Earth systems • Climate and thermal sheath • Biodynamic agriculture • One health • Goethean science

1 Introduction: Development of Scientific Consciousness and Destruction of the Environment

In the past, humans lived embedded within the earthly-cosmic rhythms of nature. They revered nature as a goddess, as the highest good, and worked with nature (Steiner, 1924–1925). With the development of modern natural science and new technologies, humans began to exploit nature in order to live more comfortably. This development, along with industrialisation and urban living, caused a separation between humans and nature. The result was widespread destruction of nature, as humans primarily thought about what they could exploit from it, rather than how they could shape processes with it (Cardinale et al., 2012; Caro et al., 2022; Hanski, 2005).

However, this separation did not only occur externally, but first of all mentally. With the development of natural science, the laws of physics, which apply to processes in the inorganic world, were increasingly transferred as causal mechanisms of chemical-physical processes to those in the organic, that is, living nature. The reductionist-analytical consciousness tried to understand organisms as a sum of individual parts, losing sight of the interconnectedness between the parts in the organism. The wholeness of living beings, which holds the parts together as a higher entity, shapes them meaningfully, and changes them over time, was excluded from scientific description. The organism is not a closed system but lives in active interaction with its environment. All of these relationships faded into the background in human consciousness, while humans focused more and more on smaller and smaller parts (Rosslenbroich, 2018a, 2018b, 2023). Furthermore, in most schools, the theory of selection, which was presented by Charles Darwin (1809–1882) in his work On the Origin of Species in 1859 with the subtitle "By means of natural selection or the preservation of favoured races in the struggle for life," is still taught to young generations as the basis of evolutionary biology. It should be clear that he did not derive the concept of the "struggle for existence" from observations in nature, but from reading the economist Thomas Robert Malthus (1766–1834), specifically his work An Essay on the Principle of Population, which was published in 1798 during the rise of capitalism.

This led to the study of the living under one-sided assumptions, with practical consequences. Today, it is no longer sufficient to attempt, with goodwill and isolated factual knowledge, to steer nature back onto a healthy path using techniques developed from these limited insights. We must achieve a fundamental transformation in our deeply disturbed relationship with nature and establish new foundations for scientific thinking that foster a life-appropriate approach to all organisms, grounded in the observation of nature. For the adopted modes of thinking build the basis of future actions (Endlich, 1985; Schnell, 1982; Suchantke, 1982).

2 Expansion of Scientific Consciousness Through Goetheanism

The contemporary human lives in a state of inner division. Through purely materialistically oriented natural science, one directs his attention towards the processes in nature, but on the condition that he does not engage with them internally. Hence, the moral impulses of the human being lose their significance, as the researcher remains a purely external observer. The appreciation of nature has become a private, personal choice for the researcher, but it is no longer a necessity. The other path in this conflict stands in opposition– one immerses oneself in the inner world and seeks to forget the outer world. In doing so, one attempts to fill oneself with many seemingly enriching feelings. This path can even lead to hostility towards science. Nowadays, people often live in a mixture of these two approaches (Bockemühl, 1984).

Is there perhaps a third possibility, through which the human being could regain access to nature and to oneself? Can the abyss between the object of observation and the observer be overcome without falling into the subjectivity of the emotional world? The ecological crisis and the psychological crisis of modern humanity run in parallel and are closely intertwined. However, there is a way to overcome the described conflict and with presence of mind connect the inner and outer life. Johann Wolfgang von Goethe (1749–1832) laid the foundation for this approach (Goethe, 1883, 1887, 2023), and it was re-enliven, re-illuminated, and elaborated as a method of knowledge by Dr. Rudolf Steiner (1861–1925) (Steiner, 1886, 1987, 1897). Rudolf Steiner frequently presented this path and, in particular, understood it as a natural science that builds a bridge between the ideas of the recognising scientist and the ideas underlying the phenomena occurring in nature, which belong to the same world of thought as human ideas. He called this method of knowledge Goetheanism, although he also used the term in other contexts (Schad, 2001). A number of natural scientists have developed in a timely manner this approach, still little known in the collective consciousness, into a meaningful and sustainability-promoting basic and applied natural science (Bockemühl, 1985; Schad, 2001; Suchantke, 1982, 2009).

This approach also lives on in many responsible scientists, even though they may not be consciously aware of it. Today, every individual is called upon to become aware of it. But how? Let us resolve to consciously re-educate our thinking and emotional world, initially through attentive, empathetic, yet alert listening to nature. Goethe sought to follow the natural phenomena of the world, whether colours from the inorganic nature or living beings, in such a way that they themselves express their idea, meaning that through the phenomena they produce, they reveal their inner nature.[1]

In this process, we do not use thinking as an arbiter, "a judge", as it is often called, but as our educator (Steiner, 2010). At the same time, we consciously cultivate our experience of nature, ensuring that when observing, we do not project our emotions onto it, but rather consciously withhold our sympathies, antipathies, and everything we already know, with our will. In this way, we create a free space for the natural phenomena, which express themselves in diverse and multilayered contexts. For we wish to first listen to nature in dialogue, purposelessly and selflessly, in order to allow new content to flow into our soul (Bockemühl, 1993).

3 Transformation of the Researcher by the Goethean Methodology

The primary aim of the Goethean-expanded methodology is to enliven abstract-theoretical and nature-detracted thinking, and to gather new concepts from nature, in accordance with life. If we wish to understand life in this way and, in the next step, also foster it through action, we must first learn to think in a living, organic-processual way. Life is always in process and transformation. Many exercises have been developed to educate our thinking in this direction, methodologically adapted to the object of observation, which will be exemplified below. For mineral, plant, animal, and human are already very different objects of observation, both chemically-physiologically and biologically.

A plant, for example, grows and develops not only as an object in space, but also in time; it has a specific temporal body (Bockemühl, 1967a). In connection with the annual cycle and the earthly-cosmic influences, it undergoes many transformations, metamorphoses (Bockemühl, 1964; Bockemühl & Suchantke, 1995; Goethe, 1992). It is the best teaching tool for practising living thinking. It performs the primary process of the living, photosynthesis, and its form is therefore, unlike ours, composed of carbohydrates. In contrast to animals and humans, who continually reshape their form from proteins through oxidation, or combustion processes when inhaling oxygen (O_2), the plant can shape its life only by "inhaling" carbon dioxide and "exhaling" oxygen, constantly producing new parts. This ultimately means that its form is simultaneously its biography, as almost all of its metamorphoses in time remain visible in space (Bockemühl, 1966; Grohmann, 1974, 1990).

We can create a so-called leaf series from the first to the last green leaf, the one closest to the flower. We notice that the form of the leaves changes during development. We must take into account that these leaves are exposed to various

[1] Original quote: "Man suche nur nichts hinter den Phänomenen, sie selbst sind die Lehre" (Goethe, 1829).

elemental and other influences at the location and throughout the annual cycle, and we can mentally metamorphose the transition from one leaf form to another in our consciousness (Bockemühl, 1966). In this way, we can practise an exact sensorial imagination (Goethe, 1985) based on the phenomena in the organic world. Thinking no longer remains purely static; instead, it becomes more fluid and gains many new qualities: it is no longer detracted from natural phenomena but connects with them, developing organically and understandably, yet still precisely, and remains artistically open to new impulses from the future. As a result, thinking acquires a pictoral character, and we gradually develop a new ability, which Goethe called perceptive power of judgment (*anschauende Urteilskraft*) (Goethe, 2023) (Figs. 1 and 2).

In the realm of feeling as well, we are invited in Goethean observation of nature to experience the transformations of the plant, which express themselves in a living image of its development as a language of gesture. We become "one" with the beings and processes of the world that we observe. This participation was instinctive in humans in earlier times, but now it must be developed in a new, conscious way. We can follow the various qualities of the colour changes, the taste, and the scent of its plant parts throughout the annual cycle. We can inwardly perceive the

Fig. 1 Natural arrangement of the leaves of common poppy (*Papaver rhoeas L.*) from bottom to top on the plant (Bockemühl, 2010, p. 73)

Fig. 2 The same leaves of common poppy arranged in a leaf series to mentally practise the transformations of the leaf form from one leaf to another (Bockemühl, 2010, p. 72)

specific relationship it has to light and darkness, to forces of heaviness and lightness, to elements such as earth, water, air and warmth, which also play a crucial role in its occurrence and appearance at its location (Bockemühl, 1985). In this way, human feeling is awakened to the world phenomena.

The realm of the human will is also addressed in this task, in order to enter into conscious and life-promoting action. First, a lively interest in another being must be awakened. This door is opened through wonder and the appreciation of all living beings. In the next step, one must create an inner free space in which the natural phenomena can reveal themselves. Only then can one selflessly immerse oneself in another being and attempt to follow all changes in time and space as accurately as possible.

4 Comprehensive Understanding of the Interconnections in a Landscape

In recent decades, awareness of the values of nature has been revived, leading to the development of instruments such as environmental planning, nature conservation, and species protection. However, the environment is still not described as a whole, but instead broken down into so-called protected goods: humans, animals, plants, soil, water, climate, air hygiene, landscape, culture, and material goods, as well as the interactions between these, but again considered separately from other protected goods (BGBl. I, 2021). Within the protected goods of landscape, beauty, character, and landscape diversity have been recognised as a value to be protected alongside biological diversity (BGBl. I, 2009). This raises many questions: How does one methodically approach the evaluation of, for example, the beauty of a landscape? Again, analytically or purely subjectively? Despite many well-intentioned efforts, global species extinction continues unabated. Why? The decline in species reflects our still predominantly abstract thinking, shaped by individual interests, with which we usually view a landscape. How can we create an understanding without losing sight of the interconnections from a holistic perspective?

The holistic approach with an aesthetic approach to landscape observation was the lifelong focus of Jochen Bockemühl (1928–2020) and is carried on by his students. He describes how we can train our recognition and experience of natural interconnections in the landscape image in three ways (Bockemühl, 1984): First, with the development of *the sense of facts*, which can still be seen as part of the analytical consciousness and perceives and defines individual things. From this stage, utility thinking is developed.

In the pursuit of beauty, we can, as a continuation, develop *the sense for holistic interconnections*. This stage includes three different aspects. (1) *The interconnection of phenomena in the landscape*, which refers to the connection of appearances in the present as the arrangement and harmony between forms in the light radius. (2) *The rhythmic interconnection of natural processes*, which involves the transformative connection in the annual cycle and the biography of a place. (3) *The interconnection of life processes in the landscape*, which is the connection of life as the characteristic feature of a place, making the landscape unique. In this final connection, the intentions of the involved humans are also included. We educate ourselves to perceive how the elements of earth, water, air, and warmth interact in the landscape and leave their impact on living beings. When we preserve the sense of facts in this process, the door opens to recognising the developmental possibilities of natural interconnections.

Third, we develop *the sense for the interests of the other being*; this means that we create an awareness not only of our own needs, but also of the needs of our "fellow beings" in these interconnections.

From this overall impression, we can create the connection of each object or living being to the whole. For in every part, the whole expresses itself in some way (Bockemühl, 1967a, 1967b). In earlier times, humans cultivated beautiful landscape images from within themselves. We have lost this ability. Now, we place many landscapes under protection and see that the beautiful and harmonious image cannot be preserved in this way when humans withdraw from it. However, when one connects holistically with a place and looks at what wants to become, one gains a share in the life of that place. In this process, one can recognise *the idea of a place* for which one may wish to take responsibility. The place then becomes something that can increasingly come into existence through human involvement. In this way, the Goethean approach of holistic perception of a landscape can lead to the individual shaping of a cultural landscape. Only where one engages in a

Fig. 3 Flower-rich permanent meadow (semi-dry grassland) at the forest edge in central Sweden (Bockemühl, 1984, p. 75)

dialogue of understanding from freedom and selects actions and technical means in harmony with this, does healing arise. This simultaneously means positioning oneself within a social context with others. Where one interacts with nature in such a way that its own being and forces come to the fore, beautiful, healthy landscape images are again created. One is also increasingly called, both in recognition and action, to that which makes life worth living (Fig. 3).

5 The View of the Earth and Its Earth Spheres

Just as a landscape has a characteristic relationship between the elements of earth, water, air, and warmth—such as a grazed, dry, and warm steppe having a different relationship than a shaded, moist meadow—so too does the Earth as a whole have a characteristic relationship between these elements, which forms the basis for life in the biosphere. The Earth's outer shell, or lithosphere, can in turn be divided into rock layers (metamorphosphere, stratosphere) and soils (pedosphere), covering about 30% of the Earth's surface. In the Earth's water layer (hydrosphere), seas, lakes, and groundwater formations can be grouped as the aquasphere, which covers about 70% of the Earth's surface. Other forms of water, such as ice deposits (cryosphere) or atmospheric water content (hydro-atmosphere), can also be counted as part of the hydrosphere. The air envelope or Earth's atmosphere covers both the litho- and aquasphere and is divided into the troposphere, stratosphere, mesosphere, thermosphere, and exosphere based on its vertical temperature layers (Huggett, 2024).

The Earth's outer shell, water layer, and atmospheric envelope, along with their respective subspheres, such as the soil cover (pedosphere), which includes the lithosphere as well as the hydrosphere and atmosphere through air and moisture inclusions, are often interpenetrated and interact with each other—this is a prerequisite for the biodiversity of the biosphere. In surface-near areas, where the biosphere exhibits its highest species diversity and biomass, the global volume ratio of water to solid components roughly corresponds to the surface coverage ratio of 70% water to 30% land, which also roughly applies to humans. When we look at these ratios, it can be said that living beings reflect the conditions of their environment, and vice versa.

From a Goethean perspective, in relation to the elements of earth, water, air, and heat, which are also reflected in the states of matter of a substance (aggregation states), the absence of a heat layer or heat sphere around the Earth is noticeable. This sphere, in turn, would permeate the other Earth spheres and encompass the processes that lead to the Earth's surface temperature. The global mean temperature of the Earth's surface has fluctuated by only about ± 4 °C over the past five million years compared to its current value (Hansen et al., 2013). It can therefore be considered a state of equilibrium (homeostasis), existing within certain limits, in the Earth's dynamic heat balance (Lovelock & Margulis, 1974).

Against this background (which could only be briefly outlined here), holistic and Goethean approaches often view the Earth's spheres not merely as parts of a whole, but as organs within an organism (Wachsmuth, 1980), in which these spheres are so closely intertwined and mutually formative that the failure of one organ often leads to the death of the organism or living being.

6 The Earth's Thermal Balance in the Cosmos

From the perspective of physical optics, the Earth's thermal balance appears surprisingly straightforward: the Earth is surrounded by a vacuum, and all heat exchange with the cosmos occurs through thermal radiation. The Earth is also in a dynamic, thermodynamic equilibrium with the cosmos (disregarding hysteresis effects): the light and heat energy continuously received by the Earth from the Sun is continuously emitted back into the cosmos in the form of light and thermal radiation. However, how this dynamic equilibrium determines the Earth's surface temperature is by no means straightforward. It depends on the atmosphere, the hydrosphere, the Earth's vegetation, and many other factors, as well as their interactions with one another.

Let us imagine the Earth without the hydroatmosphere, that is, without the interpenetration of the water and atmospheric layers. In such a scenario, the average surface temperature, or global mean temperature, would be approximately – 6 °C.

If the atmosphere were beside water vapour also devoid of carbon dioxide—whose concentration results from the carbon cycle through the interaction of the lithosphere, hydrosphere, atmosphere, and biosphere—and other greenhouse gases, the theoretical global mean temperature would drop even further, to around −18 °C (Kondratyev & Moskalenko, 1984; Roedel & Wagner, 2011, p. 44).

Instead, we observe a global mean temperature of the near-surface atmosphere of approximately + 15 °C, a phenomenon referred to as the natural greenhouse effect. From a Goethean perspective, this can be viewed as part of the Earth's warmth sphere as a "thermal sheath". This "warmth sphere" is understood as the sphere surrounding the Earth where these processes occur, shaping the Earth's thermal balance. It shares a mutually conditioning relationship with the biosphere and hydrosphere, similar to the relationship between the thermal organism of a living being and its life processes: on the one hand, these processes are only possible due to body heat; on the other hand, body heat itself arises from these processes (Lovelock & Margulis, 1974).

The warmth sphere can also be understood as forming a cover, akin to a skin organ, which separates a warmer interior from a colder habitat while simultaneously connecting the interior to its environment. Within this cover, distinct sublayers can be identified: those more transparent to thermal radiation, such as the stratosphere, which can be optically described as "window," and others that scatter or absorb radiation, re-emitting it isotropically, such as the moist air layers of the troposphere. These latter layers can be optically characterised as semi-transparent "mirrors" that retain heat. Using a weather balloon equipped with sensors to measure the Earth's thermal radiation emitted into space, as well as the atmosphere's re-emission and backscattering towards the Earth, these layers of the thermal sheath can be traversed and experimentally characterised as a vertical profile of the radiation balance within the thermal sheath (Philipona et al., 2020; Rang, 2025).

The optical properties of these atmospheric layers are not directly related to air pressure or air temperature (which decrease monotonously and predictably with altitude in the troposphere). However, they strongly depend on the interpenetration of the atmosphere with other Earth spheres, such as the hydrosphere. For this reason, it seems justified to understand them not as layers of the atmosphere but as layers of the warmth sphere, which, in the sense described above, represents an interpenetration of the Earth's spheres (Rang, 2025).

The Earth's Warmth Sphere and Its Connection to the Rock, Water and Atmospheric Layers

The warmth sphere is not limited to the interactions of these layers within the atmosphere: the fact that 70% of the Earth's surface is covered by oceans is a crucial factor in the warmth sphere. Due to their low albedo, the oceans absorb nearly all sunlight and thermal radiation from the Sun yet they warm only slightly. This is possible because of water's exceptionally high specific heat capacity and volumetric heat capacity (water appears to have the highest volumetric heat capacity among all naturally occurring substances).[2] Additionally, when releasing heat, its temperature drops significantly less than that of other substances. As a result, temperature fluctuations, such as those between day and night, are effectively balanced. Without the high ocean coverage and the unique properties of water, the Earth's warmth sphere could not maintain temperature stability in a comparable way.

Through phase transitions of water into other states of matter—solid, liquid, gaseous and plasma (in the thermosphere)—numerous processes occur such as rainfall from clouds, evaporation over water surfaces and freezing, which over relatively long periods in the past have enabled temperature regulation and the homeostasis of the mean surface temperature. At the same time, these phase transitions can only occur due to temperature differences. Warmth therefore plays a decisive role in the transformation of Earth's spheres both in the Earth's interior and on its surface and atmosphere.

The global water cycle is simultaneously a thermal cycle which has not been characterised in the same way as other cycles in natural science but could serve as a further foundation for describing the Earth's warmth sphere.[3] From the perspective of the intrinsic connection between Earth's sphere, such a warmth sphere (as a fifth primary sphere) could also act as a unifying element between the inorganic spheres and the organic biosphere.

The Goethean Approach to the Warmth Sphere and Sustainable Development Goethean methods, through their anticipatory nature, can contribute to renewing a positive relationship between us as humans, the Earth and nature, which is crucial in our current time (Amrine et al., 1987; Bembé & Rosslenbroich, 2021; Bockemühl, 1984, 2010). As described for landscapes, a Goethean approach to the Earth's spheres can foster our awareness of these spheres and our engagement with them. Due to the alienation caused by civilisation, many people have limited awareness of the Earth's rock, water and air layers, which cannot be significantly strengthened by the media's portrayal of environmental issues, as this often reflects the problematic aspects of our relationship with the environment and Earth. Through a Goethean approach, grounded in everyday phenomena, natural

---

[2] In very thin layers with microstructures on the scale of a few nanometres, higher volumetric heat capacities have been observed. However, these do not occur naturally and disappear in larger volumes or at greater layer thicknesses, as they involve surface effects (Tian, 2011).

[3] A warmth sphere or a thermal sheath in this sense should not be confused with the thermosphere as currently defined, which, as a sub-sphere of the atmosphere, plays a negligible role in surface temperature and in the temperatures and temperature gradients within the lithosphere, hydrosphere, and atmosphere in the biosphere. Originally, the thermosphere was introduced as the lower atmospheric region, up to approximately eight kilometres in height, where atmospheric processes driven by heat occur (Sen, 1924). However, this term has been redefined and established in its modern usage since the 1950s (Huggett, 2024).

processes in our environment and the environment itself come into focus. This can lead to a more conscious perception of the Earth's rock, water and air layers.

For the warmth sphere, however, the issue arises that it has not been sufficiently characterised scientifically and is not widely recognised or accepted as one of the Earth's major elemental sphere.[4] As a result, the warmth sphere remains largely unknown to contemporary people, making awareness of it nearly impossible. Yet it is precisely within this thermal sheath —reciprocally related to our awareness—that the problems of the climate crisis manifest most dramatically. Thus, we see an essential task in learning to better describe this sheath , particularly through Goethean, phenomenological and holistic approaches that can also make its integration into the Earth's whole clearly visible.

Raising awareness, however, is only a first step toward a more sustainable use of natural resources. As described in Sects. 1–4, the Goethean approach not only builds awareness but also serves as a path of personal development for those who practise it. As such, its contribution to overcoming the intention-behaviour gap cannot be overestimated.

7 The World of Microorganisms as a Field of Practice for Goetheanism

In many examples of Goethean nature observation, the object of study is a phenomenon that can be sensually perceived, leading to an essential understanding, such as a plant, an animal or a landscape. This requires the precise study of observable phenomena using the sensory organs but can and should, when possible, also include a thorough analysis of environmental conditions, plant constituents, the anatomy and physiology of animals, and the soil, plants and animals of a landscape for a more comprehensive investigation. Only the synthesis—the inner picture—of all observations and analyses reveals to us, as observers, the meaningful context of the whole. In this way, we can integrate our thoughts with the conceptually comprehensible principles and laws of the observed phenomena, as these reside on the same conceptual level where our cognition builds this bridge.

The Goethean approach differs somewhat when we attempt to study the world of microorganisms. Sensory observation as a starting point is absent, which is why relatively few such studies have been undertaken so far (Simon, 2008). Microorganisms, particularly prokaryotes—single-celled organisms without a nucleus, including bacteria and archaea—escape direct observation within their living context. They do not appear to the senses. They can only be made visible as cells through the use of a microscope, but this requires their removal from their living context and preparation for microscopic examination. Under the microscope, we see only the simplest cells, approximately one micrometre (0.001 mm) in size,

[4] For comparison, one can refer to the depictions of the Earth in school textbooks, encyclopedias, or even scientific standard works, which list four Earth layers (lithosphere, hydrosphere, atmosphere, biosphere); see, for example, German et al. (1988) and Graff (2010).

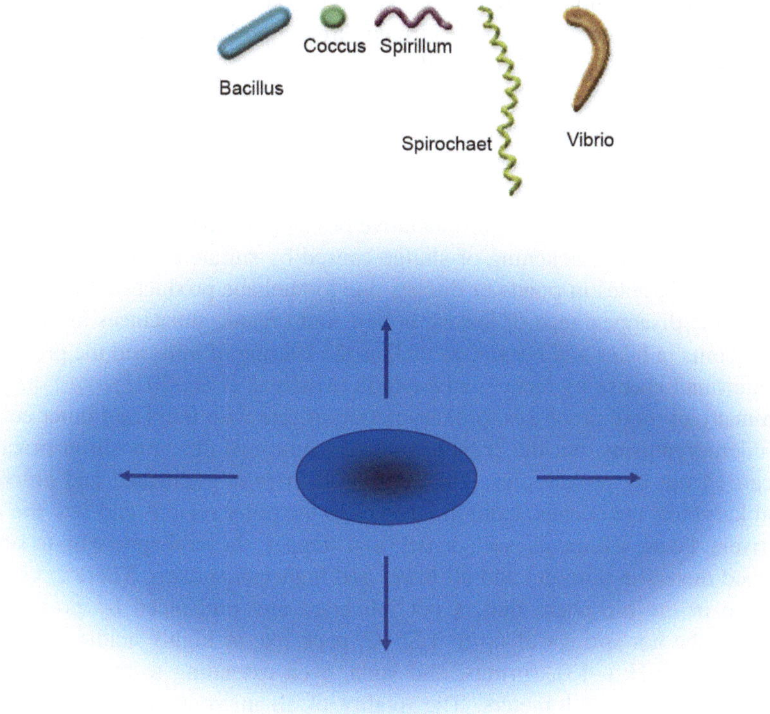

Fig. 4 Basic forms of prokaryotes (top) and the gesture of prokaryotic metabolic activities going from the cell to the environment (bottom) (Simon)

consisting of just four basic shapes: straight and curved rods (bacillus, vibrio), spheres (coccus), and corkscrew-like spirillae and spirochaetes (Fig. 4). In rare or specialised cases, these cell forms may also appear as chains or aggregates (Simon, 2008). However, these forms reveal nothing about the type of prokaryotes they are, to which species they affiliate, their metabolic processes, or their living context. This represents a gap that we must bridge through a process of knowledge formation, one that does not exist with sensory-perceptible higher organisms.

In the case of microorganisms, we cannot say in Goethe's sense that we perceive our ideas directly through our eyes.[5]

The very simple forms of prokaryotes, visible only under a microscope, stand in stark contrast to the nearly unlimited metabolic diversity of the entirety of prokaryotes. All substances and compounds formed within the Earth's biosphere can be

[5] Expressed in the already cited quote "Man suche nur nichts hinter den Phänomenen, sie selbst sind die Lehre" (Goethe, 1829).

processed and decomposed by prokaryotes, with the exception of oxidic and sulfidic rocks, which themselves were formed through prokaryotic activity during the early history of the Earth (Knoll et al., 2016).

Unlike higher organisms, the metabolism and the cycling of matter of prokaryotes are entirely directed toward the environment, impacting and shaping it in the process (Fig. 4). We perceive this cycling of matter in various ways: for example, the transformation of organic waste into humus in compost, the foul smells produced by anaerobic, fermentative prokaryotes that generate reduced sulfur compounds, the blackening of sludge caused by sulfate-reducing bacteria, the formation of bubbles in ponds or ditches due to methane production by a group of archaea, the disappearance of leaves, pollen, and organic substances at the end of algal blooms in lakes and oceans, as well as the souring of milk and the production of yogurt and cheese by lactic acid bacteria (Madigan et al., 2021).

In many of these processes, prokaryotes cooperate with fungi and other eukaryotic microorganisms, but the critical transformation and decomposition processes are carried out by prokaryotes. The entire biosphere is permeated and enlivened by prokaryotes: the oceans, from the tropics to the polar regions and into the deep sea; rivers, lakes, sediments, soil, deserts and steppes; the atmosphere up to several thousand meters in altitude; and all lower and higher organisms. There is virtually no place in the biosphere that is not inhabited and influenced by prokaryotes, whether it be hot springs of up to 122 °C, pack ice, or highly acidic or alkaline water bodies (pH 1–13), springs, and soils, even extending to the Earth's crust at depths of several kilometers. The totality of all microorganisms in a habitat, ecosystem or even in higher organisms, where they interact as living communities, is referred to as the microbiome (Berg et al., 2020).

In most habitats, there are between half a million and one billion prokaryotes per milliliter or gram of mass. The number of prokaryotes in the biosphere exceeds that of all other organisms by several orders of magnitude, estimated at 10^{30} (Whitman et al., 1998). The diversity of prokaryotes is immense and far from being fully understood. This applies to their taxonomy and phylogeny as well as to their metabolic diversity. However, not all prokaryotes in a habitat are metabolically active at the same time; they become active only when conditions suitable for their metabolism and growth are present. Some prokaryotes may be almost always active if the necessary conditions occur frequently. Others may become active only rarely, during a specific event that might happen just once a year, such as the appearance of certain algae in water or the flowering of a specific plant.

Prokaryotes are an integral part of many ecosystems and of higher organisms, with which they form a holobiont (Baedke et al., 2020). They also play a critical role in the global cycles of carbon, nitrogen, oxygen, sulfur and hydrogen (Madigan et al., 2021). This highlights the fully environment-oriented metabolism of prokaryotes, which contrasts sharply with the metabolism of higher organisms that is increasingly directed towards the development of the physical form and the autonomy of the organism (Rosslenbroich, 2014). The environmental context of a prokaryotic community can include a lake, the soil of a specific meadow, the rhizosphere of a plant or even a higher organism, including humans. Ultimately, the meaningful context of all prokaryotes is the entire Earth, which, in a Goethean sense, can be appropriately regarded as their host organism and, together with the prokaryotes, as a holobiont.

The Study of Cultivable and Non-cultivable Prokaryotes from a Goethean Perspective

If we want to learn more about the specific characteristics of individual prokaryotes, there are two methodological possibilities to their further investigation, both of which can be understood entirely in the Goethean sense.

(I) We attempt to enrich and isolate the prokaryote from its habitat and cultivate it on suitable nutrient media under conditions similar to those of its habitat. Appropriate nutrient media for enrichment include, for example, yeast extract, while the growth conditions reflecting the habitat include factors such as temperature, the presence or absence of oxygen, salinity and pH. If isolation is successful and a pure culture has been obtained, further investigations must determine which individual substrates the organism can metabolise, which carbon, nitrogen and sulfur compounds it can utilise, and which biopolymers such as proteins, polysaccharides or lipids it can break down. It must also be determined whether the organism is motile, whether it requires vitamins for growth and, if so, which ones, and under which oxygen and temperature conditions it grows at all and grows best. Growth itself consists solely of an increase in cell number, so the characteristic to be determined is its rate of cell multiplication (Madigan et al., 2021).

The formation of a visible form does not occur, and our attention is entirely focused on what happens in the environment of the prokaryote within the culture medium and how it changes over time due to its metabolic activity. For precise taxonomic and phylogenetic identification of the prokaryote, sequencing of an appropriate gene—typically the universally present 16S rRNA gene—should be carried out, and ideally, a sequence and genome analysis of the entire genome should also be performed (Madigan et al., 2021).

By synthesising all these characteristics, along with potentially additional ones, and conducting comparative analyses with other prokaryotes, we can form a fairly detailed picture of this prokaryote's living conditions. Our inner focus is primarily directed towards the environment, the life milieu, of this prokaryote. In natural settings where similar conditions exist, we can reasonably assume the presence of this prokaryote, though always in association with many other organisms.

Numerous studies have shown that only a small fraction of the prokaryotes present in nature can be isolated and studied as pure cultures (Staley & Konopka, 1985). The vast majority of prokaryotes defy isolation (Hug et al., 2016). This can be due to a variety of reasons, all of which point to something characteristic of prokaryotes: their living conditions are very different from the conditions under which attempts are often made to isolate them. Prokaryotes depend on specific, yet unknown, substances and/or metabolic intermediates for their metabolism and growth, which they cannot produce themselves and must instead obtain from other organisms in their environment. Alternatively, they may live in symbiotic relationships, often in such a way that there is mutual or even multi-sided interdependence in the supply of nutrients, metabolic intermediates, and growth factors such as vitamins (Kost et al., 2023; Wienhausen et al., 2024). They may also inhabit environments completely different from those used in enrichment and isolation attempts (Thorup et al., 2021). This demonstrates that our concrete, imaginative efforts to align with the living conditions of these prokaryotes, as well as our capacity to empathise with their living conditions and recreate them in the laboratory, are insufficient to understand them well enough to cultivate and study them in detail.

For this reason, a second approach is now widely used to study so far uncultivated or truly uncultivable prokaryotes within their communities in their natural habitats and locations. This approach forgoes enrichment and focuses directly on the taxonomic identification and analysis of the metabolic potential of the prokaryotes present within the community (Madigan et al., 2021).

(II) The prokaryotic community of a habitat—whether in soil, on the leaves of a plant, in the sea, or in the human gut—can be analysed in great detail at the genomic level regarding its taxonomic composition and the metabolic potential of its individual members. To do this, the DNA of prokaryotes must be extracted from a sample taken from the habitat and subjected to a sequence analysis of a universal marker gene, usually the 16S rRNA gene, or the genomes of all prokaryotes present in the sample. Thanks to the relatively low cost of sequencing, specialised computer programs, highly efficient computers with large storage capacities, and extensive databases for comparative analyses, such studies have become routine in many investigations. This method provides a deep and highly detailed insight into the composition of the prokaryotic community and its metabolic potential (metagenomic analyses), not only as a whole but, with careful data processing, also into individual and even rare members (Dlugosch et al., 2022; Milke et al., 2023).

How little we know about these uncultivated, but also about cultivated prokaryotes, becomes clear by the fact that up to half of the genes found in these analyses cannot be assigned to a function. Thus, the world of prokaryotes still hides an unknown potential of metabolic functions. Further, one danger of these investigations lies in relying on the abstract representations of results generated by standardised computer programs without exploring the relationship to the environmental conditions of the prokaryotic community's habitat. To prevent the findings from remaining abstract and existing only as conceptual impressions, the environmental conditions of the habitat from which the community originates must also be studied as precisely as possible and formed into an inner picture. Only in this way can the results of metagenomic analyses be placed into a meaningful context with the environment of the prokaryotic community.

Microorganisms and Sustainability
With both of the methodological approaches described, we must mentally immerse ourselves entirely in the life processes shaped by these prokaryotes, their environment, and the higher organisms that coexist with them. This requires building a direct bridge to the world of ideas underlying the life processes of prokaryotes. Unlike Goethean studies of higher organisms such as plants, animals, or landscapes, our inner focus here is far less directed towards the sensually visible object. Instead, it is much more focused on the surrounding environment, the living conditions of microorganisms, and particularly of prokaryotes. If this approach succeeds, we can even shape environments in such a way that they particularly promote the growth of specific prokaryotes. This has been known for millennia in many fermentation processes, such as in the production of wine and lactic-acid-fermented products like sauerkraut and cheese.

Recently, the promotion of soil prokaryotes that either bind or prevent the formation of climate-damaging gases or support plant growth has been demonstrated in biodynamic agriculture (Milke et al., 2024; Skinner et al., 2019). This illustrates how an appropriate and Goethean-inspired approach to agriculture, aligned with the essential nature of its processes, can sustainably foster the soil microbiome for a future-oriented Earth. Similarly, in the One Health concept—which views the health of animals and humans in a completely new and holistic way in the context of the environment—microorganisms and the entire microbiome play a central role (Berg et al., 2020). They are, in essence, included in the context in a Goethean sense.

8 Conclusion: What Can Goetheanism Contribute to Sustainable Development?

The examples presented illustrate how it is possible to bridge the gap between the ideas that form within us through our perceptions of the world and the thoughts and ideas underlying these perceptions.

The Goetheanistic approach seeks to make visible, through repeated observations of the research subject, the thoughts and ideas that fundamentally underlie both, one's own conceptions and perceptions.

In this way, Goetheanism can overcome the separation between perception and conception, which is one of the causes of our irresponsible and unsustainable ways of acting and engaging with the external world, and transform it into a participatory relationship with nature and the environment. The practice of engaging with a developing plant form or landscape formations, the meaningful perception of all of Earth's layers and spheres, and the conscious participation in the life processes of microorganisms—imperceptible to the senses—both in nature and within the human being, can be of great assistance in fostering a way of relating to the world that strengthens sustainable development.

References

Amrine, F., Zucker, F. J., & Wheeler, H. (Eds.). (1987). *Goethe and the sciences: A eeappraisal. Boston Studies in the Philosophy of Science 97*. D. Reidel.

Baedke, J., Fábregas-Tejeda, A., & Nieves Delgado, A. (2020). The holobiont concept before Margulis. *Journal of Experimental Zoology Part B: Molecular and Developmental Evolution, 334*(3), 149–155. https://doi.org/10.1002/jez.b.22931

Bembé, B., & Rosslenbroich, B. (2021). Goetheanismus und Pädagogik. In *Jahrbuch für Goetheanismus 2021* (pp. 113–133). Pädagogische Forschungsstelle beim Bund der Freien Waldorfschulen.

Berg, G., et al. (2020). Microbiome definition re-visited: Old concepts and new challenges. *Microbiome, 8*, Article 103. https://doi.org/10.1186/s40168-020-00875-0

BGBl. I. (2009). Bundesnaturschutzgesetz vom 29. Juli 2009, das zuletzt durch Artikel 5 des Gesetzes vom 3. Juli 2024 (BGBl. 2024 I Nr. 225) geändert worden ist. Bundesgesetzblatt, 2542.

BGBl. I. (2021). Gesetz über die Umweltverträglichkeitsprüfung vom 18. März 2021, das zuletzt durch Artikel 13 des Gesetzes vom 8. Mai 2024 (BGBl. 2024 I Nr. 151) geändert worden ist. Dieses Gesetz dient der Umsetzung der Richtlinie 2011/92/EU des Europäischen Parlaments und des Rates vom 13. Dezember 2011. Bundesgesetzblatt, 540.

Bockemühl, J. (1964). Der Pflanzentypus als Bewegungsgestalt: Gesichtspunkte zum Studium der Blattmetamorphosen. *Elemente der Naturwissenschaft, 1*, 3–11. https://doi.org/10.18756/edn.1.3

Bockemühl, J. (1966). Bildebewegungen im Laubblattbereich höherer Pflanzen. *Elemente der Naturwissenschaft, 4*, 7–23. https://doi.org/10.18756/edn.4.7

Bockemühl, J. (1967). Das Ganze im Teil. *Elemente der Naturwissenschaft, 6*, 1–8. https://doi.org/10.18756/edn.6.1

Bockemühl, J. (1967a). Äußerungen des Zeitleibes in den Bildebewegungen der Pflanzen. *Elemente der Naturwissenschaft, 7*, 25–30. https://doi.org/10.18756/edn.7.25

Bockemühl, J. (1985). *Toward a phenomenology of the etheric world: Investigations into the life of nature and man.* Anthroposophic Press.

Bockemühl, J. (1984). *Sterbende Wälder—Eine Bewusstseinsfrage.* Verlag am Goetheanum.

Bockemühl, J. (1993). Kann ein holistischer Ansatz im Umgang mit Landschaft formuliert werden? *Elemente der Naturwissenschaft, 58*, 28–36. https://doi.org/10.18756/edn.58.28

Bockemühl, J., & Suchantke, A. (1995). *The metamorphosis of plants.* Novalis Press.

Bockemühl, J. (2010). Lebt die Welt in mir? Wahrnehmungs- und Besinnungsübungen zum Entwickeln von Verantwortungsfähigkeit im täglichen Leben. Forschungsinstitut am Goetheanum und Petrarca, Europäische Akademie für Landschaftskultur.

Cardinale, B., Duffy, J., Gonzalez, A., et al. (2012). Biodiversity loss and its impact on humanity. *Nature, 486*, 59–67. https://doi.org/10.1038/nature11148

Caro, T., Rowe, Z., Berger, J., Wholey, P., & Dobson, A. (2022). An inconvenient misconception: Climate change is not the principal driver of biodiversity loss. *Conservation Letters, 15*(1), Article e12868. https://doi.org/10.1111/conl.12868

Dlugosch, L., Poehlein, A., Wemheuer, B., Pfeiffer, B., Badewien, T. H., Daniel, R., & Simon, M. (2022). Significance of gene variants for the functional biogeography of the near-surface *Atlantic Ocean microbiome. Nature Communications, 13*, Article 456. https://doi.org/10.1038/s41467-022-28128-8

Endlich, B. (Ed.). (1985). *Der Organismus der Erde: Grundlagen einer neuen Ökologie (Beiträge zur Anthroposophie Nr. 4).* Verlag Freies Geistesleben.

German, K., Warnecke, G., & Huch, M. (1988). *Die Erde: Dynamische Entwicklung, menschliche Eingriffe, globale Risiken.* Springer.

Goethe, J. W. von. (1829). *Wilhelm Meisters Wanderjahre*, 2nd book, 11th chapter. Cotta'sche Buchhandlung.

Goethe, J. W. von. (1883). *Naturwissenschaftliche Schriften* (Vol. 1, Edited by R. Steiner). Deutsche National-Literatur (Vol. 114).

Goethe, J. W. von. (1887). *Naturwissenschaftliche Schriften* (Vol. 2, Edited by R. Steiner). Deutsche National-Literatur (Vol. 114).

Goethe, J. W. von. (1985). *Sämtliche Werke nach Epochen seines Schaffens* (Münchner Ausgabe, Vol. 12). C. H. Beck.

Goethe, J. W. von. (1992). *Die Metamorphose der Pflanzen.* Verlag Freies Geistesleben.

Goethe, J. W. von. (2023). Anschauende Urteilskraft. In *Essay in Hamburger Ausgabe* (Vol. 13, pp. 30–31).

Grohmann, G. (1974). *The plant: A guide to understanding its nature.* Rudolf Steiner Press.

Grohmann, G. (1990). *Metamorphosen im Pflanzenreich.* Verlag Freies Geistesleben.

Graff, P. V. (2010). *Sphere of Earth: An introduction to making observations of Earth using an Earth System Science approach.* NASA/ARES Education Program. https://ares.jsc.nasa.gov/interaction/eeab/pdfs/SOE/SpheresOfEarth_StudentGuidev3.pdf

Hansen, J., Sato, M., Russell, G., & Kharecha, P. (2013). Climate sensitivity, sea level and atmospheric carbon dioxide. *Philosophical Transactions of the Royal Society A: Mathematical, Physical and Engineering Sciences, 371*(2001), Article 20120294. https://doi.org/10.1098/rsta.2012.0294

Hanski, I. (2005). Landscape fragmentation, biodiversity loss and the societal response. *EMBO Reports, 6*(5), 388–392. https://doi.org/10.1038/sj.embor.7400398

Hug, L. A., et al. (2016). A new view of the tree of life. *Nature Microbiology, 1*, Article 16048. https://doi.org/10.1038/nmicrobiol.2016.48

Huggett, R. (2024). Earth's spheres: Conceptual and definitional debates. *Progress in Physical Geography: Earth and Environment, 48*(5–6), 651–670. https://doi.org/10.1177/03091333241275465

Knoll, H., Bergmann, K. D., & Strauss, J. V. (2016). Life: The first two billion years. *Philosophical Transactions of the Royal Society B: Biological Sciences, 371*(1707), Article 20150493. https://doi.org/10.1098/rstb.2015.0493

Kondratyev, K. Y., & Moskalenko, N. I. (1984). The role of carbon dioxide and other minor gaseous compounds and aerosols in the radiation budget. In J. T. Houghton (Ed.), *The global climate* (pp. 225–235). Cambridge University Press.

Kost, C., Patil, K. R., Friedman, J., Garcia, S. L., & Ralser, M. (2023). Metabolic exchanges are ubiquitous in natural microbial communities. *Nature Microbiology, 8*, 2244–2252. https://doi.org/10.1038/s41564-023-01511-x

Lovelock, J. E., & Margulis, L. (1974). Atmospheric homeostasis by and for the biosphere: The Gaia hypothesis. *Tellus, 26*(1–2), 2–10. https://doi.org/10.3402/tellusa.v26i1-2.9731

Madigan, M. T., Bender, K. S., Buckley, D. H., Sattley, W. M., & Stahl, D. A. (2021). *Brock biology of microorganisms* (16th ed.). Pearson.

Milke, F., Meyerjürgens, J., & Simon, M. (2023). Ecological mechanisms and current systems shape the modular structure of the global oceans' prokaryotic seascape. *Nature Communications, 14*, Article 6141. https://doi.org/10.1038/s41467-023-41909-z

Milke, F., Rodas-Gaitan, H., Meissner, G., Masson, V., Oltmann, M., Möller, M., Wohlfahrt, Y., Kuligs, B., Athmann, M., & Fritz, J. (2024). Enrichment of putative plant growth promoting microorganisms in biodynamic compared with organic agriculture soils. *ISME Communications, 4*, ycae021. https://doi.org/10.1093/ismeco/ycae021

Philipona, R., Kräuchi, A., Kivi, R., Peter, T., Wild, M., Dirksen, R., Fujiwara, M., Sekiguchi, M., Hurst, D., & Becker, R. (2020). Balloon-borne radiation measurements demonstrate radiative forcing by water vapor and clouds. *Meteorologische Zeitschrift, PrePub.* https://doi.org/10.1127/metz/2020/1044

Rang, M. (2025). Experimental characterization of Earth's heat sphere. *Elemente der Naturwissenschaft, 123* (in print).

Roedel, W., & Wagner, T. (2011). *Physik unserer Umwelt: Die Atmosphäre* (4th ed.). Springer.

Rosslenbroich, B. (2014). On the origin of autonomy: A new look at the major transitions in evolution. *Springer*. https://doi.org/10.1007/978-3-319-04141-4

Rosslenbroich, B. (2018a). Eigenschaften des Lebendigen—Schritte zu einem eigenständigen Begriff des Organismus. In *Wittener Kolloquium Humanismus, Medizin und Philosophie* (Vol. 5, pp. 13–54).
Rosslenbroich, B. (2018b). *Entwurf einer Biologie der Freiheit: Die Frage der Autonomie in der Evolution.* Verlag Freies Geistesleben.
Rosslenbroich, B. (2023). *Properties of life: Toward a theory of organismic biology.* The MIT Press.
Schad, W. (2001). Was ist Goetheanismus? In *Tycho de Brahe Jahrbuch für Goetheanismus* (pp. 23–66). Tycho Brahe Verlag.
Schnell, G. R. (Ed.). (1982). *Waldsterben: Aufforderung zu einem erweiterten Naturverständnis.* Verlag Freies Geistesleben.
Sen, S. N. (1924). On the distribution of air density over the globe. *Quarterly Journal of the Royal Meteorological Society, 50*(209), 29–51. https://doi.org/10.1002/qj.49705020908
Simon, M. (2008). Aspekte zu einer goetheanistischen Mikrobiologie. In D. Pestel & W. Schad (Eds.), *Naturwissenschaft heute im Ansatz Goethes* (pp. 134–154). Verlag Johannes Mayer.
Skinner, C., Gattinger, A., Krauss, M., Krause, H. M., Mayer, J., van der Heijden, M. G. A., & Mäder, P. (2019). The impact of long-term organic farming on soil-derived greenhouse gas emissions. *Scientific Reports, 9*, Article 1702. https://doi.org/10.1038/s41598-018-38207-w
Staley, J. T., & Konopka, A. (1985). Measurements of in situ activities of nonphotosynthetic microorganisms in aquatic and terrestrial habitats. *Annual Review of Microbiology, 39*, 321–346.
Steiner, R. (1924–1925). *Anthroposophical leading thoughts: Anthroposophy as a path of knowledge* (G. Adams & M. Adams, Trans.). Rudolf Steiner Archive. https://rsarchive.org/Books/GA026/English/RSP1973/GA026_index.html
Steiner, R. (1987). *Goethean science* (W. Lindeman, Trans.). Rudolf Steiner Archive. https://rsarchive.org/Books/GA001/English/MP1988/GA001_index.html
Steiner, R. (1897). *Goethe's world view* (W. Lindeman, Trans.). Rudolf Steiner Archive. https://rsarchive.org/Books/GA006/English/MP1985/GA006_index.html
Steiner, R. (1886). *A theory of knowledge* (O. D. Wannamaker, Trans.). Rudolf Steiner Archive. https://rsarchive.org/Books/GA002/
Steiner, R. (2010). *The world of the senses and the world of the spirit* (J. Collis, Trans.). Rudolf Steiner Press. (Original work published 1911–1912).
Suchantke, A. (1982). Skizzen zu einer ökologischen Ethik. In W. Schad (Ed.), *Allgemeine Biologie* (Goetheanistische Naturwissenschaft Band 1). Verlag Freies Geistesleben.
Suchantke, A. (2009). *Metamorphosis: Evolution in action.* Adonis Press.
Thorup, C., Petro, C., Bøggild, A., Sneibjerg Ebsen, T., Brokjær, S., Nielsen, L. P., Schramm, A., & Bjerg, J. J. (2021). How to grow your cable bacteria: Establishment of a stable single-strain culture in sediment and proposal of Candidatus Electronema aureum GS. *Systematic and Applied Microbiology, 44*(6), Article 126236. https://doi.org/10.1016/j.syapm.2021.126236
Tian, H. (2011). *Volumetric heat capacity enhancement in ultrathin fluorocarbon polymers for capacitive thermal management* [Master's thesis, University of Illinois at Urbana-Champaign]. IDEALS. https://www.ideals.illinois.edu/items/25097
Wachsmuth, G. (1980). *Erde und Mensch: Ihre Bildekräfte, Rhythmen und Lebensprozesse* (4th ed.). Verlag am Goetheanum.
Whitman, W. B., Coleman, D. C., & Wiebe, W. J. (1998). Prokaryotes: The unseen majority. *Proceedings of the National Academy of Sciences of the United States of America, 95*(12), 6578–6583.
Wienhausen, G., Moraru, C., Bruns, S., Tran, Q. D., Sultana, S., Wilkes, H., Dlugosch, L., Azam, F., & Simon, M. (2024). Ligand crossfeeding resolves bacterial vitamin B12 auxotrophy. *Nature, 629*, 886–892. https://doi.org/10.1038/s41586-024-07396-y

Beyond Boundaries: Thoughts on a Science of the Living

Craig Holdrege and Jon McAlice

Abstract

What does it mean to be alive? Where do we draw the line between the living and the non-living? Is it possible not only to change our mind about where to place the border, but to transform our minds into a living thinking beyond borders? In this essay, we strive to think about life itself as a kind of belonging, which means also coming to experience how thinking itself belongs with life. We take time to exactly imagine the way a plant is, how it lives into its environment and environment lives through it. We then precisely imagine how animals live in their world in a radically different way and yet interpenetrate with plants. Finally, we see how humans are not divorced from the world but live in it as thinking beings. As living thinking, scientific research might move beyond drawing boundaries and into participating in nature's creative unfolding.

Keywords

Living thinking • Goethean science • Holistic science • Nature and consciousness • Participatory science • Phenomenology • Rudolf Steiner • Johann Wolfgang von Goethe • Gaia theory • Life processes • Organism and environment • Ecological belonging • Exact sensorial imagination • Earth as a living organism • Animal perception • Human-Nature relationship • Interconnectedness in nature • Scientific methodology • Sustainability

C. Holdrege (✉) · J. McAlice
The Nature Institute, Ghent, NY, USA
e-mail: craig@natureinstitute.org

J. Kronenberg and E. T. Lammerts van Bueren (eds.), *On the Earth We Want to Live*, World Sustainability Series, https://doi.org/10.1007/978-3-031-98758-8_17

1 Earth Alive?

We typically think of life in terms of living beings—plants, microbes, animals or humans—and tend to focus our inquiries into the nature of life on organisms. This makes sense. Yet, are the boundaries of these creatures at the same time the boundaries of life? Does life stop where organisms end? And where do organisms end?[1] There has been much discussion since the Gaia theory was formulated over fifty years ago whether it is justified to consider the Earth as an organism, or more broadly, the Earth as alive.

In an intriguing essay, plant ecologist Stan Rowe criticises the view that life is "possessed" by organisms and that organisms in a sense provide the parameters within which one can speak about life. He writes:

> Just as the living parts of an organism depend on the vitality of the whole [organism], so living organisms depend on the energetics of Planet Earth from which they evolved and by which they are maintained. From an ecological viewpoint Planet Earth, the inclusive supra-organic Ecosphere, is the best and most logical metaphor for "life" in its largest sense. (Rowe, 2006, p. 141)

His point is that just as DNA, a neuron or a liver cannot be considered alive without the context of a whole organism, the life of organisms is in turn wholly dependent on their ecological context. It is an abstraction to separate organisms from their contexts and to view "life" as only encapsulated within them. This leads him to say that we cannot separate the notion of life from that of the planet as a whole. And in the end, this wholeness would have to include at least the vital importance of the sun and the solar system as being part of this living Earth.

Rowe's perspective can help us to move beyond the limitations of a habitual way of thinking that "life equals organisms and organisms equals life." He points to the all-important insight that any time we think of an organism as enclosed and bounded, we are dealing with only a partial aspect of its reality, since every living being exists only by virtue of its relatedness to what we call its environment, or more broadly its life context.

In the early twentieth century, Rudolf Steiner outlined an approach to understanding life on Earth as a process of coevolution of the planet and the expressions of life that appear over the course of time (see, e.g., Steiner, 1997). Steiner's characterisation of the primacy of relatedness throughout the universe resounds strongly with Rowe's conception of life as a "supra-organism." Yet, while Steiner explicitly includes the human being in the wholeness of life evolving, Rowe makes no mention of human consciousness as something integral to the ecology of life. What would a practice of science look like through which we, as thinking human beings, come to recognise and embrace our role within the context of the living world? If we attend from the outset to this relatedness, to belonging together, how does our understanding of nature, life, organisms and earth grow and deepen?

---

[1] See: "Where Does an Animal End" (Holdrege, 2021a, 2021b).

In the following, we explore different aspects of such a practice. In the first two sections, we consider ways of attending to plants and animals that are responsive to how they bring themselves to expression in the world. In the third section, we look more closely at how our experience of and desire to understand the world can be the basis of a more heightened sense of the wholeness of life of which we are part.

2 Belonging—Plants and Earth

Imagine a windbreak of large white pines. It was planted years ago. The trees have grown tall, their branches wide. They shade the ground beneath them and block our view of the hills beyond. While we can consider them as individual trees, each rising in its own way into the sky, we can also attend to how they are united with the earth. Their trunks broaden as they disappear into the soil where their roots extend down and out into the Earth. An uprooted tree is no longer in any essential sense a tree. It has lost its connection to the soil—to its anchoring, to the water that courses through it, to all the soil minerals and organisms that are essential to its existence. Just as trees are rooted in the soil, so also are they "rooted" in the air and light without which they could not exist. Their green leaves (needles in the case of the pines) spread out into light, take in air through pores, and form chlorophyll and carbon-based substances through the light, air (carbon dioxide) and water. Leaves, light, air and water are inseparable. Goethe once wrote, "Were the eye not sunlike, how could we behold the light?" (2002, p. 324[2]). So too with leaves. Were they not of light, air and water, how could they develop through light, air and water? Just as earth is part of a plant through light, air, water and minerals, so are plants part of earth in the mantle of unique forms and substances they create around the globe.

This plant mantle of the earth is rich, varied and highly dynamic. In grasslands the fabric is short and dense, in a tropical rainforest voluminous and many-tiered and in extreme deserts and high mountains it apparently disappears. Yet, if we think of the oxygen or moisture that plants release into the atmosphere as the extension of plants, then this plant mantle is nowhere bounded or interrupted. When we expand our understanding of plantness to include their activity, then as oxygen and water vapor, they circle the globe and in this respect are not limited to a specific location. Plants also take up carbon dioxide out of the atmosphere. The high uptake of carbon dioxide by land plants in the northern hemisphere from late spring into fall can be recorded not only locally, but also in the weather stations in Hawaii in the middle of the Pacific Ocean.

[2] The source for all the quotations from Goethe in this chapter is one German volume that contains most of his writings on science (see Goethe, 2002, in references). All the translations from the German are ours.

The presence of this extended plant activity is not confined to the land. Phytoplankton consists of the floating plants of the oceans and other water bodies. In the oceans they form, to the naked eye invisible, a kind of "leaf" that is much denser in certain ocean areas close to coasts and almost non-existent in mid-ocean "deserts." The roles these tiny organisms play in the life of the earth are complex and many-faceted. We find the traces of their past activity imprinted deep in the geology of the earth and its soils. The extended plant activity is apparent as a temporal as well as a spatial reality.

In only one phase of its life does a flowering plant appear as a discrete individual entity—in the seed. Compact, dry and usually hard-shelled, the seed contains a whole plant in its most contracted form. A small germinal primary root (radicle), cotyledons and a growing point of the shoot (apical meristem) are all enclosed within the seed coat. They formed while the seed was developing. As long as it remains in seed form, a plant is isolated from most activity of the world surrounding it.

Depending on the species and the conditions it finds itself in, a seed can remain alive for a long time before germinating—in sacred lotus over 1000 years. Its potential comes to realisation when it begins relating to specific qualities of the environment. Some seeds need to go through the cold of winter, the heat of a wildfire or the stomach of an animal before they can germinate. Many seeds require warm, moist soil. When the conditions are right, within in a few days further development can begin. The seed takes in moisture from the soil, tissues inside swell, and the firm seed coat breaks open and the seedling grows and develops.

Its primary root grows downward—toward the centre of the Earth—while the primary shoot (stem) grows upward, away from the centre of the Earth. In taking root, the plant becomes a creature of a specific place, and at the same time relates in its growth to the whole of the earth. Imagine all taproots on earth and all upright stems and tree trunks as radii of our spherical earth. It is a powerful image. We see here again the plant's belongingness to the larger earth as home. This "Earth" includes the heavens above and the soil below. In its polar orientation of shoot and root, the plant expresses its connection to these contrasting realms of Earth.

From the perspective of a given environment, we can view the abundance of seeds each plant harbors as a potential. Most seeds do not develop into mature plants. Many decompose and become part of the soil. Seeds and seedlings feed animals. Those that germinate will often die as young plants. In these ways, plants give themselves over to the earth and its life, and in this modified form become part of, and sustain, a place and its organisms.

Typically we say: A plant grows out of the seed. We focus on the necessary presence and activity of living germinal tissue in the seed. We can follow how the seed coat breaks open, the radical emerges and grows into the soil, while the shoot and leaves grow skyward. Clearly, the plant develops out of the seed. This is a centred perspective—thinking from a centre outward. It honours the potency of life inherent in every seed.

Yet for the seed to swell and the substance transformations that lead to germination and growth to begin, the seed must come into relation with qualities of the environment that support and in part merge into its very existence. We can, therefore, take a peripheral perspective—thinking from a broad life context towards a centre. This context has plant-forming potential. We can think of water, warmth, soil, light and air as potentials that become "plant active" in the seed-environment relation. Water, in the soil, enters the seed or the roots and becomes an integral part the plant's physiological processes, substances and structures. From the peripheral perspective, we can justifiably say that the environment develops through the seed into the plant.

This strange formulation can help us from yet another angle to move beyond thinking of life as being enclosed within bodies. What we conceive of as life processes are thinkable only when we have in mind their fundamental belonging to qualities that we habitually picture as "outside of" and "other than" organismic life.

All other life forms—animals, humans, fungi, bacteria—weave in and through the plant mantle of the earth, each in its own way of belonging. Most all of these organisms build up their bodily substance and structures on the basis of plants. Either they feed on plants directly or they eat organisms that themselves feed on plants.[3] Fungi and bacteria are to us humans hardly visible members of Earth. They are all-important in their activity of breaking down organic substance that has been shed—think of leaf litter, dead trees or animal and human corpses. And they are essential members of plant and animal microbiomes. Here again, we see the interpenetration of what we often consider separate organisms. In the words of early twentieth century physiologist J.S. Haldane:

> Just as the life of any individual organism exists only as including within itself what are often called its physical environment, so does its life, when more widely interpreted, include within itself the lives of other organisms, so that these lives are not outside of its own life, though for practical purposes we usually regard separately what we can most readily treat as individual lives. (Haldane, 1931, pp. 30–31)

3 Animal Belonging

When we walk here in upstate New York through a meadow—a small sea of diverse plant life—in the early morning or before dusk, we often encounter numerous white-tailed deer. They are easy to pick out as individual animals. The same holds for the swallows flying overhead and the bumble bees moving from flower to flower.

[3] There are also some bacteria that can, similar to plants, create their living substances without feeding on plants, so-called autotrophic bacteria.

We cannot really say that we are seeing individual plants in the way we see and can count the 22 deer that are grazing on the field. The meadow's plant life is more of a unity. Surely, we can see many individual stems with leaves and perhaps flowers, but they all emerge out of the earth. Or, for my vision: disappear into the earth. Plants appear as creatures of the earth out of which they have grown and with which they remain united. That we do not speak of herds of trees or grasses points perhaps to our subliminal awareness of plants as outgrowths of Earth.

Animals, on the other hand, have a clearly circumscribed body, and as a whole move from place to place. The deer flee, white tails raised and waving, as I approach them; the swallows swoop to and from, feeding on the insects they find in their airy world; the bumble bees fly from flower to flower sucking up nectar and gathering pollen.

The vital organs of plants (think of leaves and roots) are embedded in the world. In contrast, animals develop vital inner organs enclosed within their skin. They cannot live from air, water, light, warmth and minerals alone as plants do. They are lacking capacities that plants have. Animals must feed on plants or other animals and breathe in oxygen. In this way the life of plants extends into animals and belongs to their life.

Animals, however, are open to the world through organs of sense and motion that mediate a new form of belonging to Earth. The internalisation of vital organs goes hand in hand with what we can call animal inwardness, which comes to expression as attentiveness, sensations and drives. This inwardness manifests in all the animal's doings and, from this perspective, we can think of animals as centred. The deer turns its head to face me, ears upright and oriented in my direction. On the other hand, we need to ask: Where is the deer when it sees me and then bounds away? We see it looking at us—something we'd never say about the grasses and clovers in the meadow. While we cannot know exactly what it perceives as "us" with deer eyes, ears or nose, we can know that its attention is with us. As a sentient being, an animal is as expansive as its attention reaches.

There is a quality of immediacy apparent in animal perception that is rich with significance. The deer notices us and then flees. Other deer may flee with it, not having perceived us, but rather the fleeing of their fellow. When we are distant enough and lose significance, their attention can turn again to feeding on plants, shaking their ears, scratching their body with a hoof and scattering for a moment the ever-present flies they attract.

Now think of a red fox in what for us is the "same" meadow. It trots along low to the ground. It stops for a moment at a slightly raised spot on a path that goes through the meadow, lifts its leg and urinates. Then it moves on again. Again, it stops, but this time very differently. It crouches and begins turning its head slightly from side to side. Then it leaps and dives nose first to the ground. Occasionally, it comes up with a vole dangling from its mouth. It moves on, carrying its prey.

The meadow has particular significance or meaning for each animal passing through or inhabiting it. Each animal extends beyond its physical boundary into its perceptual world, a world in which it is also active. We could call this behavioural world of animals a soulscape to emphasise that it is a world of specific meanings and not a world of physical things. James Hillman hints at this meaning-filled perceptual field of animals when he writes: "The oriole doesn't see a branch, but an occasion for perching; the cat doesn't see a thing we call an empty box, it sees safe hiding for peering. The bear doesn't smell honeycomb, but the opportunity for delicious feeding" (Hillman, 1996, p. 86). An animal's soulscape is meaning-filled and rich. It expands and contracts, and the animal is continuously responsive to its perceived surroundings.

Each animal species develops its particular soulscape, centred in its uniquely structured organism, peripherally intersecting with the soulscapes of other animals. The more we form pictures of how different kinds of animals participate in the world, the more we gain entry into what we can call the ensouled life of the planet (see Holdrege, 2021b).

There can be weeks in the winter in upstate New York during which the ground is covered by snow. The temperature can be minus 20 degrees Celsius. The deer enter the meadow and with their hooves scratch a patch, freeing it from snow. There they find some dry grass and forbs that they feed on. They move slowly. At night they tend to congregate under red cedar trees that have a dense crown and are relatively snow-free at their base. The deer feed on their needles and on those of other conifers—plants they would not seek out during other times of the year. How these creatures simply go about their lives in what we consider such harsh conditions is impressive. We never get the sense that they are unsatisfied with their existence, jealous of the bears that are hibernating, or of the birds that flew south in the late summer. They live out their lives in their ways without striving beyond or questioning their existence. This brings us to the riddle of being human.

4 Human Belonging?

In a fragment penned in the 1780s, Goethe speaks of the quandary that underlies our relationship to nature.

> Nature! We are surrounded by her and entwined with her—powerless to leave her, and powerless to enter her more deeply. Unasked and without warning, she takes us up in the whirl of her dance, and hurries on with us till we are weary and fall from her arms. (Goethe, 2002, p. 45)

Biologically, humans are nature beings. We are immersed in nature, caught up in the 'whirl of her dance'. As with plants and animals, there is an immediacy in our relationship with nature. As living organisms, we are dependent on air, warmth, water, light and on plants and animals.

As thinking beings, however, nature appears as a riddle, difficult to understand. Our way of being in nature is always accompanied by the urge to understand what it is we are a part of.

Goethe's work stands at a crossroads in humanity's striving to understand the natural world. He believed deeply that nature had something to teach us. In his view, scientific research was not a question of finding explanatory mechanisms but of learning from the phenomena as they present themselves to the perceiving, thinking human soul. Insight into the creative unfolding of the phenomenal world appeared as "an experience of a higher order" in which conceptual experience illuminates the perceptual in new, often unexpected ways and what is experienced inside appears congruent with what is perceived "outside". In the experience of congruence, the quandary arising from the dual aspects of our relationship with nature is momentarily resolved: When our thinking learns to dance nature's dance, we become creative participants in her unfolding.

At a time when scientific understanding was turning away from the phenomena towards theoretical explanations, Goethe turned towards the phenomena. In doing so he modelled an attitude of scientific inquiry that is rooted in respect for the natural world and based on an unspoken recognition that an inherent continuity exists between human consciousness and the world of sense appearance. This attitudinal divergence—turning away from or turning towards the phenomena—is significant. It marks the emergence of a certain quality of questioning concerning the relationship of the human being and the natural world. It is a questioning that rests on the recognition of the self as a source of individual agency, at once both separate from and connected with the experienced world. In the introduction to his *Ideas for a Philosophy of Nature*, the German philosopher Friedrich Schelling speaks of the conundrum that arises:

> So long as I myself am identical with Nature, I understand what a living nature is as well as I understand my own life; I appreciate how this universal life of Nature reveals itself in manifold forms, in progressive developments, in gradual approximations to freedom. As soon, however, as I separate myself, and with me everything ideal, from Nature, nothing remains to me but a dead object, and I cease to comprehend how a life outside me can be possible. (Schelling, 1988, p. 36)

The young Schelling (he was 22 when the first edition of the *Ideas* was published in 1797) recognised a relationship between an individual's inner experience and the ability to apprehend Nature as a living whole. If this relationship is severed, the living whole is replaced by a disconnected array of dead objects. The loss of an inner sense of connectedness blinds us to the living wholeness around us of which we too are part. The Canadian philosopher, Jan Zwicky, points out that the experience of wholeness lies beyond the realm that can be communicated through language "whose aim is to block gestalt comprehension and to foster awareness of the world as a collection of discrete entities whose relations are ordered in causal and logical sequences" (Zwicky, 2015, p. 252).

In a short passage from his essays on morphology, Goethe describes a further aspect of the experience of identification with nature that was later elaborated by Rudolf Steiner to provide the basis for future-oriented approaches in the natural sciences:

> When we are called to lively observation and begin to hold our own in the struggle with Nature, we first feel a strong drive to master the phenomena. It does not take long, however, before they approach us with such force, that we rightly feel the need to acknowledge their power and revere their influence. As soon as we become convinced of this reciprocal effect, we become aware of a double infinity: In the phenomena, the manifold nature of being and becoming and also living, weaving relationships; in ourselves, the possibility of an infinite development, honing our sensibilities and discernment to generate ever new forms of receiving and responding. (Goethe, 2002, p. 52)

In turning towards the phenomena and engaging with them, Goethe recognised a depth in the world of appearance that not only harboured unexpected qualities of living presence or beingness, but also an invitation to expand one's own capacities of comprehension: The invitation to grow with the phenomena.

As we work to develop a Goethean approach further in the twenty-first century, we too recognise the invitation, yet time and again must acknowledge how difficult it is to take up this invitation. Intellectual thought lets the world appear to us as an aggregate of discrete entities and often overshadows our ability to sense the living relationships that exist between them and ourselves. We are forced to address the question: Does the exercise of intellectual thinking, rooted as it is in formal structures and causal relationships, remove us from an experience of the world in which life reveals itself as a fundamental aspect of reality? That by leading us to focus on what has become, it leaves us unable to grasp the qualitative gestures through which life comes into being? That by casting a stark light on the parts, it blinds us both to the presence of the whole and the role we play within it?

That the way we think about nature changes our sense of relatedness with the natural world has been recognised since the early twentieth century. Max Weber spoke of the sociological effects of the "disenchantment" of nature in 1917 (Weber, 1946, pp. 129–156). Later Martin Wagenschein, the German science educator, posited that the more we know about the world scientifically, the less we feel at home in it (Wagenschein, 2018). Others were more explicit. Simone Weil, the French philosopher and activist, described the scientific endeavour as an alienating and uprooting experience (Weil, 2024; Roth, 2014). And in the words of Hannah Arendt, "Earth alienation became and has remained the hallmark of modern science" (Arendt, 1958, p. 264).

5 Elements of Practice

Since it is not possible to lose what one never had, the phenomenon of alienation poses the question as to the primacy of separateness or connectedness as the basis for our relationship to the natural world. Our work at The Nature Institute has increasingly been guided by the question: Is it possible to develop a thoughtful scientific relationship to nature that allows us to bring to appearance and articulate the wholeness of life in its many manifestations?

We have realised how important it is to attend to and become more clear about the nature of our lived experience in observing the natural world. This lived experience is the basis for all of our scientific pursuits, yet we hardly attend to it. This awareness is central in the approach and provides the ground for developing its practice. We must, to paraphrase Siri Hustvedt, learn to look at ourselves looking at the world (Hustvedt, 2016). Consider the following.

In the middle of spring the meadow buttercups (*Ranunculus acris*) begin to flower. Here, on the eastern edge of the mid-Hudson Valley, they appear at roughly the same time as the dandelions (*Taraxacum officinale*) and the bird's-foot trefoils (*Lotus corniculatus*), both of which also bear yellow flowers. Although each of the plants has green leaves, flowers at roughly the same time in the same place, they are easily distinguishable in the arrangement of their leaves, the form and structure of their flowers and their growth habits. Walking along the road or the edge of the meadow, we can recognise each of them and have no doubt that the perceived characteristics—the particular hue of yellow, the form of the leaves, the growth habit—belong to the plant. We do not experience these individual characteristics as disconnected singular impressions that must be integrated to give a sense of the plant as a whole. Rather, they are different sensory expressions of the whole plant. As long as we meet the plants with full-bodied and attentive experience, we trust that our inner picture of the perceived plant corresponds with the plant itself. If we move around the plant, bend down to look at it more closely or glimpse it from a distance, the plant shows itself differently. Depending on which perspective we take, it is apparent that our moving around the plant and varying the way we attend to it allows us to see it in new ways, or—if we take the plant's perspective—allows the plant to show itself to us in a variety of ways.

By varying the ways in which we attend to the plants, we become aware of details we had not noticed before. Smells, textures, the feel of the stem, qualities of stiffness, flexibility, leaf details vary between the three plants. The flowers of each invite closer attention. We can seek out the plants in different settings—at the edge of the road, amidst tall grass, at the edge of a wooded area. In each setting or habitat, their growth varies, yet each buttercup is clearly a buttercup, each trefoil a trefoil. At no point in these explorations do we have cause to question whether what we are discovering belongs to the plant or not. In fact, the more we discover, the more closely we find ourselves paying attention to the specific appearances of the plant both in space and over time.

Natural science as it has developed over the past 450 years presupposes, yet rejects as subjective, and thus inadmissible, what lies at the beginning of any striving to understand nature: our lived experience of the world. The world that arises through experience is rich with colour and shape and sound. There is the smell of fresh mown hay, the taste of asparagus, birdsong before the sunrise. The earth is firm beneath our feet, the sun warm upon our faces. It is the experienced world that moves us to wonder, it is the experienced world that presents riddles and invites us to understand. What appears to be outside us comes to life in us. It moves us, awakens in us the urge to know it better, to understand. If we take the dialogue seriously that unfolds when we turn attentively towards the world, it becomes quite difficult to draw a clear boundary that separates us from the world we are growing to understand. The world we experience and engage with shapes our sense of it and informs how we relate to it going forward.

If we turn our attention to the dialogue that unfolds when we engage intentionally with nature, we discover the reciprocity inherent in it. It is in some ways like the unfolding of a friendship. If we are able to live attentively into another's way of being in the world, it becomes possible for them to show us the essential nature of that way of being. Something similar occurs when we turn our attention to the other-than-human beings with whom we share the life of our planet. We can begin to know them in ways that bring them closer to us.

The experience of reciprocity in this dialogue—the sense of growing together with the phenomenon—is deepened through the practice of exact sensorial imagination, the conscious inner recreation of the experienced plant. In the case of the trefoil, for example, we choose a specific specimen, a trefoil we observed growing at the rambling edge of a hedgerow. It has grown quite tall, born up by the unmown grass around it. Now no longer in the presence of the plant, we give our entire attention to recalling it with as much exactitude as we can muster. In imagination, we live back into the encounter with this trefoil and picture it as vividly as possible. In doing so, we let the observed plant guide the movement of our picturing. This practice, which accompanies us at every stage of research into the nature of an organism, grows increasingly complex, multifaceted and sculptural as familiarity and a sense of intimacy with the plant deepens. When we return to the plant, we find that we "see" it differently and, perhaps in response to the shift in our seeing, that the plant shows us things about itself that we were not able to apprehend before.

The intentional intensification of the dialogue between the trefoil and ourselves through the practice of imagination also gives birth to new questions. While some of these pertain to the nature of the plant, there are others that give rise to a growing sense of responsibility. We can begin to wonder whether the way we come to know something is of consequence to the thing that is seen and known. Whether by estranging the act of knowing from the life world, we have lost something that is essential for its wholeness and vitality. The American naturalist Barry Lopez put it this way:

> The determination to know a particular place, in my experience, is consistently rewarded. And every natural place, to my mind, is open to being known. And somewhere in this process a person begins to sense that they themselves are becoming known, so that when they are absent from that place they know that place misses them. And this reciprocity, to know and be known, reinforces a sense that one is necessary in the world. (Lopez, 2022, p. 70)

6 Thinking with Reality—Conclusion

Lopez's intuitive recognition of a deep relationship between human attentiveness and the living reality of the world casts light on what is perhaps the most significant challenge in the evolution of scientific thought. The more closely we attend to expressions of inter-relatedness in the natural world and how we as living, thinking beings engage with and participate in them, the more difficult it becomes to take separateness as a fundamental aspect of reality. Earth, plant, animal, human—we are a single living whole. Each of us extends into and takes up the other in our own specific ways. It is difficult to draw the line where one ends and another begins. If we consider earth from this perspective, it is evident that she is as alive as the organisms her presence makes possible.

Rudolf Steiner was among a number of modern thinkers who questioned the validity of excluding human consciousness from an understanding of the wholeness of nature. In his autobiography he writes of the challenge of developing a practice of thinking that corresponds to reality (Steiner, 2008, Chap. 22). This is an intentional practice that proceeds from turning towards the phenomena to growing with them through focused attentiveness and finally to bringing them to life within oneself through imagination. The imaginative internalisation of phenomena of life awakens within us a world that is ever dynamic and in constant transformation, a world of interrelated movement rich in meaning. Our thinking begins to move with the phenomena. They in turn move us.

In upstate New York, the last of the trefoil blossoms disappear in the middle of September. Their leaves and stalks dry out and turn brown. With the coming of autumn, they disappear from our sense experience of the landscape. Yet they remain very much alive in us as a specific quality of plantness in our experience of the spring and summer. What begins as bodied experience of our natural surroundings lives on as an inner, conceptual experience. With some effort, we can bring the trefoil to life again through our imagination. We can let it grow inwardly in a variety of contexts and, guided by the trefoils we have come to know, experience how trefoil in its wholeness participates in the specific aspects of its extended life world. In this process, the concept of the trefoil takes on sculptural gestural qualities. We can experience it as generative activity.

Developing a thinking that corresponds to the reality of what is alive in the world requires a radical shift in what we might call knowledge practice. As long as we continue to approach nature as though it were "a collection of discrete entities whose relations are ordered in causal and logical sequences" (Zwicky, 2015, p. 252), life will remain a riddle. Life is dynamic, responsive, flexible. It requires a thinking that, in Goethe's words, is "as mobile and malleable as nature herself," a thinking that does not impose itself on life, but is willing to let itself be guided by the way life manifests.

References

Arendt, H. (1958). *The human condition.* University of Chicago Press.

Goethe, J. W. (2002). *Goethes Werke: Band XIII, Naturwissenschaftliche Schriften erster Teil.* Verlag C. H. Beck.

Haldane, J. S. (1931). *The philosophical basis of biology.* Hodder and Stoughton.

Hillman, J. (1996). *The soul's code: In search of character and calling.* Random House.

Holdrege, C. (2021a). Where does an animal end? The American bison. *In Context, 45*. The Nature Institute. https://www.natureinstitute.org/in-context-45/craig-holdrege/american-bison

Holdrege, C. (2021b). *Seeing the animal whole—And why it matters.* Lindisfarne Books.

Hustvedt, S. (2016). *A woman looking at men looking at women: Essays on art, sex, and the mind.* Simon & Schuster.

Lopez, B. (2022). *Embrace fearlessly the burning world: Essays.* Random House.

Roth, W.-M. (2014). Enracinement or the Earth, the originary ark, does not move: On the phenomenological (historical and ontogenetic) origin of common and scientific sense and the genetic method of teaching (for) understanding. *Cultural Studies of Science Education.*https://doi.org/10.1007/s11422-014-9606-z

Rowe, S. (2006). *Earth alive: Essays in ecology.* NeWest Press.

Schelling, F. W. J. (1988). *Ideas for a philosophy of nature: As introduction to the study of this science* (2nd ed.). Cambridge University Press.

Steiner, R. (1997). *An outline of esoteric science* (G. & M. Adams, Trans.; German Bibl. Nr. 13). SteinerBooks.

Steiner, R. (2008). *Autobiography: Chapters in the course of my life (1907–1925)* (R. Stebbing, Trans.; German Bibl. Nr. 28). SteinerBooks.

Wagenschein, M. (2018). *Teaching to understand: On the concept of the exemplary in teaching.* The Nature Institute. https://www.natureinstitute.org/article/martin-wagenschein/teaching-to-understand

Weber, M. (1946). *From Max Weber: Essays in sociology* (H. H. Gerth & C. Wright Mills, Trans. & Eds.). Oxford University Press.

Weil, S. (2024). *The need for roots: Prelude to a declaration of duties towards mankind* (A. Wills, Trans.). Penguin Classics.

Zwicky, J. (2015). *Alkibiades' love: Essays in philosophy.* McGill-Queen's University Press.

Key Attitudes to Understand and Guide Sustainable Development

Kees Zoeteman

Abstract

The personal interest and commitment of individuals is the real driving force behind progress towards a more sustainable world. To measure this progress, the concept of sustainability attitudes is introduced, demonstrating five key attitudes: (1) pure egoism, (2) authoritative morality, (3) smartness morality, (4) value-driven morality, and (5) conceptual intuition morality. Examples are provided to illustrate how these attitudes translate into practical actions for businesses, governments, and NGOs. Methods for individuals to test their personal attitudes are also explored. The Earth is presented as comprising seven elementary spheres, ranging from the solid and liquid to the gaseous, warmth, electric, magnetic, and radioactive spheres. The higher the elementary sphere, the greater its influence on the lower spheres and the more significant the moral challenges humankind faces when influencing it. Earth mirrors the moral challenges we must confront, highlighting that finding sustainable development solutions for the fourth elementary state of existence—climate change—is merely the beginning of what lies ahead.

Keywords

Sustainability attitudes • Sustainable development • Environmental ethics • Rudolf Steiner • Sustainable governance • Earth's elementary spheres • Gaia hypothesis • SDGs • Attitude-based sustainability • Holistic sustainability measurement • Policy and environment • Multilevel sustainability

K. Zoeteman (✉)
Sustainable Development, Tilburg University, Tilburg, Netherlands
e-mail: bcjzoeteman@gmail.com

J. Kronenberg and E. T. Lammerts van Bueren (eds.), *On the Earth We Want to Live*, World Sustainability Series, https://doi.org/10.1007/978-3-031-98758-8_18

1 Introduction

We learn through experiencing shocks, and the path to sustainable development is replete with such moments. Notable examples include Rachel Carson's *Silent Spring* (1962), which exposed how Monsanto's pesticide DDT was killing birds, and the Club of Rome's study "Limits to Growth" (Meadows et al., 1972). Even more remarkable is that these reports prompted a serious policy response at the UN level, aiming to integrate economic and environmental goals.

The report of the UN Brundtland Commission on Environment and Development (1987) and the Earth Summit in Rio de Janeiro (1992) provided positive shocks—introducing new visions and hopes for improving life on Earth. This progress was made possible because the shadow of the Cold War had dissipated, allowing inspiring new leadership from figures such as Gro Harlem Brundtland, Prime Minister of Norway, and Maurice Strong, Canadian industrialist and Secretary-General of the Rio Summit (Zoeteman & Tavenier, 2012).

The results of the Rio Summit were unprecedented and included, among other achievements, the adoption of the notion of sustainable development, three international agreements on biological diversity, climate change, and forest management, as well as Agenda 21, which outlined the UN Programme of Action for achieving sustainable development in the twenty-first century. Following the Rio Summit, a wave of further initiatives emerged, such as the World Business Council on Sustainable Development (1995) in Geneva, championed by Stephen Schmidheiny, CEO of the Swiss Eternit Group, and the Earth Charter, launched at the Peace Palace in The Hague in 2000, facilitated by Dutch Queen Beatrix and Prime Minister Ruud Lubbers.

However, subsequent high-level summits failed to maintain the same momentum among governments, as other global issues, such as the financial crisis of 2008, demanded attention. In particular, the lack of financial support for developing countries became a major source of disappointment. Additionally, the balance of geopolitical power was shifting, reducing the influence of European leaders and amplifying the impact of Asian and South American nations, which prioritised different issues.

It was not until 2015 that another landmark year for multilateralism and international policy occurred, marked by the adoption of the 2030 Agenda for Sustainable Development with its 17 Sustainable Development Goals (SDGs) and the Paris Agreement on Climate Change (UNOSD, 2024). Since then, the SDGs have proven to be a valuable tool for actively engaging businesses and local governments in the sustainable development movement, even as national governments shifted their attention to other issues, such as the COVID-19 pandemic of 2020.

Moreover, the SDGs clarified that sustainable development is not merely a synonym for environmental policies but also encompasses social and economic goals, which must be pursued in a balanced manner. In a sense, sustainable development represents the modern embodiment of ideals that date back to the French Revolution (1789–1799): Liberty! Equality! Fraternity! These ideals led to much more than the replacement of monarchies by democracies (Andress, 2015).

A century later, Rudolf Steiner sought to address societal imbalances and the dominance of economic interests by introducing what he called the "threefolding of society" (Steiner, 1919). He applied the ideals of the French Revolution to specific sectors of society, envisioning freedom in spiritual life, equality in legal life, and fraternity in economic life. After assassination attempts, Steiner was forced to cease these activities in 1922. Yet, a century later, his ideas re-emerged in the new guise of sustainable development.

The personal interest and commitment of human beings are the true driving forces behind the move towards a more sustainable world. At the heart of sustainable development lies the decision of each inhabitant of this planet to adapt their behaviour to meet the needs of both local and planetary ecosystems, as well as those of fellow human beings. Achieving a more sustainable world demands an attitude aligned with these principles. Without this, international agreements will remain hollow promises.

In the following paragraphs, a concept will be discussed that enables us to define and measure sustainable development, particularly the capacity of humankind to improve itself: the attitude of individuals and communities towards sustainability. It will also be shown how the Earth reflects back to us the attitudes needed to address or prevent sustainability distortions. A distinction will be made between different levels of organisation on our planet, outlining a future agenda for the moral choices we will face—individually and/or collectively.

2 Monitoring Sustainable Development

Achieving sustainable development is a complex endeavour and lacks a solid scientific foundation. It aims to create a positive impact across economic, social, and environmental dimensions. Consequently, numerous practical approaches to measure it have been developed over the decades (Zoeteman, 2012a). Several measurement approaches for nations gained recognition during the late twentieth century, such as the Human Development Index and the Wellbeing Index. For corporations, other indices, such as the Dow Jones Corporate Sustainability Index, came into use. However, one of the significant flaws of these indicators lies in the arbitrary methods of integrating economic, social, and ecological data, as well as the inability to compare scores objectively—for example, between countries or business sectors.

Max-Neef (1995) and Bossel (1999) stressed the need to develop integrated indicators that go beyond the economic national measure of Gross Domestic Product (GDP). Prescott-Allen (2001) proposed a method to integrate the three pillars of sustainability by first calculating the average of socio-economic indicators, then combining this result with the average value of ecological indicators. However, the combination of such fundamentally different aspects has never gained general acceptance within the scientific community. An objective unit to measure the overall sustainability of a governance area or an organisation has yet to be discovered.

The most reliable alternative has been Prescott-Allen's method of defining long-term targets (Prescott-Allen, 2001), measuring the distance to these targets, and expressing this difference as a percentage of the target, which serves as a measure of sustainable development.

Since the introduction of the authoritative SDGs in 2015, actors worldwide have been encouraged to identify which of the 17 goals and related indicators are relevant to them and to improve their performance on these goals over time. Nations and larger businesses have been motivated to take sustainable development more seriously and report annually on their progress. An example of outcomes for the EU, based on voluntary reporting by member states over a five-year period, is shown in Fig. 1 (Eurostat, 2025). The figure reveals that in 2024, around five of the 17 SDGs moved further away from their targets—for instance, "Affordable and Clean Energy"—while most others showed slight improvement, including "SDG 1: No Poverty."

The SDGs, of course, reflect the ongoing tension between achieving sustainable development and adhering to the paradigm of exponential economic growth. As illustrated in the figure, "SDG 8: Decent Work and Economic Growth" is categorised as "significant progress," whereas this is far from the case for "SDG 15: Life on Land."

The newly installed European Sustainability Reporting Standards (ESRS) are an attempt to take a next step in these efforts.

3 Sustainability Attitudes

A way to overcome part of the complexities of measuring sustainable development is to look beyond thematic goals and related indicators and address a higher level of concern: the attitude that governs the actions of stakeholders. Attitudes can be uniformly applied to any actor or organisation, as attitudes are, in principle, universally dominant factors steering behaviour.

Attitudes of actors are defined as the revealed mindsets or mental codes that potentially steer all physical behaviour of individuals or organisations. They are the manifested willpower to apply normative principles while satisfying needs. Beck and Cowan (1996) called such mindsets value systems, which, as organising principles, pervade decision structures and cultural expressions.

Attitudes have four characteristics that make them valuable as overall indicators for sustainable development. Firstly, attitudes are more pervasive than other indicators of the mental conditions of persons (Osgood et al., 1957). Secondly, attitudes are a selective force in perception and memory. People selectively seek information that agrees with their attitudes while avoiding information that does not (Festinger, 1957). Thirdly, attitudes predict behaviour. Attitude and behaviour are correlated when the attitude is based on direct experience with the attitude object, and to the extent that the attitude is cognitively accessible (Byrne & Baron, 1997). The fourth reason to choose attitudes as indicators is that they serve various psychic functions. The attitude of persons has a function in utilitarianism, social

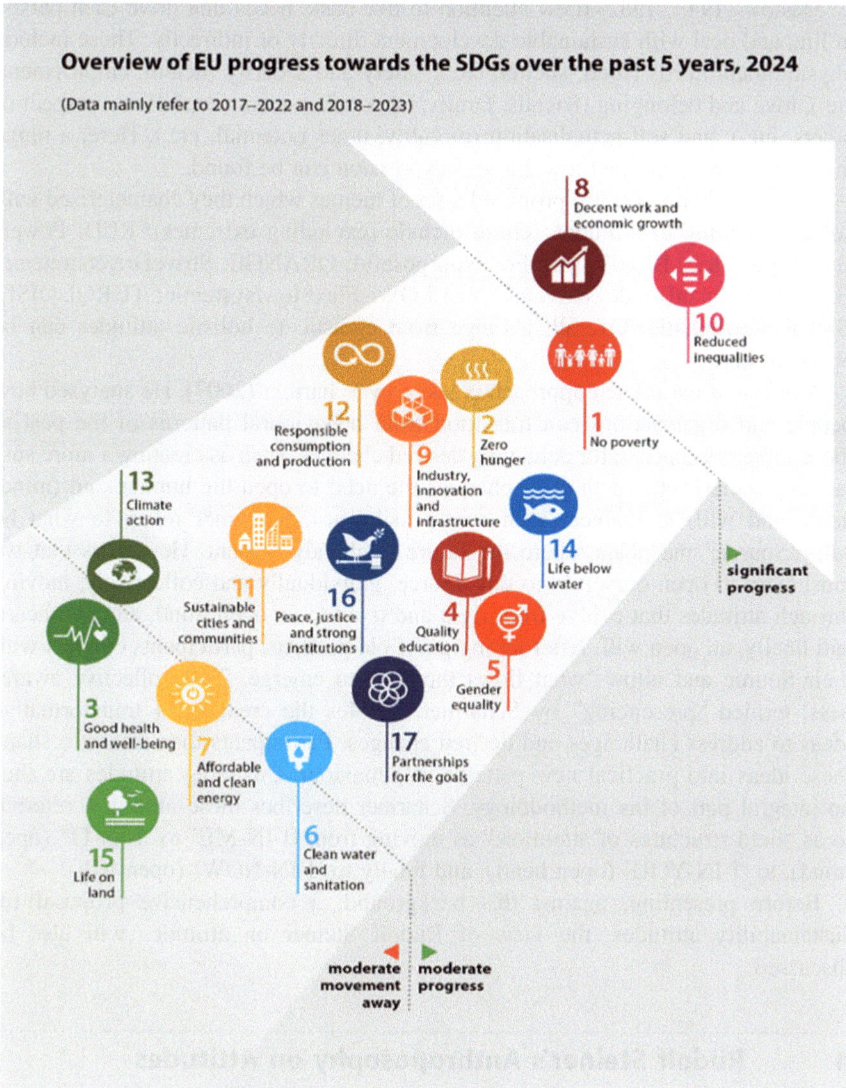

Fig. 1 EU progress towards the SDGs over past 5 years, 2024 (Eurostat, 2025)

adjustment, object appraisal, knowledge, value expression, and ego-defense (Katz, 1960).

Attitudes can represent different levels of commitment to sustainable development, from ignoring the need for it to being in favour of any action that can contribute to it. In general, attitudes can be arranged according to increasing moral complexity, from pure egoism to extreme forms of altruism. Several sets of attitudes have been proposed, of which three will be briefly discussed.

Maslow (1943, 1982) drew attention to five basic needs that drive each person in life, and deal with sustainable development directly or indirectly. These include physiological needs (food, shelter, etc.), safety and security (health, employment, etc.), love and belonging (friends, family, etc.), self-esteem (confidence, respect of others, etc.), and self-actualisation (morality, inner potential, etc.). Here, a range in attitudes from survival towards self-expression can be found.

Beck and Cowan (1996) proposed a set of memes which they characterised with colours, relating to attitudes. These include (excluding extremes): RED: PowerGods/egocentric; BLUE: TruthForce/purposeful; ORANGE: StriveDrive/strategic; GREEN: HumanBond/relativistic; YELLOW: FlexFlow/systemic; TURQUOISE: GlobalView/holistic. Overall, a range from egoistic to holistic attitudes can be observed here.

A different but related approach is taken by Scharmer (2007). He analysed how people and organisations can transition from behavioural patterns of the past to those more appropriate for achieving desired changes, such as creating a more sustainable society. His method emphasises the need to open the human soul (mind, heart, and will) to connect with new possibilities. Scharmer refers to what he calls 'Source,' the place where the future is already present. He argues that we must learn to open ourselves to this Source, individually and collectively, moving through attitudes that evolve from rigid and fixed to an open mind, an open heart, and finally, an open will. After letting go of old patterns, participants connect with their Source and allow 'what flows through' to emerge. This collective awareness, termed "presencing" by Scharmer, enables the creation of transformative ideas to address challenges and desired changes. Participants then return to shape these ideas into practical new patterns of behaviour. Changing attitudes are thus an integral part of his methodology. Scharmer describes these attitudes, referred to as "field structures of attention," as moving from 'I-IN-ME' to 'I-IN-IT' (open mind), to 'I-IN-YOU' (open heart), and finally to 'I-IN-NOW' (open will).

Before presenting, against this background, a comprehensive proposal for sustainability attitudes, the view of Rudolf Steiner on attitudes will also be discussed.

4 Rudolf Steiner's Anthroposophy on Attitudes

Steiner's descriptions of attitudes can be found in his ground-breaking philosophical work of 1893, *The Philosophy of Freedom* (Steiner, 2024, Chap. 9). He explains that mental pictures act as motives of the will:

> The principle of producing the greatest quantity of pleasure for oneself through one's action, that is, of attaining individual happiness, is called egoism. The attainment of this individual happiness is sought either by thinking ruthlessly only of one's own good and striving to attain it even at the cost of the happiness of other individuals (pure egoism). Or by promoting the good of others, either because one anticipates a favourable influence on one's own person indirectly through the happiness of others, or because one fears to endanger one's own interest by injuring others (smartness morality). (Steiner, 2024, p. 183)

Steiner's "pure egoism" attitude corresponds to Scharmer's I-IN-ME, Beck and Cowan's RED/egocentric meme, and Maslow's physiological survival needs.

Another attitude, according to Steiner, may result from moral principles, such as family habits or state laws, in the form of abstract concepts that regulate the individual's moral life without them questioning the origin of these concepts:

> In that case, we simply feel that submitting to a moral concept in the form of a commandment overshadowing our actions is a moral necessity. The establishment of this necessity we leave to those who demand moral subjection from us, that is, to the moral authority that we acknowledge (authoritative morality). (Steiner, 2024, p. 184)

Steiner describes a moral advance when a person no longer simply accepts the commands of another authority as the motive for their actions but instead seeks to understand the reason behind a particular maxim of behaviour.

> This is the advance from morality based on authority to action out of moral insight. At this level of morality (virtue-driven), a man will try to find out the requirements of the moral life and will let his actions be determined by the knowledge of them. Such requirements are the greatest possible good of mankind purely for its own sake or the progress of civilisation towards ever greater perfection. (Steiner, 2024, p. 184–185)

An example of this fourth level of sustainability attitude is Gro Harlem Brundtland's call, "Now it is time to take a giant leap forward in the upgrading of civilization," at the Toronto Conference on the Changing Atmosphere (Brundtland, 1988, p. 22).

Steiner identifies the highest conceivable moral principle as one that, from the outset, does not refer to particular experiences but instead springs from the source of pure intuition, seeking reference to life only later.

> Here the decision as to what is to be willed proceeds from an authority very different from that of the foregoing cases. If [...] no other ground for decision claims more than second place, then conceptual intuition itself comes first and foremost into consideration. All other motives now give way, and the idea behind an action alone becomes its motive (Steiner, 2024, p. 186).

In this latter case, *conceptual intuition morality* aligns with Scharmer's I-IN-NOW, Beck and Cowan's TURQUOISE/holistic attitude, and the self-actualisation proposed by Maslow.

5 Levels of Sustainability Attitude Illustrated for Different Societal Actors

A summary of sustainability attitudes, based on the above considerations, is presented in Table 1. A total of five attitudes is outlined. While further differentiation into sixth and seventh levels is theoretically possible, it currently has little

Table 1 Levels of sustainability attitude and societal actors (see also Zoeteman, 2012a)

Sustainability attitude level/Type of morality	Attitude level of societal actors		
	Businesses	Governments	NGOs
1. Very unsustainable/ pure egoism	Exhausion of resources	Ad hoc use of power	Getting organized
2. Unsustainable/ authoritative morality	Reluctantly complying with laws	Top-down law-making	Polarising actions
3. Nearly sustainable/ smartness morality	Anticipation of new laws	Voluntary agreements	Joining licensing processes
4. Sustainable/virtue driven	Anticipation of global consumer needs	Broad consensus policies	Supporting sustainable enterprise
5. Super-sustainable/ conceptual intuition morality	Voluntary projects serving societal goals	Facilitation of private initiatives	Co-organizing sustainable business initiatives

practical relevance. The five attitude levels range from very unsustainable to super-sustainable. The type of morality dominating each level is specified following Steiner's framework, showing strong similarities with other approaches discussed earlier. Table 1 also illustrates how these attitudes manifest across three societal sectors: businesses, governments, and non-governmental organisations (NGOs).

At the first attitude level, actors prioritise their own interests and are survival-oriented. They remain unaware or indifferent to other interests and focus on the short term. A positive interpretation of this attitude is that it is manifestation-oriented, as seen in the case of parents building a home for their family or an artist working tirelessly on a painting or sculpture. However, it can also be observed in the behaviour of the classical capitalist entrepreneur who maximises profit at the expense of employees and nature. During wartime, collective survival also becomes a central issue at this level.

At the second attitude level, self-interest remains dominant, but as actors expand their sphere of influence, they begin to acknowledge other interests and accept certain restrictions. This marks the establishment of public or common interest, with actors reluctantly starting to respect laws. Classic Western films vividly depict the transition from the first to the second attitude level, reflecting the societal changes during the Wild West era of American history. Businesses at this level may attempt to weaken laws before they are enacted. Governments at this stage adopt a command-and-control approach to drafting and enforcing legislation. NGOs, which are often poorly organised in societies at the first attitude level, begin to play a role at the second level by opposing the powerful through aggressive actions to bring their issues onto the public agenda.

After the polarised second-level attitudes, a third level can emerge in which actors with seemingly opposing interests begin to listen to each other and achieve mutually beneficial consensus through negotiation. Such negotiations are voluntary and may involve a multi-year perspective. Covenants between governmental authorities and business sectors exemplify activities associated with level 3 attitudes in the EU. NGOs operating at the third attitude level often pursue their interests by appealing in court against licences, which is a more civilised way of opposing industry than, for example, engaging in sabotage. Alternatively, they may negotiate a deal with industry and withdraw their case.

The fourth attitude level represents the heart of sustainable development, as actors at this level voluntarily internalise other interests in their decisions. Economic interests are no longer prioritised over social and ecological ones, and the needs of others are treated as equivalent to one's own. Like at the previous level, parties are proactive and inclusive, carefully considering the consequences of their actions. Designing creative solutions that benefit economic, social, and ecological values is characteristic of this level. A prime example of this attitude is the Corporate Social Responsibility (CSR) approach adopted by businesses. Governments at attitude level 4 strive for broad consensus policies on ambitious goals and ensure transparency in their communication. NGOs with this attitude openly support sustainable corporations.

The fifth attitude level is referred to as the level of super-sustainability. This level reflects a mental state where individuals and organisations are willing to do more than is necessary to achieve a balance between the three pillars of sustainable development. Actors at this level commit to realising intuitively perceived collective higher goals for all stakeholders and adopt new methods of value creation. Such common goals represent the potentially best possible outcomes for all concerned.

To better understand the five attitudes, it may be helpful to reflect on the attitude we personally align with. Figure 2 provides a simple test to assess where we stand in relation to five illustrative sustainability aspects. This exercise gives an indication of our average attitude level at a given moment.

As a reference, attitudes measured using a more extensive and objective methodology in 2002 showed values for airlines ranging from 1.6 (Egypt Air) to 4.1 (Lufthansa) (Zoeteman, 2012c). During the same period, national scores varied from 1.4 (Afghanistan) to 2.6 (Brazil); 2.9 (China); 3.3 (France); 3.4 (USA); 3.8 (Germany); and 4.0 (Switzerland) (Zoeteman et al., 2012).

6 Dealing with the Fall or Rise of the Attitude Level

Becoming aware of the five sustainability attitudes can be immensely helpful in understanding what is wise to do when faced with a rise or fall in attitude levels. This concern has existed since ancient times, as described in the Biblical *Sermon on the Mount* (St. Luke 6:28–39) and the even older *Tao Te King* by the Chinese philosopher Lao Tse, who was born around 600 BC in Chu, Henan.

Simplified test personal sustainability attitude

circle for each aspect the level most similar to your action or likely action at this moment

Aspect	Level 1 I first / Pure egoism	Level 2 Authoritative morality / follow the law	Level 3 Smartness morality/ negotiation	Level 4 Virtue driven	Level 5 Serving higher cause /conceptual intuition	Total average score
Energy handling at home	I open a window when it is to warm	I use cheap fossil fuel as long as possible	I buy solar panels if they are sufficiently subsidised	I participate in local energy initiatives	I organise local initiatives and support societal innovations	
Handling social conflict	I continue till I have won	I only comply when unavoidable	I am open to come to new arrangements	I try to understand, act pro-actively	I serve what brings us further together	
Purchasing food for household	Lowest price only important	Only buy quality when special offer	Prefer certificate if not expensive	Only buy biological / similar products	Actively promote bio-products and sector	
Holidays	Just fly away and enjoy	Safety counts, not environment	CO2 compensation is a nice extra	Chose sustainable touroperator	Support local initiatives at destination	
Handling household waste	Buy plastic bags at all shopping occasions	All waste ends in my container	Only seperation of waste when receiving a deposit	Voluntary waste separation also if delivery is elsewhere	Purchasing takes waste recycling into account	
Number of circles per column	..x1=	..x2=	..x3=	..x4=	..x5=	Sum:5 =

*The test requires encircling one box for each horizontally listed sustainability aspect that most closely matches our practices. After completing the selection for all five aspects, the number of encircled boxes in each vertical column is counted. Subsequently, the five calculations prescribed at the bottom line are performed, and the resulting five numbers are added together. The sum, divided by 5, represents the average attitude level.

Fig. 2 A simplified personal test of our sustainability attitude (Zoeteman, 2012b)*

Lao Tse referred to the highest level as "Tao," a principle of unity with the divine, which "acts by inactivity and accomplishes all." Achieving this level has also been described as being in "flow" (Csikszentmihalyi, 1975) or experiencing "synchronicity" (Jaworski, 1996). In the *Star Wars* films, George Lucas gives a materialistic interpretation of this concept with "The Force," drawing parallels to Lao Tse's Tao.

In ancient times, Lao Tse wrote (par. 17) that people at the highest level of awareness knew no more of their rulers than their mere existence, which he associates with attitude level 5. Later, rulers who were cautious and discreet managed matters so that people felt free to arrange their lives according to their inclinations, corresponding to attitude level 4. Subsequently, rulers conducted themselves in ways that earned them honour and praise (level 3). After this came rulers who made themselves feared (level 2), followed by those who provoked contempt among their people (level 1). Lao Tse concludes that this decline in attitude proceeds from Tao to Virtue, from Virtue to Humaneness, from Humaneness to Devotion-to-duty, and from Devotion-to-duty to Decorum. This serves as a reminder of what happens when the connection to Tao—or what Scharmer calls "Source"—is lost.

Lao Tse also offers wise insights on how to avoid the traps of lower attitude levels and contribute to improving attitudes. His fundamental advice is to avoid reaching the first attitude level of survival—characterised by "I first," a world of hollow phrases and superficial decorum—because once this level becomes established, it is difficult to change. *Prevent chaos before it breaks out!* At the second attitude level, in a still polarised society where authorities rely on command-and-control techniques, Lao Tse advises leaders: *Do not subject others to violence; if a battle is necessary to force a decision, keep it short and do not boast afterwards.* To improve attitude level 3, from Humaneness to Virtue, he recommends: *Do well without expecting reciprocity; limit your interests and desires.* For improving attitude level 4, he offers advice that remains relevant today: *Let your heart be as deep as an abyss, facilitate and do not dispute.* Improving attitude levels is not an easy task for leaders. It is, however, a task that each of us must embrace if we wish to contribute to a more sustainable society.

7 The Elemental Characteristics of the Earth

The long-term goal of sustainable development is to adapt our behaviour in order to ensure the survival of humankind on our planet, Earth. With the increasingly powerful techniques we are developing—such as nuclear explosives, ocean mining, geo-engineering, space travel, and genetic modification of cells—it is possible to disrupt life on Earth. Our activities are causing changes to the physiology of the planet, such as the poisoning of fertile soils, alteration of the reflective characteristics of the skies, slowing down of ocean currents, accumulation of plastic pollution in waterbodies, extinction of species, increasing desertification, and the build-up of space debris. To a large extent, the financially rich have passed on the burden of these problems to the poor. In a sustainable world, this is unacceptable. We need to understand each challenge and act ourselves accordingly, adopting the appropriate attitude. Different challenges require different attitudes, as we will explore in this chapter. In a way, the Earth is teaching us, step by step, how to solve sustainability issues that require increasingly evolved attitudes.

When viewed from space, Earth gives the impression of a blue planet, partially covered by white clouds, standing in stark contrast to the grey, desert-like appearance of the Moon, as illustrated in Fig. 4. Both celestial bodies are subject to constant bombardment by solar wind, which consists of heat, electrons, protons, and rapidly moving magnetic fields. When standing on the Moon, none of these flows are visible to the human eye, yet they would have a deadly impact on life. This radioactive radiation would also kill life on Earth were it not for the magnetic field of our planet, which deflects most of the solar wind, as shown in Fig. 3.

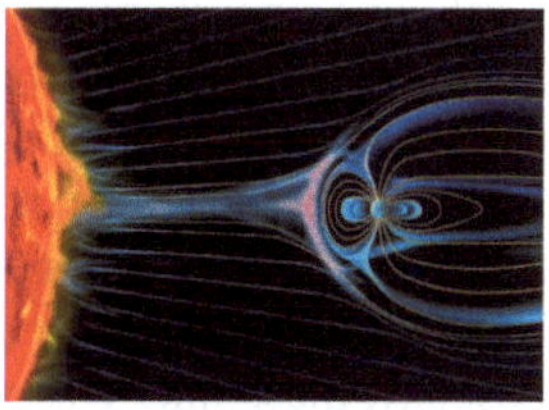

Fig. 3 Artist Impression of impact of solar wind on the magnetic field of the Earth (ESO/José Francisco (josefrancisco.org)

Fig. 4 Earthrise taken by Apollo 8 crewmember Bill Anders on December 24, 1968 NASA

Fig. 5 View by night from ISS on the red sea region (NASA, 2015)

At the Earth's magnetic poles, some of the solar wind penetrates the atmosphere, creating the two so-called Van Allen radiation belts. The outer belt primarily consists of electrons, while the inner belt is made up of accumulated protons. These radiation belts are hazardous, and spacecraft carrying astronauts avoid passing through them. Some particles from the solar wind travel closer to Earth in regions near the magnetic North and South poles. Here, they create the colourful auroras by exciting nitrogen and oxygen atoms in the atmosphere. The thin green layer visible above the Earth's surface in Fig. 5 indicates where the aurora borealis occurs and also highlights the remarkably thin layer of air that envelops our planet.

Descending from the flexible outer Magnetosphere, which marks the boundary between the deadly radioactive solar wind and Earth with its Van Allen belts, situated above 4000 km, we reach the Ionosphere below 1000 km. In this layer, light is emitted by ions of nitrogen and oxygen when excited by residual solar wind particles. Below 200 km, the temperature rapidly decreases from 1000 °C to − 90°. This layer is called the Thermosphere. Beneath it, at altitudes below 90 km, lies the Mesosphere, where the mixing of air layers begins. Below 50 km, we find the Stratosphere, home to the ozone layer, which absorbs harmful UV radiation from the sun. The Stratosphere contains air—initially at low pressure—in the composition that we breathe. Finally, below 12 km, we encounter the Troposphere and the Hydrosphere. The latter contains water vapour and liquid water, facilitating cloud formation, precipitation, and the rivers and seas on Earth's solid surface. The solid Earth, or Lithosphere, extends upward to 8.8 km through the highest mountains.

Steiner (1910) observed that the Earth develops through a gradual downward densification, transitioning from warmth to solid. However, he also characterised the sun's influence as not solely negative. While the sun emits solar wind, it also provides warmth and light. Steiner described additional, non-physical energy aspects flowing from the sun to Earth, which positively influence life. In anthroposophy, these finer energies are referred to as *life-forces*, playing a crucial role in vital processes but perceptible only through higher forms of perception. Steiner classified these life-forces or ether-forces within the realm of the Greek elements, distinguishing between light-ether, sound-ether, and life-ether. He further explained:

> It would be better therefore to say that there are seven states of elementary existence. The first, the most rarefied state of life-ether, constitutes and brings about life (*Weltenleben*); then what we call number- or sound-ether gives a certain shape to beings (*Weltgestaltung*); then light-ether; then warmth ...; then we have air, or the gaseous element, the watery element and finally the earthy or solid element ... In the Earth we have to look for the solid, fluid and gaseous elements; warmth is shared by both Earth and sun. (Steiner, 1910, p. 139)

According to Steiner, these ether qualities, with their life-giving potential, can also manifest in a different state, referred to as "fallen" or "broken" (Steiner, 1911). In this materialistic state, they become more perceptible: light-ether transforms into electricity, sound-ether into magnetism, and life-ether into radioactivity or matter-disintegration. In their fallen (dead) state, these ether qualities are more easily integrated into Earth's organisation.

In summary, we can define seven spheres for the elementary Earth as presented in Table 2. These seven states of elementary existence are related, in the final column of Table 2, to how human beings influence them through their activities. An extensive description of these influences can be found in Zoeteman (1991, 2009). It is important to recognise that the higher states of elementary existence influence the lower ones. For instance, changes in Earth's magnetic field can affect the position of the Van Allen Belts, the electrical currents in the Ionosphere, the temperature distribution and air movements in the Thermosphere, the weather phenomena in the Stratosphere, ocean circulation, climate change, and ultimately life on Earth.

The concept of elementary spheres was first taken seriously in Dutch national policymaking in 1988 (Nijpels, 1989). The National Institute for Public Health and Environment (RIVM) (Langeweg, 1988) used the first five states of elementary existence (ranging from local issues to planetary concerns) as a framework for an environmental inventory. This was documented in the report *Caring for Tomorrow* (*Zorgen voor morgen*, in Dutch).

8 Appropriate Sustainability Attitude for a Certain State of Elementary Existence

We will now explore the hypothesis that the way human beings influence a certain state of elementary existence of our planet requires a sustainability attitude corresponding to its nature, using typical examples for illustration.

The mineral state of existence of our planet is where most human activities take place—whether it is a location made of rock, sand, stone, or wood. Here, we work and live, build our homes and factories, mine resources, and dump waste materials. We believe we have the right to live our lives as we wish, provided we comply with

Table 2 States of elementary existence of planet earth and related human influences (Zoeteman, 2009)

States of elementary Existence	Position	Spatial unity	Influences by human activities such as:
Minerals	Lithosphere	Field, landscapes, city	Mining, erosion, soil pollution by chemical fertilizers and pesticides, spatial disorders, solid waste disposal
Liquid	Hydrosphere	Drainage area of river and related troposphere, ocean conveyor belts	Water pollution, overuse of fresh water quantity, floods, river dams
Gaseous	Stratosphere	Continental air circulation, jet streams	Air pollution such as acid rain, smog
Warmth	Thermosphere	Planetary thermosphere	Greenhouse gas emission, built environment heat reflection, ozone-layer depletion, geo-engineering
Light-ether/electricity	Ionosphere	Planetary ionosphere and electric current	High frequency radiowave injections, ionosphere manipulation, internet, social media
Sound-ether/ magnetism	Exosphere with Van Allen belts	Planetary proton and electron belts	GPS and other satellites, military experiments
Light-ether/ radio-activity	Magnetosphere	Earth-bound magnetic field	High earth orbit satellites, nuclear bomb testing, nuclear power generation, genetic engineering

reasonable norms. The right to survive and meet our basic needs (Attitude Level 1) is placed above other considerations. Elders prioritise protecting their families, farmers fight for their land, and artists focus on their work to create. As long as the impact of our mineral activities remains confined to our own territory, the effect on others is limited. The challenge here lies in transforming our actions into something meaningful, creating places of art and spiritual fire that pass through us as creators.

The liquid state of existence forces us to transcend earlier mineral boundaries and develop new moral qualities. Taking care of ourselves is no longer sufficient. For instance, when we produce wastewater, it flows downstream to our neighbours, who may demand protection against its pollution. Authorities hold those responsible for such impacts accountable. A sense of solidarity begins to emerge, as people can identify those responsible for the problems they face. The water sphere is an excellent teacher for cultivating Attitude Level 2.

In the gaseous elemental sphere, unsustainable situations affect larger groups of people over greater distances, often on a regional or continental scale. Social connections between those involved become more abstract but are often united through governance structures involving collaborations among metropolitanian areas or regions. These frameworks enable negotiation and the establishment of enforceable rules to address transboundary air pollution. Problems in the gaseous elemental sphere can be effectively managed with Attitude Level 3.

Elemental issues of planetary scale, such as climate change, remain largely unresolved by humanity, despite significant efforts. The treaty to stop ozone-layer depletion is a notable success, as it was relatively straightforward to implement. However, tackling climate change is far more complex, involving nearly all aspects of the economy. Nations have conflicting priorities, ranging from preserving oil production as a primary income source to protecting island states from rising sea levels. Negotiations alone are insufficient to address the problems of the warmth sphere. True solutions require reaching Attitude Level 4—the level of compassion and the voluntary internalisation of neighbours' concerns. This moral approach is essential for achieving sustainability in the warmth state of elemental existence.

When it comes to the higher states of elementary existence, humanity faces a paradox: we already have the capacity to influence their sustainability, yet there is minimal societal incentive to adopt an appropriate Attitude Level 5. At these higher realms of the planetary atmosphere, a "Wild West" attitude of expansion, lawlessness, and exploitation currently dominates.

The fifth or light-ether state of existence has been significantly influenced since the advent of communication technologies that leverage the reflective properties of the Ionosphere and the deployment of satellites into lower space, enabling communication across the globe. In 2015, more than 200 objects were launched into space annually, a number that increased tenfold by 2023, with the United States as the primary contributor (UN Office for Outer Space Affairs, 2024). This rapid expansion is only the beginning, as future plans suggest much more to come (Fig. 6).

The issue is scarcely noticed in smaller nations, which have limited awareness of or involvement in the space race among the USA, China, Russia, the EU, Japan, and India.

Even higher moral attitudes will be required to achieve a sustainable way of operating in the sixth and seventh elementary states of existence, which relate to the Earth's position in the solar system and the conditions that make spiritual life possible on Earth. A key feature of these higher attitudes will involve abstaining from the use of certain technologies until society possesses the moral capacity to handle them responsibly. However, the temptation to employ technologies such as genetic modification of plants—to make them resistant to specific pesticides that kill weeds—or to produce energy with nuclear power stations to mitigate climate

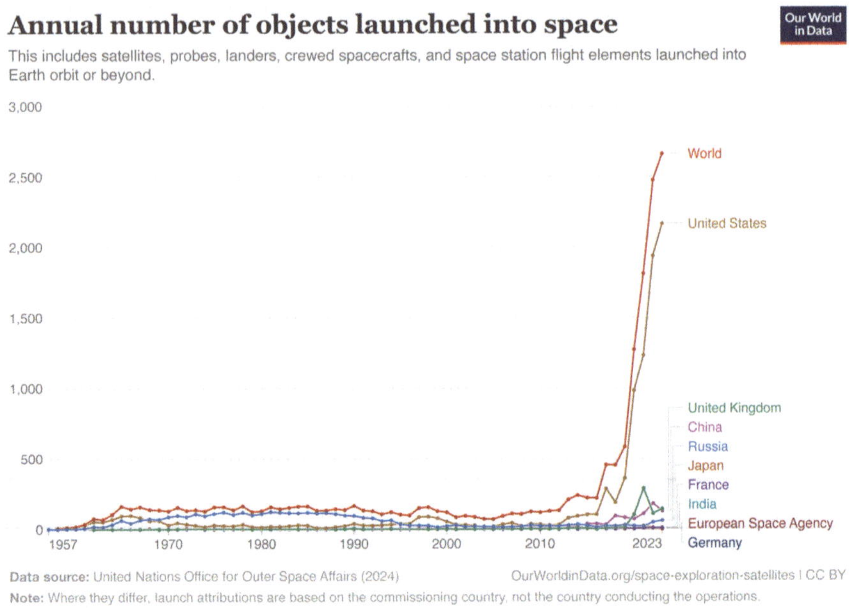

Fig. 6 Annual number of objects launched into space 1957–2023 (UN Office for Outer Space Affairs, 2024)

change is significant. These technologies will inevitably shape the way our planet can exist in the distant future.

All major geopolitical powers conducted nuclear bomb testing in the atmosphere after World War II, and further military experiments in the higher states of elementary existence cannot be ruled out in the future. Even a "common enemy" like the COVID-19 pandemic failed to unite nations. It seems likely that an "enemy" from beyond Earth would be required to unite us as one humankind (Fig. 7).

The lesson from this paper's exercise is that the five sustainability attitudes provide valuable guidance in leading us towards a more sustainable society on our planet. Earth mirrors the moral challenges we must confront, demonstrating that finding sustainable development solutions for the fourth elementary state of existence—climate change—is only the beginning of what lies ahead. We must urgently learn to act based on the higher Attitude Levels.

9 Future Challenges

Future developments will impact all seven states of elementary existence. The way technological innovations alter our lives is most directly perceived in the mineral state: the appearance of our homes, the way we eat, how we work, and the

Fig. 7 Operation Ivy, The KING Event was detonated on 15 November 1952, Enewetak Atoll (National Nuclear Security Administration, Nevada, USA)

machine-supported facilitation of handling objects. The more we become independent from the support of others, the more we may be tempted to lead isolated lives. A danger lies in the lack of challenge to our sense of self, reducing opportunities for creative expression and personal control.

At the liquid state of elementary existence, we will need to become more aware of our dependence on water. Centralised public water and wastewater systems developed in the twentieth century may partially transition to smaller-scale supplies, requiring our attention and cooperation with neighbours. Knowledge of water quality's effects on health and ecosystems, as well as making informed personal choices, will grow in importance.

Liquid issues are easier to manage than those of the gaseous state, as water first exists outside our bodies and requires a conscious act to consume it. Risks such as flooding can also be predicted and mitigated. By contrast, air inhalation cannot be delayed, and polluted air directly impacts health. Although air quality in many cities has improved over the past decades due to regional sanitation programmes, climate change will significantly affect urban air. Private air-conditioning systems will play a greater role, enhancing temperature control and air quality in homes. This could lead to spending longer periods indoors, presenting challenges such as indoor pollution and the omnipresence of social media.

Although it is easy to find arguments why humankind is failing to realise sustainable development of the higher elemental spheres, such as warmth, we can learn from the successes so far, including the reversal of ozone layer depletion and the ban on atmospheric nuclear testing. Both highlight where humankind stands today (Britannica, 2024; UN, 2009). Preventing radioactive contamination in the atmosphere that spreads all over the globe was an issue of shared global interest. The first Nuclear Test Ban Treaty, promoted by the USA and the Soviet Union, was signed in 1963 and entered into force in 1970, with over 100 countries joining, excluding France and China. This treaty prohibited atmospheric testing but allowed underground tests.

To halt all further nuclear bomb development, the Comprehensive Nuclear Test Ban Treaty was proposed. Negotiations began in 1977 but achieved limited progress. By the 1990s, countries such as Russia, the USA, and the UK declared unilateral moratoriums on testing. France and China ended their tests in 1996, while India and Pakistan conducted their first nuclear weapons tests in 1998, followed by North Korea's test of a nuclear device in 2006. In 2007, the Comprehensive Nuclear Test Ban Treaty—passed by the UN General Assembly in 1996—was signed by all nuclear powers and countries with nuclear reactors, except India, Pakistan, and North Korea. However, ten signatories, including the USA and China, had not ratified the treaty.

This situation demonstrates that even with high stakes, some nations are unwilling to accept a higher authority above themselves. This reluctance means they prioritise their own interests at the expense of their populations and inevitably all humankind. Outer space activities are inherently global in impact and will confront society with new forms of unsustainability. For example, the accumulation of space debris increasingly threatens spacecraft launches due to collision risks, potentially resulting in thousands of fragments floating in orbit. Another example is the race by major powers to establish themselves on the Moon in the coming decades, aiming to mine resources such as Helium-3 as a "clean" energy source for Earth (Kuipers, 2021).

Sustainability in the higher elementary states of existence requires higher moral attitudes. Rudolf Steiner proposed the qualities of freedom, justice, and brotherhood as foundations for sustainable development. Or, to put it another way: striving for truth, sharing love, and devoting ourselves to the higher cause of humanity's future.

10 Conclusion

Sustainable development was established as a policy theme at the UN Rio Summit in 1992, but its roots can be traced back to principles such as those of the French Revolution. Although the UN defined 17 Sustainable Development Goals (SDGs) in 2015, developing a practical approach to measure these goals individually and collectively has remained a challenge. A useful alternative is to investigate the attitudes of actors in relation to sustainability. Here too, many methods of attitude assessment have been proposed. Steiner provided clear criteria for distinguishing attitudes based on principles such as freedom and morality. Using these and other criteria, five attitudes have been defined and applied to different actors and situations.

It has been shown that these attitudes can be linked to the different elementary spheres of the Earth's ecosystem, ranging from the solid sphere of the landscape to the water sphere of river drainage areas and ocean currents, the gaseous sphere of continental air circulation, the warmth sphere of the global troposphere, and the higher spheres that envelop our planet. The magnetopause, marking the boundary between the magnetosphere and the plasma of the solar wind, represents the outermost manifestation of these seven elementary spheres. Humankind is increasingly capable of influencing even these higher spheres beyond the Earth's solid surface.

However, effective practices to prevent sustainability distortions in the higher elementary spheres are still lacking, as illustrated by the challenges of addressing climate change or achieving universal agreement on the Comprehensive Nuclear-Test-Ban Treaty. Large, powerful nations, in particular, remain reluctant to relinquish autonomy and join a global brotherhood of humankind. Yet, this is the essential challenge we must confront to ensure that Earth remains a place where humankind can continue to develop.

References

Andress, D. (Ed.). (2015). *The Oxford handbook of the French Revolution.* Oxford University Press.

Beck, D. E., & Cowan, C. C. (1996). *Spiral dynamics: Mastering values, leadership and change.* Blackwell Publishing.

Bossel, H. (1999). *Indicators for sustainable development: Theory, method, applications.* A report to the Balaton Group. International Institute for Sustainable Development.

Britannica (2024). Nuclear Test-Ban Treaty. Retrieved December 1, 2024. From https://www.britannica.com/event/Nuclear-Test-Ban-Treaty

Brundtland Commission (1987). *Our common future.* World Commission on Environment and Development. Oxford University Press.

Brundtland, G. -H. (1988). In *Proceedings World Conference on the Changing Atmosphere,* Toronto, Canada, 27–30 June, WMO, No. 710, pp. 16–30. file:///C:/Users/Eigenaar/Downloads/World_Conference_Changing_Atmosphere.pdf

Byrne, D., & Baron, R. A. (1997). *Social psychology* (8th ed.). Allyn & Bacon.

Carson, R. (1962). *Silent spring.* Houghton Mifflin.

Csikszentmihalyi, M. (1975). *Beyond boredom and anxiety: Experiencing flow in work and play.* Jossey-Bass.

Eurostat (2025). Retrieved January 2, 2025. From https://ec.europa.eu/eurostat/web/sdi/key-findings#expandable-example-content

Festinger, L. (1957). *A theory of cognitive dissonance.* Stanford, CA: Stanford University Press.

Jaworski, J. (1996). *Synchronicity: The inner path of leadership.* Berrett-Koehler Publishers.

Katz, K. (1960). The Functional Approach to the Study of Attitudes. *The Public Opinion Quarterly, 24*(2), 163–204.

Kuipers, A. (2021). We kunnen niet zomaar vertrekken als onze planeet niet meer leefbaar is: We moeten het hier met elkaar zien te redden. In M. Enthoven & K. Zoeteman (Eds.), *Keerpunt in zicht: Hoe verder met milieu en duurzaamheid* (pp. 100–105). Mauritsgroen.

Langeweg, F. (1988). *Zorgen voor morgen: Nationale Milieuverkenning 1985–2010.* Samsom H.D. Tjeenk Willink.

Maslow, A. H. (1943). A theory of human motivation. *Psychological Review, 50,* 370–396.

Maslow, A. H. (1982). *Toward a psychology of being.* Van Nostrand Reinhold.

Max-Neef, M. (1995). Economic growth and quality of life. *Ecological Economics, 15,* 115–118.

Meadows, D. H., Meadows, D. L., Randers, J., & Behrens III, W. W. (1972). *The limits to growth.* Universe Books.

NASA (2015). Astronaut Scott Kelly views the Nile at night from the International Space Station. *SciTechDaily.* Retrieved November 29, 2024. From https://scitechdaily.com/astronaut-scott-kelly-views-the-nile-at-night-from-the-international-space-station/

Nijpels, E. H. T. M. (1989). *Nationaal milieubeleidsplan: Kiezen of verliezen.* Tweede Kamer, vergaderjaar 1988–1989, nrs. 1–2.

Osgood, C. E., Suci, G. J., & Tannenbaum, P. H. (1957). *The measurement of meaning*. University of Illinois Press.
Prescott-Allen, R. (2001). *The wellbeing of nations: A country-by-country index of quality of life and the environment*. Island Press.
Scharmer, C. O. (2007). *Theory U: Leading from the emerging future*. Berrett-Koehler Publishers.
Steiner, R. (1910). *GA 122, Die Geheimnisse der biblischen Schöpfungsgeschichte* (p. 139). Retrieved November 30, 2024. From https://wn.rudolfsteinerelib.org/Lectures/GA/GA0122/19100823p02.html
Steiner, R. (1911). *Das esoterische Christentum und die geistige Führung der Menschheit* (GA 130). Rudolf Steiner Verlag.
Steiner, R. (1919). *Die Kernpunkte der Soziale Frage in den Lebensnotwendigkeiten der Gegenwart und Zukunft* (GA 23). Anthroposophic Press (Trans. F. C. Heckel, 1972).
Steiner, R. (2024). *The philosophy of freedom* (GA 4, Chap. 9). Retrieved November 25, 2024. From https://archive.org/details/the-philosophy-of-freedom-cw-4/ThePhilosophyOfFreedom_(CW4)
UN. (2009). https://legal.un.org/avl/ha/ctbt/ctbt.html. Accessed 1 December, 2024
UN Office for Outer Space Affairs (2024). Retrieved November 30, 2024. From https://ourworldindata.org/grapher/yearly-number-of-objects-launched-into-outer-space
UNOSD (2024). *Sustainable Development Goals (SDGs)*. Retrieved November 25, 2024. From https://unosd.un.org/content/sustainable-development-goals-sdgs
Zoeteman, K. (1991). *Gaiasophy*. Lindisfarne Press.
Zoeteman, K. (2009). *Gaia logica*. Uitgeverij Christofoor.
Zoeteman, K., Kersten, W., & van de Voort, L. (2012). Ranking of nations (Chap. 7). In K. Zoeteman (Ed.), *Sustainable development drivers*. Edward Elgar.
Zoeteman, K., & Tavenier, J. (2012). A short history of sustainable development (Chap. 2). In K. Zoeteman (Ed.), *Sustainable development drivers*. Edward Elgar.
Zoeteman, K. (2012c). Understanding differences in business sectors (Chap. 9). In K. Zoeteman (Ed.), *Sustainable development drivers*. Edward Elgar.
Zoeteman, K. (2012a). Can sustainable development be measured? (Chap. 4) In K. Zoeteman (Ed.), *Sustainable development drivers*. Edward Elgar.
Zoeteman, K. (2012b). Leadership and sustainability attitudes in personal life (Chap. 5). In K. Zoeteman (Ed.), *Sustainable development drivers*. Edward Elgar.

Sustainable Natives—Stories of Sustainable Development

Introduction

This chapter explores the contributions of anthroposophy to sustainable development through its practical applications and the stories of what we introduce as *Sustainable Natives*.

To do so, we present 26 enterprises and organisations actively engaged in ecological, social, and economic sustainable development, either founded upon or inspired by anthroposophy. Beyond the three pillars of sustainable development, their work distinctly integrates cultural-spiritual perspectives. Many of these organisations have existed for decades—some for a century—and are pioneers in their respective fields. Their contributions span various sectors, including agriculture, banking, education, medicine, and cultural and artistic initiatives, forming what we refer to as *Sustainable Natives* due to their early engagement with socio-ecological challenges.

These organisations are based in twelve countries across five continents and operate globally. Some employ thousands of people, while others consist of small teams of co-workers and volunteers. Some are considered a multinational, others as intentional community. They embody and practise diverse cultural and religious traditions, integrating them with anthroposophical principles and applications. Despite their differences, they share a common foundation: without anthroposophy's principles and practices, many of them might not have come into existence. A broader range of organisations was initially invited to participate. However, some were unable to contribute for various reasons, and a few had to withdraw during the process.

The enterprises were asked to write contributions at a professional practitioner level. These stories are neither advertisements nor sustainability reports; rather, they serve as windows into their founding histories, daily work, endeavours and perspectives on anthroposophy and sustainable development. Some organisations had scientific staff who could write their contributions, while others had to craft their narratives without this expertise. Some had more time, others less. Although all contributors received a semi-structured guideline with a set of questions, they had the flexibility to highlight what they deemed most important in answering the question of their engagement with sustainable development and anthroposophy.

Our intention is to showcase different understandings and applications of both sustainable development and anthroposophy, recognising that, naturally, contradictions may arise between the enterprises in terms of definitions and perspectives—something inherent to the discourse on sustainable development. Some organisations may place a stronger emphasis on the ecological dimension, while others focus primarily on social or economic sustainability challenges. With this collection, we aim to present a nuanced and pluralistic contribution to the ongoing discourse on sustainable development.

Biodynamic Federation Demeter International—BFDI

Jean-Michel Florin and Maja Kolar

Abstract

The BFDI was founded to bring together all farmers' associations practising biodynamics worldwide. This method was developed in 1924 by a multidisciplinary team led by Austrian philosopher Rudolf Steiner. Since then, it has evolved and been enriched by numerous research projects and the practical experience of farmers around the world.

Biodynamic Federation Demeter International with the vision is to create a new level of inclusivity, where not only certifying organisations, but all biodynamic organisations can take actively part in decision making, activity and development of the movement on the international level.

The purpose of BFDI is to enable farmers, processors and traders to practise agriculture for the future and for all people connected to biodynamic agriculture to take responsibility. We want to:

- Inspire and build awareness through exchange, dialogue, research, training and advice.
- Provide services such as market development, trademark protection, local capacity building, training and advice, and certification.
- Provide and guarantee the standards for working with our trademarks.

J.-M. Florin (✉)
BFDI, Bonn, Germany
e-mail: jean-michel.florin@goetheanum.ch

M. Kolar
BFDI, Bonn, Germany

J. Kronenberg and E. T. Lammerts van Bueren (eds.), *On the Earth We Want to Live*, World Sustainability Series, https://doi.org/10.1007/978-3-031-98758-8_20

Keywords

History of agriculture • Biodynamic farming • Agricultural Development • Certification • Sustainable agriculture

Features:

(a) Founding year	2020 (Demeter certification since 1928)
(b) Employees	20 (50+ in supporting commissions)
(c) Location	Germany, 88,603 Deggenhausertal
(d) Active in which countries	Licences in more than 40 countries worldwide
(e) Sector	Sustainable Agriculture
(f) Other	Federally governing the first organic agriculture certification lable

1 Birth of the Biodynamic Impulse

The origin of the biodynamic farming impulse can be traced back to a particular event: a series of eight lectures followed by questions, answers and tours given in June 1924 by the Austrian philosopher Rudolf Steiner. These lectures, known as the Agriculture Course (Steiner, 1999) were given on an agricultural estate at the cutting edge of agricultural progress at the time, in Koberwitz, not far from Breslau in Silesia. They were given in response to the questions and concerns of the participants, which—even at the time—related to issues of sustainable development (Fig. 1).

Indeed, the participants were concerned about the effect on the soil of the "new" chemical fertilisers (mainly synthetic nitrogen), the degeneration of plants, the weakening of animal vitality and increases in disease. Participants also wanted a more global approach to agriculture, integrating an understanding of the influences of the cosmos on plants and the Earth. Added to this were questions about the farmer's attitude to the land, including, for instance, the role of meditation in agricultural work.

Steiner's lectures placed the question of sustainable development in a very broad macrocosmic context, evoking the fact that the Earth was heading towards

Fig. 1 Logo BFDI

a decline in its vitality and that it was now the role of human beings to maintain and even regenerate its vitality. To achieve this, he called for a paradigm shift.

During this founding congress, a research circle was created, bringing together all the farmers present. The aim of this association was twofold: on the one hand, to support the practical application of agronomic indications from the lectures, so that they could be validated in the field; and on the other, to set up trials to scientifically verify the method. This research group worked in close collaboration with the Natural Science Section of the Freie Hochschule für Geisteswissenschaft (School of Spiritual Science) at the Goetheanum, and in particular with its research institute, directed at the time by Rudolf Steiner's close collaborator, Ehrenfried Pfeiffer, today considered one of the pioneers of the worldwide organic farming movement.

Today's biodynamic farming is the result of the initial impetus given by Steiner, combined with all the field experience and research carried out over the last 100 years around the world. It is a method in constant evolution, enriched by the contribution of each participant (Bind and Hurter, 2023).

Biodynamic farming developed rapidly on an ever-increasing number of estates, some of them very large. In 1928 the 'Demeter' trademark was registered so that consumers could identify the products of biodynamic agriculture, taking the name of the Greek goddess of earth fertility—and the first standards were drawn up which defined the practice. At the time, the significant development of biodynamic farming worried the agrochemical lobbies who, in turn, saw it as competition and launched smear campaigns.

2 Organic and Biodynamic Agriculture

Biodynamic agriculture is one of the components of organic agriculture, as its name suggests. As agricultures of life, organic and biodynamic agriculture have in common that they are organic agricultures inspired by the principles of life, such as the notions of the organism (the whole is greater than the sum of its parts), metamorphosis and evolution. They adapt both their practices to the terroir rather than moving towards soil-less farming. On a day-to-day basis, the differences are not very significant, because 80–90% of what an organic and biodynamic farmer does is identical, and their specifications are similar. The biodynamic farming specifications have specific features relating to the link with the soil, limiting the purchase of external inputs, stricter respect for the integrity of the animals (e.g. a ban on dehorning animals), the importance of biodiversity. The major practical differences are the use of specific biodynamic preparations inserted in the compost or sprayed on the soil or crops. However, a number of practices that were still specific to biodynamic agriculture 30–40 years ago, such as the use of nettle and horsetail teas and respect for planetary positions (not compulsory in biodynamic agriculture), are now widely used in organic and sometimes even conventional agriculture.

In practice, the boundaries between these two concepts are not very clear, as many organic farmers are inspired by biodynamic agriculture and, conversely, many organic farming innovations are incorporated by biodynamic agriculture. Furthermore, each biodynamist, while adhering to shared principles and the practices outlined in the brand's standards, develops a personalised approach to biodynamics, adapting it to their specific terroir to create a unique farming estate. In this context, the use of biodynamic preparations remains indispensable.

It is in their overall concept that we can attempt to distinguish between organic and biodynamic agriculture. The word 'dynamic' in biodynamic refers to the forces (dynamis = force in Greek) that are very important in biodynamic agriculture. The aim is essentially to stimulate the forces of life, their health and even their resilience. In practical terms, this means activating the health and vitality of crops using preferably open pollinated varieties, biodynamic preparations and other plant extracts. The same goes for animals, which are reared with respect for their specific nature and on high-quality feed. The primary aim is to support health rather than fight disease and parasites.

Biodynamic agriculture is based on a vision of the world that goes beyond the physical and chemical realities studied by the quantitative approach of experimental science, to include the more subtle planes of the forces of life. In this way, we work with the visible and invisible planes, most often expressed through life forces and processes. Further, Biodynamic agriculture is also based on an attitude that takes account of the inner nature of each living being, which can never be reduced to a machine or an object. This vision leads to a relationship of participation and cooperation with nature, while respecting its intrinsic values (Foyer et al. 2020).

3 The Structuring of the Worldwide Biodynamic Farming Movement

After the troubled period of the Second World War, biodynamic farming began to spread to German-and English-speaking countries in the 1950s and 60 s. From the 1970s and 1980s, it expanded to France, Spain, Italy and other nations. Gradually, from the 1990s, and especially between 2000 and 2010, biodynamic farming reached a global scale. Today, it is practised in over 50 countries on every continent (except Antarctica) by an estimated 40,000 farms, 7000 of which are certified Demeter farms.

In the 1980s, the need arose to create an organisation to bring together the various associations certifying the Demeter trademark which had already been founded in 1928. The Demeter International (DI) federation was established in 1997 in Colmar, France. However, numerous biodynamic farming associations in other countries were active in biodynamics but not involved in certification. To unify these organisations, the International Biodynamic Association (IBDA) was founded in 2002. This association took ownership of the Demeter brand, while DI continued to manage and certify it. In 2010, a process of regular meetings between the three international biodynamic farming bodies—DI, IBDA, and the Section for

Agriculture at the Goetheanum—was initiated. This collaboration led to the creation of a new federation; the Biodynamic Federation Demeter International. This federation brought together all members of DI and IBDA under a shared vision to create a new level of inclusivity, where not only certifying organisations, but all biodynamic organisations can actively participate in decision-making, activities, and the development of the movement at the international level.

The purpose of the Federation is to enable farmers, processors and traders to practice *agri-culture* for the future and for all people connected to biodynamic agriculture to take their responsibility. In order to do so we want to: inspire and build consciousness by exchange, dialogue, research, training and advisory; provide services such as market development, trademark protection, local capacity building, training and advisory, and certification; provide and guarantee the standards for working with our trademarks and provide quality assurance in our certification system.

4 Our Vision Today

Since its origins, biodynamic farming has been based on a holistic approach to agriculture that includes all aspects of working the land and aims to develop agriculture that is sustainable in every sense of the word. The international biodynamic farming movement has worked together to crystallise its vision, which is expressed as follows (Fig. 2).

In its practical application, this approach to agriculture is based on the following principles:

- Encouraging the expression of the potential of the agricultural domain and of living beings (plants and farm animals) instead of exploiting them by reducing them to objects. So, instead of simply making use of land, plants and animals, the aim is to understand them in their deepest nature (their spiritual nature) to support them in expressing their own potential. One of the key elements is to consider the agricultural estate as a living unit, an agricultural organism, practising 'circular' agriculture linked to the soil in the same way as the circular economy seeks to do. This means that biodynamic farms strive for self-sufficiency in fertility and fodder, support biodiversity and wildlife as vital parts of the farm, and work to integrate the farm into its ecology and community.
- Favouring health enhancement and prevention of disease through an attitude of 'care' and support rather than control.
- Offering a place to each being on each farm. In the event of a problem, the approach is to regulate the presence of so-called "parasites" rather than eradicating or destroying them. At the overall level of the farm—considered as a living organism—we will seek to encourage diversity by offering each living being a place in the farm.

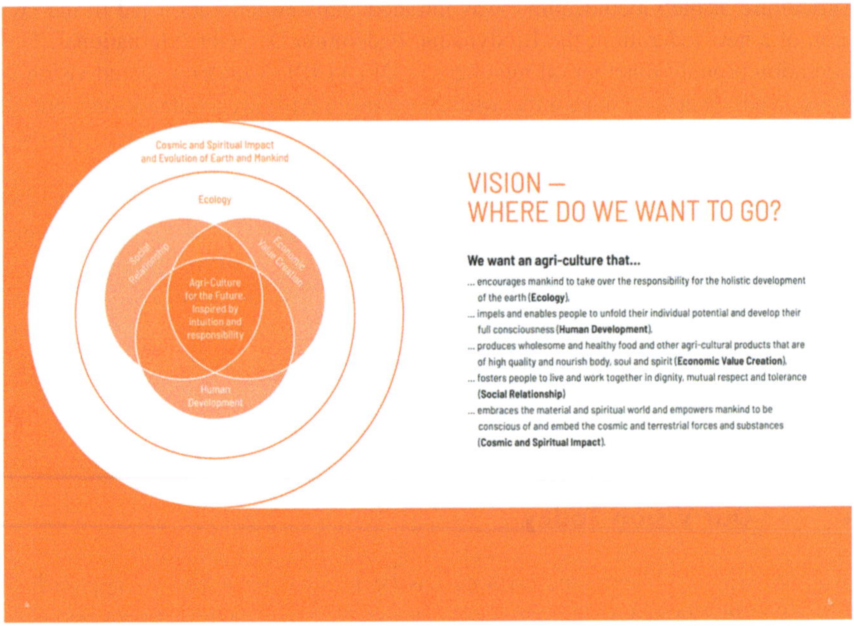

Fig. 2 BFDI Vision

- Pursuing the evolution of living things by individualising places, plants and animals (creation of new plant varieties, animal breeds, etc.), rather than standardising (cloning, genetic manipulation, etc.).

5 What We Are Doing

The Biodynamic Federation is active in bringing the work of its members into connection and integration. Working with the principle of subsidiarity, we and our members are involved in the following work:

Ecology—creating living soil and lasting fertility.

1. Creating a living context within which human beings, animals and plants can thrive and develop.
2. Advancing the continued evolution of domestic animals and cultivated plants.

Human development

1. Finding innovative social and technical solutions to the challenges we face and developing a new comprehension of nutrition and food quality with adequate methodologies to evaluate it.

2. Entering into dialogues with those working with traditional, organic, spiritual and alternative growing methods as well as with those engaged with conventional agriculture.
3. Educating consumers and raising awareness of their responsibility.
4. Enabling farmers, processors and traders to practice agriculture for the future.
5. Further developing the vision, mission and principles with all interested parties.

Economical value creation

1. Handling resources with care during the processes of both production and consumption.
2. Developing a cooperative approach throughout the production chain and creating partnerships with supporting businesses and civil society groups.
3. Making use of technology in a conscious and focused way to serve our objectives.
4. Supporting producers in our movement to recognise and adapt to increasing market demands and encouraging consumers to understand the importance of seasonality and regional production.

Social relationship

1. Encouraging individual communities to pursue and develop agriculture in their specific context.
2. Actively campaigning to create the conditions needed for sustainable and holistic development.
3. Formulating guidelines and standards that include social values and developing appropriate ways of assessing them to ensure that our brand is protected and remains trustworthy.

Spiritual impact

1. Developing spiritual abilities, supporting the dignity of the farmers and seeking knowledge.

6 The Anthroposophical View of the World and the Human in Our Work

As Biodynamic farming originated in the series of lectures given by Rudolf Steiner in 1924 (Agriculture Course), biodynamic farming has its roots in the anthroposophical movement. However, biodynamic farming evolved ever since. Practitioners developed and adapted these first impulses by putting them to use in daily farming practice and by observing the results.

The anthroposophical view brings a specific understanding of the role of farmers on Earth. It is also committed to restoring ecological balance, increasing biodiversity and developing other forms of cooperation and marketing of agricultural products, drawing on the multitude of 'acupuncture points' represented by biodynamic farms on every continent.(Roquebert, 2018)

Since the anthroposophical vision places the human being and their evolution at the centre of the process, it could be said that the sustainability of a process is fundamentally linked to the responsibility taken by people who make a personal commitment to accompany a place and then pass it on, ensuring its continuity. To achieve this, farmers first need to develop an attitude of cooperation and co-creation with the living beings they work with, rather than one of simple exploitation.

It is also necessary to take legal action. This has led the biodynamic farming movement, along with other social movements, to explore new forms of collective land ownership (Terre de Liens, Bodenfonds, etc.), as well as seed ownership (seeds as commons) (Wirz et al., 2022).

7 BFDI and Sustainable Development

From an agricultural perspective, our definition of sustainable development goes beyond the three commonly used pillars (ecological, social and economic dimensions) by extending it to include two other values: human development and cosmic and spiritual impact (see Sect. 4: Our Vision).

The ecological dimension is evident in our practice. As the Demeter brand is owned by the biodynamic community worldwide, we have the capacity to determine what the brand stands for in the form of the standards which are informed by the research, experience and inspiration of member organisations, farmers and processors all over the world. Rather than being implemented in any one country's law (as organic is), the International Demeter Biodynamic Standard is reliant on the consultation, collaboration and democratic voting of the biodynamic community. This has resulted in our standard being the strictest organic production and processing standard in almost all aspects.

The social dimension is expressed within BFDI by the development of exchange networks amongst all members. We base our work on a bottom-up approach, enabling everyone concerned to get involved. We are an active 'learning community', i.e. a place for exchange and development with all our members and allies (associations, licensees, trainers, advisors and researchers). We aim to be a place that promotes synergies and develops networks for cooperation, exchange and learning at both the national and international level. By sharing resources and experience, we can accompany and encourage farmers to develop and improve their practice and, in the end, to strengthen their commitment to the biodynamic impulse beyond certification.

For instance, to create a supportive network of biodynamic farming advisors, the Section of Agriculture of the Goetheanum and the Federation have collectively drawn up a charter freely signed by the advisors who then recognise each other through a peer-review system. In this way, everyone gets to know their colleagues, and a climate of collaboration and knowledge-sharing is created instead of one of competition. A conference is organised annually for these advisors to meet, discuss, exchange and develop.

Similarly, to encourage the development of biodynamic farming research, biodynamic farming organisations have created a platform for exchange between researchers. Conferences are organised collaboratively every other year to share research, discuss results and methods and facilitate communication amongst the researchers involved.

For a movement like biodynamic farming, which began in Europe and then spread throughout the world, there is always a risk of adopting a 'neo-colonial' attitude. This could manifest, for example, in prioritising the conversion of estates in countries of the Global South to meet the demands of European markets, while neglecting the development of local markets. We have clearly identified this risk and are working actively to support the training of local advisors, training in each country and above all—the greatest challenge—local markets. Some flagship biodynamic farming initiatives, such as Sekem in Egypt have managed to avoid this pitfall from the outset by developing a local market while exporting certain products to support the continuation of the project (SEKEM, 2023).

This brings us to the economic dimension, which represents one of the major challenges. How can a fraternal economy—whether associative or solidarity-based—be developed within a capitalist system characterised by competition and the survival of the fittest? This challenge is further compounded by the fact that agriculture is often treated as just another production sector, despite its fundamental differences.

From the point of view of the internal structure of the federation, we rely on the following values to develop an agile, efficient and horizontal structure that encourages the empowerment of each member and farmer.

- Solidarity—the Federation is a service organisation, supporting members worldwide and relying on income mainly from the added value in established markets.
- Equality in the right to participate and decide—each country with a member (or members) in the Federation has one vote—not related to size, history or market.
- Subsidiarity—the Federation does not aim to centralise power or money but works to support development at the local and country level.
- Clarity—by designing a structure that is both simple and modular the Federation can adapt itself to new situations while not leading to structural exemptions out of the enormous diversity in the world.

8 Successes and Challenges of BFDI

From the outset, despite phases of challenge and metamorphosis, the biodynamic farming movement has always managed to essentially develop from the bottom up and through concrete initiatives in the field. It has also remained united overall: Even if different tendencies have developed over time, there has never been a total split.

It is surprising that 100 years later, this series of conferences given in Koberwitz in front of 130 people (Steiner, 2004) has resulted in a vast network present in more than 50 countries, with a great diversity of farms, gardens and parks all working according to the same principles and at the same time adapted to different soils, climates and socio-economic contexts. Today, biodynamic farming is practised in many parts of the world in a wide variety of ecological contexts, by small farmers in Asia and Africa, by large mixed farming and wine-growing estates, in public parks and gardens, in shanty towns, and so on. These farms, which from the outset have been the places where biodynamic farming has been developed, form a vast network of farmer research, the results of which the Federation—in collaboration with the Agriculture Section—is seeking to gather for the benefit of the entire network. These farms have not only innovated in the agronomic and ecological fields, but also in the legal field (with the development of collective forms of land ownership, work on the farms, etc.), the social field (education and community activities on the farms, social inclusion, etc.) and the economic field (participation in the creation of CSAs, and other forms of associative economy). In fact, since its inception, the biodynamic farming movement has been a leavening agent for the whole organic farming movement, to which it often provides inspiration and innovation.

The greatest challenge today lies in addressing the major ecological crises that were identified 100 years ago. These challenges not only persist but have significantly grown in scope. In addition, many people have become even more distanced from nature and agriculture and those working in agriculture have become increasingly disempowered. The ecological and social services provided by sustainable and regenerative agriculture are increasingly acknowledged. However, the economic value of these contributions is rarely recognised or adequately rewarded for those who provide them. We also face the challenge of having solutions to offer for the widespread crises in the world, yet we must find effective ways to ensure these solutions are adopted on a larger scale.

9 Evolution and Metamorphosis

As mentioned above, one of our fundamental concepts is that of evolution, in the sense that the manifestation of life on Earth is always in transformation, in metamorphosis. Similarly, the Earth itself is always evolving.

So, whether in agriculture or in the development of biodynamic farming organisations, our role is not to maintain a fixed system but to accompany a process that is always evolving. We can take the example of the metamorphosis of the plant described by Goethe to show how biodynamic agriculture is a process of change. "But if we look at all forms, especially the organic ones, we find that nowhere is there anything enduring, anything at rest, anything complete, but rather it is far more the case that everything is in continuous motion and flux" (Goethe, 1807/2009).

10 Empowerment—Dignity of the Farmers

As a Federation of associations, BFDI wants to strengthen the empowerment of each association and each biodynamic farming actor by respecting the following principles:

- Solidarity—BFDI is a service organisation, facilitating decision-making amongst members.
- Equality—Every member has the right to participate decide.
- Autonomy and Subsidiarity—The aim of the Federation is not to centralise but to make development at farm, local and country level wherever and whenever possible.
- Clarity—By employing a structure that is both simple and modular, it can adapt to new situations without creating structural challenges, even when confronted with the immense diversity present in the world.

All of this work leaves us with the following existential questions and aims in our work:

- How to support all farmers around the world with training, advice, certification, marketing: a major project which we hope will enable many more smallholders in many countries to be included in the future.
- How to support and inspire a younger generation for meaningful work with agriculture and nature. We need to facilitate building a conscious and purposeful relationship with life, nature, food and the Earth.
- How can we communicate with the general public to raise awareness of biodynamic farming products?
- The future of the biodynamic movement depends on individual empowerment and collaboration. Today, we must prioritise fostering a sense of inner certainty that drives meaningful action. By nurturing a deep connection to the living world, individuals can authentically engage in farming, gardening, breeding, research and food production. This inner certainty is essential for meeting life's challenges and ensuring the sustainability of our biodynamic practice.
- How to create powerful alliances with all the organisations developing agriculture that respects the Earth and living beings, in order to bring about a global paradigm shift. Individual situations may be different, but on the whole most companies and most politicians are continuing with the methods that really belong in the past. We are on the cusp of a paradigm shift, and we need to support its emergence.

11 Conclusion

As we have seen, the theme of sustainable development has been one of the fundamental objectives of biodynamic agriculture since its creation 100 years ago. The aim was to develop an agriculture for the future that would enable humans and all living things to continue to live on Earth in good health. Since then, the growing network of biodynamists has explored many avenues and opened up paths in many areas to ensure sustainable agricultural development (farmers' seeds, collective forms of land ownership, quality food, experience-based agricultural training, farmers' research, etc.).

The worldwide biodynamic movement has also demonstrated its resilience, creating its own structures (Demeter brand, network of associations, network of formations and advice, etc.) which have enabled it to grow through all the trials of the last 100 years. One of the latest stages in this structuring process was the recent founding of the Biodynamic Federation Demeter International.

References

Bind, R., & Hurter, U. (2023). *Biodynamisch*. Verlag am Goetheanum.

Foyer, J., Hermesse, J., & Hecquet, C. (2020). Quand les actes agricoles sont au care et au compagnonnage: L'exemple de la biodynamie. *Anthropologica, 62*(1), 93–104. https://doi.org/10.3138/anth.2018-0103.r1

Roquebert, C.-I. (2018). *La critique écologique radicale dans le contexte de l'entreprise : L'ambivalence des démarches de RSE dans les entreprises en Biodynamie* [Doctoral dissertation, Université de Nantes, Audencia Group].

SEKEM. (2023). *Annual report 2023*. SEKEM. https://sekem.com/wp-content/uploads/2024/06/Annual-Report-2023-small.pdf

Steiner, R. (1999). *Geisteswissenschaftliche Grundlagen zum Gedeihen der Landwirtschaft (GA 327)*. Rudolf Steiner Verlag.

Steiner, R. (2004). *Agriculture course: The birth of the biodynamic method* (C. Creeger, Trans.; CW 327). Rudolf Steiner Press. (Original work published 1924)

Wirz, J., Hurter, U., & Kunz, P. (2022). *Seeds as a commons: Breeding as a source for real economy, law, and culture*. Sektion für Landwirtschaft am Goetheanum.

Bio-Stiftung Schweiz—Shaping a Fruitful Future Together in the Service of the Living

Mathias Forster and Christopher Schümann

Abstract

When the Bio-Foundation Switzerland was established by Dr Rainer Bächi in 1987, a love for the Earth, nature and humanity drove the development and realisation of its specific ideas. Anthroposophy, with its expansive perspective on the world and humanity, provided the founder of the Bio-Foundation with both inner and outer guidance. Rainer Bächi was a cosmopolitan figure, untouched by national egoisms. His concerns were for the welfare of all people, regardless of the regions or parts of the world in which they lived. With this perspective, he sought to ensure that small farmers in developing regions could access markets in Europe and North America for their organic produce.

Features:

(a) Year of foundation	1987 by Dr Rainer Bächi
(b) Employees	5
(c) Location	Basel, Switzerland
(d) Active in the which countries	Switzerland, Germany, Austria, Liechtenstein, Slovenia, Serbia, Mozambique, Egypt. In principle, the Bio-Foundation Switzerland can carry out its activities worldwide
(e) Sector	Organic farming, sustainable forms of economic activity, soil fertility, biodiversity, ecology as a whole

M. Forster (✉) · C. Schümann
Bio-Stiftung Schweiz, Bordeaux-Strasse 5, Basel 4053, Switzerland
e-mail: m.forster@bio-stiftung.ch

J. Kronenberg and E. T. Lammerts van Bueren (eds.), *On the Earth We Want to Live*, World Sustainability Series, https://doi.org/10.1007/978-3-031-98758-8_21

Fig. 1 Logo Bio Stiftung Schweiz

1 Introduction

When the Bio-Foundation Switzerland was established by Dr Rainer Bächi in 1987, a love for the Earth, nature and humanity drove the development and realisation of its specific ideas. Anthroposophy, with its expansive perspective on the world and humanity, provided the founder of the Bio-Foundation with both inner and outer guidance. Rainer Bächi was a cosmopolitan figure, untouched by national egoisms. His concerns were for the welfare of all people, regardless of the regions or parts of the world in which they lived. With this perspective, he sought to ensure that small farmers in developing regions could access markets in Europe and North America for their organic produce (Fig. 1).

Although his primary interest lay in biodynamic agriculture, developed from anthroposophical principles, his commitment extended across a variety of associations. The promotion of organic farming and the sustainable consumption of organic food was largely realised through the founding and management of the Institute for Marketecology (IMO), which still provides certification services for organic enterprises today, though it is no longer under the management of the Bio-Foundation Switzerland.

Profits from IMO's activities were donated to the Bio-Foundation Switzerland, a charitable organisation, and used for philanthropic purposes. From its inception, the Bio-Foundation served as a bridge not only between countries and cultures but also between various organic associations. The purpose of the Bio-Foundation Switzerland is to support life in its full complexity and wholeness: healthy soil, healthy plants, healthy animals, healthy people and a healthy planet.

2 Background of the Bio-Foundation Switzerland: An Expanded Concept of Art and Anthroposophical Impulses in Service of Life

From its inception, art has played a significant role in expressing and fostering the creative potential of the human being. Rainer Bächi passed away in an accident in 2010, yet the recognition of human creativity as a vital source of development, and the conviction that it serves as a renewing force capable of addressing today's most urgent challenges, remains a guiding principle of the Bio-Foundation Switzerland. This is where Rudolf Steiner's anthroposophy, with its threefold view of the human being (see Chap. 2), continues to exert a crucial influence today.

This threefold view posits that humans are not merely creatures but also creators; alongside body and soul, they have a spirit. The ways in which this individual human spirit acts, moves and produces are matters of individual freedom (Steiner, 1904/1994). Here, Joseph Beuys's concept of an expanded art form comes into play. 'Everyone is an artist,' Beuys famously stated, implying that every individual holds an innate source of creativity, and that its activation constitutes an artistic act in itself. In this sense, we regard ourselves and our work as acts of art.

It is evident that the future evolution of Earth and human culture ultimately depends on the quality of human consciousness and the inspirations we choose to embrace. When we seek to understand the causes of the ecological crisis, which we ourselves have brought about, we realise that this understanding is itself an act of creation and thus an exercise of free will. Insights do not emerge naturally but are the products of the human spirit's often painstaking labour. 'Thinking makes the soul, with which even the animal is endowed, into spirit,' Hegel once wrote (Hegel, 1827/1991). Today, precise observation and active thinking may be more necessary than ever to gain a broad and profound understanding of these problems and to devise sustainable, harmonious solutions that fit seamlessly.

In this way we can create the conditions for that which has become ill through us to become healthy again. These are great tasks that can only be solved if we not only cooperate with each other in a positive way, but also associate with each other in order to promote the living on Earth. And not only the life of nature, but also the life of culture. For the development of both is intimately connected. In every foodstuff, on every farm, in every landscape, human formative impulses have taken on external forms. These external forms are each expressions of a particular spiritual culture, of a spiritual consciousness.

By fostering these conditions, we enable what has been made ill by human activity to heal again. These are immense tasks that can only be achieved through positive cooperation and association aimed at promoting all forms of life on Earth—not only natural life but also cultural life. For the development of both realms is intricately linked. In every food item, on every farm, in every landscape, human formative impulses taken on external forms. These forms reflect specific spiritual cultures and states of consciousness.

Industrial agriculture embodies a different cultural spirit from that of biodynamic agriculture. The outward forms, which express ways of thinking, convey a clear message. This can be observed, and each person can form their own perspective. What does the cowshed look like? What about the pigsty? What do the animals eat, and how are they? How close are they to living in accordance with their nature? Can a rich diversity of plant and animal species thrive on the farm? How are the soil and soil organisms doing? And how do people treat each other on the farms and beyond, along the value chains in cooperation with traders and consumers? How do politicians treat farmers? What prices are consumers willing and able to pay for high-quality food? And how should agriculture be funded to facilitate sustainable agricultural development?

Depending on the underlying philosophy and view of humanity, responses to these questions can vary significantly. Artist Joseph Beuys (1921–1986), who was familiar with Rudolf Steiner's anthroposophy, coined the term 'social sculpture' through his active engagement with it. We are all invisibly connected within this social sculpture—as creative and productive beings, as artists. This sculpture is continually evolving because we, as human beings, shape it, consciously or not. It changes from moment to moment, reflecting the consciousness from which we act. This process produces both ugly aspects and beautiful ones; it generates suffering as well as joy and fulfilment. The whole of this is in constant evolution. In which direction is the social sculpture currently moving in relation to agriculture? It appears to be moving in very different directions simultaneously, shaped by varied perspectives on the world and humanity.

3 Industrial Agriculture and Organic Farming

We wish to highlight two developmental directions here due to their significance. The first is industrial agriculture, which is based on a materialistic worldview. This approach emerged alongside the industrialisation of various aspects of life and has brought progress and prosperity to many areas of human life. However, it is becoming increasingly evident that the forms of thought and action shaped and aligned with industrial processes can, when taken to extremes, do more harm than good within the field of agriculture (Forster & Schümann, 2020).

The second worldview can generally be described as holistic, clearly distinguishing between inanimate objects and living plants, animals and human beings, each of which has its intrinsic value and characteristics to be observed and respected in agriculture. The biodynamic approach to thinking and working, rooted in Rudolf Steiner's teachings, has been a pioneering movement within the innovative and life-promoting development of agriculture. It opposes harsh materialism and overcomes it by demonstrating that an agrarian culture grounded in respect for and mindfulness of life is not only possible but urgently necessary, especially in light of the worsening ecological crisis.

In the extreme forms of industrial agriculture, this materialistic worldview manifests itself as physical–chemical reductionism. All problems and challenges are believed to be solvable through chemical substances such as nitrogen fertilisers and synthetic pesticides, or with physical interventions and other technical solutions, such as 'precision farming'. Since this worldview recognises only matter and denies the existence of soul or spirit, human beings, animals, plants and all other living beings are inevitably reduced to objects and economic goods.

This reductionist view has led to the creation of meat factories, which produce cheap meat. However, upon closer inspection, this meat is not truly inexpensive when one considers the considerable economic and ecological costs stemming from this materialistic approach to living beings in agriculture. The sole focus is on maximising the production of meat, milk and eggs in terms of quantity and weight, while minimising costs for the individual farm.

In line with this logic, animals are confined to the smallest spaces, and, as a result of the stress imposed by such husbandry, they inevitably fall ill. To counter this, they are given antibiotics as a preventive measure, ensuring they stay alive until they are ready for slaughter. No consideration is given to the animals' souls, which manifest clearly in their capacity to suffer and the suffering they endure, as their sentience is overlooked. This is because such a worldview refuses to acknowledge any differences between animals and inanimate objects or economic goods.

4 The Healthy 5

The holistic approach to agriculture stands in contrast to the materialistic view and is the one that we, as the Bio-Foundation Switzerland, support and promote. In this perspective, all living beings on Earth are interconnected in a symbiotic relationship. This also includes the Earth itself and the entire cosmos. There are living relationships and dependencies among all living beings, which necessarily means that disease phenomena occurring in one area almost inevitably result in similar phenomena in other areas. Conversely, it also means that holistic and sustainable health promotion must consider all areas simultaneously.

We use a "healthy 5" as a guide, but in a different sense than is commonly understood. By this, we mean: healthy soil, healthy plants, healthy animals, healthy human beings and a healthy planet. Only when all these spheres are in balance and thus healthy is the entire living organism truly healthy.

5 The Soil Fertility Fund of the Bio-Foundation Switzerland

In discussions with Christian Hiss, a biodynamic vegetable gardener and the founder of Regionalwert AG and Regionalwert Leistungen GmbH, we became aware of the alarming decline in soil fertility worldwide. It became evident to us that soil fertility is also deteriorating within organic farming. This means that the Earth, on whose fertility the life of all plants, animals and human beings is existentially dependent, is slowly losing its capacity to ensure healthy nutrition for its inhabitants. In response, we developed the idea of establishing a fund to foster new forms of cooperation within society and to support a select number of farms over a defined period in enhancing the fertility of their arable land and promoting life in the soil.

The project extends over a period of seven years. We work with biodynamic farms, but we are also open to collaborating with organic farms affiliated with other associations. A total of around thirty participating farms from Switzerland, Germany, Austria and Liechtenstein are involved in the project as partner farms. As the project concludes in 2024, we can confidently state from our own experience that the process of declining soil fertility is reversible if the problem is understood in its full complexity and effective solutions are developed and implemented simultaneously at several levels (Bio Stiftung Schweiz, n.d.).

However, this requires financial flows and sources that recognise their return on investment in increased soil fertility. We identify the primary cause of declining soil fertility as the tendency within our society to exploit soils and farmers, that is, to take more from them than we are willing to give back for their regeneration. Moreover, promoting and maintaining healthy soil fertility is an inherently complex issue. Even highly experienced soil experts frequently affirm that they are continuously acquiring new knowledge in this field. Since it is primarily farmers who are ultimately responsible for soil fertility on land used for food production, they must be given the opportunity to continuously exchange and develop their knowledge and skills through further training and shared experiences, given the complexity of the topic. This requires both time and financial resources, which farms, under economic and time pressures, often lack.

This is where the concept of the Soil Fertility Fund of the Bio-Foundation Switzerland comes in, along with our desire to provide concrete support. We regularly organise further training and networking opportunities for the farming community and offer financial support to enable them to implement additional measures for soil development. However, the Bio-Foundation Switzerland is not an awarding foundation with a large endowment from which the funding could be sourced. This meant that we had to secure the necessary financial resources through fundraising activities, which was an extremely challenging task. However, to our great joy, it was a complete success.

Through this process, we connected with other foundations, companies and private individuals, with whom we have developed a trusting relationship over the years. We are convinced that a forward-looking approach to developing and implementing innovative solutions lies in the formation of communities of institutions, organisations and human beings who share common goals and are committed to achieving them together. In this way, each participant becomes an essential contributor, adding their unique value to the success of the whole. In this context, the quality of the relationships that emerge from cooperation is important to us. They should be characterised by warmth.

We are convinced that the future lies in free cooperation between free human beings who develop and pursue common goals together. This is also how the concept of the Soil Fertility Fund came about. In numerous discussions with various stakeholders, we have endeavoured to understand the complex issue of declining soil fertility and to develop and implement solutions that we believe will deliver the best results. We also cultivate the quality of freedom in our relationships with the farmers involved in the project.

For example, we leave it to our partner farms to decide which measures they want to implement. However, we expect them to report annually, both to themselves and to us, on whether they have implemented what they set out to do at the beginning of the year and, if not, to explain the reasons. In this context, we work with the concept of 'free space'. In this free space, suggestions can be made, for example, by the soil expert Dr. Ulrich Hampl, who visits our partner farms twice a year, where he examines the soil together with the farmers and is available for advisory discussions. But what is actually done on the soil is decided by the farmers themselves. The aim here is to revitalise the soil to a depth of about thirty centimetres.

Today, we can see from soil samples and documentation carried out over the entire period that much can be achieved in terms of soil structure over a period of seven years when organisations and people join forces with farmers in new forms of association (Bio Stiftung Schweiz, 2022). However, it is also clear that care for the soil must be continuous and must never diminish, because the soil is a living being whose life and development depend on the care we human beings give it when we use it for agriculture.

6 Calculating Correctly in Agriculture

During his time as an independent entrepreneur, biodynamic gardener, economist and innovative thinker, Christian Hiss realised that the agricultural industry does not do its sums properly. This is because the macroeconomic costs incurred by agricultural production methods are not passed on to the farms that cause them. These macroeconomic costs arise, for example, in the areas of air and water pollution, soil contamination and the resulting additional health costs, loss of biodiversity, flood damage, etc. If the substantial economic costs caused by agricultural businesses were to be invoiced to them, many industrial agriculture businesses would probably have to file for bankruptcy immediately. However, since these costs are not included in the product price today, and the damage is instead borne by the general public through taxes, the illusion can be maintained that so-called 'conventional' food is cheaper than organic products. Furthermore, the services that farmers provide for the common good are also not recorded and are not adequately remunerated through subsidy payments. This is mainly due to an accounting system that is not designed to capture such services and value creation.

In response to this, Christian Hiss developed an extended accounting system and corresponding instruments with which the performance of farms in the areas of ecology, social affairs and regional economics can be realistically recorded and evaluated. He has thus created the conditions for a performance-related remuneration and subsidy policy. We were convinced by this innovative approach, and we consider it to be an important key to moving more and more in the direction of sustainable development in the agricultural sector. In the second pilot phase of our soil project, we incorporated Hiss's performance calculator into the concept and had it tested and expanded several times by our partner farms. As a result, it is now considered the most comprehensive and practical tool that best reflects the realities of farming. Our participating partner farms were enthusiastic, but over time they also became frustrated because they could now see in monetary terms what they were doing for the common good, but also how much of this was not being remunerated by society.

7 Further Projects

The diversity of our efforts to make Earth a vibrant place is also reflected in our internal and external project funds.

Internal Project Funds
In addition to the Soil Fertility Fund, we have set up the BodenKunstFonds (Soil-Art-Fund). The intention here is to bring fertile soil, sustainable agriculture and art closer together. In this context, art should be able to serve the soil and become a strong voice for it. The highlight of these efforts to date was the field concert in the summer of 2023 at the Rengoldshausen estate, where the Nuremberg State Philharmonic Orchestra played Beethoven's 6th Symphony in front of 500 guests.

Another internal fund is the Stiftungsfonds (Foundation Fund). The purpose of the Stiftungsfonds is to create a space for the as yet undefined and emerging in our work and to enable free development for the future. We often deal intensively with current issues, where the first step is to understand the often complex interrelationships. Only then can we answer the question of whether we can contribute something fruitful to these issues. Such work cannot be assigned to any existing projects, which is why we have set up a separate fund for it.

External Project Funds (A Selection)
Erde und Kultur Fonds (Earth and Culture Fund): This fund aims to promote and support the development of responsible relationships between human beings and nature, especially in the fields of agriculture and environmental education.

'Samenfest' Fonds (Seed Proof Fund): This fund supports consumer demands regarding seeds. Its aim is to promote the demand, cultivation and sale of seed-proof (reproducible) biodynamically developed varieties.

Fonds Teil der Lösung (Part of the Solution Fund): This fund aims to contribute to a sustainable society by organising events for people interested in sustainable lifestyles and economic systems.

Regenerative Agroecology Fund Mozambique: This fund is committed to the dissemination of agroecology in Mozambique.

SANER NaturFonds (SANER Natur Fund): This fund was set up by SANER Project AG in recognition of the fact that they take plants from nature to produce remedies. With the fund, they want to give something back to nature (Fig. 2).

Fig. 2 Farmers workshop Bio-Stiftung Schweiz

8 Future

The potential of the performance calculator will certainly keep us busy. In addition, we will take a thorough look at the question of how farms should and could be integrated into the economic, legal and cultural context in the future, so that life on them, and in society as a whole, can develop in the healthiest way possible. Through practical projects, we will continue to show that many things are possible when human beings work together in a fruitful way.

The tasks are many and great. We can achieve almost nothing alone. We all depend on each other, and we can only achieve a prosperous future for all living beings on this planet, and for the planet itself, by adopting a broader concept of the division of labour. By extending the division of labour to include everything that lives and works, i.e. soil, microorganisms, plants, animals and human beings.

This happens most effectively through consciously shaped interaction in which everyone can contribute the best of what they have developed for the benefit of the greater whole. We are open to shaping and developing our further development together with those human beings who find their way to us on this path of joint development. Whether it be by supporting and enhancing our work and activities with freely donated money, founding a fund with us to enrich and expand the family of funds, or even by supporting us inwardly and taking a warm interest in our work and how we do it. We are deeply grateful to all the people who have supported us in this to date and will continue to support and promote us in the future.

If in a global world

We take responsibility for the well-being of all

And if this becomes the principle of our action,

Our unique development will play an integral part

In a colorful social, economical and ecological harmony.

In a global world

The outreach of our actions is universal.

The more we anticipate the well-being of the community

And of its individual members in our doings

The better it will be for the future of humankind

And in return for ourselves.

—Dr. Rainer Bächi, Gründer der Bio-Stiftung Schweiz

References

Bio-Stiftung Schweiz (2022, December 16). *Bodenentwicklung auf den Partnerhöfen des Bodenfruchtbarkeitsfonds im Jahr 2022* [*Soil development on the partner farms of the Soil Fertility Fund in 2022*]. https://www.bodenfruchtbarkeit.bio/blog/konzepte-berichte/bodenentwicklung-auf-den-partnerhoefen-des-bodenfruchtbarkeitsfonds-im-jahr-2022/

Bio-Stiftung Schweiz (n.d.). Abschlussbericht Pilotphase 1: Bodenfruchtbarkeitsfonds [Final report pilot phase 1: Soil fertility fund]. https://www.bodenfruchtbarkeit.bio/aktuelles/jahresbericht/abschlussbericht-pilotphase-1-bodenfruchtbarkeitsfonds/

Forster, M., & Schümann, C. (Eds.). (2020). *Das Gift und Wir* [*The poison and us*]. ISBN: 978-3-86489-294-3. Westend Verlag

Hegel, G. W. F. (1991). *Enzyklopädie der philosophischen Wissenschaften im Grundrisse* (2nd ed.). Felix Meiner Verlag (Original work published 1827)

Steiner, R. (1994). *Theosophy: An introduction to the supersensible knowledge of the world and the destination of man* (GA 9) (C. E. Creeger, Trans.). Anthroposophic Press (Original work published 1904)

Egyptian Bio-dynamic Association—How Can an Economy of Love, Inspired by SEKEM, Contribute to a System Change?

Justus Harm

Abstract

Economy of Love is a holistic sustainability standard inspired by SEKEM in Egypt. It supports biodynamic agriculture while integrating ecological, social, cultural, and economic efforts within the value chain. Through training, certification, and an innovative nature credit system, EoL enables farmers to improve soil health, increase biodiversity, plant trees, produce compost and earn additional income by reducing CO_2 emissions through solar pumps. Since its launch, over 30,000 farmers have adopted EoL practices, transforming desert areas into productive farmland and contributing to climate mitigation. The initiative promotes fair trade, strengthens rural communities, and emphasizes education and cultural development. By aiming to include 250,000 farmers by 2028 and introducing biodiversity and water credits, EoL offers a practical and values-driven model that can be scaled in Egypt and beyond.

Features:

(a) year of foundation	1994
(b) Employees	96
(c) Location	Egypt
(d) Active in which countries	Egypt and India
(e) Sector	Agriculture, standard, ecosystem services
(f) Other	Gulbenkian Prize for Humanity 2024

See Fig. 1.

J. Harm (✉)
EBDA, Cairo, Egypt
e-mail: justus.harm@economyoflove.net

J. Kronenberg and E. T. Lammerts van Bueren (eds.), *On the Earth We Want to Live*, World Sustainability Series, https://doi.org/10.1007/978-3-031-98758-8_22

Fig. 1 Logo EBDA

1 Introduction

In the heart of the Egyptian desert, where barren, arid soils predominated just a few decades ago, a vision began to flourish in the late 1970s: SEKEM. Founded by Dr. Ibrahim Abouleish, this initiative embodies a profound commitment to biodynamic and regenerative development, deeply inspired by the principles of anthroposophy. SEKEM is far more than an agricultural project; it serves as a pioneering model for communal living that harmonises nature, economy, culture and the social dimension. Over the years, SEKEM has established extensive networks not only within Egypt but also internationally, becoming a lighthouse of holistic sustainable development.

Within the framework of SEKEM's economic model, the concept of the "Economy of Love" (EoL) emerged and has evolved, since 2019, into a comprehensive production and carbon credit standard. This standard is seen as a pioneering model that embodies a holistic approach to certifying and promoting biodynamic practices. It integrates a CO_2 component that promises higher incomes for farmers contributing to ecological improvement. Unlike conventional sustainability certification methods, this standard incorporates the ecological, cultural, social, and economic performance of farmers. Its foundation in regenerative and biodynamic principles provides a unique perspective and depth, making it highly relevant in today's discourse on sustainable development.

2 From Desert to Flourishing Habitat

The story of SEKEM began in 1977, when Dr. Ibrahim Abouleish, inspired by the teachings of Rudolf Steiner, envisioned transforming desert soil near Cairo into fertile, thriving land and laying the foundation for a vibrant community where people could work, learn, and realise their individual potential. At a time when the environmental crisis was not yet firmly rooted in global consciousness, Abouleish recognised that the solution to many of humanity's most pressing challenges lay in a sustainable and holistic approach to nature.

The central challenge was clear: how could desert land, traditionally considered barren, be converted into agriculturally viable land? Although the vision of creating a sustainable community in the desert seemed absurd—even impossible—to many in the late 1970s, Abouleish was joined by a group of like-minded people. Together, they began implementing biodynamic farming methods based on anthroposophical principles. These methods, which treat soil as a living system and avoid chemical fertilisers and pesticides, quickly yielded impressive results. Within a few years, the desert turned green, and SEKEM began producing biodynamic food that found markets not only in Egypt but also internationally.

SEKEM's impact, however, extended far beyond agriculture. Dr. Abouleish understood that the holistic development of society could only be achieved by equally addressing ecological, social, cultural and economic dimensions. Thus, SEKEM began establishing schools, a University, medical centres and cultural institutions to promote the potential unfolding of people in the community. This approach, which places the human being at the centre of development, became part of SEKEM's identity and inspired the creation of the Economy of Love programme.

3 A World in Balance

The vision of SEKEM and the Economy of Love approach is simple yet powerful: a future in which every human being can realise their individual potential; where humanity lives in social structures that reflect human dignity; and where all economic activities align with ecological and ethical principles. This vision is also deeply connected to the anthroposophical philosophy, which views the human being as an integral part of nature. In this worldview, humans are not masters of nature but stewards, bearing the responsibility to preserve, nurture, and engage with the Earth for future generations.

For SEKEM and EoL, sustainable development means more than reducing the ecological footprint; it also entails inspiring a society founded on the principles of justice, solidarity, and respect. This holistic perspective permeates all aspects of EoL's work, from promoting biodynamic agriculture to supporting cultural and social activities in rural areas (Fig. 2).

4 Scaling Biodynamic Agriculture through Economy of Love

The activities and programme of EoL are the result of years of research and development aimed at creating a model that enables farms and organisations to integrate SEKEM's holistic values into their production and processing practices.

A central function of EoL is to ensure the certification, validation and verification of agricultural enterprises that adhere to the principles of biodynamic farming. These enterprises commit not only to social, cultural and economic activities as

Fig. 2 New biodynamic farmers working in Camille field in Egyptian desert area

in associative economics but also to holistic ecological practices. These practices focus on regenerating and building soil, enhancing biodiversity and avoiding the use of chemicals.

A unique feature of the EoL standard is the introduction of an innovative carbon credit system that measures and financially rewards certain ecological contributions made by farmers. Within this system, carbon credits are allocated to biodynamic farmers who reduce their greenhouse gas emissions and engage in practices inherent to biodynamic farming such as composting, forestry, regenerative soil management and renewable energy. These CO_2 certificates can be traded on the Voluntary Carbon Market (VCM), providing farmers with an additional income stream that enables them to further develop their biodynamic and regenerative farming practices.

In addition to the certification and the validation and verification of carbon credits, EoL supports educational and training programmes aimed at raising awareness about biodynamic agriculture and the importance of protecting natural resources. These programmes target not only farmers and organisations but also the wider public, promoting an understanding of the relevance of biodynamic agriculture across all aspects of life.

5 The Impact of Agri-Culture on Earth

The holistic understanding of human beings and nature, as described in anthroposophy, profoundly shapes the work of EoL in many ways. Biodynamic agriculture is based on the principle that the agricultural organism is a living cycle with which human beings must interact harmoniously. Through this approach, nature is not merely a resource to be protected; rather, humans are seen as part of the whole organism and therefore tasked with understanding agriculture as cultural work on the Earth. By cultivating the soil, humans assume the responsibility of developing and enhancing it—a process of co-creation between humans and nature.

Another central principle of anthroposophy that plays an important role in the work of EoL is the social threefolding. This idea, developed by Rudolf Steiner, posits that society consists of three spheres: economic life, rights life and spiritual life. Each of these spheres has its own function, and they should operate independently of one another to foster a harmonious and balanced society. In practice, this means that EoL not only includes economic sustainability but also supports the social and cultural activities of certified participants, contributing to the holistic development of their communities.

6 Holistic Sustainable Development

EoL pursues a holistic definition of sustainable development that equally considers four dimensions: social, cultural, ecological and economic. This definition surpasses conventional approaches by emphasising the importance of cultural and spiritual development and scaling these elements with innovative methods.

Social sustainability is fostered through the integration of comprehensive social criteria into EoL, complemented by educational initiatives. These activities aim to strengthen the social fabric within companies and farms in rural areas by promoting a sense of community and raising awareness about the importance of sustainability. Targeted educational programmes empower people to develop and implement their own sustainable practices, resulting in more resilient and interconnected communities.

Cultural activities plays a central role in EoL. Culture is understood not only as the preservation of traditions but also as the promotion of a vibrant and dynamic cultural life that develops in harmony with the environment and the social needs of the community. Culture is viewed as an essential factor in enabling individuals to realise their potential, placing personal development at the core. Within the EoL system, each human being is seen as a unique individual whose abilities are cultivated through targeted educational programmes, artistic initiatives and spiritual practices. These approaches contribute not only to individual development but also to strengthening the community as a whole, creating an environment where cultural values and creative expression can thrive.

Ecological and regenerative sustainability is pursued through various measures aimed at protecting and restoring natural resources. A key aspect is the introduction of biodynamic farming, which enhances soil fertility, increases biodiversity and maintains ecological balance. Each biodynamic farm develops its own unique farm organism, integrating plants, animals and biological cycles to build resilience against external influences. This holistic approach strengthens the independence of farms and minimises the need for external resources. Practices such as seed cultivation, compost production and reforestation contribute to ecosystem regeneration, eliminating the need for chemicals and fertilisers. Activities such as reforestation and integrating trees into farm systems, such as agroforestry, are also actively supported.

Economic sustainability is achieved by creating new income opportunities for farmers, such as the CO_2 remuneration system for ecosystem services. EoL's carbon credit system plays a central role, offering farmers an additional income source that enables them to expand their biodynamic and regenerative practices. Moreover, Economy of Love promotes fair trade practices and facilitates access to local and international markets for biodynamic products. These economic measures ensure that sustainable practices are not only ecologically and socially viable but also economically profitable.

7 The Contribution of Anthroposophy to Sustainable Development

Anthroposophy offers a unique contribution to sustainable development by promoting a holistic approach that considers both spiritual and material elements. In the work of EoL, this approach is shown in the way ecological practices are interconnected with social and cultural initiatives to foster a comprehensive form of individual and societal development. Many of EoL's functions are rooted in elements of associative economics, which emphasises cooperation, justice and solidarity. In associative economics, the market is not viewed as an arena for uncontrolled competition but as a space where producers, traders and consumers engage in a conscious dialogue. These actors collaborate to agree on fair prices that respect the needs of both producers and consumers while taking into account ecological and social costs. This principle is reflected in EoL's practices, which focus on creating values that go beyond the purely monetary.

Another central concept of anthroposophy applied in EoL's work is the principle of "economics as a means of livelihood." This principle advocates for an economy aligned with the needs of both human beings and nature, based on the principles of justice, solidarity and sustainability.

8 Achievements, Challenges, and Lessons Learned

Since the introduction of the EoL programme, significant achievements have been made. These include the successful transition of over 30,000 Egyptian farmers to biodynamic practices, resulting in a substantial reduction in greenhouse gas emissions, increased CO_2 sequestration and improved soil fertility. Over the past two years alone, more than 260,000 tonnes of CO_2 have been avoided and sequestered. However, this transition was not without challenges, requiring considerable investment in training, equipment and farmer support.

Another major achievement is the recognition of EoL carbon credits on the international market. This recognition has enabled Egyptian farmers to not only monetise their sustainable practices but also gain international acknowledgement for their efforts. This has raised awareness of the importance of sustainable agriculture and expanded the market for sustainably produced goods which can now be sold at conventional market prices thanks to additional CO_2 revenue. SEKEM and the Egyptian Biodynamic Association (EBDA), in collaboration with the Financial Regulatory Authority and the Egyptian Stock Exchange, played a key role in establishing Africa's first voluntary carbon market. The first auction of EoL carbon credits took place in August 2024.

In July 2024, SEKEM and EoL were jointly awarded the Gulbenkian Prize for Humanity in Portugal. The award was presented by the Chair of the Jury, Dr. Angela Merkel, in a grand ceremony in Lisbon. In 2025, SEKEM was awarded the Champions of the Earth award by the United Nations Environment Programme. Alongside other international accolades and participation in UNFCCC COP conferences, SEKEM and EBDA are set to attend further major events in the coming years to market biodynamic smallholder farmers' carbon credits. However, fluctuating demand in international markets poses an ongoing challenge.

Another challenge lies in addressing the impacts of climate change on agricultural production. The increasing frequency and intensity of extreme weather events threaten harvests and necessitate innovative solutions to further strengthen the resilience of biodynamic farming. SEKEM and EBDA are working closely with research institutions and partners to develop and implement new methods that help farmers adapt to changing climatic conditions.

Valuable lessons have been learned from these challenges. One of the key insights is the necessity of continuously adapting and improving regenerative biodynamic practices to meet evolving conditions. This requires close collaboration with farmers to ensure that the solutions developed are both practical and effective.

Another important lesson is the critical role of education and awareness-raising. Experience has shown that the success of these projects largely depends on how well the communities involved are engaged in learning processes and informed about the importance and benefits of sustainable practices. As a result, EoL has expanded its educational and training programmes to ensure that farmers and communities are equipped with the knowledge and skills needed to successfully implement sustainable practices.

9 Shaping a Regenerative Future

In the coming years, EoL plans to expand its work and integrate 250,000 farmers into the EoL programme by 2028. A central focus of this expansion will be the further development of the EoL standard to fully capture the farmers' comprehensive contributions. This includes the addition of further scopes, particularly Biodiversity Credits, Water Credits, and Social Credits.

Biodiversity Credits aim to document and reward the contribution of biodynamic agriculture to promoting biodiversity. Since biodynamic farming has been scientifically proven to require less water, Water Credits will be developed to recognise efficient water use and conservation, providing compensation to farmers. Social Credits will be integrated to measure and reward the societal value generated by EoL farmers through their work. This includes creating jobs, fostering communities and improving quality of life in rural areas.

A particular emphasis will continue to be placed on strengthening farmers' resilience to the impacts of climate change. This involves developing and implementing new agricultural methods to help farmers adapt to changing climatic conditions.

EoL also plans to intensify its collaboration with other initiatives and organisations pursuing similar goals. The objective is to increase the impact and reach of EoL's work and extend the EoL standard model to other regions and countries. Promoting partnerships will play a key role in leveraging synergies and fostering the exchange of knowledge and experiences.

Another goal is to enhance the transparency and traceability of the carbon credit system. EoL is working on developing new technologies and methods, including the use of satellites, to measure and document emission reductions with even greater accuracy. This is not only critical for the credibility of the system but also for its acceptance on the international market.

EoL will also continue investing in educational and awareness-raising programmes to foster understanding and support for holistic sustainable development in society. These programmes aim to inspire a new generation of farmers, entrepreneurs and decision-makers to integrate the principles of sustainability into their work and contribute to a better future for all.

10 Conclusion: On Earth We Want to Live

The work of SEKEM and the EoL demonstrates how an inspiration through anthroposophy contributes to sustainable and regenerative development. In practice, this approach is reflected in how EoL preserves natural resources while simultaneously promoting human and cultural development. "The work is the transformation of the Earth. The work on the Earth is the force of attraction for the vision. The vision needs a piece of Earth," said Ibrahim Abouleish, who transformed the Egyptian desert into a thriving oasis, unlocking the potential of thousands of people. This statement describes the essence of holistic agriculture: the development of people and nature through their interaction.

The EoL approach promotes the idea that agriculture is more than just cultivating soil or protecting the environment. It emphasises that organic or biodynamic agriculture requires meaningful collaboration between humans and nature—a truly cultural act. Biodynamic farming, as taught by Rudolf Steiner, views the farm as a living organism interconnected with people, animals and nature. This philosophy teaches us to work with natural rhythms and enhance the vitality of the Earth without depleting it. This is central in EoL practices, which combine high productivity and efficiency with environmentally friendly methods.

In a world where industrial agriculture and forestry are among the largest contributors to climate change, accounting for nearly 24% of global greenhouse gas emissions (IPCC, 2014), EoL offers a sustainable alternative. It demonstrates that regenerative biodynamic agriculture provides solutions to many of the current challenges threatening our future. The challenges we face require a holistic perspective that goes beyond short-term food production to consider the entire value chain.

With only 1.6% of global agriculture certified as organic (Willer et al., 2022, p. 20), it is crucial to bring the knowledge and principles of biodynamic farming into the mainstream. The EoL approach offers a model that not only ensures ecological regeneration but also strengthens the social and cultural dimensions of agriculture.

Living on Earth means utilising its resources, protecting them and fostering a deep connection with nature. Guided by Economy of Love and anthroposophical principles, we can collectively create a world where regeneration is not just a word but a lived reality. EoL presents a promising vision of how agriculture can become a true co-creation with nature and society—a vision that supports the full flourishing of both humanity and the Earth.

References

IPCC. (2014). Climate change 2014: Mitigation of climate change. Contribution of working group III contribution to the sixth assessment report of the intergovernmental panel on climate change. https://www.ipcc.ch/report/ar5/wg3/

Willer, H., Trávníček, J., Meier, C., & Schlatter, B. (Eds.). (2022). *The world of organic agriculture: Statistics and emerging trends 2022*. Research Institute of Organic Agriculture (FiBL) and IFOAM—Organics International. https://www.fibl.org/fileadmin/documents/shop/1344-organic-world-2022.pdf

Eosta & Robin Food Coalition—Dream, Dance, Deliver: for Purpose, People, and Planet

Volkert Engelsman

Features:

(a) Founding year	1990
(b) Employees	100
(c) Location	Waddinxveen, the Netherlands
(d) Active in which countries	20
(e) Sector	Organic fruit and vegetables
(f) Other	Royal Award Officer in the Order of Orange Nassau (2023) European Business Award for the Environment (2019) King William I Prize for Sustainable Entrepreneurship (2018) No 1 at the Sustainability Leaders Top 100 (2017)

1 Introduction

First of all, let us pause for a moment and go back in time. If you zoom out, you will see that we are living in the middle of an interesting era that began around 1989, at the end of the last century, with the fall of the Berlin Wall, the Tiananmen Square protests in Beijing, Václav Havel becoming president of Czechoslovakia, Nelson Mandela being released from prison and Al Gore writing his book Earth in the Balance. Gore spoke not only of a growing climate problem but also of the

V. Engelsman (✉)
Eosta, Founder Robin Food Coalition, Waddinxveen, Netherlands
e-mail: volkert.engelsman@gmail.com

J. Kronenberg and E. T. Lammerts van Bueren (eds.), *On the Earth We Want to Live*, World Sustainability Series, https://doi.org/10.1007/978-3-031-98758-8_23

Fig. 1 Logo Eosta

Fig. 2 Logo robin food coalition

pollution of the atmosphere as the pollution of "Buddha's breath." He described the ecological and social crisis as a mirror in which we see the limitations of our own thinking, feeling and willing, which he was convinces was, "for lack of a better word, spiritual" (Gore, 1992, p. 12)—unusual language for a vice president of the United States (Figs. 1 and 2).

From the first UN Rio Earth Summit in 1992, from which the IPCC emerged, the Earth Charter and Green Cross also arose through a collaboration between Maurice Strong, Al Gore and Mikhail Gorbachev. That was, in fact, one of the birth moments of the sustainability movement. It was a movement that attracted great minds. Of course, the Club of Rome and figures like Gro Harlem Brundtland had been active for some time. Grandmasters of the sustainability movement appeared on the scene. A joint building of the culmination of a cosmopolitan human impulse for the Earth can be seen in these developments—a civilisation in which the thinking, feeling and willing of each individual human being can become subservient to humanity and the Earth. In the service of People & Planet, as we would now say. A transition that invites us to view everything we do in the economy in light of its moral consequences for humanity and the future of the Earth.

These impulses were behind the birth of Eosta in 1990. Al Gore's book Earth in the Balance was a direct echo of Walter Johannes Stein's book for Daniel Nicol Dunlop, which dealt with the same theme. Dunlop's thesis was that we are not going to regain order from the chaos of the global economy until we learn to understand the Earth as a living organism again (Dunlop and Stein 1937).

One of the current major thought-leaders of the sustainability movement is Swedish climate scientist Johan Rockström, known for the planetary boundaries models, which show that the economy must stay within the limits of the Earth's carrying capacity. British economist Kate Raworth (2017) adds with her 'doughnut economics' that the economy should not only stay within planetary boundaries but also within the boundaries of our social foundations (Raworth, 2017). The transition to an economy that takes responsibility for humanity and the Earth starts with a paradigmatic shift from reductionist to wholeness thinking, from abstract to living thinking.

Ecological and social development are each other's condition. As individuals, with every transaction, we participate in a global economy that fully connects us as humanity. The Earth is our common destiny. Our consciousness just does not pull that off yet. Simply put, the mission of the sustainability movement is: Do everything from awareness and responsibility for People & Planet and derive the Purpose of your humanity from that.

2 Triple Bottom Line—Three Layers of Intimacy

You could speak of three layers of intimacy. First, what is a meaningful footprint on the Earth (planet)? This is the most visible side of the sustainability movement. This involves the four natural kingdoms (mineral, plant, animal and human) in their greatest possible cosmic dimension. The second more intimate question is: How do we design our society to create freedom, equality, brotherhood/sisterhood (People)? The ultimate and most intimate third question is: How do we develop our thinking, feeling and willing into a humanity thinking, feeling and willing (Purpose)?

The core question of today's sustainability movement is: How are the Earth and social organism connected, how can they become one again? And how is that connected to the question of how we become human beings? In 'triple P bottom line' language: People, Planet, Profit, where Profit is only profit if People & Planet also 'win'. The current profit definition seems to be based on the opposite. Neo-liberalism seems to consider it normal that profit is also profit if it parasitises the Earth and society. This old profit definition is stubborn but increasingly under pressure from the realisation that in this way we are sawing away the chair legs under our own existence.

Climate change and a widening wealth gap threaten our existence and force us to reconsider our current systems and realites. This is the core question that has the potential to unite the global society. This is also the core of the Cosmopolitan School, which the anthroposophical movement has been trying to realise for 100 years. It is, as it were, the Studium Generale of humanity. For those who open their eyes, another school is also visible which seeks to solve these challenges with reductionist, technocratic solutions. For example, geoengineering to solve the climate problem, genetic engineering to make plants climate-resistant. Often such 'solutions' are offered by a handful of powerful multinationals, the dependence on whom then only increases, thus widening the wealth gap. If you look through your eyelashes, you see an almost apocalyptic divide looming between reductionist-and wholeness-thinkers—between the reductionist school and the Cosmopolitan world school on Earth. In this, anthroposophy constitutes and develops not so much "the wisdom of humankind" but the "becoming aware of our humanity".

3 Eosta—Where Ecology Meets Economy

For Eosta—Europe's multiple award-winning market leader in organically grown fruit—sustainability starts with transparency. We provide every product with a Nature & More QR code that gives access to the organic grower's unique story. That code also provides access to a Nature & More "sustainability flower", as we call it. In this way, we unlock the impact on People & Planet, providing insight into the consequences a supply chain has for people and nature. At the heart of this sustainability flower (see Fig. 3), we place the unique story of the farmer, surrounded by the three petals of freedom, equality and fraternity that represent the 'People' dimension. Then, there are four petals for soil fertility, water, biodiversity and climate representing the impact on 'planet'.

We derive meaning (Purpose) for everyone in the chain from what is meaningful for society (People) and the Earth (Planet). So we look outwards, at current developments in the world. From that, we develop the navigation for business

Fig. 3 Eosta's sustainability flower

development. This approach has kinship with the Arthurian movement that drew inspiration from the elements, from the way out, from the macro cosmos. Other Eastern spiritual currents sought that inspiration from the outset. The outside world is penetrated, so to speak, only through the stained-glass windows of a cathedral. The rather Western spiritual currents, on the other hand, were characterised by the absence of such interior spaces. The Western, Celtic movement experienced spirituality precisely in the 'inner' space of nature and cosmos. As far as we are concerned, learning to speak the language of the world is mainly an ability of Western spirituality. This also explains our connection with Dunlop. He stated that humankind will only find order in the chaos of the world economy again when we learn to understand the earth as a living organism (Dunlop quoted in (Stein, 1937). Steiner (1924) differentiates in his Karma lectures between the culture impulse from Central Europe or Eastern Europe versus the civilisation impulse from the West. When thinking of culture, one tends to think more of the arts. Civilisation involves society and the economy. Cultural and civilisational impulses converge in anthroposophy. The key question here is how an initiation impulse can become civilisational impulse. The new mysteries can be found at the train station.[1] In the temple legend, that impulse is represented by Solomon's master builder Hieram Abiff (Steiner, 1904–1914).

4 Threshold

Nevertheless, to work in the economy from a humanity (People) and Earth (Planet) perspective requires crossing some thresholds. The ecological and social crises in the world hold up a mirror to us in which we encounter the limitations of our own (small) thinking, feeling and willing.

We face a triple divide: the ecological one that separates us from nature, the social one that separates us from others, and the spiritual one that separates us from ourselves (Scharmer, 2007). That's where it gets personal! Prejudice, cynicism and fear hold us hostage. This must be overcome to develop the big thinking, feeling and willing that unlocks the future of people & planet. "I am me and my circumstances" as Spanish philosopher Jose Ortega Gasset (1983) practically summed it up. The key to transition also lies right away in those personal circumstances. Situational awareness is an art in and of itself. That implies being able to see that your own development challenges reveal themselves in your own unique circumstances, in the magic of the human encounter, in the people who cross your path. This is now expected of the economy that makes us (still) unconsciously part of Earth and human development.

[1] Interview Joseph Beuys. Die Mysterien finden im Hauptbahnhof statt. 03.06.1984. DER SPIEGEL 23/1984.

5 Dream, Dance, Deliver

With that, the transition towards sustainability becomes a personal transition, a Human Resources challenge. That is where leadership deepens into self-leadership. That is where it becomes consciousness soul development. In modern leadership language, you come across this in Peter Senge's (1999) The Fifth Disciplin, Otto Scharmer's (2007) Theory U, Joseph Jaworski's (1996) Synchronicity Theory, or Stephen Covey's (1989) 7 Habits of Highly Effective People. There are all kinds of forms in which this spiritual development path expresses itself.

Based on this, our HR chief developed a three-pronged vitality policy: Performance Vitality, Social Vitality and Mental Vitality. There is a sequence to this: if we want to leave a meaningful footprint (performance vitality), we must learn to co-create (social vitality). The prerequisite for co-creation is "being able to become who you are" (mental vitality). Later, we started calling this "Dream, Dance, Deliver", fully in line with our mission "Planet, People, Purpose."

If we wish to realise our sustainability goals (Deliver), we must unlock the magic of the human encounter, learning to apply the secret of the $1 + 1 = 3$ in our collaboration, internally and externally (Dance). This accelerates when you do so from a higher purpose that starts with self-development (Dream). In this sense, success is not self-development. It is the result of a 3D policy and of unlocking the magic of encounter through the path of a shared higher purpose. That also explains why, as a team, we were able to innovate quickly, move with market opportunities, co-create with big names like the Dalai Lama, Julia Roberts, the (then) Prince of Wales, Desmond Tutu etc. All these "guardian angels" have committed to the mission of our business. Not because they were so impressed by a simple business in Waddinxveen West but because they found themselves in the higher purpose for which we were fighting. That, of course, is a climate in which employees can develop their talents. In which they can become who they are by making a concerted effort, not only for financial, but also for "natural, social, and human capital."

It is often explained how important it is to have a dot on the horizon. The problem with that is that it is less of a dot but usually more of a vague cloud. Hence, we thought Dream was a good name. You work from an "emerging future" that you intuitively sense (Dream) and, while co-creating and within your own karmic horizon or "circle of influence," unlock action potential in an inspiring way (Dance), enabling you to create something that is greater than the sum of its parts, greater than you could ever have achieved on your own (Deliver). We always experienced a lot of support from the "serenity prayer": "Grant me the serenity to accept the things I cannot change [circle of concern], the courage to change the things I can [circle of influence], and the wisdom to know the difference."

Change always starts from a trend-setting minority, never from a following majority. That is why it is important to develop and follow your own purpose (Dream) and from there on form 'coalitions of the willing', together with the people who cross your path and inspire you (Dance). Then just get to work. What you have to do reveals itself in the doing. Intuition develops in practice (Deliver).

The market knows the work of Eosta, but of course a lot of subsidiaries and sister companies surround it, such as Soil & More, Vitalis Organic Seeds, Ecos del Agro Costa Rica, Red Sea Organics Israel, TOSC UK, Florganic, Endala South Africa and so on. This is purely the merit of a team with an unstoppable entrepreneurial ability to develop vision in action. You do not plan something like that, our companies emerge from observation, co-creation and the will to want to solve problems in the world.

6 Carbon Credits for Soil Fertility in South Africa

An example: In the Sundays River near Port Elizabeth in South Africa, we observed a growing problem with alien vegetation in river deltas. The rivers were simply growing thick, which was depleting irrigation water for local citrus crops. When we looked closely, it turned out to be the result of excessive fertiliser use. The crops grew but the soil became impoverished so fertiliser leached into the groundwater and stimulated excessive plant growth in the rivers, which sucked up the water. We got in touch with the World Bank, which had developed a 'Working for Water' programme for this purpose. Apparently, this problem occurred in several river deltas around the world, where there is usually a lot of agriculture. The programme made millions available to put people, equipped with heavy equipment, to work pruning back the excessive vegetation in the rivers and throwing it into large landfills, allowing the water to flow again. We then thought, that sounds like mopping up with the tap running, it could be more efficient. What if we use all that green waste to turn it into compost to replace artifical fertiliser. The World Bank provided the green waste, we processed it into fertile humus compost using Ehrenfried Pfeiffer's[2] composting methods. We did this together with the local citrus cooperative that represents almost 60% of South Africa's citrus crop. As a result, artificial fertilisers were replaced by compost, soils recovered, leaching into rivers stopped and farmers switched en masse to organic farming. Moreover, we managed to get carbon credits for compost. These were awarded on the basis of methane gas avoidance (no more landfills), nitrous oxide avoidance (fertiliser emission) and carbon fixation in the soil as a result of humus build-up. Farmers could now sell not only their crop with an organic-premium, but also their compost and carbon credits. En passant, we had thus developed a revenue model for not only products but also farmers' eco-system services.

Given the fast-growing demand in other parts of the world for this kind of solution, we decided to house it in a separate company we called Soil & More. Follow-up projects quickly emerged in Mexico, Costa Rica, India, Ghana, China, Argentina, Chile, and Egypt (Sekem).

[2] Ehrendfried Pfeiffer (1899–1961) was a pioneer of organic and biodynamic agriculture and closely associated with anthroposophy.

You do not plan something like that, it arises from an ideal, from the magic of the human encounter and the will to solve challenges in the world. That requires moving from leadership to self-leadership. How do you develop that? A few examples from our HR policy will follow.

7 Marvellous Mondays, Freaking Fridays & More

As is sometimes the case, I was working late in the office when I struck up a conversation with our cleaner. It turned out he was also an imam, and he talked at length about Al-Chidr, a mythical figure in Islam associated with holy Jihad. Not the outer, but the great inner Jihad, the battle we must wage within ourselves with the dragon of our little self on the way to our big self. I was impressed by how content and word related to the battle of St George, St George and the dragon!

Just before, I had a conversation with Rabbi Awraham Soetendorp, president of the European Jewish Council on a similar topic, but illuminated from the point of view of Kabbalah. You can feel it coming: the next Sustainable Thursday at our office was born. It was our custom every Thursday to invite inspiring speakers for a lunch meeting. A kind of studium generale for the entire staff team. The focus was always on humanistic themes. What a wonderful opportunity to now hear an imam and a rabbi on a universal self-development theme. The themes we covered ranged from philosophical, spiritual to near death experiences, karma and reincarnation, Sustainable Development Goals as well as Inner Development Goals.

Besides Sustainable Thursdays, we also did Marvelous Mondays. Every Monday, classical music sounded throughout the company. This stemmed from an agreement with students from four conservatoires who used our office for rehearsals. From chamber music to solo concerts and operas, everything passed by. Our team also included employees from production and storage, forklift drivers etc. They seemed to have no problem spontaneously clapping in the middle of beautiful passages. The great thing was that the conservatoire students were fine with that too. At one point, even Rachmaninov's Third Piano Concerto blared through the loudspeakers of the large storage hall. Music opens up universal worlds and can penetrate deeply, not infrequently to tears.

On Freaking Fridays, we shared our "fuck-ups." If you invite people to have the courage to take the initiative, that includes the space to be allowed to make mistakes. On Freaking Fridays, we exchanged what had gone wrong and why. Not in the form of accountability to a higher-ranking employee, but simply among ourselves, where everyone feels safe to be vulnerable and to learn from each other's mistakes in such a way that it gives people courage to keep trying again and again. Fully in line with Nelson Mandela, where the greatest glory in living lies not in never falling, but in rising every time we fall (Mandela, 1994).

We also had Terrible Tuesdays and Wild Wednesdays but that would go too far here. Most importantly, our 3D self-leadership policy is in sync with our 3P sustainability policy. The way in and the way out must be balanced. They are each other's prerequisite. Dream, Dance, Deliver is a self-leadership model that is about becoming who you are as well as co-creating and developing the courage to take initiative for a better world. Dream—Purpose, Dance—People, Deliver—Planet. This is the condition for overcoming the triple divide: the ecological one that separates us from nature, the social one that separates us from others, the spiritual one that separates us from ourselves.

Eosta is derived from Eos, resurrection, dawn, resurrection of the human being on the other side of the threshold. This is reflected in its foundation stone meditation/verse:

From the Earth,
By the Sun,
For Humankind.
So that Humankind
Becomes a Sun
For the Earth

The challenge is to consistently articulate that impulse in the language of current events. By doing so, you prevent anthroposophy from becoming solely a worldview. On the contrary, you want anthroposophy to resurrect again and again in the language of the spirit of the times. Dare to let go of everything that concerns anthroposophy, every moment. And find it again in the encounter with the other, but in their words or in their actions. Yes, and sometimes that is gambling. Sometimes it works, sometimes it does not. At first, you think that the other person doesn't understand you, but in reality you do not understand the other person. In the latter case, you often fail to speak their language.

8 Desert Club

An important humanities research forum for us was the Desert Club. A group of about 15 entrepreneurs including Götz Rehn—Alnatura, Peter Blom—Triodos bank, Thomas Jorberg—Gemeinschaftsbank, Sanjay Bansal—Ambootia, Peter Senge—OFF UK, Patrick Holden SFL UK, Ibrahim and Helmy Abouleish—Sekem, Ulrich Walter—Lebensbaum ed.

There, we developed our own views on, for example, sustainability policy, health and nutrition, the new economy and true cost accounting. Most but certainly not all of the above were or are self-proclaimed anthroposophists. What united us were the challenges posed by the zeitgeist. One aspect of anthroposophy is that you connect to the challenges in the world and articulate everything in cosmopolitan language. Anthroposophical jargon can then be a barrier. Anthroposophy is never meant to be a worldview. It is always meant to be a way of knowing which connects the spiritual in the world with the spiritual in the human being, where we have to make all the terms stand up every day in the language of the world.

In our office, we have given all the meeting rooms names of great people, such as/including Steiner, Dunlop, Pfeiffer and so on. We have always been careful with the word anthroposophy, not in order to conceal it but rather to prevent it from having a dogmatic effect. In the sense of: "yes that's fine that you're an anthroposophist, I'm a Catholic." We owe it to anthroposophy to find a language that does not exclude but connects to something generally human—not high above but embedded in everyday practice.

9 5M Sustainability Navigation

At Eosta and now at the Robin Food Coalition (an alliance of about 100 organic food SME (small and medium size entrpise) front-runners), we navigate the 5Ms to define Materiality, Measure, Manage, Market and Monetise sustainability impact. The M of Determining Materiality has been added recently, more on that in a moment.

When I stepped down from Eosta, the request came from a number of companies to also implement the 5M strategy with them. The questions came in various formulations from large multinationals but also from medium-sized organic companies. To keep it a bit workable, we decided to focus on the medium-sized organic and sustainability leader companies. Together with Paul Polman, former CEO at Unilever, Peter Blom, former CEO at Triodos Bank, Marjan Minnesma, CEO at Urgenda, Derk Loorbach of Erasmus University and a few more people, we then founded the Robin Food Coalition. A "coalition of the willing" that joins forces to accelerate the food transition and deliver prototypes for the new economy in which profit is only profit if people and nature also win. Talk about grand masters: Paul Polman is one of those. He spearheaded the Sustainable Development Goals, the first UN/humanitarian agenda for sustainable development.

We found ourselves confronted with the challenge of helping to transform the economy into a People-& Planet Proof Profit model. The language we use is that of the CSRD, the Corporate Sustainability Reporting Directive, which requires European companies to report on their impact on people & planet. If they do not comply, they will not receive an approvement from their accountant and thus no funding. Smaller companies do not yet have to comply, but they do get the sustainability reporting burden pushed backwards into the chain, which means this obligation still lands on their doorstep.

What we were concerned with was not just becoming CSRD-proof, but using CSRD to back up the impact benefits of organic food and farming with data and thereby demonstrate that sustainable products are not too expensive but unsustainable products are too cheap because the harm to people & planet are passed on to future generations. The CSRD consists of three dimensions: Enivironmental, Social and Governance. Governance also involves self-governance, stakeholder engagement, ownership to capital etc. With a little imagination, you can recognise the three Ps in there. Planet, People, Purpose. CSRD can be seen as a lot of bureaucratic compliance paperwork, but also as a potential Trojan Horse in which the new economy is driven into the old.

9.1 Materiality & Karma

CSRD starts with a compulsory Double Materiality Analysis. In other words, what impact do you have on People & Planet and what impact do changing People & Planet-related conditions such as climate change or migration have on your company. Basically, this is a sort of karma analysis. What substantial "material" impact do your actions have on the well-being of humanity and the planet. So, as a business, you no longer look away, but you want to hold all economic transactions against the light of karmic consequences for people and planet. You place your actions and transactions in the perspective of the other side of the threshold: the big Thinking, Feeling, Willing; perhaps a somewhat idealistic interpretation of "double materiality," but certainly not an unrealistic one.

This is not just an exercise you go through on your own on a rainy Tuesday afternoon. The CSRD provides concrete prerequisites, such as stakeholder consultation and documentation (European Commission, 2023). You look at your own life from the perspective of all peers and the Earth. You basically run your life through a self-induced purification process, or "kamaloka" as the Indian philosophers called it.

9.2 Measure

So much for the first M of determining the materiality of your impact. Next, the CSRD mandates measuring impact. That starts with a baseline measurement, a kind of karmic file recording. That is not so easy because how do you want to map thousands of impact data points? You cannot do that with primary data from your own observations, like 'dear producer please supply me with all your impact data for soil fertility, water use, biodiversity and climate'. What we are organising collectively is a combination of data bases, satellite data, remote sensing techniques and geo-imaging which, with the help of artificial intelligence, produces impact data. With that, we can determine down to a few metres the humus composition, water holding capacity, biodiversity, greenhouse gas emissions and carbon fixation in the soil. All this still says little about the regenerative capacity or vitality of a farm but it does map its footprint.

9.3 Managing

The CSRD prescribes that you set improvement targets based on measurement data. This is what the third M of Managing stands for. This is not just about setting targets but also about how to free sustainability from the isolation of a sustainability department and how to let it permeate the capillaries of the organisation: from purchasing, sales, operations, packaging policy, ERP systems, financial reporting, HR policy, etc.

9.4 Market(ing)

Sustainable policies are not without cost implications. When you focus not only on yield maximisation but also on preserving nature, ensuring social inclusion, and promoting health, your product inevitably becomes more expensive. Farmers then often ask: "How can I go green when I am in the red?" That is the key question behind our fourth M of sustainability Market(ing). In our coalition, we develop revenue models for ecosystem services and help organic companies cash in on their human, social and environmental impact benefits. We do so, for instance, with collective communication campaigns, carbon credits, biodiversity credits, true cost accounting and activating natural and social capital on the balance sheet.

9.5 Monetise

This is the bridge to the fifth and final M of Monetisation. True Cost Accounting is now mainstream and widely accepted. Banks increasingly apply sustainability criteria in their financing conditions. If you are not sustainable, you pay more interest and are required to keep a higher equity buffer to be able to absorb climate-related supply-chain disruptions, for instance. True Cost Accounting helps companies monetise the positive and negative impact.

For the record, nature has no price, it does not belong on the balance sheet, nor is it part of the economy, but of the commons. What you can monetise though, is what it would cost to restore the damage you cause to nature. The same goes for social value. What you can monetise is what it would cost if you paid everyone in the chain at least a living wage. That does belong in the economy as well as on the balance sheet. One of the pilots we are implementing in the coalition is about activating investment in eco-system services on the balance sheet. Now, every sustainability investment still weighs on the profit and loss account. In the future, investments in natural capital can be assetised on the balance sheet. This implies that sustainable and organic products will be cheaper for the consumer while farmers will still get their well-deserved premium for it, just not from the consumer but from capital markets. The above-mentioned pilot involves major parties such as banks, insurers, accountancies, UN departments and various organic companies providing 'proof of concept'. Obviously, it would be an absolute game-changer if this succeeded. The fifth M also provides the basis for new governance where 'people & planet' also act as co-owners and managers of the company. For this too, we are pioneering stewardship-owned governance models.

Fig. 4 Nature & More brand of Eosta with plastic free distribution method

In the past, we had to make do with an organic or Demeter certificate to get a premium price. With the CSRD, the opportunity arises to put the economy in front rather than behind the cart and cash in on the added social, ecological and health value of organic and biodynamic farms.

Complex? Sure, but change and leadership never comes from a following (and usually complaining) majority, but always from a trendsetting minority. "If the going gets tough, the tough get going." We have united that trendsetting minority in our Robin Food Coalition. In doing so, we do not hesitate to also unlock the intelligence of large mainstream parties for the smaller organic players, such as those from banks, UN departments, accounting firms, universities and NGOs.

We need prototypes for the new normal! Where there is a will, there is a detour. Dream, Dance, Deliver! (Fig. 4)

10 Conclusion

For more than 30 years, Eosta and its subsidiaries and sister companies (and recently also the Robin Food Coalition) have been providing prototypes for a new economy in which 'profit' is only 'profit' if 'people & planet' also win. The companies show that you can become market leader with this, even in an old economy where the polluter still gets away with a competitive advantage. For these companies, sustainability starts with radical transparency and taking responsibility for your impact on humankind and the environment. The 5M strategy of 'impact Materiality, Measuring, Managing, Marketing, Monetising' provides a navigation that is as simple as it is logical for both the 'future proofing' of companies and the systems transition needed to cope with the ecological and social crises. The 3D HR policy 'Dream, Dance, Deliver' provides the (self-) leadership conditions that make this possible.

References

Covey, S. R. (1989). *The 7 habits of highly effective people: Powerful lessons in personal change.* Free Press.

DER SPIEGEL. (1984, June 3). Interview with Joseph Beuys: Die Mysterien finden im Hauptbahnhof statt. *DER SPIEGEL, 23*(1984). https://www.spiegel.de/kultur/die-mysterien-finden-im-hauptbahnhof-statt-a-7e610ad1-0002-0001-0000-000013508033

European Commission. (2023). *Corporate sustainability reporting directive.* Retrieved December 12, 2024, from https://finance.ec.europa.eu/regulation-and-supervision/financial-services-legislation/implementing-and-delegated-acts/corporate-sustainability-reporting-directive_en

Gore, A. (1992). *Earth in the balance: Forging a new common purpose.* Houghton Mifflin.

Jaworski, J. (1996). *Synchronicity: The inner path of leadership.* Berrett-Koehler Publishers.

Mandela, N. (1994). *Long walk to freedom: The autobiography of Nelson Mandela.* Little, Brown and Company.

Ortega y Gasset, J. (1983). *Obras completas.* Alianza Editorial.

Scharmer, C. O. (2007). *Theory U: Leading from the future as it emerges.* Society for Organizational Learning.

Stein, W. J. (1937). The Earth as a basis of world economy. *The Present Age, 2*(7), 5.

Steiner, R. (1904–1914). *Die Tempellegende und die Goldene Legende als symbolischer Ausdruck vergangener und zukünftiger Entwickelungsgeheimnisse des Menschen.* Rudolf Steiner Verlag.

Steiner, R. (1924, August 21). *Esoterische Betrachtungen karmischer Zusammenhänge.* Rudolf Steiner Verlag.

Senge, P. M. (1999). *The fifth discipline: The art and practice of the learning organization.* Currency/Doubleday.

Banca Ética Latinoamericana (BELAT) and Fundación Dinero y Conciencia—Creating a New Economy: Ethical Finances in Latin America

Juan Bottero

Abstract

Our purpose is to create a new economy based on human dignity and care for the earth in Latin America. We aim to bring awareness to how we use our money when we consume, invest and donate, thus putting it at the service of sustainable development for the region. To achieve this we have 2 objectives: To promote the use of money as a tool for economic and social transformation and to create financial instruments for positive impact.

Features:

(a) Year of Founding	2016
(b) Number of Employees	40
(c) Location	Chile
(d) Active in which countries	Argentina, Chile, Brazil and Uruguay
(e) Sector	Finances

1 Vision and Purpose

The Foundation *Dinero y Conciencia* was established with the objective of promoting a new economic model based on the conscious use of money. Our mission is to address the challenges of our time, which include the individual, social and environmental crises. In Latin America, uncontrolled development has resulted in

J. Bottero (✉)
Fundación Dinero y Conciencia, Santiago de Chile, Chile
e-mail: jbottero@dineroyconciencia.org

J. Kronenberg and E. T. Lammerts van Bueren (eds.), *On the Earth We Want to Live*, World Sustainability Series, https://doi.org/10.1007/978-3-031-98758-8_24

the region becoming the most unequal in the world[1], with serious implications for environmental and social wellbeing.

BELAT
FUNDACIÓN
DINERO &
CONCIENCIA

Latin America is rich in culture and biodiversity. It has 31% of the freshwater reserves in the world[2], vast arable land and a shared language and history. We want to understand our region better so we can make it thriving and able to also serve the world needs and development.

Our work is guided by the principles of *social threefolding*, introduced by Rudolf Steiner in Europe between 1917 and 1922. This framework inspires our efforts to create an economy that serves the needs of humanity, the planet and the greater good. We are committed to fostering cultural renewal and promoting social inclusion. To achieve a harmonious society where individual potential can flourish, we must ensure an economy based on solidarity, a free culture, and an equal and democratic political system.

In 1919, at the conclusion of the First World War, Rudolf Steiner visited various cities and engaged with prominent figures in politics, culture, and economics. He delivered public lectures, emphasising the necessity for a new societal structure based on social threefolding to avoid a potential recurrence of the devastation witnessed. Despite the subsequent developments in Europe in the next decades, Rudolf Steiner's contributions were not in vain, they served as a source of inspiration for various anthroposophical initiatives worldwide, including the establishment of Waldorf education, biodynamic farming, and anthroposophical healthcare.

Social threefolding is not an applicable formula, but rather a set of guiding principles based on the anthroposophical understanding of the human being. It responds to the evolving needs and potential of the human being in our time and helps us to direct our work. The human being is understood as threefold: spirit, soul, and physical body; and the guiding principles are: freedom, equality, and solidarity. The central question, therefore, is: How can we co-create a healthy cultural, political, and economic life.

In order for individuals to unfold their talents and walk their life path, regardless of their circumstances, we need a free and pluralistic culture. This can enable one's

[1] It is the region of the world with the highest income inequality in the 2019 Human Development Report of the United Nations Development Programme (UNDP), released in December. The richest 10 percent accumulate almost 80 percent of the wealth. And the poorest 50 per cent, only 1 percent of the wealth in our region. SOURCE: CEPAL (ONU).

[2] Although Latin America has about 31 per cent of the world's drinking water sources, it could be one of the regions hardest hit in a climate change-induced crisis. Even so, it has high levels of water stress in cities and productive areas, limited access to safe drinking water, with 166 million people still without a safely managed service. SOURCE: CEPAL (ONU).

full potential to be realised and translated into social wellbeing. From a political standpoint, it is essential to uphold the principle of equality among all individuals, ensuring that no one is in a position of authority over another and that the rule of law is upheld. The primary role of politics is to ensure peace within society. The economy is the creation of value in products or services that serve to satisfy basic human needs. In summary, as biological beings living on Earth, we have needs that make us interdependent. It is therefore important to ensure that one's own work is aligned with the needs of others. A personal talent can be engaged in a way that benefits humanity, while allowing others to meet their own needs through the work that they do. Conversely, the broader society and community hold a responsibility to support and foster individual needs and development. As previously stated, the guiding principle of the economy is solidarity or mutual support.

We will have a healthy society when all three spheres—cultural, political and economic—are separated from each other and also when they are ruled by their guiding principal: freedom in the culture sphere, equality in political and legal sphere, and solidarity in economic sphere. On the contrary, when one sphere is imposed on another or guided by a different principle than the corresponding one, this will create individual crises, social crises and environmental crises.

Our vision and purpose are to contribute to the sustainable development of Latin America based on Rudolf Steiner's principles of social threefolding. Central to this mission is fostering awareness around the use of money, particularly in the context of investments, to align financial actions with these guiding principles.

2 What We Do

The foundation was established in response to the thought-provoking question: "Who does your money serve?" Money is a fundamental aspect of our daily lives, yet we often fail to make conscious decisions about how we use it. When a consumer purchases a product, they are effectively endorsing it and instructing the producer to create more items of a similar nature. It is important to recognise the influence that consumers can exert on producers when selecting products that have been manufactured in a sustainable manner. This line of thinking can also be applied to the financial sector. When I save or invest money, I should be aware of the impact my actions have on the world. From this conviction arose the two main objectives of the Foundation: (1) promoting a conscious use of money, and (2) financing sustainable enterprises in the region. The Foundation has been in charge of organising events, courses and generating strategic alliances to promote a new economy. Additionally, it holds the political rights of the financial group BELAT (Banca Ética Latinoamericana), which facilitates credit structuring for companies and connects them with investors seeking to give meaning to their money.

Fig. 1 August 2024 information Fundación Dinero y Conciencia

3 Ethical Banking in Latin America

The main characteristics are: (1) Real economy: it seeks to generate value creation and employment, (2) Explicit investment criteria: a 100% positive impact portfolio (every company financed by us is generating a good impact in the world), (3) Transparency: investors are informed about the destination of the money, (4) No economic incentives for workers, and (5) Governance: the identity of the project is protected by the Foundation, the equity investors only have economic rights, the political rights it is hold by the Foundation. To date, 1200 successful projects have been financed, mobilising more than US$130 million in Chile, Argentina, Uruguay, and Brazil. The money has been invested in three main areas: education and culture, social development, and the environment, each with its own main sectors. Some of the concrete results of these investments are (Fig. 1).

It is intended that the amounts of these investments will be distributed in a balanced way across these three areas to achieve a diversified impact portfolio (Fig. 2).

In this way, we put our work at the service of a new economy based on human dignity. Our impact is to add value to the production chain of companies that create a sustainable future. It is not about finding the most profitable sector, but about making a positive impact in all possible areas.

In this social art of connecting people, we pair investors with entrepreneurs who need credit, our tools to do so are: crowdlending platform, investment funds and direct contracts. This means that at the moment we are not a regulated bank working with savings (Fig. 3).

4 Our Founding Story

Inspired by Triodos Bank in Europe, in 2013 Joan Melé, then Deputy General Manager of Triodos Bank Spain and Vice President of the Triodos Foundation, was asked to travel to Argentina and Chile to share the case of this bank. This trip generated kinterest and commitment, and the Fundación Dinero y Conciencia was set up in Chile, with Joan Melé as president and Guillermo Scallan, who held a leadership position in the Avina Foundation, as vice-president.

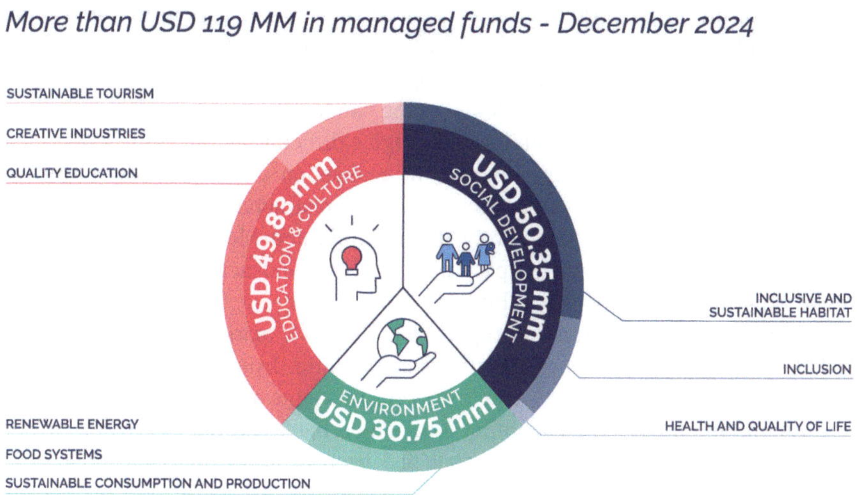

Fig. 2 Values as of August 2024, BELAT

Fig. 3 Funded projects supported by Fundación Dinero y Conciencia trade fair and an organic farm

In 2016, the first operation was carried out in Chile. In 2019, it was decided that it would be a unified Latin American project and in 2020, during the Covid pandemic, the teams from Argentina, Uruguay, and Brazil were formalised. In the following years, we faced the challenge of the pandemic and worked to accompany the companies that had taken out loans through a difficult period while, at the same time, supporting the first steps towards the realisation of new offices. Thanks to this support, the financed companies were able to restructure their loans and move forward. We also worked intensively in Colombia and Mexico with a volunteer promotion team tasked with preparing an analysis and action plan for opening future offices in these countries (Fig. 4).

In 2023, the first institutional report (SP, PT, EN), marked a milestone at the end of the first seven years of the organisation's life, from an idea to operations in four countries with over 30 employees. In the same year, it was also decided to postpone the process of becoming a regulated bank and the expansion of new offices, in favour of consolidating what had been achieved until the break-even point was reached.

Fig. 4 Our story from a social movement to financial group in 4 countries

5 Creating a Movement

As mentioned above, the Foundation promotes a new economy based on the conscious use of money, where individual ethics are put into play every time one buys, saves, invests or donates. This work is carried out through lectures, workshops, seminars and courses, both in person and online.

In 2020, in the context of the Covid pandemic confinement, we carried out online activities that allowed us to reach people all over the world. We held lecture series on anthroposophical foundations with guests from different countries, held various workshops on education, on agriculture and our connection to the Earth, as well as on responsible corporate ownership. In this work we collaborated with dozens of well-known organisations and individuals such as Gerald Häfner, head of the Social Science Section at the Goetheanum, Otto Scharmer, author of Theory U (MIT), Purpose Foundation Latin America and *Sistema B* Latin America, among others. These videos reached more than 100,000 views and generated a lot of interest. We also offered an online course called "Fundamentals for Social Renewal," taught by Joan Melé, which is held every year.

We also carry out face-to-face activities. By way of illustration, in 2022 we held two "tours" for a total of 21 weeks. We worked in Chile, Argentina, Uruguay, Brazil, Mexico, Colombia, and Peru. Highlights from the year include 71 face-to-face events attended by approximately 5,000 participants, 55 strategic meetings, and 7 press appearances. The last lecture of the year was held in Santiago, Chile, in collaboration with Helmy Abouleish, President of the International Federation of Demeter Biodynamics and CEO of SEKEM (Egypt).

On these journeys we come into contact with different type of groups: mainly anthroposophical, business, and academic. One aim is to bring the anthroposophical movement itself together, e.g. the Waldorf movement to collaborate with the biodynamic movement, etc., but it is also important to build a bridge to the world outside the anthroposophical movement itself. This dialogue, although it is a challenge to find the right language in which to convey anthroposophical knowledge, is indispensable and nurturing in both directions.

6 Anthroposophy's Contribution to Sustainability

Rudolf Steiner not only identified the "social question"—what we might call sustainable development in today's terms—as a subject for spiritual science, but it is precisely on the basis of this knowledge that the right approach to these questions can be taken. It is not a matter of looking at the social conditions that produce misery in the world and attributing to them a power that they do not have, since they are nothing more than the expression of an inner life. Instead, the circumstances in which human beings live are shaped by their own thoughts, feelings and actions, which holds the key to genuine social transformation.

The anthroposophical approach gives us an orientation to think about our connection to Earth, the cultural renewal of society, and the understanding of sustainability as the search for balance between polarities. It is a healing impulse based on self-awareness and a sense of responsibility towards the world and other human beings. Sustainability must be worked on as a search for coherence between what one thinks, feels, and does.

Inspired by the research of Bernard Lievegoed, Lex Bos, the Brazilians Jair Moggi and Daniel Burkhard wrote their book *The Transforming Spirit* (2007), in which we can see how a social organisation also has a quadri-structure, similar to the image of the human being: The social organisation has an identity, a purpose, values, and a biography (the "I"); then it is made up of people who come together in function of a common task, in this dimension there is the challenge of overcoming personal sympathies and antipathies and putting themselves at the service of the task (astral body); then there are work processes that can be more or less efficient, respond better or worse to a plan (vital body); and finally there is a physical-material expression, there is an infrastructure where the work is carried out, there are resources (physical body).

In this way, we can say that, in our case, the search for transformation towards sustainable development takes place at three levels:

1. Society: at the external level we aim to generate positive impact from the financing of projects in the three areas already mentioned: (1) Education and culture: the link of the human being with his most spiritual part; (2) Social development: the link between human beings; and (3) Environment: the link of the human being with the Earth. Before opening a new office, an in-depth study of the problems in the area is carried out, with the aim of ensuring that our work is an effective response to these diverse problems in a given context.
2. Organisation: At the level of the organisation itself, we have an area of work called Organisational Development, which promotes the continuous training of BELAT workers, renewing the connection with the purpose of the organisation itself. This work is done through dialogue tables, social dynamics, and artistic activities.
3. The individual: At the individual level we work with questions about the meaning of life and propose an anthroposophical view of the human being, their dignity and purpose on earth. This work is carried out both within the organisation and externally through lectures, seminars and courses.

7 Conclusions

Despite the efforts of global organisations, the climate crisis is progressing, and social tensions are increasing. In addition, new phenomena are emerging: the digitalisation of everyday life and artificial intelligence. In the coming years it is likely that major global challenges will get more complex and increase.

Overcoming materialistic reductionism and consumerism, which tends to deny the spiritual dimension of life, and striving for a common purpose out of dignity is key. Overcoming selfishness, which neglects community life and development, depends on each one of us, and for this it is fundamental to understand humanity in its evolutionary process and in constant development.

One of the vital leverage points is the recognition that we must not delegate our responsibilities to external entities. Instead, each individual should become the author of the transformation in their own life and environment, supported by the collective strength of the community. To know oneself, to be interested in another human beings, and to have a disposition to collaborate in the face of the needs of others are the three paths that we must follow to solve the global crisis that challenges us as humanity.

Perhaps the main task at this time is to create initiatives, projects, or organisations with ethical aims and coherent, healthy, and sustainable behaviour, which seek to have a positive contribution to their environment and support each other. These can be companies, hospitals, financial institutions, schools, etc., and their work can be translated into an education that teaches for the freedom and love, an agriculture that evolves the soil with life and produces food with vital forces, a therapy that heals the soul, an art that develops the spirit, and an economy based on solidarity in which the use of money is at the service of a larger sustainable development.

References

Burkhard, D., & Moggi, J. (2007). *El espíritu transformador* (128 p). Editorial Antroposófica.
Steiner, R. (2014). *El nuevo orden social* (160 p). Editorial Antroposófica.
Steiner, R. (2014). *Rethinking economics* (310 p). SteinerBooks Anthroposophic Press.
Steiner, R. (2012). *El aspecto interno de la cuestión social* (84 p). Editorial Antroposófica.

Freie Gemeinschaftsbank—Exploring Organic Principles in Banking

Max Ruhri

Features:

(a) Year of foundation	1984
(b) Employees	37
(c) Location	Basel, Switzerland
(d) Active in which countries	Switzerland
(e) Sector	Banking system

1 Introduction

Banks promote themselves with high interest rates and security for the investment funds of potential clients—and with low interest rates and quality advice for financing projects through bank loans. But what do we really know about these institutions? Very little. They present themselves as large, outwardly mirrored black boxes: no matter the angle, one only ever sees a polished image reflecting the clients' own interests. The box seems to return the gaze—but in a more flattering light. Its inner workings remain inaccessible. Confronted solely with one's own reflection, the only course of action is to optimise personal gain. In essence, it is an infrastructure of self-optimisation (Fig. 1).

A number of banks—particularly members of the Global Alliance for Banking on Values (GABV) (n.d.) have committed to breaking through this system

M. Ruhri (✉)
Director Freie Gemeinschaftsbank, Basel, Switzerland
e-mail: max.ruhri@gemeinschaftsbank.ch

J. Kronenberg and E. T. Lammerts van Bueren (eds.), *On the Earth We Want to Live*, World Sustainability Series, https://doi.org/10.1007/978-3-031-98758-8_25

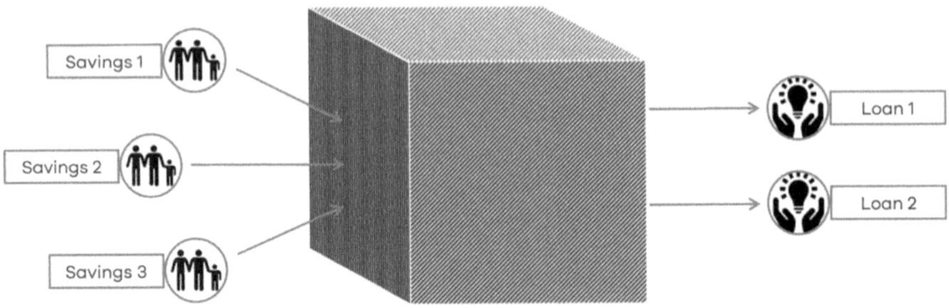

Fig. 1 Bank as a black box

of pure self-optimisation. These institutions prioritise social and environmental values, advocating for marginalised groups, environmental issues, and cultural initiatives—all with full transparency. The mirrors of the black box become transparent, allowing investors to see exactly how their money is being used, down to a detailed list of every loan recipient funded by the bank. In this way, the bank transforms into a transparent box.

Over the past 50 years, this segment of values-based banking has flourished globally, playing a significant role in ecological and economic developments. As an investor, one has the opportunity to place funds with such banks and thereby support their mission. This represents an infrastructure where personal resources are entrusted to values-driven decision-makers who carefully consider the broader consequences of financial investments for the greater good (Fig. 2).

Within the ranks of the GABV, there have been efforts to take things a step further by dissolving the very concept of the bank as a box. Instead, the box "unfolds" to open up into a financing ecosystem where investors can directly participate in project funding, thereby assuming responsibility themselves. This development is not solely about what a bank finances but also about how it operates as a bank. Such an approach holds extraordinary promise for sustainable economic development.

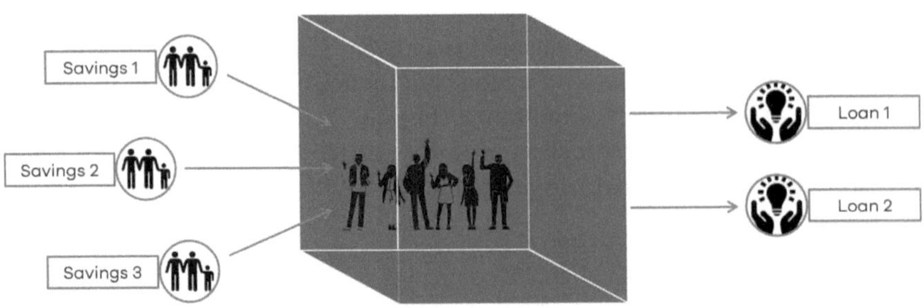

Fig. 2 Transparent box

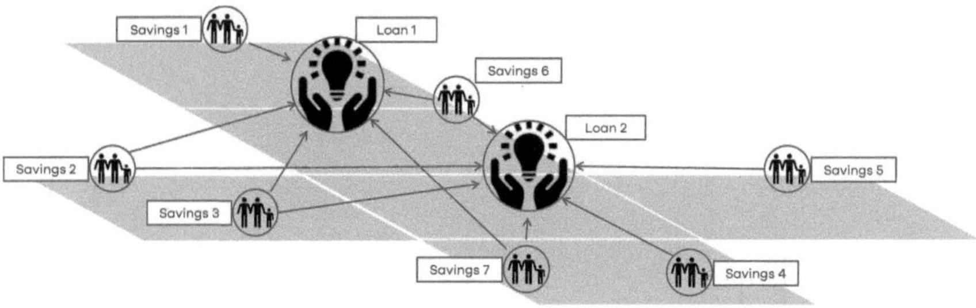

Fig. 3 Bank as an ecosystem

This essay offers initial theoretical reflections and empirical insights into this concept. A complete and comprehensive description remains a work in progress. The Freie Gemeinschaftsbank Genossenschaft in Basel (Switzerland) not only operates as a values-based bank but also positions itself as a laboratory for developing such an infrastructure (Fig. 3).

The first step presents concrete examples from the activities of the Freie Gemeinschaftsbank. We are aware that these are laboratory reports from a niche in which people have chosen a sustainable approach to money. The focus here is not on the ecological and social financing criteria that connect the Freie Gemeinschaftsbank with its clients, but rather on the bank's own methods of operation.

2 Examples from the Activities of Freie Gemeinschaftsbank

Trustee Loans—Clients Take Responsibility

Direct loans from clients to projects have held significant importance alongside traditional credit financing since the founding of the Freie Gemeinschaftsbank. These private loans were managed and administered through the bank in a fiduciary capacity, which is why they are referred to as trustee loans. The risk and the determination of terms were borne by the clients themselves. Essentially, the trustee business today operates in much the same way as it did then, though it has evolved technically and legally.

Nowadays, the bank and investors share the risk equally, supported by parity in securities and processes for interest and capital payments. Since 2022, an annual event has been held where projects are introduced to potential investors, allowing the people involved to connect and exchange ideas. Currently, the volume of trustee loans facilitated by the Freie Gemeinschaftsbank stands at CHF 32 million, representing approximately 10% of the bank's total credit volume.

The projects chosen for facilitation address significant societal needs, such as ecological agriculture, social institutions, sustainable enterprises, or housing initiatives. For each project to be financed, a maximum interest rate is agreed upon,

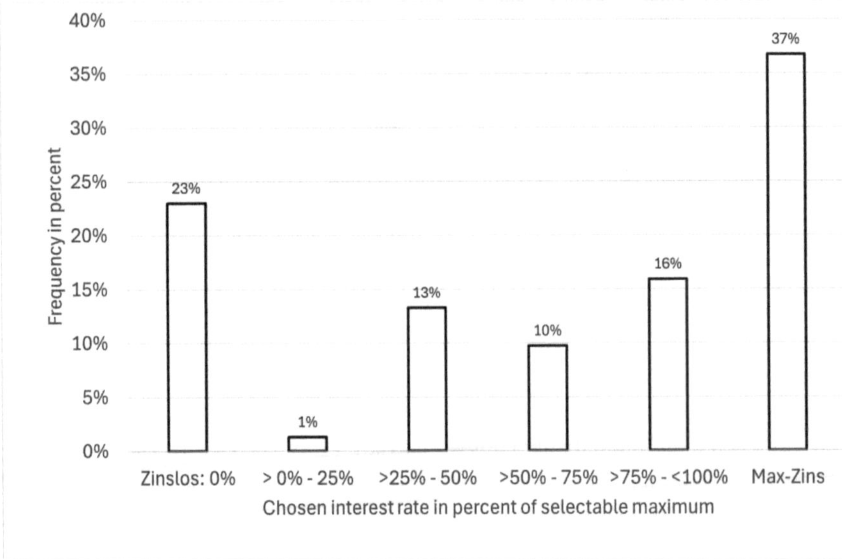

Fig. 4 Selected interest rate as a percentage of the maximum selectable rate

which, along with the bank's commission, roughly aligns with standard loan interest rates. This rate, combined with appropriate minimum terms and securities, is offered to potential investors, who then decide on their participation by specifying their desired interest rate and the amount they wish to contribute (Fig. 4).

There is a strong willingness among investors to forgo interest earnings, even though the bank maintains a neutral stance during consultations. On average, investors select an interest rate equivalent to 62% of the maximum rate offered. As shown in the diagram, 23% of investors forgo interest entirely, opting for a 0% rate. Around 37% choose the maximum interest rate, while 50% partially forgo earnings. This approach makes financing through trustee loans more cost-effective than conventional bank loans.

Defaults are extremely rare, limited to only a few isolated cases in the bank's 40-year history. However, the direct involvement of investors does carry a real risk. It is worth noting that during the financial crisis of 2007–2008, the trustee loan segment did not incur any losses, even with a trustee volume of nearly CHF 60 million at that time.

Free Choice of Account Fees

Following the significant drop in interest rates, which fell into negative territory after the 2007–2008 financial crisis, Swiss banks increasingly relied on fees to bridge revenue gaps—the Freie Gemeinschaftsbank was no exception. To emphasise both customer agency and individual responsibility, the bank undertook an extended development process involving interested cooperative members to introduce a fee system where customers could freely determine the amount they wished to pay—without the use of incentives. The underlying assumption was that responsibility entrusted to the bank's clients would be recognised and embraced by them. A significant source of guidance, alongside the bank's experience with its customers, came from Samuel Bowles of the Santa Fe Institute and his work on the function and impact of incentives (Bowles, 2016).

The new fee system was implemented on 1 July 2018. Prior to its introduction, all clients were informed that they could freely choose their fee, and a corresponding form was included with the information. Clients who did not respond were assigned a default fee of CHF 5 per month. Clients were provided with three guiding principles for determining their fee:

1. **Solidarity and financial capacity**—those with greater means were encouraged to pay more, while those with limited means could pay less.
2. **Bank service usage**—clients who required more services or consultations were encouraged to pay higher fees.
3. **Innovation and resources**—contributions would support the bank's innovation and development.

Ultimately, the decision regarding the fee amount was entirely left to the discretion of the clients. They can still chose a new fee whenever they want (Fig. 5).

By the final quarter of 2021, the service contribution level had stabilised at a relatively consistent rate: the average chosen contribution amounted to CHF 3.84. Approximately two-thirds of clients retained the standard amount of CHF 5. Around 2.5% opted for a higher contribution, with the maximum monthly amount reaching CHF 100, while about 19% selected a contribution of CHF 0 (see diagramme). The revenue target of the Freie Gemeinschaftsbank was met precisely and has remained stable since the introduction of the system. New clients particularly value the option to choose their fees, perceiving it as a gesture of trust and an opportunity to actively participate in shaping the bank's approach.

Stable Interest Rate Development

Following the 2007–2008 financial crisis, an analysis of the Freie Gemeinschaftsbank's interest rate development was conducted. The aim was to assess how independently the bank could operate from market trends. Since its inception, the

Fig. 5 Service contribution

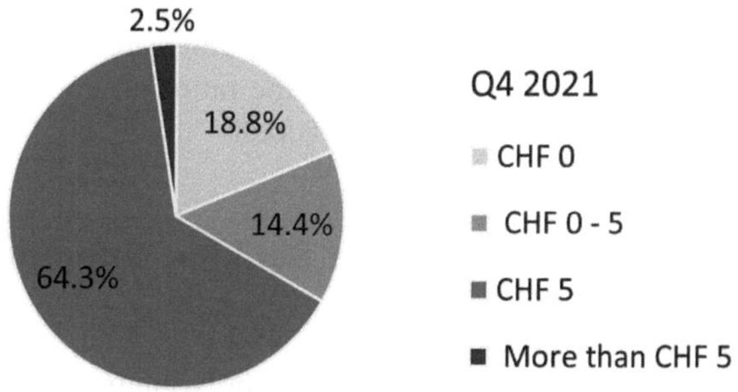

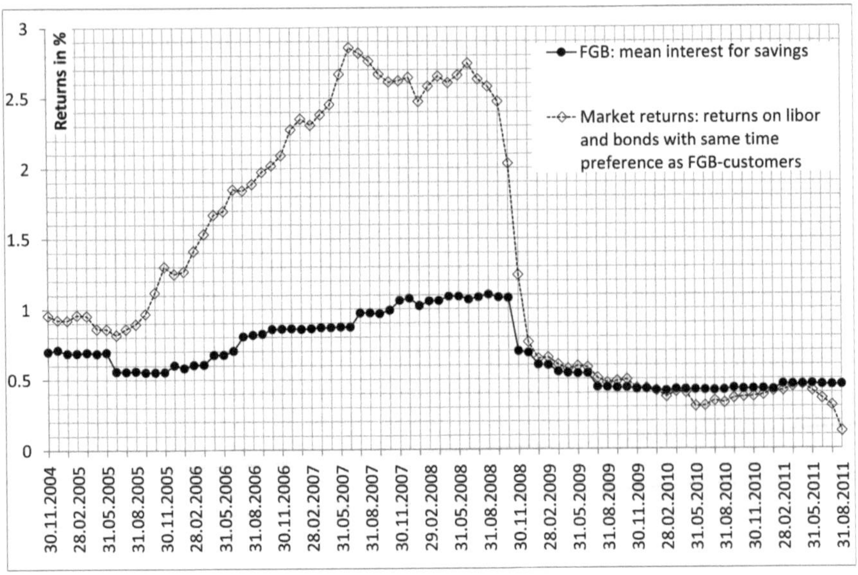

Fig. 6 Comparison of interest rate development between the market and the Freie Gemeinschaftsbank

Freie Gemeinschaftsbank's interest rate policy has been characterised by gradual adjustments and only moderate changes. When market interest rates shifted, the bank waited longer than others to adjust and adopted only modest changes.

In periods of high interest rates, the bank's rates were typically lower than market rates (see Fig. 6), while in low-interest phases, they were higher. The graphic illustrates market interest rates compared with the Freie Gemeinschaftsbank's rates between 2004 and 2011, weighted according to the composition of accounts and maturities chosen by clients for their investments. It is evident that the rise in market interest rates up to 2007 was only modestly reflected in the bank's rates, while the sharp decline in 2011 was initially disregarded.

This relationship between the market and the Freie Gemeinschaftsbank is typical and is expected to persist during future phases, such as the interest rate increase in 2022 and the expected decline in 2024. Corresponding evaluations are still pending. The bank's interest rate policy aims for the greatest possible stability, resulting in a "smoothed" interest rate curve. The bank's clients support this stabilising policy through their behaviour, as the rationale for maintaining consistent financing costs for loans is readily understood

3 Foundations of Banking Operations

Systemic Exploitation

The current economic culture comprises numerous characteristics, most of which we are scarcely aware of. I would like to highlight three aspects, as they help clarify the approach of the Freie Gemeinschaftsbank.

Capital Orientation: In most companies, particularly corporations, investors (e.g., shareholders) are the ultimate decision-makers through the general assembly. All corporate decisions must align with profit generation to retain investors. Voting rights are determined by the proportion of capital ownership.

Fragmentation: The self-interest of investors, and their focus on profit maximisation, is pursued through the "flexible combination of isolated, 'freely available' production factors that have been disentangled from relationships" (Exner, 2021, p. 287). Raw materials are viewed in isolation from their natural environments, leading to practices such as the development of monocultures and the burning of oil reserves. Communities that, for generations, lived off the yields of their own agricultural production lose these familial and local ties, moving into factories to join the industrial workforce.

This exploitation persists today because not only are resources themselves extracted from their contexts, but the value chain is also fragmented. Consumers can choose from countless products but see nothing of their social and natural origins. Thus, purchase decisions are influenced solely by taste and willingness to pay—the "black box" of consumer behaviour. Through this fragmentation, we, as consumers, participate in global exploitation.

Philanthropy: It is noteworthy that philanthropic care, in the form of gifting and charitable donations, emerged simultaneously with the self-interest-driven economic system. Through acts of care, this bourgeois morality seeks to remedy the social and ecological damage caused by the very profit-oriented economy championed by the same class (Exner, 2021, p. 252).

While the dominance of self-interest is not further explored here, the altruism, social integration, and value integrity found in various older and Indigenous cultures suggest that these qualities are intrinsic to human nature. Perhaps they were lost somewhere along the path to our modern economy (Malinowski, 1922/1979, p. 88).

How might human behaviour, banking, or the economy as a whole change if we reconnect with these lost or marginalised qualities?

Relationships Between People as the Basis of Banking Work

What allows the Freie Gemeinschaftsbank the freedom to explore new or unconventional approaches? Several factors combine to create this space for development.

Human Orientation: The cooperative legal form prioritises individuals over capital ownership. In the Freie Gemeinschaftsbank, each person has one vote at the general assembly (a "one person, one vote" principle), regardless of the size of their financial stake. Membership requires purchasing just one share, valued at CHF 300. Moreover, the bank pays no dividends on cooperative shares, and the shares are non-redeemable. Even the participation capital introduced in 2023 offers a modest and stable dividend of just 2.00%, adjusted to reflect prevailing interest rates. Since profits do not influence share value or dividends, the trading value of the shares remains equal to their original issuance price. The sole purpose of the bank's profits (aside from modest dividends on participation capital) is to build risk reserves for potential loan defaults. Profits are one among many operational conditions of the bank and do not hold a central role in its management. This freedom from the pressure of profit maximisation creates room for the development of a sustainable banking model.

Relationship Orientation: Freed from the demands of profit maximisation, the focus shifts to cultivating relationships. Through events, articles in the customer magazine, project excursions, and cooperative dialogues, the bank fosters connections and provides opportunities for participants to engage with one another. Such interactions foster recognition. While economic goods often incite competition (e.g., "if I eat the apple, you cannot"), recognition strengthens community awareness and provides a counterbalance to egoism.

The Gift in the Economic Process: Recognition and care for others must be reintegrated into economic processes. The concept of the gift represents acknowledgment within the economy rather than in charitable galas disconnected from value creation. As noted, "the gift functions as an operator for the recognition of the value of social groups and actors as such" (Caillé, 2022, p. 105). Gifts extend beyond human relationships to encompass care for nature, beginning with an awareness of natural systems and ecosystems.

The essential precondition for conscious recognition is visibility, which is largely absent in the fragmented capitalist economy. This absence eliminates any counterweight to egoism, leaving ecological and economic exploitation unchecked. The Freie Gemeinschaftsbank's experience with trustee loans shows that direct lending to initiatives is often accompanied by acts of giving, such as forgoing interest. Through relationships, people and their impulses become visible—otherwise, they would remain hidden. Transparency in this sense goes beyond mere traceability; it enables a deeper understanding of ecological sustainability. Making visible both environmental care and destruction could significantly influence consumer behaviour. In a transparent system, consumers and bank clients are empowered to take responsibility differently. Recognising human initiatives can inspire and guide them. Relationships and visibility are prerequisites for assuming responsibility.

The banking sector is one of the most highly regulated areas of the economy—and rightly so, as self-regulation in a fragmented world is nearly impossible. In today's culture, there is no reliable infrastructure to support responsible individual actions in banking and finance. The relationship-based banking model described here opens pathways for responsible action. Alongside external legal regulation, it promotes an inner, self-responsible regulation that can sustain itself. People invest in cooperative capital without expecting returns, forgo interest to support initiatives, or voluntarily pay account fees. Such behaviours stand in stark contrast to the pursuit of maximum returns.

The Bank as a Moderator

Banks are decision-makers. They determine where financing is allocated and where it is not—at least in the realm of external capital. This will remain true in many areas, alongside the role of creditworthiness assessments. However, if one takes the relationship-based approach to banking to its logical conclusion, a new role emerges: that of a moderator.

The role of moderation involves coordinating and facilitating a social framework in which projects are financed. Through this process, relationships and awareness of others are cultivated, extending far beyond mere lending relationships. Based on this responsive network (as opposed to the isolated individual relationships typical of conventional banking), purchasing decisions also begin to shift, and new donation dynamics emerge within the sphere of gift money (Stiftung Freie Gemeinschaftsbank, n.d.).

Outlook

Initial experiences suggest that risk behaviour within this financing structure differs fundamentally from that of conventional banking. In traditional banking, the concept of a risk premium is well established: higher risk, higher interest rates. This approach can exacerbate already challenging situations by making them even more difficult due to higher interest costs.

In the context of the Freie Gemeinschaftsbank's trustee financing, certain phenomena have emerged that demonstrate a different dynamic. Clients show a fundamental willingness to support otherwise healthy initiatives through reduced interest rates during difficult times, and conversely, to accept higher interest rates when circumstances are particularly favourable. While there is limited experience with this approach so far, it is plausible that this relationship-based financing system not only demonstrates inherent social and ecological sustainability but also exhibits self-stabilising tendencies. Challenging situations can be alleviated through interest reductions, while prosperous periods allow for higher interest rates.

4 Freie Gemeinschaftsbank and Anthroposophy

The Freie Gemeinschaftsbank was founded in 1984 from an anthroposophical impulse, with the support of the GLS Bank in Bochum. The preamble of its statutes states:

> The founding of the Freie Gemeinschaftsbank was inspired by the impulses of the spiritual science initiated by Rudolf Steiner. The bank is invigorated and renewed through the effort to implement anthroposophy in practice. (Freie Gemeinschaftsbank Genossenschaft, 2023, p. 3).

This preamble remains as relevant today as it was at the time of its founding. Unlike biodynamic agriculture, anthroposophical medicine, or Waldorf education, no bank was established during Steiner's lifetime. However, Steiner repeatedly expressed interest in founding a bank. For instance, during his address at the orientation meeting for the planned enterprise *Der Kommende Tag A.G.* on 11 March 1920, he referred to an earlier statement:

> far more than a theatre, I would prefer to establish a bank—not out of a love for money, truly not, but because I realised that it will inevitably become necessary to tackle the most practical aspects of life to address the needs of our time (Steiner, 1920/2020a, p. 228, translated by editor).

In addition to this and similar statements, there is an unpublished 1919 fragment titled *A Proposed Enterprise* (Steiner, 1919, 2020b). This document outlines the concept of an institution resembling a bank, many of whose key features align with those of today's banking model. More broadly, Steiner's social science works, particularly the *Nationalökonomischer Kurs* (National Economic Course) (Steiner, 2002) and *Kernpunkte der sozialen Frage* (The Fundamental Social Law) (Steiner, 1919/1976), offer important points of orientation for the work of the Freie Gemeinschaftsbank. This essay will explore three fundamental aspects central to the bank's approach.

Ethical Individualism and Diversity

In his book *Theosophy*, Rudolf Steiner states that "in spiritual terms, every human being is a species unto themselves" (Steiner, 1904/1990, p. 71). What unites this radically diverse view of humanity? In *The Philosophy of Freedom* (1894/1992)—developed by Steiner at the end of the nineteenth century—he introduced the concept of ethical individualism. Steiner posits the possibility of individual autonomy and authenticity in action: "A human being is free only to the extent that they are able, at every moment of their life, to follow themselves" (Steiner, 1992, p. 164).

This self-following is further described as: "when I follow my love for the object" (Steiner, 1992, p. 162). Acting out of love—though it may manifest differently in each person—does not, according to Steiner (1992, p. 166), lead to contradictions. Instead, this ethical individualism, which Steiner considers the highest level of moral action, does not result in arbitrariness or chaos but creates harmony among individuals.

Through the recognition described above, we perceive the spiritual dimension of the human being and experience the unifying force—a sense of enthusiasm for the perspective that others reveal from their spiritual essence.

The Social Question and Harmony in Diversity

Approximately 25 years after *The Philosophy of Freedom* and in the aftermath of the First World War, Rudolf Steiner initiated the political campaign for social threefolding. His book *The Threefold Social Order* (original German: *Die Kernpunkte der sozialen Frage*) was written in the context of this campaign. In this work, the ideals of the French Revolution—liberty, equality, and fraternity—are described as three central elements of social coexistence. Steiner viewed the cultivation (and, within the context of the campaign, the institutionalisation) of these elements as the foundation for bringing human diversity into harmony.

For Steiner, liberty applies to the individual and the sphere of cultural and spiritual life, granting individuals the opportunity to "follow themselves." In the social sphere, or the sphere of law, equality is paramount: relationships require equality and mutual respect. Yet, within relationships, the individuality of each person with their unique impulses emerges and is recognised. This recognition, in turn, fosters the cultivation of giving within economic life, enabling the element of "siblinghood" to manifest.

The principles central to a new approach to banking—interest in the cause, nurturing relationships, and recognising others, combined with the element of giving—already appear in the context of the social threefolding movement.

Organics

If we understand the economy as a complex system in which the interests of diverse individuals interact in either harmonious or destructive ways, we are immediately presented with the image of a living organism. This economic organism requires an organic approach. Organics, in this sense, encompasses many elements that differ fundamentally from the view of the economy as a machine. One such element is worth highlighting here: while non-organic conceptions of the economy often focus on creating incentives for good behaviour (such as reducing CO_2 emissions), an organic perspective seeks to understand and nurture the organism so that it produces what is good—ecologically and socially sustainable—through its own inherent forces. As Samuel Bowles' research demonstrates, incentives as an instrument are systematically unsustainable.

Despite technological progress, environmental strain and global inequality remain unresolved. This perhaps explains the dominance of dystopian visions of society in our time. To prevent negative consequences, dystopian systems rely on regulation through laws; without such regulation, ecological and social challenges cannot be managed. In contrast, utopian systems rely on positive ideas and impulses of individual people with their organic interactions to enable the sustainable good. Thus, "it is not about constructing the social organism through concepts but about understanding the conditions of life under which it forms and functions as a living entity, according to its own inherent forces." In this sense, a bank conceived organically positions itself as part of a living organism within the broader context of a similarly envisioned society and nature.

5 Conclusion

Over the course of economic development, interpersonal relationships and the connection between humans and nature along the value chain have almost entirely disappeared. This loss makes it nearly impossible to consider anything beyond individual needs in production and consumption decisions. The result is a culture of self-optimisation and exploitation of others, reflected in both economic theory and practice, with all its negative consequences for humans and the natural world.

Since its founding, the Freie Gemeinschaftsbank has been working on a banking model based on relationships between people and the visibility of their impulses. Integrating these dimensions of "visibility" and "relationship" into economic processes leads to a stabilisation of economic conditions through systemic regulation. This regulation is rooted in the recognition of people, their impulses, and nature. Through this recognition, an element re-emerges that has largely disappeared with the loss of relationship and visibility: the gift. This element transforms competition in the economy into collaboration and exploitation into cultivation.

By integrating relationships and the visibility of humans and nature into the economic process, and cultivating cooperation on this basis, the economy begins to appear as an organism. If it is to thrive, this organism must be cared for as something living and organic.

References

Bowles, S. (2016). The moral economy: Why good incentives are no substitute for good citizens. Yale University Press.

Caillé, A. (2022). Das Paradigma der Gabe. Transcript Verlag.

Exner, A. (2021). Ökonomien der Gabe. Mandelbaum Verlag.

Freie Gemeinschaftsbank Genossenschaft. (2023). Statuten der Freien Gemeinschaftsbank Genossenschaft.

Malinowski, B. (1979). *Argonauten des westlichen Pazifik* (Original work published 1922, p. 88). Syndikat Autoren- und Verlagsgesellschaft.

Steiner, R. (1990). Theosophie (GA 9). Rudolf Steiner Verlag. (Original work published 1904)

Steiner, R. (1976). *Die Kernpunkte der sozialen Frage* (GA 23). Rudolf Steiner Verlag. (Original work published 1919)

Steiner, R. (1992). *Die Philosophie der Freiheit* (Taschenbuchausgabe). Rudolf Steiner-Nachlassverwaltung. (Original work published 1894)

Steiner, R. (2002). *Nationalökonomischer Kurs* (GA 340). Rudolf Steiner Verlag. (Original work published 1922)

Steiner, R. (2020a). Ansprache an der Orientierungsversammlung über die geplante Unternehmensgründung Der Kommende Tag A. G. (1920). In *Zu sozialen und wirtschaftlichen Fragen der Gegenwart* (GA 332b). Rudolf Steiner Verlag.

Steiner, R. (2020b). Eine zu gründende Unternehmung (1919). In *Nachgelassene Abhandlungen und Fragmente 1879–1924*. Rudolf Steiner Verlag.

Freunde der Erziehungskunst Rudolf Steiners—Education is the Most Important Resource for Enabling a Future for Children

Nana Goebel

Features:

(a) Year of foundation	1971
(b) Employees	140
(c) Location	Germany
(d) Active in which countries	More than 90 countries worldwide
(e) Sector	Education

1 Foundation

In the mid-1970s, questions of sustainability, environmental protection (as it was then called), and respectful interaction with nature began to enter public consciousness, stirring many young people. Among other influences, this awakening was inspired by reading Rachel Carson's Silent Spring (1962). Naturally, for all mindful young people, the question arose: What do the looming crises of failure in relation to nature and the environment mean for one's own behaviour and sense of responsibility? What we knew with certainty was that we had to act. But in which field?

We committed ourselves to education, convinced that it was—and continues to be—the indispensable foundation for responding responsibly to the pressing challenges of our time. We believed then, and still believe now, that education underpins all human behaviour: how human beings interact with the Earth and

N. Goebel (✉)
Freunde der Erziehungskunst Rudolf Steiners, Friends of Waldorf Education, Berlin, Germany
e-mail: n.goebel@freunde-waldorf.de

J. Kronenberg and E. T. Lammerts van Bueren (eds.), *On the Earth We Want to Live*,
World Sustainability Series, https://doi.org/10.1007/978-3-031-98758-8_26

Fig. 1 Logo Freude der Erziehungskunst Rudolf Steiners

advocate for it, and how they engage with other people and society as a whole (Fig. 1).

However, it was not just any form of pedagogy that concerned us, but one that embraced spiritual dimensions. Already as young people in the 1970s, we were keenly aware that materialism, which had become increasingly pervasive and unfortunately influential in public education systems after World War II, was destined to lead to decline. Thus, what was required was an alternative approach to education—one that took the Earth as seriously as it did the cosmos.

While many of our generation and other like-minded contemporaries were deeply involved in the anti-nuclear movement or the peace movement against the deployment of intermediate-range missiles in Germany during the early 1980s—which we also considered profoundly important—our commitment was entirely focused on establishing the Friends of Waldorf Education (Freunde der Erziehungskunst Rudolf Steiners). This organisation was envisioned as a means to secure the long-term global presence of Waldorf pedagogy and to ensure that children and young people from all social backgrounds could access these educational institutions.

2 The Vision

Over the past fifty years, the Friends of Waldorf Education has been built and steadily expanded. The Waldorf pedagogical movement has since achieved global reach, with Waldorf educational institutions successfully establishing themselves in over 90 countries, particularly since the 2000s, offering a sought-after educational approach. During this time, more than 121 million euros have been raised primarily in Germany in order to support the global Waldorf movement, and to extend its impact internationally. Now, the question of vision arises anew.

During this period, we have managed to increase the number of Waldorf schools worldwide tenfold, to make the pedagogical principles of anthroposophical education known in the more remote regions (not only in Central European countries but also in places like Fiji, Senegal, Myanmar, and Guatemala), and to draw increasing attention to the question of what constitutes good education. And yet, we cannot be satisfied with this. In light of the pressing challenges of our time, as outlined, for example, by Jonathan Haidt in his book The Anxious Generation (2024), the question of vision demands a fresh perspective. Today, it can no longer be primarily about the quantitative expansion of meaningful, human-centred pedagogy—though this remains important—but must focus above all on qualitative deepening.

To be and, above all, to become fully human is undoubtedly more challenging today than in any preceding decade or century. Becoming human is a task for giants, and thus it becomes imperative to discover the giant within oneself—or at least the potential for such greatness. Supporting this endeavour is what we strive to achieve today.

The qualitative deepening of pedagogical work, in the face of the pervasive superficiality that is spreading everywhere, is one of the most critical tasks of our time. All the dimensions tied to this goal must be integrated with the ongoing quantitative dissemination of this pedagogy. Achieving this will continue to require mobilising financial resources from within society, bringing about change by investing in people—their intentions, capabilities, enthusiasm, and competence. In doing so, we hope that ever more people, in freedom and with a sense of responsibility, can utilise their lives to strengthen humanity, the Earth, and the cosmos.

3 Three Core Areas of the Friends of Waldorf Education

The Friends of Waldorf Education operate in three distinct fields:

For approximately 50 years, we have been supporting the Waldorf movement worldwide through financial aid made possible by identifying donors who share our conviction that fostering education is a vital and future-oriented initiative. The Friends of Waldorf Education are not a foundation and do not "have" money. We are an association that, primarily in Germany but increasingly in other European countries and the United States, finds individuals who share our goals—our "friends"—and are willing to contribute financially through various actions and campaigns.

To put it plainly, we acquire donations. However, this oversimplifies our work. At its core, our mission is about finding and connecting with people who want to become actively involved. It is not just about money. We seek individuals committed to promoting independent education because this model, driven by the engagement of parents, teachers, and students, holds greater potential and societal relevance.

Our vision—though it remains a dream—is that all schools in a country could one day be operated by those for whom the schools exist: parents, teachers, and students. But, as mentioned, this is an aspiration. Returning to the matter of donations: we aim to ensure that donors retain decision-making power over how their contributions are used. This principle has guided us since the beginning. Approximately 95% of donations we receive are earmarked for specific purposes, meaning we do not decide how funds are allocated. We act as intermediaries, forwarding the donations as specified. Only in cases of justified concerns about the quality of an institution would we investigate and, if necessary, withhold funds. However, this has never occurred.

In essence, the Friends of Waldorf Education function as a "donation bank"—a term that could be misleading and should be understood as an intermediary or facilitator. We serve as a bridge between donors and the institutions requiring financial support, such as kindergartens, schools, and special education facilities. One could also describe the Friends of Waldorf Education as a network of individuals deeply committed to the Waldorf pedagogical movement, who utilise our services to channel their support. In this sense, the success we have achieved thus far is entirely due to these dedicated friends.

For around 30 years, we have been supporting young people through a year-long programme offered after they complete their regular schooling. By organising voluntary service opportunities both within Germany and in countries with Waldorf educational, special education, or social therapy projects, we enable young people to gain work experience abroad. During this year, they learn to collaborate with people initially unfamiliar to them, acquire a new language, and adapt to a previously unknown environment. To date, approximately 40,000 young people have taken advantage of this opportunity, completing voluntary service either in Germany or internationally. A study on the benefits of such a year is currently underway and is expected to be available in due course. This area of work is largely subsidised by the state, making voluntary service accessible to young people from diverse social backgrounds. Such a year typically results in profound, often life-changing experiences, even though the challenges have significantly evolved in the wake of the COVID-19 pandemic.

A third area of focus is the educational support of children and adolescents in crisis situations, whether caused by natural disasters—such as earthquakes, volcanic eruptions, or floods—or by human actions, such as civil wars, wars, displacement, or impoverishment. The Friends of Waldorf Education organise educational initiatives in these crisis-affected regions, aiming to shield children from trauma and nurture the seeds of healthy self-confidence within them. This work is conducted on a global scale.

4 Anthroposophy and the Friends of Waldorf Education

The foundation of anthroposophical pedagogy—referred to by this name officially only in Israel—is a nuanced view of the human being or, to use a more contemporary term, an anthropology derived from an observation-based understanding of the human being. This perspective distinguishes between physical and psychological phenomena while also describing their interplay. It is precisely at this interplay that anthroposophical pedagogy focuses, as it varies depending on the age of the child or adolescent.

This pedagogy is therefore age-specific, employing diverse methodological and didactic tools tailored to the developmental stage of the individual. In this sense, the anthropology of Waldorf education—referred to as the anthroposophical view of the human being—plays a significant and central role in the work of the Friends of Waldorf Education. Most importantly, it must be emphasised that neither Rudolf Steiner nor any Waldorf educator considers the human "I"—whose existence is not merely postulated but recognised and experienced as a reality—to be an object of education. The human "I" is inviolable, just as human dignity is inviolable in society.

At the same time, Waldorf pedagogy is not teleological; it does not educate toward any goal external to the self-determination of the individual. Such intentions are obsolete and incompatible with the principles of this educational approach.

5 Dimensions of Social, Ecological, and Economic Sustainability

The foundations of sustainable development have been thoroughly developed and repeatedly discussed since the 1970s. They were formally agreed upon in the 2015 UN Sustainable Development Goals as benchmarks for human, societal, and industrial behaviour. Naturally, these foundations and goals also apply to us. In principle, we consistently reaffirm our full support for them and align our work accordingly. This commitment extends to the many seminars and training sessions conducted as part of the voluntary service programmes.

However, we sometimes face a conflict of action. To make meaningful decisions about funding allocations, it is imperative to visit kindergartens, schools, and special education facilities in other countries. Only a comprehensive understanding of these institutions—on pedagogical, legal, and economic levels—enables us to make informed funding decisions. Yet, undertaking these visits often requires air travel, which, from the perspective of environmental and climate concerns, should ideally be avoided.

For instance, when emergency educators plan assignments in Colombia, Ethiopia, or Indonesia, these missions are only feasible with the use of air travel. Similarly, a voluntary service placement in Latin America or Africa is almost impossible without it. Consequently, we must continuously deliberate—on all levels—where we are willing to act in ways that are harmful to the climate and consciously accept the consequences. I suspect that this dilemma is shared by all those who undertake responsibilities that extend beyond regional boundaries.

6 How Are the Dimensions of Social, Ecological, and Economic Sustainability Manifested?

Education is widely regarded as central to sustainable development, a point I believe is virtually uncontested. However, the individual dimensions of sustainability sometimes compete with one another, as has been frequently noted. In the work of the Friends of Waldorf Education, social sustainability holds a particularly prominent position.

For example, when a well-paid senior nurse at a hospital in Nairobi recounts that she grew up in a corrugated metal hut under the most impoverished conditions and attributes both the courage to pursue an education and the ability to succeed entirely to her time at the Steiner School in Mbagathi (Ongata Rongai), it is clear that the intended impact has been achieved. This is precisely what we, along with the sponsors who financed her education, hoped to accomplish: enabling a person to develop the skills and self-confidence necessary to break free from the cycle of poverty. Similarly, when an IT developer employed in Silicon Valley chooses to send his children to a nearby Waldorf school so they can grow up in a wholesome environment, it demonstrates how significant social and ecological sustainability are in influencing decisions about education.

Moreover, in many Waldorf schools—particularly those located outside major urban centres—school gardens are established and maintained by the students. This practice incorporates ecological values into daily school life. Increasingly, studies have shown that integrating green spaces into school grounds and cultivating school gardens are not sentimental traditions of overly romantic Waldorf educators but instead represent some of the best tools for fostering healthy physical and psychological development. These gardens serve as a form of anti-aggression training through collaborative work and as responsibility training through action.

The Friends of Waldorf Education have increasingly been approached for financial support for school construction projects, especially over the past two decades. This presents a wonderful opportunity to highlight the immense benefits of green and nature-oriented schoolyards and gardens, while encouraging colleagues worldwide to recognise not only educators but also architecture and schoolyard design as essential co-educators.

In the field of economic sustainability, there is significant room for improvement. Schools operating independently of the state receive government subsidies in only a few countries and are recognised as equal providers in even fewer. As is well known, this often means that such schools are funded exclusively through parental contributions and donations, making them less accessible than the Waldorf movement envisions.

The Friends of Waldorf Education are committed to ensuring that access to education is not solely determined by the economic circumstances of parents. This is the driving force behind our educational sponsorship programme. However, our resources remain limited, and high tuition fees are still a reality. Only a change in political frameworks could provide free access to education. Unfortunately, such changes are unlikely in the short term. We remain dedicated to this significant political challenge—working to establish conditions in the future where education and pedagogy are recognised as publicly funded responsibilities of those involved.

7 Putting Sustainable Development into Practice

The work of the Friends of Waldorf Education is based on anthroposophy, which lives as a motivation within the people involved. It is neither the goal nor the content of the respective tasks. For example, when I carry out a booking process, it is the bookkeeping rules established in German regulations that are decisive, not my personal convictions. This is essentially true for everything: anthroposophy exists internally as a driving force and orients a search process for addressing the most important questions.

This means that we do not aim to teach anthroposophy to the children in a Waldorf school or to the volunteers completing their service in a special education institution. They will only hear about it if they ask corresponding questions. In our environmentally and climate-conscious actions, we include the spiritual dimension of the world and strive to implement environmentally and climate-conscious behaviour as broadly as possible.

We aim to ensure that the dimensions of sustainability are understood and incorporated into individual actions. Only when understanding is followed by corresponding action can sustainability be achieved. It would be a success if, for some, anthroposophy could deepen their understanding, particularly of the relationship between Earth and cosmos, and strengthen their motivation to act accordingly. We assume that sustainable development is accessible to every healthy human understanding and can also be pursued without engaging with deeper layers.

8 Freedom and Siblinghood

These two ideals lie at the heart of our work. The Friends of Waldorf Education aim not only to support independent schools, as previously outlined, but to foster schools where a small person can grow so fully that they are equipped with a wide range of abilities, enabling them to realise their own intentions. The focus is not merely on talking about freedom but on laying the groundwork during the years of education to develop competencies that can serve as a foundation for a life guided by intention.

The challenge lies in cultivating these competencies within a classroom community, so that, alongside individual skills and abilities, those of all others can flourish as well. It is through valuing both the strengths and weaknesses of others that the profound opportunity emerges: to achieve more collectively than one could alone. Fellowship, therefore, is not merely abstract consideration but the discovery that the interplay of abilities can generate more than any single individual or the sum of individual capacities could accomplish.

This great potential can be experienced not only within the class community of a school but also in the work of volunteers. When the interplay of freedom and fellowship, in this sense, leads to an amplification of possibilities, it can be experienced as profoundly fulfilling.

9 Building Trust

In the work of the Friends of Waldorf Education, both successes and failures—quantitative, qualitative, and relational—are encountered. Both offer opportunities for learning.

For instance, when we successfully facilitate the construction of a school, as recently happened in a rural area of western Kenya near the Ugandan border, this represents a quantitative success. It means that the necessary resources were secured to provide local children with access to education—not in overcrowded classrooms, as is common in most public schools, nor in barracks-like buildings, but in an environment designed to be as beautiful as possible.

When a fourth-grade student at a small, makeshift school on the outskirts of Freetown, Sierra Leone, reverently whispers the morning verse of the Waldorf school into my ear, entrusting me with what she holds as her greatest inner treasure, this is a qualitative success. It demonstrates that the school has provided her with something meaningful that she cherishes and protects within herself.

When a faculty shares its concerns with us—and every faculty has concerns—this is a relational success. It shows that mutual trust has grown to the extent that there is no longer a need to maintain appearances in front of visitors. Trust has grown enough to allow genuine worries and challenges to be openly shared and addressed together. In this trust lies the opportunity to find a way forward. However, whether we can generate meaningful ideas in such moments remains, and will always remain, a challenge.

In fact, every personal and institutional relationship is inherently challenging if one seeks not to drift into repetition but to continuously develop and learn something new. Repetition is always tedious and ultimately serves only to preserve what already exists. A school—or any institution, for that matter—is only truly alive when it grows and evolves. Learning to embrace change is itself a challenge (Fig. 2).

10 Education for All

As long as injustices persist on this Earth—and it seems they will for the foreseeable future—there will be work for the Friends of Waldorf Education. We fully support the demand proclaimed by UNESCO as a motto in the year 2000: Education for all. In this spirit, we will continue to work in the coming years to ensure that more and more children gain access to high-quality, meaningful, and human-centred education.

Fig. 2 Class in Lutsk, Ukraine (Freude der Erziehungskunst)

We also fully endorse the demand that children must have the opportunity to grow up healthily. Achieving this requires radical measures. The dependency of young children on digital media is increasing everywhere. The magnetic pull of smartphones has been widely documented, as have their harmful effects. There is no shortage of knowledge about this issue. Waldorf kindergarten teachers and school educators were among the first to raise awareness about the consequences of digital media usage. For example, the Alliance for Childhood, initiated by Joan Almon in the USA in the late 1990s, is a significant reminder of these efforts. So, while there is no lack of knowledge, there may be a lack of understanding—or perhaps willingness to act.

Although some countries, such as Australia, France, and Sweden, have recently banned smartphones from schools to make learning possible again, much remains to be done. Much of this work is uncomfortable for the adults involved. Just as learning to embrace change is challenging, so too is learning to embrace discomfort.

Only if as many parents and teachers as possible learn to love this discomfort and enable children to grow up without digital media will there still be people in the future with healthy bodies and healthy minds who can contribute meaningfully to society. In the future, societal effectiveness for the benefit of humanity—and not its exploitation—will depend on who is willing to engage and who possesses the strength and creativity to do so. Education and upbringing are the key.

11 Education as the Foundation of the Future

Preparing for the future requires an understanding of essential values. Just as cultivating healthy soil is a prerequisite for growing healthy vegetables in the future, so too is the role of education and upbringing. If we wish to enable societal engagement for the benefit of humanity, Earth, and the cosmos in the future, we must plant the seeds for it today. This is especially true for education and upbringing.

Every mistake made in kindergarten and school has consequences—extending even into health and illness. Likewise, every healthy, constructive, and intellectually enriching experience will also have an impact. One of the most renowned educational scientists in the Netherlands considers memories of joyful childhood experiences to be among the most critical components in developing resilience. It is likely that for us, and for Waldorf schools, beyond imparting skills and sufficient knowledge, the primary focus is on fostering resilience. Those who possess resilience are better equipped to deal with life's events than those who are anxious or depressed.

Since the future always brings the unexpected—as both lived and historical experience teach us—we are shaping how the future will be approached through our actions today: whether it is met with curiosity or rejection, joy or fear. What is certain is that new things will emerge in the world, and we can choose to anticipate them with optimism if we wish. The most important question is whether we can translate what we understand and comprehend into our actions. The longest journey, as always, remains the one from the head to the hand.

12 Conclusion

For over 40 years, the Friends of Waldorf Education have been committed to fostering respectful and humane pedagogy as well as ensuring education for all children. Globally, the organisation is active in three key areas:

First, it supports the worldwide Waldorf movement by raising donations. This effort is not solely about financial support but also about finding individuals who are actively committed to the organisation's educational goals. Second, it empowers young people through voluntary service programmes, both domestically and internationally, which provide valuable, profound, and sometimes life-changing work and life experiences. Third, it offers educational support to children and adolescents in crisis regions affected by natural disasters or human-made conflicts. These efforts aim to alleviate trauma and strengthen the children's self-confidence.

Anthroposophy serves as the foundation for this work. Education and upbringing are understood as essential pillars of sustainable development.

References

Carson, R. (1962). *Silent Spring*. Houghton Mifflin.

Haidt, J. (2024). *The Anxious Generation: How the Great Rewiring of Childhood Is Causing an Epidemic of Mental Illness*. Penguin Press.

Gemeinschaftskrankenhaus Havelhöhe—A Climate-Friendly Healthcare Campus for the Future

Christian Grah, Maurizio Bär, Dirk Buchwald, Heike Breidenich, Christa Foppe, Jörg Probst, Friedemann Schad, and Roland Zerm

Features:

(a) Year of foundation	1995
(b) Employees	1000
(c) Location	Berlin, auf der Havelhöhe
(d) Active in which countries	Germany
(e) Sector	Health care

C. Grah (✉) · M. Bär · D. Buchwald · H. Breidenich · C. Foppe · J. Probst · F. Schad · R. Zerm
Pneumology and Lung Cancer Centre, Lead Climate Friendly Hospital Havelhöhe, Berlin, Germany
e-mail: christian.Grah@havelhoehe.de

M. Bär
e-mail: Maurizio.Baer@havelhoehe.de

D. Buchwald
e-mail: dbuchwald@havelhoehe.de

C. Foppe
e-mail: Christa.Foppe@havelhoehe.de

J. Probst
e-mail: Joerg.Probst@gertec.de

J. Kronenberg and E. T. Lammerts van Bueren (eds.), *On the Earth We Want to Live*, World Sustainability Series, https://doi.org/10.1007/978-3-031-98758-8_27

1 Initial Impulses in Havelhöhe

The Founding Impulses

The impulse to establish a self-managed hospital for Integrative Anthroposophical Medicine in Berlin arose in the period following the fall of the Berlin Wall, between 1989 and 1995. It originated from individuals who had spent their youth and early professional years in the divided city of Berlin and inspired by the atmosphere of new beginnings after reunification, sought to make a meaningful impact in this transformative environment. Among these founders were also people from what was then "West Germany."

A key inspiration came from the hospital in Witten/Herdecke, founded 25 years earlier, which served as a model in many respects. In particular, it provided a pivotal concept for the new hospital in Berlin/Havelhöhe: "community hospital" (Gemeinschaftskrankenhaus) (Fig. 1).

A third impulse also played a significant role, initiated by young doctors and, in some cases, even students. Their starting point was the question of how to foster a new consciousness in medical practice—one that considers the human being as interconnected with the living world, the animate planet and, ultimately, the broader solar system. This unusual perspective had already been introduced to medicine in 1924 through Rudolf Steiner's lectures, inspiring generations of young physicians up to the present day (Steiner, 1924).

In some respects, the same questions driving this group now resonate within the frameworks of Planetary Health (Whitmee et al., 2015) and One Health Medicine (UNICEF, 2008). However, guided by Steiner's lectures, these questions found a more inwardly focused starting point. During the founding period of Havelhöhe, broader Earth system science concepts, such as the "Planetary Boundaries" framework (Richardson et al., 2023), which have since become scientifically established, were still niche ideas—addressing the boundaries of Earth's biosphere was not yet mainstream.

One can therefore say that the founding impulse of Havelhöhe was shaped by three distinct elements: drawing inspiration from the fall of the Berlin Wall as a defining cultural moment of the twentieth century, embracing the idea of a healing

Havelhöhe

Climate Friendly Hospital

Fig. 1 Logo Gemeinschaftskrankenhaus Havelhöhe

therapeutic community and addressing the challenge of creating climate-resilient and climate-friendly medicine.

Healthcare Institutions in Transition—Places for Ailing Human Beings and a Diseased Planet Earth

The paradigm under which modern medicine was practised in the twentieth century was shaped by the narrative of an uncritical exploitation of nature. Nature was seen as existing solely to serve humanity. Whether in the form of highly purified phytotherapeutics, medical products derived from bacterial production (antibiotics), the cultivation of cell cultures, the use of oil, gas, rare earths, synthetic analogues, or newly engineered "antibodies," the question of whether plants, animals, elements, or even electrons "consent" to their use in human medicine was—and still is—never raised. After all, who would one ask?

The act of practising medicine for the benefit of humanity and fighting for every human life is an unassailable principle of human medicine. Implicit in this principle, however, is the idea that all other life—and indeed all matter—has no other purpose. Everything is there to serve human needs. The drive for further medical advancement remains unrestrained, aimed at overcoming disease. Disease is increasingly understood in terms of its biological functions and is treated as a "flaw" at the molecular level. This understanding is based on biochemical explanations of what constitutes "life," but the phenomenon of life itself remains unresolved (Scheidler, 2021).

This lack of understanding regarding life and our relationship with all other forms of life on this planet has contributed to the adoption of concepts like a life-hostile notion of growth in economics and politics. These ideas have led to the human-made crisis threatening the habitability of Earth. Albert Schweitzer, in the face of this incomprehensibility of life's phenomenon, called for an attitude he described as "reverence."

Given the threat to all higher forms of life, a new guiding principle could emerge among the "cultural creators" of society: declaring themselves "healers" or even "therapists" of the living planet (Latour, 2017; Richter & Bernd, 2024) (cf. Lovelock J., Margulis L. The Gaia Hypothesis, New York, 2007). They could adopt the habitability of the planet (cf. Latour B., Kampf um Gaia, 2022) as their shared maxim. Biological diversity and all planetary and biological conditions necessary for life could be designated as protected cultural zones. These zones would form the foundation for both the health of humanity and the safeguarding of life itself. And medicine? Human medicine would assume a dual role within this framework: providing care and treatment for the individual patient while simultaneously addressing the needs of humanity as inhabitants of the planet.

One might conclude that such a perspective necessitates a rejection of modern, technical medicine. Can modern technical medicine continue to exist at all if it imposes such burdens on the biosphere? At Havelhöhe, our self-conception explicitly rejects this dichotomy. We believe that the integration of environmental conservation and modern medicine is not only possible but essential. The objective of our ClimateFriendlyHospital project embodies precisely this vision.

Climate-Friendly Medicine

What concrete solutions can we propose when modern healthcare finds itself in a contradiction? On the one hand, it aims to strengthen self-regulation (seeing the human being as part of nature), yet on the other, it relies on modern technologies, materials and substances that harm Earth's living conditions. We see a way to resolve this tension between our being part of nature and the technically instrumental orientation of the medical system in cultivating a new awareness of cyclical processes, increasing the use of remedies that align with living nature for a large proportion of treatable conditions, and, above all, fostering our capacity for "moderate" action. The decisive moment of this century, with its irreversible timeframes for maintaining the planet's habitability, has led us to set our own goal: to transform ourselves and our hospital into a future-oriented healthcare institution.

Undoubtedly, this challenge can only be met with great effort. Why, then, should we associate this effort with a "friendly" relationship to the climate? A friendly approach to all other beings and things on Earth serves as a counter-image to the conquest and exploitation of nature and broadens the concept of adaptability and resilience. For this reason, we advocate the term "climate-friendly medicine" and propose extending the idea of "climate-resilient medicine" to include medicine's active role in cultural development. In this context, the healthcare system could prove to be a key driver of societal transformation and change (Grah, 2022). We aim to model resilience in the face of inevitable changes while fostering a new mood and attitude of "friendliness" toward our home planet and the Earth's biosphere. No less a figure than Aristotle, approximately 2360 years ago, articulated the qualities of a friendly relationship, which we believe can extend to the relationship between humanity and nature:

> The friendly man behaves rightly towards those with whom he associates. He pays them the respect due to them, and his actions aim at what is honourable or pleasant, so far as he can manage (Aristotle, 2009).

How the Climate-Friendly Hospital Goal Was (and is Being) Developed

As early as May 1997, just 16 months after the municipal hospital at the Havelhöhe site was transferred to the sponsorship of Gemeinschaftskrankenhaus Havelhöhe, an Ecology Working Group (Arbeitskreis Ökologie) was established. This group addressed various topics associated with environmentally conscious corporate management at the time.

In the summer of 2019, the year was marked by a powerful climate movement, driven by the climate strike calls from Fridays for Future, inspired by Swedish student Greta Thunberg, which brought tens of thousands of people to the streets for demonstrations. A local H4F group was quickly formed in Havelhöhe to build on the work of the Ecology Working Group. In 2020, the mission statement of Gemeinschaftskrankenhaus Havelhöhe was expanded to include the concept of sustainability. Consistent with its identity as a hospital community that considers the concerns of all staff, the organisation held a plenary conference to discuss and formally endorse the fundamental reorientation of the institution: transforming itself into a Climate-Friendly Hospital within ten years.

To facilitate this transformation, a steering group was established, comprising a climate manager, a managing director and two senior physicians. This group initiated key steps in the transformation process. The focus has consistently been on fostering awareness among all employees, aiming to serve as both a model and a reflection of a broader societal transformation. On the scientific and practical front, the project was further enriched through the establishment of a real-world laboratory (*Reallabor*). This initiative aims to leverage the expertise of young scientists and collaborations with universities to support the impending transformation.

Integrative Anthroposophical Medicine in the Era of Climate Change
Integrative medicine, which combines conventional medicine with complementary healing methods, has the potential to play a pivotal role in the global healthcare system, particularly in the transformation toward climate-friendly healthcare facilities. A distinctive form of integrative medicine is Anthroposophical Medicine, which regards the human being as a whole—body, soul, and spirit. It incorporates conventional scientific medical practices as well as movement-based, naturopathic, and artistic therapies, all with a focus on sustainability.

The World Health Organization (WHO), in its Strategy for Traditional Medicine 2014–2023, underscored the importance of integrating traditional and complementary approaches into the healthcare systems of all countries. This strategy aims for all WHO member states to develop integrative health systems that include both conventional and traditional healing methods. Such a vision aligns with the goal of promoting healthcare that meets the needs of populations while becoming more environmentally friendly and resource-efficient (WHO, 2013). In 2023, the WHO decided to continue this framework under the concept of "Traditional, Complementary and Integrative Medicine" (TCIM) as part of its global strategy for 2025–2034.

Anthroposophical Medicine has always seen itself as contributing to such a transformational process. It prioritises prevention, the strengthening of resilience and salutogenesis (self-healing capacities), and resource-efficient therapies. Its plant-based remedies, derived from biodynamic cultivation, and its artistic therapies—such as painting, music, or sculptural therapy—and mindfulness-based movement therapies, such as eurythmy, require fewer natural resources. At the same time, it emphasises an approach that strengthens individual responsibility, activates existing resources, and considers the social environment of the patient. This complements the predominantly medication-based approach of conventional medicine, which relies heavily on chemically produced pharmaceuticals.

The WHO Benchmarks for Training in Anthroposophic Medicine establish a framework to ensure that physicians worldwide are trained to a high standard in this holistic medical method (WHO, 2023).

Social Sustainability Through Self-governance and Integral Leadership
Our mission statement declares:

> Recognising patients as empowered partners requires that the employees of Gemeinschaftskrankenhaus Havelhöhe develop and maintain working practices in which they themselves collaborate as empowered partners. To this end, they are to be actively and responsibly involved in shaping the therapeutic community through participation in internal advisory and decision-making processes. To enable us to continuously and collectively improve our performance, we maintain open working practices that ensure the expertise of each professional group is mutually recognised and consistently considered, regardless of hierarchical position.

The hospital's departments are jointly managed across professional groups (physicians, nursing staff, therapists) because processes on individual wards often involve all these groups. This collaborative approach helps shift focus away from particular interests, prioritising high-quality patient care.

Approximately 70 leadership positions, including hospital management and administration, are assigned for a five-year term. Before renewal, leaders must report on their work and receive feedback from employees within their respective areas. Monthly plenary conferences involve all leaders in critical advisory and decision-making processes, such as the allocation of the annual investment budget. Transparency is a cornerstone of this approach, particularly regarding the salary structure. Salaries are not individually negotiated; instead, they are agreed upon with the works council and adhered to strictly. Over our 30 years of experience, this approach has proven to be highly trust-building and socially sustainable.

Our vision for the future draws on a developmental perspective that describes growth in personal and communal capabilities through stages, referred to as "first-tier" and "second-tier" consciousness. This progression aims for the advancement of social and awareness qualities toward an "integral" consciousness (cf. Beck, D. E., & Cowan, C. C. Spiral dynamics: Mastering values, leadership and change. John Wiley & Sons, 2014). Through these processes, we see ourselves on a path to making a self-directed and self-governed contribution to the coexistence of human beings, living creatures, and life-sustaining conditions on Earth (Steiner, 1918).

Economic Sustainability Through the Economy for the Common Good
As a non-profit organisation, we have embedded sustainability as a central corporate objective in our mission statement, alongside our core task of medical care. This is highlighted repeatedly and prominently throughout the text. For example, paragraph 1 in the the preamble states: "[…] All the necessary collective efforts for this purpose are guided by the idea of a free, developing, and self-determining personality and the principle of sustainability, ensuring that these developmental opportunities remain open to future generations in the spirit of intergenerational equity."

However, not only should one's own actions be sustainable, but there should also be "careful management of resources and efforts to protect the natural environment," which should be communicated to "patients as well as […] external partners in our social and economic environment," thereby extending these efforts into society. In this context, the GKH aims to "take a pioneering role in the healthcare sector."

It is clear to experts in health economics that achieving this goal within an increasingly economised healthcare system presents significant challenges. Germany's current Diagnosis-Related Group (DRG) reimbursement system provides no incentives for additional social-ethical or ecological efforts. Unlike the free market, free pricing is not permitted. Additional expenditures, for example, for sustainable procurement, must be offset by savings elsewhere. Nevertheless, we remain committed to addressing these challenges. At GKH, we firmly reject a healthcare system based on profit expectations, as it contradicts our social-ethical convictions. Instead, we focus on creating healthy and sustainable living and working conditions for patients and staff while promoting planetary health. To expand this approach to economic sustainability within the organisation, we achieved certification as a balancing company under the "Economy for the Common Good" standard in 2022 (Source: GWÖ).

The Economy for the Common Good (ECG) places emphasis not only on resource-conserving business practices but also on cooperative, socially ethical, and democratic engagement with patients, employees, and trade partners. Similar conceptual principles were articulated in Rudolf Steiner's "Threefold Social Order" (Soziale Dreigliederung) at the beginning of the last century. GKH is the first hospital ever to receive this certification.

Through the balancing process, we have made the hospital's sustainable corporate philosophy visible to employees and external stakeholders with an official label. Furthermore, it has helped us identify areas where additional efforts are needed to further improve sustainability in the future.

Ecological Sustainability Begins with Nutrition

For us at Havelhöhe, ecological sustainability means a forward-thinking and careful approach to natural resources. The healthcare sector has much to address in this regard, as its ecological footprint is significant. According to the Health Care Climate Footprint Report,[1] it accounts for more than 5% of global net emissions—equivalent to 2 gigatonnes of CO_2 per year. What can an individual hospital contribute? Here are two examples from our field of action in nutrition, which we have been actively working on since approximately 2015.

Our food system is responsible for around 30% of human-made greenhouse gas emissions, with the production of animal-based foods being the primary contributor to these negative impacts.[2] In 2017, we transformed our meal provision system to be more plant-based. Currently, 60% of our food is organic, with a daily budget of €5.40 per person for breakfast, lunch, and dinner.

[1] *Source:* https://global.noharm.org/media/4370/download?inline=1.

[2] Source: https://www.klimawandel-gesundheit.de/planetary-health/ernaehrung/.

Fig. 2 Havelhöhe next tot the Havel, from above

We made a clear decision regarding meat consumption: less, but of higher quality. Meat is still included on our menu daily but not at every meal—only at breakfast, lunch, or dinner, not all three. All our organic meat and sausage products are sourced exclusively from the Berlin/Brandenburg region. Other organic foods such as grains, bread, apples, and fish are also procured locally. From spring to autumn, the majority of our organic vegetables come from a community-supported agriculture initiative located just 4 km from our hospital (Fig. 2).

2 Development into a Climate-Friendly Healthcare Facility 2020–2030

Zero-Emission-Hospital

Hospitals worldwide account for 6.7% of CO_2 emissions. This creates a paradoxical situation: the CO_2 emissions primarily generated by energy consumption for heating, cooling, electrical systems and mobility significantly contribute to global climate change, thereby adversely affecting human health. At Gemeinschaftskrankenhaus Havelhöhe, we have decided to make a meaningful contribution in this area by becoming climate neutral within a few years.

The first step involves analysing energy flows and determining the necessary useful energy required for heating, cooling, lighting, electrical systems and devices. Since the founding of Gemeinschaftskrankenhaus Havelhöhe in 1995, and relative to the baseline year of 1990, we have already reduced CO_2 emissions from energy usage from over 6000 tonnes per year to less than 2000 tonnes per year (Source: Project Outline Zero-Emission Hospital Havelhöhe, Status 1/2021). Developing and implementing a comprehensive energy strategy for a campus like ours—a 17-hectare health campus consisting mainly of existing buildings and legacy technologies—is only feasible through a step-by-step breakdown of tasks:

- Massive expansion of photovoltaic systems
- Dynamic procurement of electricity and gas
- Establishment of an energy management system compliant with DIN 50001
- Model trial for cooling
- Installation of energy storage systems
- Import of renewable energy sources.

A representative example of the numerous technical challenges is the project for cooling existing buildings. As climate change progresses, the need for cooling in these structures is increasing. This demand will be addressed for the first time in a pilot project scheduled for 2025/2026. As part of the renovation, lightweight clay cooling ceiling panels will be installed. These panels, connected to a heat pump, will provide both heating and cooling. The key advantage of clay is its ability to regulate humidity, allowing patient rooms to be cooled without the need for a ventilation system.

Through these individual steps and trials, a comprehensive approach to energy use and management is being developed. At its core is the habitability (or "comfort") of patient rooms, application areas, and clinical spaces. Heating and cooling in these areas will be achieved through a modelled energy portfolio, integrating renewable, sustainable, and economically viable energy from both self-generation and external sources. This approach aims to combine economic efficiency with CO_2 neutrality, a goal expected to be fully realised within 10, and at most, 15 years.

Main Project UKAM (Environmentally and Climate-Friendly Pharmaceuticals and Medical Products)

Modern medicine relies on pharmaceuticals and medical products that are accessible worldwide. With globalisation, items such as antihypertensive drugs, antibiotics, cancer treatments and vaccines are widely available and used. However, the use of these pharmaceuticals has largely overlooked side effects on nature and the planet. Awareness of these impacts is only beginning to grow (Daughton & Ternes, 1999; Kümmerer, 2009; Fatta-Kassinos et al., 2016; Wilkinson et al., 2022). The production, use, and disposal of pharmaceuticals and medical products not only cause climate damage through energy consumption but also introduce pollutants into the biosphere, contributing to biodiversity loss and destabilising planetary boundary tipping points.

Environmentally and climate-friendly medicine expands its focus to a dual objective: caring for the individual patient while addressing the needs of the planet.

This dual responsibility must become central to future healthcare practices, without compromising the quality of care for individual patients.

For example, prior to the start of our project, few staff members were aware that diclofenac-based topical products (ointments) cause significant harm to wildlife, including birds such as raptors, storks, cranes, owls, and even fish. Several vulture species have already become extinct as a result, disappearing irreversibly from Earth's biosphere. Many were also unaware that psychopharmaceuticals lead to behavioural changes in fish, such as perch, and that traces of these drugs in the environment endanger the future of these species. Furthermore, it was not widely recognised that anaesthetic gases and the propellants used in inhalers for bronchial asthma cause substantial climate damage (Schiemann et al., 2024).

At Havelhöhe, we have launched projects to protect the planet by critically evaluating the use of medications and medical products. We aim to identify situations where harmful pharmaceuticals can be avoided without compromising efficacy. We also seek to highlight where overuse and misuse of medications can be addressed and where alternatives, including environmentally harmless drugs or no drugs at all, can be utilised effectively.

Projects for Climate Adaptation

In a world rapidly changing due to climate change, it is essential for every healthcare facility to prepare for future climatic challenges. Our vision is to systematically strengthen the resilience of the hospital to protect both patients and staff from the effects of climate change. Only a resilient healthcare system can ensure the population's health is effectively addressed, even in the face of future crises. To achieve this, we have initiated numerous activities and projects aimed at reducing our ecological footprint while simultaneously enhancing health outcomes:

1. Development, implementation and maintenance of a heat protection plan
2. Preparation for and regional risk assessment of severe weather events and the provision of care for affected individuals
3. Establishment of secure supply concepts to ensure resilient procurement strategies
4. Preparation for future health challenges, including building diagnostic and treatment competencies for diseases previously unknown in the region but emerging due to climate change

Our holistic approach addresses the individual needs of patients while also considering broader ecological and social contexts. In doing so, we aim to make the hospital a model for climate-resilient healthcare.

Where We Want to Be in 2030 and What We Envision for 2050

The ideal healthcare system of the future is comprehensively sustainable. It uses 100% renewable energy, implements a closed-loop system for waste management, and ensures fully transparent and clean supply chains for pharmaceuticals and medical products. This system fully embraces the concept of a climate-friendly hospital, operating not only emission-free but integrating ecological, social and economic sustainability at all levels.

Energy: Hospitals and healthcare facilities of the future are energy self-sufficient. They utilise renewable energy sources, such as solar, wind and geothermal systems, to meet their entire energy needs. Efficient building designs, including passive houses and green technologies, reduce energy consumption, while smart grids enable flexible management of energy supply and demand.

Waste Management: All materials used are either recyclable or biodegradable. A closed-loop waste management system ensures that no environmentally harmful waste is generated. Materials such as plastics, metals and glass are maintained within a circular system, where they are either recycled or used to produce new medical devices and products.

Supply Chains: The healthcare system of the future employs transparent and sustainable supply chains. Pharmaceuticals and medical products are produced with minimal environmental impact and under fair working conditions throughout the supply chain. Local production and decentralised structures reduce the need for transport and, in turn, CO_2 emissions. This aligns with the Lancet Countdown recommendations, which emphasise the relationship between climate change and global health and identify sustainable supply chains as a key element of a climate-friendly healthcare system.

Climate-Friendly Hospital: The climate-friendly hospital concept encompasses more than emission reductions. It also includes the use of environmentally friendly materials and the integration of nature into hospital operations. Green roofs, water-saving systems and ecological landscaping enhance patient well-being while reducing the hospital's ecological footprint.

Mobility, Nutrition and Presence: Staff adopt healthier and more sustainable lifestyles, including a focus on a planetary health diet, sustainable mobility and mindful living. Work schedules are designed to balance professional responsibilities with time for cultural, natural and creative activities, harmonised with healthcare duties. Car use is the exception; cycling and public transport are free for all employees. Travel for conferences on other continents is conducted online or by ship, with the time required counted as work hours. All processes—from patient and staff nutrition to hygiene and wastewater treatment—are climate-neutral, contributing to both human health and the biosphere's vitality.

Resilience and Social Justice: Such a healthcare system is not only ecological but also resilient. It is robust in the face of crises like pandemics or natural disasters because it relies on prevention and locally available resources. It guarantees equal access to medical care for all, regardless of social status. Social justice is essential for reducing health disparities and ensuring equitable healthcare worldwide.

Economic Sustainability: Economically, this system eliminates harmful incentives. It is founded on values-based financing that prioritises prevention and holistic health. Payment is not tied to the volume of treatments but to the long-term health of the population. This represents a paradigm shift, redefining health as a long-term objective rather than a commodity.

3 Conclusion

In summary, the healthcare system of the future is built on the principles of ecological sustainability, social justice and economic efficiency for the common good. It combines modern medical standards with a profound sense of responsibility for the planet and future generations, as called for by studies such as the Lancet Countdown on Health and Climate Change. The future of medicine lies in a system that heals without harming, fostering health and development for individuals while promoting a humane and free coexistence of all life and its co-creators.

References

Aristotle. (2009). *Nicomachean ethics* (W. D. Ross, Trans.; R. Kraut, Ed.). Oxford University Press.

Daughton, C. G., & Ternes, T. A. (1999). Pharmaceuticals and personal care products in the environment: Agents of subtle change? *Environmental Health Perspectives, 107*(Suppl 6), 907–938. https://doi.org/10.1289/ehp.99107s6907

Fatta-Kassinos, D., Dionysiou, D. D., & Kümmerer, K. (Eds.). (2016). *Advanced treatment technologies for urban wastewater reuse*. Springer.

Grah, C. (2022, July 11). Die große Transformation. *Deutsches Ärzteblatt, 119*(27–28).

Kümmerer, K. (2009). The presence of pharmaceuticals in the environment due to human use—Present knowledge and future challenges. *Journal of Environmental Management, 90*(8), 2354–2366. https://doi.org/10.1016/j.jenvman.2009.01.023

Latour, B. (2017). *Kampf um Gaia: Acht Vorträge über das neue Klimaregime*. Suhrkamp Verlag AG.

Richardson, K., Steffen, W., Lucht, W., Bendtsen, J., Cornell, S. E., Donges, J. F., Drüke, M., Fetzer, I., Bala, G., & Von Bloh, W. (2023). Earth beyond six of nine planetary boundaries. *Science Advances, 9*(37), eadh2458. https://doi.org/10.1126/sciadv.adh2458

Richter, H. U., & Bernd, R. (2024). *Demokratie und Revolution*. Kiepenheuer & Witsch.

Roquebert, C.-I. (2018). *La critique écologique radicale dans le contexte de l'entreprise : L'ambivalence des démarches de RSE dans les entreprises en Biodynamie* [Doctoral dissertation, Université de Nantes, Audencia Group].

Scheidler, F. (2021). *Der Stoff, aus dem wir sind: Warum wir Natur und Gesellschaft neu denken müssen*. Piper Verlag GmbH.

Schmiemann, G., Dörks, M., & Grah, C. (2024). S2k-Leitlinie Klimabewusste Verordnung von Inhalativa. *AWMF Online*. https://register.awmf.org/de/leitlinien/detail/053-059

Schmiemann, G., Dörks, M., Martin, E., & Grah, C. (2024b). Klimabewusste Verordnung von Inhalativa—Umsetzung in der hausärztlichen Praxis. *Zeitschrift Für Allgemeinmedizin, 3*, 1–6. https://doi.org/10.1007/s41996-024-00047-8

Steiner, R. (1918). *Die Philosophie der Freiheit*. Philosophisch-Anthroposophischer Verlag.

Steiner, R. (1924). *Meditative Betrachtungen und Anleitungen zur Vertiefung der Heilkunst: Vorträge für Ärzte und Studierende der Medizin*. Rudolf Steiner Verlag.

UNICEF. (2008). *Contributing to One World, One Health: A strategic framework for reducing risks of infectious diseases at the animal-human-ecosystems interface*. https://www.unicef.org/documents/contributing-one-world-one-health-strategic-framework-reducing-risks-infectious-diseases

Whitmee, S., Haines, A., Beyrer, C., Boltz, F., Capon, A. G., de Souza Dias, B. F., Ezeh, A., Frumkin, H., Gong, P., & Head, P. (2015). Safeguarding human health in the Anthropocene epoch: Report of The Rockefeller Foundation-Lancet Commission on planetary health. *The Lancet, 386*(10007), 1973–2028. https://doi.org/10.1016/S0140-6736(15)60901-1

Wilkinson, J. L., Boxall, A. B., Kolpin, D. W., Leung, K. M., Lai, R. W., Galbán-Malagón, C., Adell, A. D., Mondon, J., Metian, M., & Marchant, R. A. (2022). Pharmaceutical pollution of the world's rivers. *Proceedings of the National Academy of Sciences, 119*(8), Article e2113947119. https://doi.org/10.1073/pnas.2113947119

World Health Organization. (2023). *WHO benchmarks for training in anthroposophic medicine.* https://iris.who.int/handle/10665/366645

Grupo Cultiva

Guillermo Scallan

Features:

(a) Founding year	2000
(b) Employees	Over 300
(c) Location	Chile
(d) Active in which countries	Primarily Chile
(e) Sector	Environmental and social regeneration, education, ecological restoration, sustainable development, urban transformation

1 Grupo Cultiva

Grupo CULTIVA consists of a non-profit corporation established in the year 2000, a B Corporation[1] founded in 2010, and a nursery founded in 2012. The corporation holds a majority stake in the company and the nursery, which gives the Group its first distinguishing feature. Everything we do is guided by a vision centred on the well-being, care, and evolution of both human beings and the Earth, while also ensuring the economic soundness of the corporation and the financial profitability

[1] A B Corporation (B Corp) is a certification granted by B Lab to companies that meet high standards of social and environmental performance, accountability, and transparency. It reflects a commitment to balancing profit with purpose and is recognised in many countries worldwide.

G. Scallan (✉)
Director of Social Innovation at the AVINA Foundation, Ñuñoa, Chile
e-mail: Guillermo.Scallan@avina.net

J. Kronenberg and E. T. Lammerts van Bueren (eds.), *On the Earth We Want to Live*, World Sustainability Series, https://doi.org/10.1007/978-3-031-98758-8_28

CULTIVA

SEMBRANDO CONCIENCIA

Fig. 1 Logo Grupo Cultiva

of the company and the nursery, which allow us to sustain our positive impact over time (Fig. 1).

2 Corporation

The corporation was founded within a Waldorf school with the aim of bringing this pedagogy to all children and young people in the capital city of Chile. We set out to contribute to solving the city's primary environmental issue: pollution. This involved actively engaging with the understanding of the problem (thinking), the harm and pain it causes to the population and the ideal that motivates us to reverse it (feeling), and finally, the firm and demanding commitment to transform the problem through concrete actions (willing).

Our commitment and research led us to understand that a significant part of the issue stemmed from the deforestation of the foothills and hills within the city, which caused mud to flow into the streets of Santiago and later become suspended in the atmosphere as fine dust due to vehicular movement. From this, we also realised that trees could help us—they would enable us to prevent this process by regenerating the ecosystem (stabilising the soil, absorbing rainwater, and supporting the wildlife of flora and fauna). Furthermore, they would help capture particulate matter already floating in the city with their leaves.

The findings were undeniable: the foothills of Santiago alone lost 47,000 tonnes of soil annually due to rainfall, and coincidentally, 47,000 tonnes of dust floated

Fig. 2 Planting trees with Cultiva with Grupa Cultiva

in the city's atmosphere over the same period. ACTION WAS NEEDED, and with this, we sought to show ourselves—children, young people, and adults alike—that circumstances challenge us to act, that nothing is immutable, and that the human spirit can transform anything. It is worth noting that, up until then, being born in Santiago was equivalent to having to live in a polluted city (Fig. 2).

Dream, Act, Transform!

3 The First Ten years: Changing the City with Children and Young People

We planted native trees with students from various schools in Santiago, both public and private, creating a green belt along much of the foothills at an altitude of 1000 m, while also reforesting entire hills scattered throughout the city.

The young people turned this task into a guiding ideal in their lives, and soon, thousands of us were working together, transforming spaces and creating life. We surprised even ourselves when the country's most-watched television and radio programmes described what was happening as "one of the initiatives that mobilised the most students in Chile." In a very short time, more than 20,000 young people formed a community dedicated to bringing life to the surroundings of this beautiful city.

4 A Job Well Done and the Wisdom of Anthroposophy

The young people worked with enthusiasm, love, and hope, and their efforts began to make a tangible difference. While other reforestation projects in the area, supervised by the National Forest Corporation (CONAF), achieved survival rates of only 25–30%, CULTIVA's plantations boasted a remarkable 98% survival rate. This achievement soon caught the attention of the Ministry of the Environment and CONAF, prompting visits from officials to the plantations, as they struggled to understand the stark difference in species survival.

Undoubtedly, the love and care that the young people devoted to each tree played a significant role in this success. However, another factor was the unwavering dedication of the CULTIVA Corporation team and the application of biodynamic preparations in the plantations. The wisdom imparted by Steiner through biodynamic agriculture and anthroposophy has a concrete and direct impact on all measurable indicators, including the survival of the species.

5 The Creation of CULTIVA EMPRESA in 2010

Faced with evidence of the results, state authorities challenged us to create a company that could extend the impact of reforestation efforts beyond the educational role guaranteed by the Corporation. This company would enable the scaling up of reforestation projects to help offset greenhouse gas emissions generated by industries, as well as the dust and vegetation loss caused by real estate developments. In 2010, we established CULTIVA EMPRESA as the first company within the group, with CORPORACIÓN CULTIVA becoming its majority shareholder. The company's dividends fund a significant portion of the Corporation's programmes.

CULTIVA EMPRESA was founded as a B Corporation, focusing on ecological regeneration by designing and implementing reforestation and restoration programmes in alignment with the Native Forest Law and the Metropolitan Region's Decontamination Plan. A decade later, in 2020, its activities expanded to include green area maintenance and the measurement and offsetting of carbon footprints. Both organisations, Corporación Cultiva and Cultiva Empresa, share a common purpose: to make a positive contribution to society through actions that promote environmental care, education, collaborative work, and community strengthening.

Over the course of our 24-year history, Cultiva has developed projects in reforestation, urban transformation, environmental education, and educational gardens. In each initiative, we prioritise community participation, engaging students and volunteers while forging partnerships with public and private institutions such as schools, municipalities, government authorities, and companies. To date, we have collaborated with more than 35,000 students and volunteers, reforested over 700,000 native trees across 860 hectares, and led more than 150 projects.

Today, Grupo Cultiva has a team of over 300 employees dedicated to making a positive impact in the world through their daily work in social, environmental, and economic spheres. These individuals are united by their conviction that they can transform the world for the better, redefining the ways we act and interact with our environment.

6 New Business Units

We have a nursery for native trees and shrubs that has been producing plants for operations since 2012, and has recently begun selling in both domestic and international markets. We are also developing, inspired by geographical medicine, the production of medicinal herbs, with plans to launch a project in the fields of cosmetics and medicine.

7 The Dignity of the Human Being and the Dignity of the Earth

We aim to raise awareness, grounded in commitment and experience, of the interconnectedness between human beings, all other life forms, and the Earth. Our goal is to uphold both the dignity of humanity and the dignity of the Earth. We strive to create a world in which the well-being of human beings and the environment lies at the heart of development.

8 In the Face of so Much Despair, We Invite You to Change the World, for It is What Connects Us to Hope

For 24 years, we have worked alongside people to positively transform the world with nature through the regeneration of native forests, educational gardens, and the creation, maintenance, and improvement of public spaces. For over two decades, we have put our hands to the earth, reforesting and driving transformative action, so that children and young people can be at the forefront of change; so that companies can expand their vision of their impact on the planet and the society in which we live; and so that we all commit to the future. Twenty years of promoting socio-environmental regeneration.

We have collaborated with public, private, and community organisations in dozens of spaces, schools, gardens, plazas, urban parks, and degraded areas, fostering the common good in connection with nature.

9 While Climate Agendas Seek to Generate Paralysis and Anxiety, CULTIVA Aims to Foster Commitment and Hope on a Global Scale

In 2019, in response to the overwhelming sense of climate despair and the emergence of the diagnosis of "eco-anxiety," we, along with a group of allies, decided to lead the most significant climate action chain in history. Together with the Avina Foundation, we led the creation of the 6D IT'S NOW campaign, aiming to prevent humanity from being restricted by concepts like 15-min cities, movement limitations, or similar measures, and instead, encouraging us to commit our will to healing what we have harmed. We initiated this movement, which, as seen in the following chart, took place on a single day (December 6, 2019) across all five continents, with over 300 major milestones dedicated to healing the Earth.

10 Anthroposophy is at the Heart of Our Work

Anthroposophy, with its vision of the human being and the cosmos, has been central to Cultiva since its inception. In recent years, however, we have been working on the challenge of creating a new practical application of anthroposophy, one that integrates Waldorf education, agriculture, therapies, and emergency pedagogy to address the increasingly complex realities of modern cities, where the primary human challenges of our time converge. We have reflected deeply on the question: How can anthroposophy be brought to modern cities? How can we think of biodynamic agriculture beyond the traditional farm? What and how can we adapt Rudolf Steiner's agricultural principles to the complex context of large urban centres? What human actions can serve as therapeutic practices in cities? And how can Waldorf pedagogy be adapted, beyond the classroom, to transform urban relationships?

One might think that the first 25 years of Grupo Cultiva were focused on demonstrating the value of anthroposophy's various fields in the world. The next 25 years could be dedicated to developing a new "branch" of anthroposophy, one that allows us to tackle emerging human challenges. To achieve this, we will need considerable support—not only in terms of knowledge and skills but also financial resources. Yet, we are convinced that by delving deeply into these questions, we can uncover a new dimension to our organisation's work. The intuition is there; now we must cultivate the imagination and techniques necessary to bring it to life.

Biodynamic Agriculture becomes part of Chilean Culture

From its very beginnings, Biodynamic Agriculture has been present. This form of agriculture is the one that unites, the one that interweaves the different aspects and areas of Cultiva. It extends from the Earth to the Cosmos in a vertical sense, and widely covers the sphere of the social in a horizontal sense.

In 2019, as part of the IT'S NOW initiative and in cooperation with the Chilean Biodynamic Association (ABDChile AG), we applied biodynamic preparations to the reforestation efforts on Cerro Renca. We were even able to experience, in awe, a light summer rainfall after one of our applications in December—an unusual occurrence in Chile. Each gesture guided and reaffirmed the path we were taking. This endeavour involved donation and collaboration, yet it deeply resonated within our organisation. Through Biodynamics, we seek to practise agriculture that works with both soil and sky—an agriculture that contributes, endures over time, remains sustainable, regenerates, nourishes, and heals. We do so by fostering collaboration with all beings of nature—the plant world, animals, and human beings—ensuring that they all exist in harmony and develop in the best possible way.

Since 2021, we have taken another significant step towards Biodynamic Agriculture. We have intensified the application of biodynamic preparations and begun producing our own biodynamic compost at the Cultiva Nursery. Once again, in a collaborative effort, we invited Cultiva members to take part in this initiative. We started working along these lines to strengthen the nursery's crops, fostering plants that are resilient and wise from their very origins, capable of carrying these forces into plantations and growing into strong, resilient trees. We saw the effects firsthand—sickly plants recovering quickly upon contact with our compost, the surrounding environment revitalising, and the nursery becoming more vibrant. We continue working to sustain and expand these efforts. We have begun keeping records and have recently started conducting chromatography analyses in the hope of contributing our experiences to the biodynamic community in the future.

In keeping with this approach, we have continued deepening our collective understanding of the anthroposophical principles underlying Biodynamics. Today, part of the team leading different areas is undergoing formal training in this field. Our commitment to both Biodynamic Agriculture and the social dimension was further strengthened in 2023 when we became members of the Chilean Biodynamic Association (ABDChile AG). As a result, and in alignment with these principles, the Cultiva Nursery is now part of ABDChile's Participatory Guarantee System (PGS) certification programme, moving first towards organic certification and ultimately towards biodynamic certification.

Biodynamic Agriculture has the power to bring people together. Today, through it, we are forging subtle connections between the various physical locations where our plants grow and where we apply preparations. In doing so, we cultivate greater awareness among participating groups, for Biodynamics requires not only that we practise it but also that we unite and work together towards this shared ideal. This year, we are planning biodynamic preparation applications within a programme involving 14 municipalities across Greater Santiago. The initiative encourages citizen participation, including local residents and schools, and features workshops that allow children and young people to engage in activities for all beings of nature, expressing their gratitude in a tangible way.

11 Transitioning from a Trend to the Essence of Sustainable Development

Sustainable development has steadily gained prominence in our culture. From our perspective, this growth is linked to a heightened awareness that, in many cases, still lacks the depth necessary to bring about the profound change required. Such transformative change must be rooted in the evolution of the human being, moving from competition to collaboration, fostering communities of free and conscious individuals who harmonise with each other and their relationship with the Earth.

12 Three Transformative Projects Shaping Santiago Entrusted to the Group

Together with the government of Santiago, we are implementing the BROTAR project, which will add 35,000 new trees to the city, benefiting 500,000 people. These trees will not only enhance Santiago's green spaces but also help mitigate pollution and reduce the city's temperature. This initiative is complemented by the "pocket forests" project, also led by Cultiva, which is beginning to establish 13 forests within the city.

Two of Santiago's largest projects have been entrusted to an anthroposophical organisation, which has also been recognised with the Bicentennial Seal by the Government of Chile for its contribution to the values the country sought to promote during its bicentennial celebrations.

The third major project currently led by Cultiva is the creation of what will become the city's largest park, located in some of Santiago's most impoverished areas. Cultiva has spearheaded a collaborative effort involving local communities to plant 70,000 native trees and develop the necessary infrastructure to establish this park as a reality.

13 The Future that Wants to Emerge: Everything Must Be Sustained from Within

We are convinced that we can attune ourselves to perceive the future that wants to emerge in two very concrete ways: through young people, who bring new impulses from their recent presence in the spiritual world, and by overcoming fear, despair, and the paralyzing effects of uncertainty. We must allow what is ready to emerge to take shape, as fear and despair close the doors to this vital dimension necessary for all spiritual initiatives. For this reason, over the past 24 years, we have prioritised working with new generations, helping them overcome fear, taking action to strengthen ourselves, and fostering hope in tangible ways.

The new challenges of extreme temperatures, wildfires, and droughts demand a new inner strength in humanity, one that can sustain everything from within. "What we build today may not remain tomorrow, taken by the fire," as we have experienced with some of our plantations. Yet, we must remind ourselves: this forest may not provide ecosystem services to the current generation, but it has undoubtedly planted a seed for our next planetary incarnation. Through our actions, we construct the future, heal the soul, and heal the Earth.

14 Conclusion

Over the course of its 24-year journey, Grupo Cultiva has emerged as a pioneering organisation dedicated to addressing the interconnected challenges of social, environmental, and economic sustainability. Rooted in anthroposophy, its initiatives have combined education, ecological regeneration, and community engagement to create lasting impact. From reforesting native forests to transforming urban landscapes and fostering socio-environmental regeneration, Cultiva's work exemplifies the power of collaboration, creativity, and hope in overcoming global challenges.

Looking ahead, Cultiva remains committed to exploring new dimensions of anthroposophy, empowering younger generations, and sustaining its transformative vision. Through partnerships, innovative projects, and a steadfast focus on inner development, it continues to pave the way for a more harmonious relationship between humanity and the Earth.

Kongsi Co-op—Towards Individual Development and Community Wellbeing

Siu Hou Chong, Paw Lan Tai, Chen Yee Teh, Siao Hoong Sam, Ning Hii, Jia Yi Tan, M D Isa M. Saffiuddin, Jia Hao Lau, Chee Yuan Ng, and Chiou Woan Leng

Abstract

Koperasi Kongsi Selangor Berhad (Kongsi Co-op), a Malaysian consumer co-operative society, is dedicated to address social, environmental, and economic challenges by championing sustainable living and community empowerment. Guided by its core principles of inclusiveness and collaboration, the Co-op connects local producers and consumers through community-supported agriculture and community-supported education. Its efforts promote organic and sustainable practices, ethical business, and lifelong learning, all aimed at fostering both individual development and community wellbeing. The Co-op prioritizes organic community development and authentic relationships, overcoming financial and resource challenges by building a strong culture and resilient network. By expanding sustainable consumer choices ad raising environmental awareness, Kongsi Co-op is paving the way for more responsible, conscious communities and a better sustainable future.

Features:

(a) Founding year	2021
(b) Number of employees (as of August 2024)	8 and 116 members as co-owners
(c) Location	Kuala Lumpur, Malaysia
(d) Active in which countries	Malaysia
(e) Sector	Consumer Cooperative Society

S. H. Chong (✉) · P. L. Tai · C. Y. Teh · S. H. Sam · N. Hii · J. Y. Tan · M. D. I. M. Saffiuddin · J. H. Lau · C. Y. Ng · C. W. Leng
Koperasi Kongsi Selangor Berhad (Kongsi Co-op), Kuala Lumpur, Malaysia
e-mail: siuhou@kongsi.coop

J. Kronenberg and E. T. Lammerts van Bueren (eds.), *On the Earth We Want to Live*, World Sustainability Series, https://doi.org/10.1007/978-3-031-98758-8_29

1 Introduction

In today's fast-paced society, we are facing unprecedented challenges that have reshaped how individuals and communities are formed. The multifaceted challenges, that includes rapid technological advancements, geo-political instability, economic injustice, environmental degradation and materialism development of a society, have exacerbated existing challenges and created new ones in a rapid rate (Gyamfi et al., 2021; Martin et al., 2020; Qin et al., 2023; Söderholm, 2020; Zhang et al., 2022). However, despite the growing awareness of these challenges, there remains a significant gap in addressing the root causes of these matters (Henderson & Loreau, 2023; Shittu, 2020). While mainstream thinking has constantly pressed for these challenges (e.g., socio-economic and environmental issues) to be resolved on national policies or academic levels, these actions are deemed reactive, and they are impossible to resolve without tackling the underlying systemic problems (Camilleri, 2020; Fan & Fang, 2020; Shittu, 2020).

Kongsi Co-op, as a consumer cooperative society in Malaysia, saw grassroots movements as huge opportunities to approach these challenges together. The Co-op realised the contemporary socio-economic challenges unconsciously fostered economic inequality, environmental exploitation, and cultural implications. For example, the variation in economic development within Malaysia has created socio-economic development inequality (Sulaiman et al., 2021; Yusof & Kalirajan, 2021); the attempt for globalisation has encroached into and compromised local cultural practices for capitalistic gain (Ullah & Ming Yit Ho, 2021); the rights to food especially quality food is becoming less accessible to most Malaysians (Sulaiman et al., 2021). The Co-op is inspired to explore how Rudolf Steiner's social threefolding order namely, the interconnectedness of economy, rights, and culture can be implemented in the hopes of addressing the above-mentioned challenges. Through grassroot movement with holistic solutions that promote both social justice and ecological sustainability. Thus, the Co-op hopes to co-create a lasting sustainable model that strives towards individual development and community wellbeing.

2 Founding Story

Drawing inspiration from Rudolf Steiner's concept of the threefold social order which emphasises the inter-relationship of economic, cultural and social life, Kongsi Co-op (The Co-op) has since been exploring optimal models to meet the needs of its members (Lamb, 2010a; Steiner, 1999, 2003).

The initiative's roots trace back to 2012 when Dr. Joean Oon, a homeopath who introduced the International Postgraduate Medical Training (IPMT) to Malaysia, providing the local community with opportunities for holistic learning. As a result, the local community was inspired and eager to find ways for uplifting individual development and building community wellbeing. In 2013, some of current founding members had the opportunity to attend the Anthroposophic Business Workshop facilitated by Dr. Michaela Glöckler, during the IPMT programme held in Malaysia. In the 7-days workshop, Dr. Glöckler introduced Rudolf Steiner's concept of "Social Threefolding", a new imagination of the nature of economic and social life that takes into consideration the whole human being—body, soul and spirit. Besides, the ideal of "Solidarity" (fraternity) in the economical life of the social threefolding order mentioned by Dr. Glöckler has set a base working framework for the Co-op. This early endeavour laid the foundation for the principles that now guide Kongsi Co-op in deepening the connection between economic, social and ecological spheres.

It was a decade later that the idea of a community-owned enterprise sprouted. The Co-op was initiated in 2021 by a group of concerned individuals to address current social challenges, such as:

1. Socio-economic impacts between consumers and producers especially on the uneven distribution and access to quality food.
2. Relationship between consumption behaviour and environmental wellbeing.
3. The contribution of individual development towards community wellbeing.
4. Fostering connections between like-minded producers and consumers committed to supporting the local circular economy.

Thus, Kongsi Co-op was officially incorporated in March 2022, with forty-three founding members. As of August 2024, the Co-op has 116 members, who are called "Co-owners". It hopes to generate and reintroduce a sense of belonging and warmth into how people interact with each other, with our soil and nature again.

3 The Co-op's Vision, Mission, and Principles

As mentioned, Kongsi Co-op started humbly with a group of concerned individuals who wanted more organic and biodynamic products and services to be available to fill the gap in the Malaysian market. The Co-op has since been looking for an optimal operating model to fulfil the group's everyday needs. Furthermore, the Co-op is concerned with relationship and community building as much as with factors of production, consumption, labour and economy. The relationships mentioned should include not only our relations with fellow human beings, but also of those with nature. Besides that, we believe that a healthy economic model should be imbued with genuine social responsibility towards the community. Mainstream economic models are often detached from real-world economic phenomena where modern mainstream economists are still struggling in resolving the paradigm crisis especially after the global economic and financial crisis since 2018 (Xie, 2019). Thus, Kongsi Co-op hopes to reach out to wider producers and consumers who share the same values that are in favour of supporting local sustainable economic development. Upon establishing the Co-op, we have been striving to build a harmonious society of three-folding through our vision, missions and principles.

Vision

Kongsi Co-op aims to serve as a model co-operative society by fostering a profound connection between people, soil and nature, drawing on the significant impact of consumer choices on sustainable economies and communities. The relationship between individuals, the environment and local producers plays a critical role in sustainable practices. By consciously navigating consumer spending toward local producers, individuals contribute not only to the environment but also to the strengthening of community ties. This is done through various initiatives such as Kongsi Community Market which we described as an 'inconvenient market', Kongsi Food Bag programme, co-owners orientation programmes and the community-based business mentoring programmes.

Kongsi Co-op further advocates for conscious consumerism via Kongsi Academy, the education arm of the Co-op, with its vision—to cultivate a conscious community by nurturing a home gardener, caregiver, nutritionist, an educator, and an economist in every household. It is active in incubating local producers and tutors and offering anthroposophy-inspired courses.

Mission

As a community enterprise, the Co-op mission statement reflects the core purpose and vision in building our community. Our mission is to co-create, connect, collaborate and build a sustainable community.

Our commitment to co-create means to create a platform with the local community to support farmers and producers. The platform created by Kongsi Co-op could serve as community points to bridge the relationship between producers and consumers. Hence, its purpose is providing better understanding between these two parties to address actual social needs.

Upon the creation of the platform, the Co-op aims to connect people with resources. This can pair up different communities with different needs in optimising resources and building sustainable communities. We are committed to promoting practices that could benefit both communities and the environment through better resource utilisation.

Our focus lies on collaborating with producers and consumers in enlarging the sustainable food system in organic and biodynamic products and services. We prioritise fostering partnerships with local suppliers to grow organic and biodynamic food supplies, while creating new organic and biodynamic market opportunities through raising consumer awareness.

At the core, the Co-op wants to support communities to become culturally, environmentally and economically sustainable through education and community services (Fig. 1).

Seven Sustainable Living Principles

The Co-op is guided by seven principles to ensure ethical business and operation practices, foster community collaboration and promote transparency for stakeholders. These principles provide a framework that empowers the Co-op in

Fig. 1 The first Kongsi Co-op Annual General Meeting in 2023 with local organic farmers and artisans at Kongsi House, Kuala Lumpur

making decisions and collaborating with different stakeholders. As such, the principles reinforce the Co-op's sustainability and responsibility towards the broader community. The seven principles are as follows:

(1) *Earth-friendly, organic, and biodynamic food*

This is the most fundamental and core principle of the Co-op. It is also the main reason for establishing Kongsi Co-op. The Co-op supports farming methods that are not harmful to people, soil and nature. This practically commissioned the farming methods to enhance soil health, protect biodiversity and prevent the use of synthetic chemicals or fertilisers. Also, this principle focuses on advocating for food to be grown naturally and it integrates with the ecological balance to produce food sustainably in order to support both environment and community wellbeing. The principle also refrains from genetically modified produces and growing methods that are in contradiction with natural law. The Co-op strongly believes that it is food with life force that can truly uplift the development of the body, soul and spirit of an individual. The nutrition dynamic of food allows us to reconnect with our spirituality as put forth by Rudolf Steiner (Rudolf Steiner, 2008).

(2) *Products and services friendly to the people and environment*

Not only is the Co-op diligent about how food is farmed, but the Co-op is also careful about how products are processed, and how services are rendered. The Co-op prioritises sustainable practices to minimise environmental impacts and ethical business efforts for a better and healthier economy.

The Co-op works closely with its partners to ensure fair trade, respect for rights and sourcing of products and services that cause as little environmental impact as possible. This also comes in the form of services where the providers enjoy working on our projects and do not just do so to make profit, thereby contributing to a greater sustainability effort. We want to think that what is friendly should also adhere to the providers' intentions and principles. For us, contentment is also a key criterion in determining who we work with and why.

(3) *A green consumption model with minimal wastage of resources*

Within current consumption behaviour, consumers have taken for granted the conveniences of the readiness and availability of a product or service. Most of the time, consumers may not realise the implications for the convenience prepared for them. For example, single-use plastic bags and shopping bags are common in Malaysia's local market with local consumers heavily reliant on this packaging, especially during grocery purchasing. Excessive consumption has created an undesirable collection of shopping bags which mostly end up in landfills and choking waterways. Thus, this principle guided the Co-op to focus on dealing with unnecessary wastage and how we as consumers and producers could optimise the usage of current resources without creating additional waste by working with producers to forbid single-use packaging of any forms. Furthermore, we ensure that goods available in the Co-op are in bulk form with minimal or reusable packaging, and we encourage consumers to purchase according to their needs only, abstaining from excessive purchases and using containers already in possession. By avoiding unnecessary waste creation, we were able to create a green consumption model that benefits all parties, including nature.

(4) *Doing business without exploiting others and nature*

In the current business context, the pursuit of profit has become a customary culture to many businesses. Often, such profit-driven culture comes at the expense of labour rights, environmental rights and social justice. In our Co-op, we recognise the need to expel such exploitative business models and the need for businesses to change towards an ethical business model on every level along the supply stream. This principle anchors on promoting fair trading and transparency between consumers and producers as well as environmental stewardships.

(5) *Supporting local small-scale farmers and artisans*

Through knowing our suppliers, especially the local small-scale farmers and artisans, we can reinforce a trust-based economy. Supporting local farmers and artisans fosters a sense of community and warmth while providing high-quality products at the producers' capacity. Meanwhile, consumers can share production risks with the producers, such as natural risks in support of long-term producers' production, thus, fostering healthier and connected communities. This also promotes long-term sustainable development for the consumer-producer market while strengthening local socio-economic resilience as well as preserving cultural heritages by buying locally.

(6) *Providing diverse and holistic education in human development*

Genuine actions or behavioural changes can only come from the awakening of our will, and it takes holistic education for that to happen. Thus, we have incorporated activities, such as arts, gardening, parenting, and cooking classes, that encompass the thinking, willing, and feeling into all our learning contents, with the goal of bringing strength into individual development to make the suitable choice which we believe will always come back to the community—we are essentially wired to give altruistically.

(7) *Creating a sustainable community with associative economics*

By achieving the previous six principles, we foresee that the vision of a sustainable habitat can surface and prevail for generations to come. We are inspired by the thought that our work today will be valued by our descendants for up to six generations. This creates a sense of belonging and ownership in what we are doing.

4 What Does Kongsi Co-op Do?

We believe that true sustainability is about community, cooperation and mindful consumption. Therefore, we operate based on Rudolf Steiner's threefold social order and the Co-op's seven principles to achieve sustainability. We view the threefold social order of Culture, Rights and Economics through the strong arm of Community-Supported Agriculture, Community-Supported Education and the Co-op's governance.

Community-Supported Agriculture
The Community-Supported Agriculture (CSA) model exemplifies a successful integration of business and environmental stewardship (Bisht et al., 2020; Enthoven & Van den Broeck, 2021; Lamb, 2010c; Landwehr et al., 2021; Volz et al., 2016). The core of the CSA model is based on the two-prong approach of having a community to get fresh produce from known or neighbourhood farmers while farmers get good conditions and produce for the community they know (Volz et al., 2016). The Co-op adopted this model to connect consumers directly with local farmers through farm purchases.

As our focus is food-centric, we can explain it this way: By understanding the importance of life forces in food and sustainable farming practices, consumers can make informed choices. The CSA model also offers mutual benefits for both farmers and consumers.

The farmers gain financial stability from consumers' direct food subscription as farmers can receive upfront payment and therefore ensure cash flows throughout the growing periods. This also shares the production risks with consumers for better farming plans, thus lessening the burden on farmers during challenging seasons.

On the other hand, the consumers enjoy fresh and locally grown produce. Instead of relying solely on external recommendations, the consumers can actively support farming methods that ensure a just environment for farmers while also acquiring fair-priced food.

Tied to the CSA model, the Co-op also holds Kongsi Community Market (formerly known as Kongsi Recycle Day and Market) regularly. One of the main intentions is to provide direct marketing support to local organic and biodynamic producers and artisans. The market can connect directly with consumers while also encouraging environmentally friendly consumption patterns with minimal packaging, minimising wastage by shortening the supply stream.

By meeting each other's needs, we can establish pricing that is not solely determined by conventional supply and demand mechanisms. Instead, our pricing reflects the true dynamics of a just market. This model not only strengthens the local food supply and promotes sustainable agricultural practices but also fosters community ties. Besides that, consumers engage in more than just purchasing food; they invest in their health, the local economy and environmental sustainability. Participation in a CSA reduces carbon footprints by eliminating long-distance food transportation and supporting small-scale, local farmers, reflecting a conscious commitment to both ecological and community wellbeing.

Community-Supported Education

We believe that self-initiative is what motivates a person to carry forward a self-directed act, and that is cultivated through holistic education. For those of us who have passed the age for Waldorf education, our education pillar—the Kongsi Academy, offers an opportunity of continuous learning through the principles of Anthroposophy. Here, we navigate and develop a just and equitable culture that supports our rights, honours everyone's unique gifts, and brings new meaning to life in alignment with our time.

By understanding our innate roles, we can contribute more effectively to the community as a whole and meet each other's needs. Also, by equipping individuals with knowledge about their rights and responsibilities, education empowers them to make choices that benefit both themselves and the environment. An informed individual is more likely to engage with local initiatives, support community-driven efforts and contribute to a collective move toward a sustainable future. We envision this as the foundation for a healthy threefold community that fosters thinking, feeling, and willing. As a result, the Co-op has derived a community-supported education (CSE) model to support the community's learning journey. The CSE model is modified based on the CSA model by looking out for subjects that are essential to and enriching for adult learning as well as individual's personal development and community development, which have often been sidelined by mainstream education. We invite tutors and participants with similar intentions to curate and support these holistic learning topics so to keep the warmth going. We often ask the question "Does the subject or the topic serve our members, learners, and communities' inner and outer development?". We believe every individual has gifts to contribute to the community, and that we are constantly learning and can be cultivated through a healthy education that takes in all aspects of a whole individual's development. Therefore, a community supported education system emphasises holistic lifelong learning focusing on livelihood essentials instead of analytical prowess for mainstream working environment.

Co-op Governance

The threefold social order transpires into Kongsi Co-op's seven principles that guide our decisions regarding partnerships, product selection and business support. Our common goal is to promote a sustainable way of life that aligns with our people, planet and prosperity paradigm. Governed by Rochdale's seven cooperative principles, we are bound by the cooperative values of self-responsibility and caring for others, among others, as our core pillars. Thus, we believe that through the collective support of our members we achieve economic independence while achieving for-purpose initiatives in serving the needs of our members as well as fulfilling our duties to our community. Through self-sustainability, we can strengthen our community and build a more resilient future.

5 Contribution of Kongsi Co-op Towards Individual Development and Community Wellbeing

The philosophy of anthroposophy plays a foundational role in the Co-op initiatives and activities. It helps the Co-op create a supportive environment that fosters a more humane, sustainable and interconnected world. The Co-op supports sustainable practices in agriculture, education, nutrition, healthcare and economics—this includes promoting biodynamic agriculture, holistic education, associative economics practices, as well as anthroposophy-based nutrition and homecare. Through major activities like the Anthroposophic Conference Malaysia, regular essential skill workshops and Kongsi Community Markets, it serves as a platform for learning and sharing anthroposophical ideas. These initiatives aim to inspire participants to apply the principles in their personal and professional lives, contributing to individual and community wellbeing.

As a consumer co-operative in Malaysia, Kongsi Co-op emphasises the importance of community building by bringing together people from diverse backgrounds to create a supportive environment where mutual respect and understanding can flourish. By integrating diverse cultural and spiritual backgrounds, the Co-op implicitly promotes the anthroposophic ideal of understanding and respecting individual differences, practised through its day-to-day affairs. The social threefolding model as introduced by Rudolf Steiner aims to create a more equitable and sustainable society by ensuring that economic activities are aligned with social and environmental wellbeing.

Social Threefolding of Kongsi Co-op

These are the two verses by Rudolf Steiner which the working team will cite during the start and end of the team meetings or before carrying out any working tasks:

> May our feelings penetrate to the centre of our hearts
>
> And seek in love to unite with those who share our goals
>
> And with the Spirits who look down benevolently
>
> On our hearts' earnest strivings
>
> Sending us strength from realms of light
>
> As they illuminate our love.
>
> —Rudolf Steiner, *Verse for America*

***

> A healthy social life is found only when,
>
> In the mirror of each soul,
>
> The whole community finds its reflection,
>
> And when, in the whole community,
>
> The virtue of each one is living.
>
> —Rudolf Steiner, *The Motto of Social Ethics*

Among many references through our journey with Kongsi Co-op, we draw inspiration from the book "Associative Economics—Spiritual Activity for the Common Good" by Gary Lamb which we encountered in 2021. We had the privilege of inviting the author for two online exchanges and discussions on how to apply associative economics. Associative economics is concerned with and depends on building and sustaining healthy relationships. It is the wish of the founding members of the Co-op to eventually be able to implement the concept of "Associative Relations Audit as a Means of Transformation" as stipulated in Chap. 16 of this book. This audit focuses not on dollars and cents but on assessing the quality and types of relationships that are a part of a business or farm, to promote team spirit, integrate idealism into daily operations and enable an initiative to be more productive.

Kongsi Co-op also wishes to apply Steiner's Fundamental Social Law as expressed in the contemporary terms as follow (Lamb, 2010b):

> The more that individuals work for the benefit of society or the needs of others,
> and the more that each person is supported by others to lead a dignified existence,
> the greater the well-being and overall prosperity of the society will be.
> —Gary Lamb, Associative Economics

We consider this an ideal that must be practically applied in all aspects of business to achieve true social responsibility in economic life, which, in turn, will support the cultural life of society.

Journey Reflection—Awareness and Transition into Organic Lifestyle

Since the establishment of the Co-op, we feel strongly that our efforts in raising awareness are considerably successful on three aspects, namely, the consumer, producer and retailer sectors. We have witnessed participants or learners in our activities seeing organic food and organic lifestyle in a new light and this comes from within them.

Furthermore, our persistence in zero-waste and zero-packaging policies has also prompted producers and retailers to relook at their packaging and provided us with alternative bulk purchases specially for our consumers. Also, our guiding principles have given much more confidence to producers and retailers in terms of knowing that they are backed up by an entity that is supportive as long as their intentions are aligned. Example of this include doctor-turned farmer Dr. Azman converting his conventional rice field into practising organic farming; Centainnel Agriculture into providing bulk packaging for their processed sweet potato noodles. These are some of the successes that we have observed and that motivate us in the work we are doing. We might not be able to quantify it, but we are noticing retailers or producers realigning their intentions after our "Beyond Organic Retailing" workshop for local organic businesses.

Slow and Organic Growth vs Financial Viability (Diversity in Opinions, Inclusiveness)

Our biggest challenge was understanding that what we are doing is an organic process which takes time, while also needing to maintain financial stability to truly realise our vision of meeting each other's needs through job opportunities that safeguard the long-term health of our people, soil and nature. By being unapologetic about the lifestyle, we believe is necessary for building a harmonious ecosystem, we may encounter resistance from some, leading to inconveniences and disagreements among members. However, we have learned that this is a natural part of building a diverse and inclusive community. As this model is uncommon in Malaysia and most of us have no prior experience with it, we are bound to encounter obstacles and disagreements about what is being done and what needs to be done. Nonetheless, we are grateful for the clarity we have gained through each struggle we have faced.

Our priorities may differ, but one outstanding example is the restoration of our current headquarters, a 60-year-old wooden stilted Malay house with decent soil space, which is becoming increasingly rare near cities. Our core team felt the need for a grounding spot to gather hearts, minds and hands. While some may consider investing in digital trading platforms more financially sound, for those of us actively involved in the work, refurbishing the headquarters has served as an invaluable way to connect with people and conduct actual work. The digital space is effective but lacks the warmth and connection that we have experienced time and again through in-person activities, sweating together and enjoying good meals, creating ensouling memories.

As the Co-op enters the third year of operation in 2024, like many startups, it is still facing financial instability and a shortage of manpower. We have encountered numerous opportunities, challenges, and learning experiences throughout our journey, which have helped us realise the importance of building a strong culture to attract the right people to co-create and shape this path together. Every situation has its two sides, and we share some of the experiences that have defined Kongsi Co-op.

Kongsi Co-op faces several challenges in its growth, including a shortage of manpower across all levels, from administration to daily chores. Despite efforts to reach a wider target audience seeking alternative lifestyles, the community is not yet large enough to relocate to a larger space. While there are numerous ideas for expansion, a lack of space currently limits their execution. The challenge of competing with greenwashing tactics is also prevalent. Land ownership, attracting new members with diverse perspectives, and ensuring everyone is aligned with the Co-op's goals are additional hurdles. Education plays a crucial role in addressing these challenges, as diverse intentions among members can hinder progress. Matching people to suitable roles is another factor to consider.

6 Concluding Remark

Kongsi Co-op's platform is designed to facilitate the development of like-minded individuals and communities by fostering opportunities for collaboration and shared objectives. Each initiative undertaken by the Co-op is strategically aimed at enlarging the "consumer" pie and appetite towards earth-friendly way of living, which involves broadening the spectrum of available consumer options and enhancing awareness.

This approach enables consumers to explore a more diverse range of choices, thereby informing and enriching their decision-making processes. By expanding the visibility of sustainable and locally sourced products, Kongsi Co-op not only supports more informed consumer behaviour but also contributes to individual development. This process of offering varied and sustainable options encourages personal growth and self-awareness, thereby promoting a more engaged and knowledgeable community.

Despite the challenges, Kongsi Co-op has made significant strides in its mission to foster a sustainable and interconnected community. The Co-op has demonstrated resilience and adaptability. By focusing on building a strong culture, attracting the right people, and continuously learning from experiences, Kongsi Co-op has positioned itself for continued growth and impact. Each challenge has presented an invaluable opportunity for us to realign ourselves.

References

Bisht, I. S., Rana, J. C., & Ahlawat, S. P. (2020). The Future of smallholder farming in India: Some sustainability considerations. *Sustainability (Switzerland), 12*(9). https://doi.org/10.3390/su12093751

Camilleri, M. A. (2020). European environment policy for the circular economy: Implications for business and industry stakeholders. *Sustainable Development, 28*(6), 1804–1812. https://doi.org/10.1002/sd.2113

Enthoven, L., & Van den Broeck, G. (2021). Local food systems: Reviewing two decades of research. In *Agricultural systems* (Vol. 193). Elsevier. https://doi.org/10.1016/j.agsy.2021.103226

Fan, Y., & Fang, C. (2020). Circular economy development in China—Current situation, evaluation and policy implications. *Environmental Impact Assessment Review, 84*(April), Article 106441. https://doi.org/10.1016/j.eiar.2020.106441

Lamb, G. (2010a). Part two: The threefold nature of social life and associative economics. In D. Mitchell (Ed.), *Associative economics: Spiritual activity for the common good* (pp. 36–78). Waldorf Publications.

Lamb, G. (2010b). The fundamental social law: A new foundation for economic life. In D. Mitchell (Ed.), *Associative economics: Spiritual activity for the common good* (pp. 49–62). Waldorf Publications.

Lamb, G. (2010c). Trans-sector economic associations: Earthly and spiritual potentials. In D. Mitchell (Ed.), *Associative economics: Spiritual activity for the common good* (pp. 81–84). Waldorf Publications.

Gyamfi, B. A., Ozturk, I., Bein, M. A., & Bekun, F. V. (2021). An investigation into the anthropogenic effect of biomass energy utilization and economic sustainability on environmental degradation in E7 economies. *Biofuels, Bioproducts and Biorefining, 15*(3), 840–851. https://doi.org/10.1002/bbb.2206

Henderson, K., & Loreau, M. (2023). A model of sustainable development goals: Challenges and opportunities in promoting human well-being and environmental sustainability. *Ecological Modelling, 475.*

Landwehr, M., Engelbutzeder, P., & Wulf, V. (2021, May 6). Community supported agriculture: The concept of solidarity in mitigating between harvests and needs. In *Conference on Human Factors in Computing Systems—Proceedings*. https://doi.org/10.1145/3411764.3445268

Martin, A., Armijos, M. T., Coolsaet, B., Dawson, N., Edwards, A. S. G., Few, R., Gross-Camp, N., Rodriguez, I., Schroeder, H., Tebboth, G. L. M., & White, C. S. (2020). Environmental justice and transformations to sustainability. *Environment*, *62*(6), 19–30. https://doi.org/10.1080/00139157.2020.1820294

Qin, M., Su, C. W., Umar, M., Lobonţ, O. R., & Manta, A. G. (2023). Are climate and geopolitics the challenges to sustainable development? Novel evidence from the global supply chain. *Economic Analysis and Policy, 77*, 748–763. https://doi.org/10.1016/J.EAP.2023.01.002

Steiner, R. (1999). Finding real solutions to social problems. In *Towards social renewal: Rethinking the basis of society* (4th ed., pp. 38–62). Rudolf Steiner Press.

Steiner, R. (2003). Culture, law and economy. In *Social threefolding: Rebalancing culture, politics & economics*. Rudolf Steiner Press.

Steiner, R. (2008). *Nutrition: food, health, and spiritual development*. Rudolf Steiner Press.

Shittu, O. (2020). Emerging sustainability concerns and policy implications of urban household consumption: A systematic literature review. *Journal of Cleaner Production, 246*, Article 119034. https://doi.org/10.1016/j.jclepro.2019.119034

Söderholm, P. (2020). The green economy transition : The challenges of technological change for sustainability. *Sustainable Earth, 3*(6). https://doi.org/10.1186/s42055-020-00029-y

Sulaiman, N., Yeatman, H., Russell, J., & Law, L. S. (2021). A food insecurity systematic review: Experience from Malaysia. *Nutrients, 13*(3), 1–41. https://doi.org/10.3390/nu13030945

Ullah, A. K. M. A., & Ming Yit Ho, H. (2021). Globalisation and cultures in Southeast Asia: Demise, fragmentation, transformation. *Global Society, 35*(2), 191–206. https://doi.org/10.1080/13600826.2020.1747992

Volz, P., Weckenbrock, P., Cressot, N., & Parot, J. (2016). *Overview of community supported agriculture in Europe*. http://urgenci.net/the-csa-research-group/

Xie, A. (2019). The paradigm crisis of modern mainstream economics. *Axiomathes, 30*(1), 37–48. https://doi.org/10.1007/s10516-019-09435-0

Yusof, Y., & Kalirajan, K. (2021). Variations in economic growth across states in Malaysia: An exploratory analysis. *Journal of Economic Studies, 48*(3), 699–719. https://doi.org/10.1108/JES-06-2019-0279

Zhang, L., Xu, M., Chen, H., Li, Y., & Chen, S. (2022). Globalization, green economy and environmental challenges: State of the art review for practical implications. *Frontiers in Environmental Science, 10*(March), 1–9. https://doi.org/10.3389/fenvs.2022.870271

Kufunda Learning Village

Maaianne Knuth and Trevor Button

Features:

(a) Founding year	2002
(b) Employees	31
(c) Location	Ruwa, Zimbabwe
(d) Active in which countries	Zimbabwe with outreach into other countries
(e) Sector	Community Development, Community and Social Profit Organisation
(f) Other	Hildur Jackson Eco-village award year

1 The Start

Our story began out of a meeting between Denmark and Zimbabwe. Maaianne Knuth, founder of Kufunda is a woman of Danish and Zimbabwean origin, born in Kenya, raised in Denmark and returned to Zimbabwe in her teenage years with her family (Fig. 1).

Later, in university back in Denmark she was deeply inspired by a Danish educator, Grundtvig whose work had to do with supporting people in coming more fully into their humanity and freedom. He was the founder of the modern Folk high

M. Knuth (✉) · T. Button
Kufunda Learning Village, Harare, Simbabwe
e-mail: maaianne@gmail.com

J. Kronenberg and E. T. Lammerts van Bueren (eds.), *On the Earth We Want to Live*, World Sustainability Series, https://doi.org/10.1007/978-3-031-98758-8_30

Fig. 1 Logo Kufunda learning village

school movement in Denmark, which is alive and strong still today, focusing on adult education for life.

For Grundtvig to be human is to be a member of the School for Life. Initially, he sought to support young adults with a farming background to discover their own dignity and their innate capacity to learn and grow, out of a sense of wonder. He was a priest, a teacher, author, historian and hymn writer. He worked out of a way of seeing the world as an interconnected microcosm/macrocosm. For him, our time on earth is a time of creative fulfillment of divinely-granted human potential. In many ways this is similar to what Rudolf Steiner brought with anthroposophy. This inspiration from Denmark—led to Maaianne's return to Zimbabwe to co-create Kufunda in 2001.

Initially Kufunda worked with young people, many of whom had dropped out of school, supporting them in becoming agents of their own destiny, waking to their potential and possibility. Our work strove to challenge the colonial script in which they were seen as poor and less educated, in order to enable them to find their way into who they truly are, and what they have to bring out of their own culture, experience and imagination.

This grew into Kufunda Village. Kufunda means 'learning' in Shona, the local language. Kufunda is a learning village in which we are learning our way into what it takes to create a healthy, vibrant community as whole, becoming individual creative human beings.

Over time we have moved beyond a focus on young people to focus on people who are part of community life, wherever they may be, working now with communities across Zimbabwe and with relationships into the world, in particular in Brazil, India, Denmark, the US and several African countries.

2 Our Vision

Our vision has to do with human awakening. Our vision is one of awake, alert human beings, working together in community in service of life. It relates deeply to Steiner's social ethic:

"The healthy social life can be found.

When in the mirror of each soul

The whole community finds its expression,

And when in the whole community

The soul strength of each one is living."

It is thus a movement beyond purely community focus, towards living and learning how to cultivate community in which the freely-offered gifts of each person are given and received. It is a community in which we are learning how to practice sisterhood and brotherhood.

Our vision is of people with pride and dignity in who they are, confident and therefore able to co-create innovative and even beautiful solutions to everyday challenges, strengthening the shared ability to respond to what the future brings, even when the context is deeply challenging. Together, we are learning how to build the future, through organic, collaborative, co-creative practices.

3 Working with Healing Out of an Appreciative Embrace of What is

Our work is threefold: Healing in community, Healing with the land; and bringing a Healing Education for the young. We speak of healing as a return to wholeness. It is a continuous living process of becoming whole.

(1) **Healing in Community**

We work with participatory, collaborative community development. We work out of appreciative inquiry and seek to meet communities as mysteries to be discovered and uncovered, rather than problems to be solved.

When we engage with communities (and through this how communities then engage themselves), we enter into a collective search, for the good, the true and the beautiful. We look for beauty. We look for goodness. We look for essence. And out of that, we ask the question, what does this inspire? How might we grow this? In this way, we support communities to gather around the best of what is, around what they care about, and to find a way to mobilise around it.

These kinds of explorations are strengthened in direct correlation with the wakefulness and commitment of those participating. Therefore, included in much of what we do are aspects of inner work. We seek to invite people into a sense of possibility, of being active participants in their own life's unfolding.

This naturally means including artistic exploration, living in nature, and other creative, spiritually connected pursuits. We undertake study of spiritual science together with those who wish. We seek to use all these ways that connect people with something beyond the purely physical to find inspiration from within, in connection with that which is without, including the community itself.

(2) **Healing with the Land**

For many years, healing in community was our primary work: enabling communities to reconnect with the pride in themselves and to organise around what they cared to affirm from the emerging future. Out of this we evolved our practical farming work, which is so essential to the life of the rural communities in Zimbabwe, into a path of biodynamic farming. Biodynamic farming is the agricultural expression of our purpose–a return to health and wholeness for the earth, through the preparations that restores the earth, and strengthens the individual diverse aspects of the farm, into an emerging healthy whole. Biodynamic farming is best practiced in community, although it needs clear insights of the individual farmer to tend to the cycles and the timing. Our work with biodynamic farming has woven us into a deeper kinship with nature and the land, and we are teaching what we are learning to growing numbers of community farmers in Zimbabwe.

(3) **Healing Education**

Similarly, our work with communities led us to see that we had to work with education, as much of our work with youth began with a need for deep unlearning from an education that was in fact deadening.

And so we work with education, with the young child, and with supporting others to be able to work with children in an alive way, supporting their becoming. We are inspired by Waldorf to bring a form of education that supports the spirit of the human being to land here in a good way, each one bringing their gifts from beyond the threshold, and learning what is needed to be able to navigate and operate in this earth sphere. The village has come to life in a different way since we included a school and we have been called upon to bring the foundations of Waldorf education to nearby teachers (Fig. 2).

Fig. 2 Community activities in Kufunda

4 Kufunda and Anthroposophy

Anthroposophy has been an important way of strengthening what we had already begun doing and living at Kufunda. Perhaps the most important premise is that everything that is materially here, exists in the spiritual realm: Life and the physical material reality is an expression of the living creative spirit.

Anthroposophy situates us in a bigger unfolding. As a learning village engaged in supporting the waking and emerging expression of becoming human beings, anthroposophy is a key: Anthroposophy is about becoming human, and Kufunda is about the becoming human in community.

It is important to state that anthroposophy is not only terminology and concept. It is spirit in thought, feeling and deed. In living anthroposophy, we are dealing with truth, beauty, and goodness. In thinking, we deal with the truth, in feeling with beauty and in doing goodness. We bring this into our work and into our lives. These living ideals form a foundational core of our approach in meeting people, nature and communities. It is the collaborative search we spoke of earlier, seeking the good, the beautiful and the true. Naturally, there are limits in how we express this, because as individuals, we all have our limitations, but the essential thing is the striving to realise the ideals.

In the second part of Goethe's Faust[1] the angels say that they can redeem the one who is constantly striving. In other words, the outcome of our work does not necessarily have to be successful, nor does it have to be radiantly glowing, shining examples of human activity. For the angels, the striving is the essential thing, and here in Kufunda, we are striving. To achieve these ideals of the expression of divinely-inspired human potential, within the social ethic, we are doing something that the angels can take up and, so to say, redeem through their work in turn.

5 What Is Sustainable Development to Us?

Nature sustains itself through cycles. For example, in the plant world, from the seed to the sprout, the leaf, the bud, the flower, the fruit, the dying away and back to the seed again. All of this is like a breathing in nature.

What sustains the breathing is the spirit. The spirit is the force and the source of the system of sustainability. Sustainable development for us therefore has to do with development that is in resonance with natural laws, originating in spirit. The gift of biodynamic farming to Kufunda was opening us to a way of farming that gave a way to be in this resonance, and to experience how we work in the material with spirit. It was, and still is, an experience of working with grace in a very practical way.

For us as humans, it is the same. We breathe in the cycles of life and time; earth and death; day and night; spring, summer, autumn and winter; taking in, digesting and casting out. All of these and others are rhythmic processes in which we live, and the sustaining element behind all of these processes is spirit.

So the true ingredient of sustainability is spirit, which exists beyond the outer rhythms of life. So in human development, the sustainable factor, or the factor that gives sustainability, is spirit.

That may sound lofty, but actually, it becomes very practical. Sustainable development is a development that is enhancing what is here now, allowing for an unfolding of Spirit in life. We practice and implement sustainable development through continuous learning—in resonance with life. Continuous learning from ourselves, from our communities, from nature. Kufunda means learning. We are a learning village. That is our way of living the cycles of life–being in learning and working with emergence. We never arrive, we are constantly on the journey of learning, which is spiritual development. Sometimes we make mistakes—our mistakes teach us important lessons. Being aware that we are on a spiritual journey in the physical collective experience gives a strength, a capacity to listen more deeply to the nuances, to guidance beyond our own intellectual knowing.

[1] Goethe FAUST Part 2. Verse 11,936.

Ecology as a science of the balance of nature and human activity is taken into account within the work done at Kufunda, for example, through biodynamic farming, organic architecture and building using renewable products as much as possible, living close to nature and using minimum mechanisation. These are the ecological factors that we work with here in Kufunda, and which touch also on the economic, the social and on the cultural/spiritual realms of our community.

6 Living the Future Now

One of our key mottos as Kufunda is 'Living the Future Now.' We were inspired by a poem long ago that asked, 'What if the future is waiting for you?' We live with the assumption, that the future needs us for it to come into being. The future is future and now at the same time, and so we seek to live the future we long for, and that is calling to us, right here and now.

It is as if we are composing today with the music coming to us from tomorrow. We are living with that music in our daily work–at the school, with the children who are learning, with the biodynamic farmers, with the community engagement work–allowing the emerging future to shine through and into the emerging work. Everything is a process of learning. We are working on this cusp of actually meeting life right here and now, as an emerging piece of music or art. We are being together, learning, struggling, striving, evolving.

And so every day is a practical step towards the future. We are not just fumbling about with vague ideas and lovable people helping each other. It is far more than that. It is that, but it is far more than that, because we are building a kind of understanding. Our learning is also an understanding of how human beings can work and live together today, as well as in the future.

We are learning and providing something that communities elsewhere in the future can use. It can be applied. What we are doing here is our application in our situation, but it can be applied and modified in infinitely many ways. Depending on where people come together and want to work together in this way, what we are living here has endless possibilities.

Key to our work has been the transformation that has taken place in people and their communities. There are so many examples of personal transformation, combined with community transformation. We are not a place for individual development on its own, nor a place for community development only. We work with this interface between individual and collective identity and transformation, as two-sides of a coin.

Sometimes individuation leads to separation from community life, and yet something is happening here, where people are waking and transforming and connecting to their deeper gifts and dreams and staying connected in the web of community; finding their way to actively create together.

This is significant. Through our work over the years we have seen the power of individuals waking to who they are—which often happens in community. We see inner confidence grow through the work—leading to individuals stepping into bigger community roles—youth without a high school diploma becoming community leaders, councillors, social workers, starting their own enterprises, etc.. We have seen communities gathering around their collective vision and–without big external funding—making their way tangibly and practically—building community centres, creating libraries, youth programmes, community gardens, bee keeping initiatives, planting trees, creating community water points, and more. We have seen politically polarised communities finding the place in the middle where people who used to be pitted against each other politically have been able to connect around their tangible dreams and initiatives. All of this points to individuals waking to who they are and what they care about *in community*. Out of this practical collaborative work happens for their shared future. In the face of the socio-economic challenges of Zimbabwe it sometimes feels like a miracle to experience what can be unleashed in communities with very little material wealth—and yet people are taking charge of their own destinies.

7 What Are We Learning that is Relevant for the World?

Our way of working—with inner and outer, self and collective, out of appreciation, is deeply relevant for the world. In our work through our Communiversity Programmes with communities, we are saying to communities, based on our own lived experience 'you can take control of your own lives, as we have in Kufunda'. This is not only about projects but also about governance. Kufunda is a self-governing community. We make our decisions as regards our income, expenses, our visitors and all the things that happen here. This is something that in the near future is going to be an increasing need amongst smaller communities. So Kufunda village is a kind of example of where smaller communities can govern and lead themselves in an unfolding expression of truth, beauty and goodness–in striving, always a striving. This is probably going to be one of the biggest things where we can make a difference in the near future, not only in Zimbabwe, but all over the world.

People can come and see and learn and say "That works. Let's try it in Kenya, or let's try it in Germany, or in South America."

Recently we hosted a big pan-African gathering on agro-ecology at Kufunda, with people from almost 30 African countries. The feedback was overwhelmingly positive, even if it was not so clear to the attendees what made Kufunda a special place–"*We can't really put our finger on it, but this is magical; something is alive here.*"

Kufunda is not a big fancy NGO. We are a group of people, living in community seeking our way into the future that is beckoning, always inviting others to join us. Out of that, we are working with Life and Spirit, and finding something is alive here that has more value than the material.

That is what we are working towards, not only for the next 10 years, but for future. If we were to summarise some of our key lessons around what it takes to live the future now, we could say:

> Within everyone, regardless their background, lies possibility and potential which can be activated without disconnecting from community. It can in fact be strengthened in community.

We are learning that it takes warmth to nurture this potential. At the heart of our work lives gratitude and appreciation. Individual and community riches can be unlocked through gratitude and grace. We are learning the power of appreciation to bring deep transformation.

We are remembering that respectful involvement is key–an involvement that honours the value of each one, and invites their participation. People support what they have been part of creating. That has always been the starting point of Kufunda, but it is also an ongoing learning to live it. Involving each voice takes much more time, and yet it builds sustainable solutions that belong to the people involved. This lesson also means that there are aspects of our community life that some people choose not to join, and that is okay.

We have learned of the need to be willing to prioritise. When we come together as a community around a key focus, we galvanise momentum and movement. We used to try to do everything–and being a learning village working with communities, land and children made for a very diluted focus. We have learnt that we can have the everyday broader work, <u>and</u> that we can galvanise the future through shared clear sharp intention. Focusing spirit into shared action! Recent priorities include, creating a community youth hub, building new classrooms for the children, renovating and upgrading the village, hosting two pan-African gatherings and each year coming together to prepare for the upcoming growing season. Short term clear shared priorities galvanise people to come together in support of the whole.

We have been learning about the need to connect with the world, with others who work like us, seeking the potential for mutually beneficial, expansive expression. We are never actually alone–and we can always be enriched by finding ways to connect and weave our work together with others, just as we are doing in this publication.

8 Our Key Challenges?

Leadership in the community is a question we are always working with. Spiritually, each one of us is a "King" or "Queen." When the Book of Revelations speaks about kings, what is meant is individual human beings who are striving to find their higher meaning in the world. The more we know ourselves in the higher sense, the more we become our own leader, our own guide. In the way we work—through shared responsibilities and individual responsibilities—we discover our own leadership roles within the community. The more this can be developed, the more mature our community will become. In a collective culture it takes courage for each to step into their king- and queen-ship, but it is part of learning of how to live in that place between individual and collective culture and expression.

As a social initiative, our biggest challenge comes out of issues around finance and money. Almost every problem we are confronted with today stems from a misunderstanding of what money really is. Often we live from a place of scarcity which his not helpful! We are attempting to embark on a way to better understand what money is, how we can work with it and thereby begin to solve and confront the real challenge that money poses. From struggling with fundraising since our inception, we are growing into a self-recognition of the value we bring, through doing the work we love and are good at, and discovering ways to attract money for it. So something is shifting in our relationship with money, and that is bringing its own new money challenges around who gets what and how we let the money flow and be shared in the community. As we said above, this is a striving—and we are very much still on the way.

9 The Coming Years

Waking to a higher consciousness is key for what needs to emerge for society in the coming decades.

In our experience, this means softening, becoming more inward and willing in our encounter with the world. It has to do with cultivating a stance of appreciation—which is actually love—through which we can then embrace and face what comes to us.

For us at Kufunda this means embracing what is, and seeking out and acknowledging the gift of each one (person, community, situation), regardless of how it presents. What goes into the compost heap does not look pretty and yet it is living gold for the earth through the process of transformation. The overwhelmingly intellectual materialistic tendency in the world so often seeks to control the world. What is needed is a way of being with the world in which we can allow ourselves *to take in*, and move with what is, as it is, and in that movement we can transform and be transformed. We then become a channel for an arising evolving future.

10 Conclusion

To summarise the above, one could say that Kufunda Village unites, through its international appeal, people and impulses of various kinds, in an African—European setting which is typically Zimbabwean.

This "outer" circumstance is supported by an "inner" attitude fed by Spiritual Science. The inner spiritual stream works quietly—by many visitors totally unnoticed—but nevertheless tangibly in the community. Just as a true Waldorf teacher brings into the classroom inspiration out of an inner life which enriches the lessons for the children, but which remains remains in the background, so too does the classroom called Kufunda Village have an inner living stream flowing through it as an underlying, invisible, source of inspiration.

This inner source is the true Sustenance of the outer work and activity. It is really what makes the village-community ultimately sustainable.

L'Aubier—Freedom, Commitment, Responsibility

Aline Amrein and Christoph Cordes

Abstract

L'Aubier began in 1979 with the enthusiasm of two young people who decided to take over a farm in the French-speaking region of Switzerland, and convert it to biodynamic agriculture. Through a collaborative relationship with customers and supporters, numerous additional ventures have emerged over the past 45 years: Today, in addition to the Demeter-certified farm with an on-site dairy, L'Aubier comprises a 100% organic and Demeter-certified restaurant, a bio-hotel with seminar rooms, an organic shop, an intergenerational residential community, an academy and a café-hotel. Solutions for the ownership of a company as well as for the internal organisation have been developed that could serve as models. Supported by over 2000 partners and sympathisers, L'Aubier stands as a pioneering enterprise in associative economics, applied ecology, and innovative social structures.

Keywords

Associative economics • Applied ecology • Social structures • Company property • Entrepreneurship • Anthroposophy

A. Amrein (✉) · C. Cordes
Montezillon, Switzerland
e-mail: aamrein@aubier.ch

J. Kronenberg and E. T. Lammerts van Bueren (eds.), *On the Earth We Want to Live*, World Sustainability Series, https://doi.org/10.1007/978-3-031-98758-8_31

Features:

(a) Founding year	1979
(b) Number of employees	around 50
(c) Location	Montezillon and Neuchâtel (CH)
(d) Active in which countries	Switzerland
(e) Sector	Biodynamic agriculture, hospitality, intergenerational living, education, associative economics

1 Introduction

L'Aubier (Fig. 1) began in 1979 with the enthusiasm of two young people who decided to take over a farm in the French-speaking region of Switzerland, located in Montezillon above Lake Neuchâtel, and convert it to biodynamic agriculture. Through a collaborative relationship with customers and supporters, numerous additional ventures have emerged over the past 45 years from what was originally a small farm. Today, in addition to a 40 hectare Demeter-certified farm (Fig. 2), featuring one of Switzerland's first free-range barns for horned cows and an on-site dairy, L'Aubier comprises a 100% organic and Demeter-certified restaurant, a bio-hotel with six seminar rooms (recognised in 1994 as Switzerland's first eco-hotel), an organic shop, an intergenerational residential community, an academy, and a café-hotel, also 100% organic, with its own coffee roasting facility, located in the city of Neuchâtel itself.

At its Montezillon site, heating throughout the premises is provided using wood from the surrounding forests, as well as through the recovery of waste heat from refrigeration units and the ventilation system. Rainwater is collected and utilised for laundry and toilet flushing. Purchases are made with deliberate consideration of the products' origins, production and marketing methods, and their social and ecological impacts. Electricity is generated on-site, and any additional electricity purchased is 100% certified from renewable energy sources. The construction materials used are of natural origin: paints, lime plaster, wool carpets, and floors

Fig. 1 Logo L'Aubier

Fig. 2 Cows on their way to the pasture in L'Aubier

made of wood, stone, or linoleum. Hotel duvets and pillows are made of pure sheep's wool, with covers and sheets crafted from organic cotton. Background music is intentionally omitted, allowing the senses to be enriched through authentic experiences of colour, scent, taste, and texture.

Guided by the current team of eight managers overseeing various areas and supported by over 2000 partners and sympathisers, L'Aubier stands as a pioneering enterprise in associative economics, applied ecology, and innovative social structures.

2 Three Essential Gestures

What unites all these areas? And what distinguishes L'Aubier from other enterprises operating in similar fields? Three defining gestures can be identified.

A first gesture, present from the inception of L'Aubier and recurring in all its activities, is the will to engage in practical action. Initiative holds the strength and potential for change to occur and for something new to come into the world. This active engagement of human beings is seen as essential at L'Aubier, something that must be protected and encouraged.

A second gesture involves an expansion of self-interest. The goal is not the realisation of one's own idea or the pursuit of profit, but the harmonisation of individual commitment with the needs of the world. The guiding principle is the well-being of the whole—whether it concerns customers, financial partners, employees, animals in the barn, or the cultivated fields. This widening out of self-interest can be seen as a fundamental gesture of ecology and associative economics.

A third gesture centres on nurturing self-awareness and an understanding of the world, and the ability to place one's thoughts and actions within a broader context. As a path of insight—into what it means to be human, one's relationship with others, and one's connection to the world—and as a cultural impulse to overcome today's materialistic worldview, anthroposophy serves as both an inspiration and a unifying element for the daily work of those responsible at L'Aubier. When we observe and engage differently with the fields of activity through anthroposophy, our relationship to them changes. A personal, practical connection emerges, creating a bridge that allows insight not to remain a mere idea but to be imbued with thought, feeling, and will, thereby enriching one's activity. This benefits everyone as the quality of the work is continually developed and enhanced.

The attempt to harmonise these three gestures gives rise to an ongoing search for coherence and authenticity, an approach that can be described as free, committed, and responsible.

Obstacles and challenges can stimulate creativity, leading to the discovery of new paths and the ability to "do things differently". New activities often arise from the intersection of an initial idea, the fertile ground of a specific opportunity, and the initiative of an individual who connects with it.

As the enterprise grows and the number of active participants increases, two main areas require particular attention: establishing a structure for the management of property, and the internal organisation. For these, pioneering solutions have been developed that could serve as models.

3 Who Owns the Company?

An initiative can begin without money, but to grow and develop, it must access the necessary capital. This raises the question of what money actually is: it is an expression of relationships between human beings, and it is up to us to shape those relationships. L'Aubier's approach is based on trust and partnership, both in the financing of the enterprise and in its relationships with business partners.

In terms of financing, L'Aubier's approach is to use a limited company combined with an association. The limited company issues two types of registered shares with equal voting rights in a specific ratio. For a capital amount of 100,000 CHF this consists of 90 shares at 1,000 CHF each available to the public, and 100 shares at 100 CHF each, which carry the majority vote and are held by an association whose purpose is to uphold L'Aubier's ideals and ensure the company's ethical orientation. If the company needs to grow—requiring increases in capital—the selling of shares at higher prices finances the 100 CHF shares held by the association, which thus retains its voting majority. By this, the shareholders ensure the company's adherence to its ethical orientation. This arrangement enables those who carry the company in their daily activities and in its development to make responsible decisions without being dependent on the goodwill of an owner or investor. At the same time, they do not need to provide their own capital to take the initiative.

This structure removes decision-making authority from monetary control. It is founded on the belief that human beings can govern and correct themselves. As a result, L'Aubier does not separate the strategic and executive levels; the board of directors and management work as one.

Thanks to this structure, generational transitions and other transfer processes are facilitated. The outgoing generation does not face the dual concern of ensuring the continuation of the enterprise while also striving to sell their shares as profitably as possible. Moreover, the company cannot be inherited, and the incoming generation does not need to raise significant capital to join and continue the enterprise. The focus is on individual human beings with the will to act and the ability to commit their talents to L'Aubier. The company's future is entrusted to such individuals.

With this approach, existing legal structures are adapted to ethical goals, and constructs that are incompatible with human initiative are overcome. This means that responsibility and engagement have as much weight as capital. The company becomes unsellable and, in effect, owns itself.

4 Freedom and Responsibility

Within L'Aubier, activities are independently organised to grant entrepreneurial freedom to the individuals responsible. Each area operates with its own profit and loss statement, functioning like a small, autonomous business with a clear view of its financial situation. To maintain cohesion and consistency across the enterprise, weekly management meetings are held. Every two months—or monthly if necessary—a broader exchange on business progress takes place. During these sessions, the various financial reports are shared and discussed. This transparency offers an objective overview, fosters awareness, and creates understanding through the reflective process of observing others' perspectives. A similar approach is followed for investments: discussions take place, but the final decision and responsibility rest with the individual in charge.

The approach differs slightly for decisions that affect L'Aubier as a whole. In such cases, the interests of individual areas must give way to a broader perspective. For example, while discussions may involve the farmer and the hotelier, they do so not as representatives of their respective areas but as individuals who carry shared responsibility for L'Aubier as a whole. Decisions of this nature are made by consensus, ensuring they are collectively supported and sustained over the long term.

The enterprise functions as a living organism, evolving in accordance with its own inherent principles. Its various areas form a flexible network of relationships, interacting with a focus on mutual well-being and the overall health of the whole. These interactions are driven by genuine interest rather than mere necessity. The areas are not detachable elements that can be arbitrarily added or removed; they

Fig. 3 Sowing the Future at L'Aubier

are integral parts of a greater whole. Denis de Rougemont, a thinker held in high esteem at L'Aubier, described the principle of individual freedom combined with responsibility for the whole as a 'federalist attitude', which characterises well L'Aubier's working ethos. This applies both internally and in its relations with the wider world.

5 Agri-Culture

The sustainability of biodynamic agriculture—with its living farm organism, spray and compost preparations, respect for the living world, and the long-term build-up of humus through its farming practices—is self-evident. It is one of the cornerstones of L'Aubier. Within this framework, and extending beyond it, the initiative "Sowing the Future!" was developed by L'Aubier in 2006 in collaboration with the biodynamic cereal breeder Peter Kunz to promote a genetically modified organism (GMO)-free agriculture. Since then, this event has taken place annually at L'Aubier and on many other farms around the world (Fig. 3).

A group of people, young and old, gathers in a field, each holding a cap full of seeds. After an introduction to the gesture of sowing, they walk across the fertile soil and, with a generous movement, sow their grains together, living grains—capable of being sown again—grains for the bread of tomorrow. This is the archetypal gesture of agriculture, carried out collectively by many people. It represents a commitment to a sustainable future for farming and is, at the same time, a cultural act of renewal. The driving force behind this initiative was the feeling that one should counter the development of genetic engineering, which impoverishes life and creates dependency among farmers.

In a single act, ecological, social, economic, political, and cultural dimensions converge. Such an act resonates: it raises awareness, opens up new perspectives, connects generations, and forges relationships between farmers, consumers, plant breeders, and farms.

Sustainability is not a system or a switch that can simply be flipped—it requires a shift in consciousness. Existing practices must be re-evaluated in light of their consequences and transformed with forward-looking impulses. In this way, sustainability goes beyond conserving resources to include a developmental aspect: becoming creatively active through personal commitment to the common good.

6 On the Path to the Future

Within L'Aubier, every person—whether employee, customer, visitor, or financial partner—is regarded as a whole human being, capable of taking initiative, committing to an ideal, and assuming responsibility for one's actions.

The future potential of L'Aubier lies in human beings who are open to the needs of the times and equipped with the will to respond to them. However, this capacity requires protection and encouragement, as our times bring numerous challenges. Society increasingly faces weakening will-power and declining self-initiative. Instead of cultivating a healthy emotional life, instinctive behaviours are becoming more dominant. A lack of commitment and a fixation on personal well-being are becoming more prevalent, as is a diminishing ability to think independently.

In response to these challenges, in 2016 L'Aubier launched a nine-month full-time training programme "Oser l'indépendance!" (Daring Independency!), which has been offered annually from October to June ever since. The programme focuses on the archetypal gesture of taking initiative: independence. It is delivered by the responsible persons at L'Aubier who share their extensive experience from years of entrepreneurial and idealistic engagement, as well as their inner motivations, in an environment that fosters initiative. Based on anthroposophy, the training is divided into courses, artistic workshops, and practical work.

The courses cover an understanding of the world and humanity, perspectives for the future, and practical tools such as accounting and legal structures, with the goal of empowering participants to actively and entrepreneurially shape the future. In workshops, such as groundwork with horses or engaging with language, participants focus on developing and strengthening their own personality and abilities. The practical work within the company serves a dual purpose: it funds the training, which is offered free of charge, and allows participants to immerse themselves in the reality of a business driven by ideals. This programme, whose courses and workshops are also open to all L'Aubier employees, makes the inner essence of L'Aubier—its vision of the human being and the ever-present ideals—tangible and explicit. Independence of head, heart, and hands is cultivated, enabling participants to bring forth new initiatives and solutions to pressing global challenges.

Behind the three gestures outlined at the beginning, it is the human being, with its potential for insight and freedom, who stands as both the starting point and the ultimate aim of L'Aubier's endeavours.

7 Conclusion

The core impetus of L'Aubier lies in seeking answers and demonstrating practicable approaches to address the urgent questions of our time. This is undertaken with the understanding that success can only be achieved through collaborative partnerships. This principle applies equally to internal processes and external relationships with customers, suppliers, and financial partners. L'Aubier is founded on a living network of interactions and relationships, which can only thrive through constant care and attention, requiring inner wakefulness and ongoing engagement.

L'Aubier's approach is the antithesis of a rigid system in which processes, once defined and established, operate autonomously. It requires human beings with the skills, potential, and will to engage, particularly in today's world. However, such qualities can no longer be taken for granted. L'Aubier therefore strives to provide a ground where the capacity for initiative can develop and thrive. Currently, this is evident in areas such as agriculture, food processing, hospitality, gastronomy, and training. In principle, however, there are no limits to the areas of work.

Monte Azul—A Beacon of Hope and Education in the Favelas of São Paulo

Renate Keller, Susanne Charlotte Rotermund, and Ute Craemer

Features:

(a) Founding year	1979
(b) Employees	2.026
(c) Location	Monte Azul Community/Horizonte Azul Community/ Jardim São Luis Neighborhood—South Zone of São Paulo
(d) Active in which countries	Brazil and internationally through an independent association of former volunteers in Germany, in Japan and on request in other countries and contexts
(e) Sector	Education and Culture, Integrated Health, Social Assistance, Environment, Institutional Development
(f) Other	The "Prêmio Itaú-Unicef", Prize for Education and Participation of Unicef. And Order of Merit of the Federal Republic of Germany 1987 (Ute Craemer) (Fig. 1)

R. Keller (✉) · S. C. Rotermund · U. Craemer
Associação Comunitária Monte Azul, São Paulo, Brazil
e-mail: renate@montezul.org.br

J. Kronenberg and E. T. Lammerts van Bueren (eds.), *On the Earth We Want to Live*, World Sustainability Series, https://doi.org/10.1007/978-3-031-98758-8_32

Fig. 1 Logo Monte Azul

1 The Associação Comunitária Monte Azul as a Social Workshop School

Origin and Development

45 years ago, Ute Craemer, Renate Keller Ignacio and Paulo Roberto Ignacio started a unique social project in the favela Monte Azul (São Paulo, Brazil) that was to profoundly change the lives of thousands of people.

Renate Keller Ignacio trained women from the favela to become Waldorf kindergarten teachers. This created a network of small kindergartens and nurseries between and in the wooden huts of the favelas, which in a joint labour effort together with the parents were embellished in a child-friendly way. These became beacons of light for the people living in the muddy, desolate situation of the favela.

Paulo Roberto Ignacio shared his many years of experience as a carpenter with young people and helped many of them to become self-employed and make the leap out of the favela.

Ute Craemer, a Waldorf teacher, fought tirelessly for financial resources and exemplified an unshakeable belief in the good of people to the growing workforce, even when it came to drug-addicted youths who did not shy away from any act of violence.

The Origins of Work

The work began in 1975 at Ute Craemer's home. When the kept children begging at her doorstep the Waldorf teacher did not want to just give bread or clothes. She wondered what these children actually needed and asked her seventh-grade students. They were enthusiastic about the idea of organising play afternoons in Ute's home with the favela children. They collected wax crayons, watercolours, wool scraps, flutes, modelling wax and much more, and came in groups after school to Ute's house, which was very close to the favela Monte Azul. Thus, a first bridge between the economically rich and poor was built. The children of the Waldorf School taught the children of the favela how to knit and crochet and read them stories. They also learned to admire the favela children, as they showed them what it was like to take responsibility for their younger brothers and sisters at an early age or to manage the household of a family of many. They played theatre, sang and danced together. When more than a hundred children were playing in Ute's garden, it was time to look for other premises. at Ute's home. It became clear that every social work sets a process in motion in which everyone gives and takes, teaches and learns: a first step towards sustainability.

The Social and Spatial Dynamics of Favelas

Brazilian slums, called 'favelas', emerged in the 1970s. Attracted by the growing industry in São Paulo and displaced from the persistent drought areas in north-eastern Brazil, thousands of people migrated to the city and built makeshift shelters on open slopes, on the outskirts of the city or next to the multi-lane highways. This is how Monte Azul, Peinha and Horizonte Azul came into being, with which the development work of the association became deeply connected during the 1980s and 1990s.

Integration and Social Commitment

The staff at Monte Azul included people from the favelas, many of whom had little or no formal education but were educated by their hard experiences of life. They were joined by socially committed people from Brazil's middle class who had been fortunate enough to have learnt a trade, and who were able to contribute to the success of the project. Every year, young volunteers from Germany, Japan, the United States and many other countries came to share their lives with the favela residents for a year and, transformed by this experience, gave their own lives a new meaning.

The Concept of the Social Workshop School

We gradually realised that this colourful community is a school in which everyone can be a teacher and a student. We call this school the 'Social Workshop School' with the motto: 'Learning by Working—Working by Learning', where the teacher may also be an illiterate person. In this way, every co-worker can develop not only professionally but also personally and artistically. To this end, we offer annual workshops, study groups, themed festivals and courses.

The Monthly Integration Day

The highlight is our monthly integration day: On this day ~200 co-workers meet in the cultural centre or on the sports field in the centre of the favela or in Horizonte Azul in the shade of large trees, to meet and work on the theme of the year. In 2023, for example, the theme was 'Respecting differences.' In 2024 it is: 'Anthroposophy in the actions of Monte Azul'. We usually start with a short presentation, then groups are formed to reflect on what they have heard and/or work on it artistically. At the end, each group presents what they have worked on to the others. Over a shared lunch, people from different work areas can get to know each other better.

As Monte Azul has a cooperation agreement with the Ministry of Health, the last integration day took place in the 17 health posts that serve a region of about 280,000 people. Thanks to this cooperation, about 1800 employees have gained insights into the practice of anthroposophy.

Mainumby

The Mainumby training course arose from the necessity of training social workers, educators and teachers. Recognised by the state as a training course since 2024, it is based on practice- and experience-based learning.

Many of the topics currently discussed intensively in Brazil are part of the history of Monte Azul, such as racial equality, recognition of diversity and an interest in the many cultures in Brazil. This also includes the study of the diverse contexts and backgrounds of Brazilian history. A deeper understanding and sense of belonging should take the place of the traditional one-sided Eurocentric view of humanity.

2 Current Activities of the Associação Comunitária Monte Azul

Today, the Associação Comunitária Monte Azul (ACMA) works in 21 locations with different priorities in the areas of education, health, culture, environment and research.

- **Infancia Querida (Beloved Childhood)**: Nurseries for children up to 4 years of age. In friendly rooms that invite free play, the Waldorf kindergarten teachers take care of the children while their parents work.
- **Nossa Ciranda (Our Circle Dance)**: After-school care for school children with artistic activities and plenty of free space to play. All children receive two healthy meals and, if necessary, are monitored by our medical-therapeutic outpatient clinic.
- **Vocational Preparation Courses**: Carpentry, fashion design, bakery, IT and administration—these courses are available to young people, but also to adults up to the age of 59, who can gain new impetus for their professional careers.
- **Music School**: Since 2012 ACMA offers a music school in the centre of the favela Monte Azul, where children and young people learn string instruments, play in the orchestra and use their music to pacify and cheer up the atmosphere in the favela. The orchestra is often invited to cultural events in cultural centres and companies, which in turn builds bridges between different social realities.

- **Casa Angela**: Birth centre where 30–40 babies a month are born through a humanised birth process. The birth centre also trains midwives from all over Brazil and conducts research into humanised birth.
- **Medical-Therapeutic Health Centre (Outpatient Clinic)**: A multi-professional team offers consultations and treatments in anthroposophic medicine to the economically disadvantaged population of Monte Azul and the neighbouring peripheral residential areas. The success is so great that this anthroposophically enhanced medicine is even gaining increasing recognition from the cities Health Department. We are in the process of concluding a cooperation agreement that will enable staff in the family health programme learning to practise external treatments.
- **Cultural Center**: A cultural centre offers artistic workshops in the evenings and at weekends. Here dance, theatre, capoeira, paint course, yoga practices and language courses are offered. There are regular artistic presentations and theatre festivals.
- **Caminhando Juntos (Walking Together)**: A project for young people and adults with disabilities that seeks their integration into social life and accompanies them to become increasingly independent in their lives.
- **Pontinho de Cultura (Culture Point)**: A project that takes place in the centre of the favela Monte Azul in the afternoons and evenings. Two educators play with children and juveniles, helping to maintain a positive atmosphere in the favela and protecting children from the destructive effects of drug trade.
- **Escola de Resiliência**: A full-time Waldorf School in the Horizonte Azul district with two kindergarten groups and five school classes. The school cultivates a large organic vegetable garden where students plant and harvest the vegetables for lunch, experiencing the different cycles of nature. This school is financially supported by the Rudolf Steiner School, a result of the bridge once built by Ute's students and our ongoing efforts to make Waldorf education available not only to financially privileged families. These efforts, which have been ongoing since the 1980s, concluded in many Waldorf schools in the poorer areas of Brazil and have now been officially recognised by the Brazilian Federation of Free Independent Waldorf Schools in the so-called Integration Commission.
- **Research, Education, Publications and Lectures**: Since the 1980s, books and other publications on social, educational and anthroposophical topics have been written. They are available in Portuguese and other languages, which provide food for thought both in and outside the anthroposophical context. In addition, there exists a worldwide programme of lectures and seminars, including at the Goetheanum.

Growth of Monte Azul

Together with the volunteers from around the world, we operate with almost 300 co-workers, most of whom come from the favelas. Around 20,000 families benefit from ACMA's services, and 1350 children visit the facilities every day.

Health Family Programme "Estratégia Saúde Da Família" (ESF)
ACMA manages the public family health programme 'Estratégia Saúde da Família' (ESF) with 1750 employees in the 300,000-strong municipality of São Luis. This programme was introduced throughout Brazil in the 1990s and in São Paulo in 2001. It involves a health team consisting of a doctor, a nurse, two nursing assistants and six health agents being responsible for every 1000 families. The health agents are themselves part of the 1000 families and are tasked with visiting these families every month, advising them on health issues. In doing so, they determine whether there are old, bedridden people, pregnant women, small children, people with disabilities, etc. Those who need it, are referred to the appropriate therapeutic facilities. This programme is financed entirely from state funds.

Monte Azul and Social Responsibility
Through its work over the past 45 years, Monte Azul has shown that commitment and responsibility can significantly improve the quality of life in favelas. Some of the original residents have left Monte Azul and now live in better neighbourhoods. They have managed to get out of the favela through courses and training opportunities ACMA offered. But they have not forgotten their roots and often return to help others and support the next generation. There are many stories of people who, thanks to Monte Azul, have found a better future. This is the result of work based on trust, love and faith in the good in people (Fig. 2).

3 Mission, Sustainability and Anthroposophy

The mission and purpose of Monte Azul is to reduce social injustice and contribute to environmental balance through education, culture and health. The organisation strives to ensure that all living beings have the right to live and that everyone can contribute to improving the world. This understanding of sustainability is not only practised in Monte Azul but is furthermore spread throughout the world. For us, the question of sustainable development begins with the social question: the existence of every human being, their path of development and their potential. We see the formation of communities that embrace this as a social sustainability task that we strive for. Once this foundation has been laid, an individual can better take responsibility for the broader community, become effective, giving something back to their fellow human beings and to earth, our basis of life.

The economic realm should serve this purpose, rather than placing an additional burden on social life. A sustainable economy can therefore only emerge if it is at the service of human development and the care of the earth.

In this respect, Anthroposophy has been a fundamental element and a source of inspiration for Monte Azul from the very beginning, primarily through Waldorf education and the art of healing. Addressing the potential for development in every person and in every community and creating the conditions for this to unfold on a physical, soul and spiritual level are the guiding principles of our work. Taking joint responsibility for this is at the heart of Monte Azul. In this respect, we

Fig. 2 Socio-educational community activity, Monte Azul

understand sustainable development as the unfolding of potential in the individual, in the community and in the foundations of life, the earth and its inhabitants, on which we live.

Global Impact

Monte Azul's values and experiences are shared by staff and volunteers in other regions of Brazil and around the world. Through international conferences, seminars and the co-founding of other social initiatives, Monte Azul inspires and educates other social institutions. Several similar initiatives have been founded in Brazil and other countries by former volunteers who brought their experiences from Monte Azul back to their home countries. These initiatives follow the same principles and work to improve the lives of people in precarious situations and open up new perspectives for them.

One example of this is the ongoing work with UNESCO-inspired Waldorf schools in Japan and the efforts of the Pindorama movement, which researches Brazilian culture in the light of anthroposophy (www.pindorama.art.br). It is co-founded by Monte Azul. We initiated a two-way process asking: What can the anthroposophical movement learn from Brazilian culture and how can Brazilian culture be fertilised by the human and world view of spiritual science?

Monte Azul wants to be a melting pot for sustainable and humanising impulses that radiate out into the world and enable people to live together in love.

4 Conclusion

Monte Azul emerged from a personal encounter with the social needs of São Paulo's favelas in the context of anthroposophical social and educational work. Over the past 45 years, this led to thousands of people in the favelas of São Paulo not only receiving massive structural support for their basic needs, but also everyone being recognised and supported as an individuality with potential for development and fulfilment. Many educational, medical and training support programmes, cultural activities and the sense of community are constantly building on this ideal and have succeeded in creating hope and prospects in one of the most economically disadvantaged regions of Brazil. Through this work, the understanding of (social) sustainable development is complemented by the dimension of recognising and supporting each individual and their community in their potential so that this can unfold. The book *A Sustentabilidade do Eu* (Sustainability through the I), written by us, describes this process.

The Nature Institute

Craig Holdrege and Ryan Shea

Abstract

Located in Ghent, New York, The Nature Institute is one of only a handful of organizations worldwide dedicated to a rigorous application and further development of the Goethean method in ongoing scientific research and training (natureinstitute.org). These efforts are expressed practically in research projects, educational programs, and publications. We work to understand the living world as a dynamic, interconnected whole. Wholeness includes the scientist as a participant in a respectful dialogue with the wisdom that is inherent in nature. By researching, uncovering, and communicating the interconnected nature of things, we offer ways to overcome the destructive tendencies of fragmentation and isolation that dominate our age. We strive to embrace the complexity of life in all its appearances and awaken the capacity to care for the planet we share.

Keywords

Goethean science • Participatory science • Phenomenology • Living thinking • Goethe • Rudolf Steiner • Sustainability as transformation • Adult education • Experience-based learning • Transformative learning • Holism • Wisdom of nature

C. Holdrege (✉) · R. Shea
The Nature Institute, Ghent, NY, USA
e-mail: craig@natureinstitute.org

R. Shea
e-mail: ryan.shea@natureinstitute.org

J. Kronenberg and E. T. Lammerts van Bueren (eds.), *On the Earth We Want to Live*, World Sustainability Series, https://doi.org/10.1007/978-3-031-98758-8_33

Features:

(a) Founding year	1998
(b) Employees	9
(c) Location	Ghent, New York, USA (natureinstitute.org)
(d) Active in which countries	Talks, Workshops, and Courses in U.S.A., Australia, Canada, Brazil, UK, Germany, Switzerland
(e) Sector	Not-for-profit Organisation
(f) Other	Research, adult education, publications; only institute dedicated to the Goethean approach to science in the English language with a worldwide reach

1 Introduction

In his early 20s, Craig Holdrege formed the germinal idea of founding an institute to develop and practice Goethean phenomenology of nature in the United States (Fig. 1). It seemed to him at the time in 1975—and still does—that this approach can lead to healthy insights into the deeper nature of nature. It can do so when grounded in careful attention to the sense world and just as careful attention to the way we participate as knowers in our efforts to understand the world. The practice thereby avoids the contrasting pitfalls of materialistic interpretation and sense-fleeing spirituality that are so prevalent in the US and elsewhere.

The idea had a long incubation period. During this time, Henrike Holdrege, Steve Talbott, and Craig Holdrege were preparing, each person in his or her way and with varying individual emphasis, the ground for the institute they would—unbeknownst to them until much later—found together and carry into the present. They were studying the works of Rudolf Steiner, Goethe, or Owen Barfield, and

Fig. 1 Logo the nature institute

learning in person from important teachers—Jochen Bockemühl, Ernst-Michael Kranich, Manfred v. Mackensen, Georg Maier, Wolfgang Schad, Andreas Suchantke, and Frank Teichmann. Craig and Henrike developed Goethean practice as Waldorf high school teachers and in areas of research. Steve worked for many years as an editor and writer.

In these and other ways, more than twenty years of diverse activity created the basis of an institute. In 1998, the intentions and biographies of the three founders coalesced and The Nature Institute came into being in Ghent, New York. The founding vision is articulated in our mission statement:

> Nature around us is whole and interconnected. Though we are part of nature, we do not yet fathom nature's depths, and our actions do not embody her wisdom. A fundamental shift in our way of viewing the world is necessary if we would contribute to nature's unity rather than dissolution. At The Nature Institute, we develop new qualitative and holistic approaches to seeing and understanding nature and technology. Through research, publications, and educational programs we work to create a new paradigm that embraces nature's wisdom in shaping a sustainable and healthy future.

Our work of original research and educational workshops is primarily aimed at helping to bring about this fundamental shift in our way of viewing the world. For that reason, in order to pursue sustainability as an institutional goal, it is crucial for us to turn towards nature herself and see what she reveals.

2 Sustainability as Transformation

To our mind, we need to see the issue of sustainability and sustainability education in terms of transformation. What does it mean when we say we want to sustain something?

To sustain means to nourish or to give support to something. In any organic process there is activity in sustaining; it does not just happen. We need to grow and prepare the food we eat, and our digestive organs actively break down and transform food to make the substances our body needs to sustain itself. This is quite different from the more static capacity of columns of a Greek temple to sustain the weight of an architrave. The columns have not needed, over the past 2500 years, periodic infusions of nutrients to hold the weight they carry. But our bones do, and so do the branches and leaves of a 2000-year-old Sequoia. The ability to sustain, viewed from an organic perspective, is dynamic. What is sustained is active and what helps to sustain is active. Think of the sun's energy-generating capacity that the plant utilises to create its own substances. An animal must constantly change, take in new food, metabolise it, to stay alive. At a first level we see that to maintain or sustain is constant activity.

A cow, for example, does not merely stay the same. She changes, grows and develops. She is born as a calf and grows into a cow, but even before that she starts as a fertilised cell (zygote) that must undergo radical transformations to become anything remotely resembling a calf. All organisms live in time, have their own particular life cycle, and go through stages of development. Young mammals need the milk of their mother to sustain them and then shift to a species-specific diet as they mature; they need to shift again when they no longer grow. Like other ruminants, mature cows feed on grass, which is remarkably hard to digest. Cows ruminate. They chew, swallow into their rumen, regurgitate, and chew again. However, this is not enough to break down and digest the cellulose-dense grass. The cow's rumen is populated by microorganisms that do much of this work. In turn, the alkaline saliva helps to keep the microorganisms alive, as they would die in a too acidic environment. The cow would not be able to live without her microorganisms, and vice versa (Holdrege, 2021).

An organism's microbiome reveals the dance of different kinds of life—bacteria, viruses, fungi and more—that all work together in and on the body of any living being. They also mediate between organisms. So, in order to learn what sustains an organism, we must carefully study its life through time and its relations to both its internal and surrounding environments.

The life of an individual organism may last for minutes (in some bacteria), years (in animals), or centuries (in trees). But the life of any species extends beyond the life of the specimen. Each individual is borne of the substance of the previous generation and in death it dissolves into the larger field of earthly substances. When an organism propagates (sexually or asexually), it brings forth one or more progeny that are members of that species. The life of a specimen is thereby a moment in the life of a species, which can span thousands or even millions of years. This extends again our notion of what it means to sustain an organism. It means providing not only the conditions for the co-becoming of the individual, but also for it to multiply.

And species evolve. This evolution cannot be predicted, although in retrospect we can discover evolutionary patterns. The evolution of a species is grounded in its own characteristics and potentials and by changes in the world that sustains it—what we often glibly call the environment. No organism lives in a vacuum, and its environment consists of the life of other organisms, of the changes in weather patterns or seasons and, over long periods of time, of geological transformations. All evolution is co-evolution. There is an ongoing interaction and evolution of diverse organisms within the larger context of the changing ecosphere. We can say simply: the earth evolves.

When we see an organism, a community, or an ecosystem as "states" we try to understand to some degree the complex array of substances and forces that sustain them in the present, as they are now. Any picture of them as existing in the future will be a picture of the present extended into the future. But the more deeply and broadly we understand organisms, communities, or ecosystems, the more the image of a "state" recedes into the background. We recognise them as being dynamic, evolving, and fluid. Our limited picture opens up and becomes less rigid. This is due to the inherent plasticity of living partnerships, the unpredictable nature of evolution and—to make matters worse (that is, more real!)—the fact that we human beings cannot separate ourselves and our activity from this evolution, since we ourselves are part of the web of life on Earth.

For these reasons, we cannot know in the way of a set of rules what is needed to sustain evolving life in the future. This is not to say that we cannot know anything about what is needed. But we want to emphasise that inasmuch as we acknowledge transformation and evolution in planetary life—and that we are active participants—we enter a realm of freedom. Because we are dealing with dynamic processes and interactions that we can engage with in a variety of ways, we need to conceive of sustainability not as a set of programmes but as an evolving process. We need to ask: What kind of capacities do we require for engaging in such an open-ended process and how can we develop these capacities?

If we take the transformational, evolutionary nature of life seriously and design programmes and plans of action, we need to be keenly aware of our ignorance. Socrates' insight that wisdom consists in awareness of our ignorance has lost no pertinence over nearly two and a half millennia. Awareness of what we do not know makes us more cautious, circumspect and observant. It infuses us with a readiness to change our ideas and course of action (Vitek & Jackson, 2008). It is a key feature of a real ability to evolve.

At a fundamental level, sustainability hinges on evolving human capacities and the awareness that how we engage can also evolve in as yet unknown ways in the future. As an attitude of mind, sustainability would entail wholeheartedly embracing this journey (Fig. 2).

3 The Nature Institute: Cultivating Transformation

For us at The Nature Institute the primary question is: How can we, in the present, begin developing an awareness of the dynamic and transformational nature of life that can increasingly inform our own thinking and doing, and that can enable us to become more conscious and responsible participants in an evolving Earth?

This question is our guiding question, the one we carry with us, and the one that carries us along. Every aspect of what we do at The Nature Institute engages this question. The work is in the doing, not the having done. Our guiding motto comes from Goethe, who said that "If we want to attain a living understanding of nature, we must follow her example and become as flexible and mobile as nature herself" (Goethe, 1817/2002, p. 56).

First and foremost, the staff at The Nature Institute strives to practise a phenomenology of nature. In order to develop a Goethean, qualitative, contextual and holistic science, we need to practise. A core feature of this endeavour is learning to attend to phenomena of the world with patience and persistence. We could speak here of practicing faithfulness to the concrete appearances (see our other chapter in this book). This fidelity to the phenomena—this careful and disciplined striving to leave behind our preset plans and goals, and our pre-given judgments and prejudices—all this is not easily done. It needs to be developed and cultivated.

We are a small group, with currently nine full and part-time staff. The institute is in the middle of a transition as the founders work together with a new generation that will take the work into the future. As an organisation, we are currently

Fig. 2 Nature Institutes summer course

right in the thick of questions concerning institutional sustainability and transformation. Our attempt is to not separate the organisational (social) and economic aspects of the institute from its core scientific and educational work. The founders are fully aware that they cannot replace themselves with younger colleagues who merely continue what has already been done. The work will metamorphose. We are keenly aware and deliberate in confronting these challenges with the conviction that sustainability begins at home.

Much of our research and publications have focused on living beings and the nature of life. For example, Craig Holdrege has published many whole organism studies, culminating in two books: *Thinking Like a Plant: A Living Science for Life* and *Seeing the Animal Whole: And Why It Matters*. Steve Talbott has been working for several decades on a project titled "Biology Worthy of Life," which has culminated in his book *Organisms and Evolution: Agency and Meaning in the Drama of Life*. We have also, however, investigated other realms. Henrike Holdrege has focused on phenomenology of the visual world, the four elements and projective geometry, on which she published a two volume projective geometry workbook titled *To the Infinite and Back Again*.

Each person's practice has a twofold commitment: (i) to be with the phenomena, to slow down enough to learn the teachings of nature and (ii) to cultivate a living thinking, to help our thinking become "as flexible and mobile as nature herself." In this commitment our efforts are inspired by Rudolf Steiner's work. In articulating the need for Goethean phenomenology of nature, Steiner said that "science must awaken in a Goethean sense… We need to learn to get beyond our dead abstract concepts and move into living, concrete concepts" (Steiner, 1919/1979, p. 103). And it was Goethe who "brought the petrified formal concepts into movement. That was his great central act" (Steiner, 1914/1980, pp. 17–18). By bringing our thinking into movement we can open ourselves more fully to the things themselves and develop what Steiner once called a "phenomena-immanent science" in which the task is "to take the field of phenomena as it is and learn to read its own significance" (Steiner, 1922/1997, pp. 25–26).

Part of our research work takes up mainstream science and current hot topics of contemporary research. For example, Craig and Steve spent many years investigating and writing about genetics, both genetic reductionist explanations and technological genetic engineering. Steve has spent much of his time over the last several decades researching mainstream accounts of evolution and materialist explanations of organisms. Craig and Jon McAlice sought out a more holistic understanding of viruses during the COVID-19 pandemic. Three of the staff are currently in a multi-year project looking at intelligence in nature, starting with the popular current topic of "plant intelligence." Although each researcher at the institute has their own individual phenomenological practice, we do not work in isolation. We come together for weekly collaborative inquiry and work together in our education programmes.

In all our interaction with mainstream science, our guiding task is to continue our twofold activity of learning from the phenomena themselves and cultivating living thinking. To do this requires different responses, for not all science is created equal. Sometimes we are called to give a strident critique of a particular scientist's claims or more generally held current scientific ideas or paradigms when we find that they ignore or deny the phenomenal realm or that they deaden our thinking and thereby blind our vision. This is the case, for example, in genetic reductionism, wherein an organism is seen as the result of genetic mechanisms. Here we must clear the roadblocks that skew our understanding. However, sometimes what is called for is to point out that the phenomena uncovered by mainstream scientists allow for new and more adequate ideas. Much of Steve Talbott's work has been to show that the actual empirical findings of contemporary biology can be seen in a much more qualitative, holistic, and dynamic light than the prevailing habits of thought and theories allow. And, of course, there are those mainstream researchers who share our core orientation and with whom we become collaborators, either through direct communication or simply by being mutually aware travellers on the same path.

As a further expansion of our core practice, we offer a variety of workshops and courses. They vary in length from a half day immersion into the world of color, to three day workshops investigating the world of plants, to our fifteen month, low residency "Foundation Course," which is attended by people from around the globe. We offer courses tailored to Waldorf teachers, biodynamic farmers, teacher trainers, Camphill workers and homeschooling parents. We are also asked to give talks, workshop,s and courses in many different countries. In every course and workshop we introduce participants to Goethean phenomenology by doing it.

For example, we have taken a group of twenty participants out to a red maple tree right next to our main classroom, where they are guided in encountering the tree in its immediacy with all their senses. The group then sat in a circle—each with a different leaf from the same tree. They were invited to go round the circle and each person described an aspect of their leaf. Sometimes we can go round the circle two or three times and still make new discoveries.

This is a simple exercise, but a powerful one. A transformative one. The Nature Institute provides a place for people to be unrushed, to slow down, to quiet their chatting and their ever harried minds, and to simply look, listen, think carefully and learn. Time and time again, participants state how transformational this practice is for them. It is particularly important that they are working together as a group. Time and again we experience the truth of what Goethe wrote: "It is true here as in other human endeavors that only the interest of many focused on a single point will generate something excellent" (Goethe, 1792/2010, pp. 19–20).

Through these courses, we want to help participants gain insights into qualities of living thinking and and develop skills that can come to the unique expression in their biography, their profession, their specific community, and their other daily commitments. We are not trying to "make" Goethean researchers. We want to give participants the basis to take up, work out of, and develop the Goethean approach in their own ways and within their own life contexts. This is a facet of our efforts to stimulate sustainability as a transformational process. And this includes our commitment to continue to work together with those individuals and communities that strive to develop their work further.

A final expansion of the practice comes in our publications. In them, we are doing yet again the same thing, the twofold core practice of turning to the phenomena and cultivating living thinking to enable us to learn from the appearances. We strive to write in such a way that the phenomena considered—skunk cabbage, the blue of the sky, bison, sloth, cow, or the appearance of buds and more—come vividly to the reader's imagination. Our hope is that our portrayals help bring readers into a more living understanding of the natural world. Even more, our hope is that the publications inspire readers to attend more carefully themselves to the kind of natural phenomena we describe and to make their own discoveries. Our website, natureinstitute.org, contains hundreds of articles by our staff and other authors. It is a world-wide resource on Goethean science in the English language.

Many of our publications also directly address problematic and fraught topics in current mainstream scientific research, such as sustainability, genetics and evolutionary theories. When our writings take up critique of deadening ideas in the mainstream, we still strive to orient ourselves and the reader so that it is not so much us, but the holistic, living world itself, that does the critique. Taking nature as our guide, the deficiencies of reductionist or anthropocentric ideas become obvious by contrast.

4 Conclusion

There is no doubt that we human beings are in many ways a powerful destructive presence on earth today. This recognition often leads people to feel that the best solution to problems of sustainability involves removing human influence. The focus on our negative impacts can easily lead to despair and resignation. However, are we by necessity only taking, consuming and destroying? Can we also be giving, building, restoring, and healing? Can the human-earth relation evolve? In our work, we take up the task of developing and increasingly engaging our whole-human capacities in turning toward and learning from the wisdom of the natural world. This is one contribution to a much needed and promising adventure—the potential healthy co-evolution of humanity and Earth.

References

Goethe, J. W. v. (1792/2010). The experiment as mediator of subject and object. *In Context #24, Fall 2010*. https://www.natureinstitute.org/article/oethe/experiment-as-mediator-of-object-and-subject

Goethe, J. W. v. (1817/2002). Zur Morphologie: Die Absicht wird eingeleitet. In Goethes Werke, Bd. *XIII (Naturwissenschaftliche Schriften I)*. München: C. H. Beck, 2002. Translation by C. Holdrege.

Holdrege, C. (2021). The dairy cow and our responsibility to domesticated animals. In *Seeing the animal whole and why it matters by Craig Holdrege* (pp. 259–289). Lindisfarne Books.

Steiner, R. (1914/1980). *Der Menschliche und der kosmische Gedanke*, lecture on January 20, 1914; German Bibl. Nr. 151. [English translation: *Human and Cosmic Thought*; Rudolf Steiner Press, 2024].

Steiner, R. (1919/1979). *Die Erziehungsfrage als soziale Frage*, lecture on August 17, 1919; German Bibl. Nr.296. Translation by CH. [For a published English translation see: *Education as a Social Problem*; Anthroposophic Press, 1984.].

Steiner, R. (1922/1997). *Erneuerungsimpulse für Kulture und Wissenschaft*, lecture on March 6, 1922; German Bibl. Nr. 81. Translation by CH. [For a published English translation see: *The Impulse of Renewal for Culture and Science*; Rudolf Steiner Publications, 2018.].

Vitek, B., & Jackson, W. (Eds.). (2008). *The virtues of ignorance*. The University Press of Kentucky.

NatureTex—The (R)Evolution of Egypt's Textile Industry

Konstanze Abouleish

Abstract

NatureTex, a SEKEM subsidiary in Egypt, combines biodynamic agriculture with the four dimensions of sustainability: ecology, economy, social life, and culture. Guided by anthroposophical principles of human consciousness development. Founded in response to a pesticide crisis in textile production, it has proven over 30 years that high-quality cotton can be produced without chemicals while fostering social justice, cultural vitality, and economic fairness. This model shows how philosophy and practice can merge to achieve lasting sustainability in the textile industry.

Keywords

Sekem • Egypt • Cotton • Textiles • Sustainability • Consciousness • Dimensions

Features:

(a) Founding year	1994
(b) Employees	200
(c) Location	Cairo/Sekem
(d) Active in which countries	6+ (Country count for production: only Egypt)
(e) Sector	Textile production
(f) Characteristics	One of the first organic cotton producers

K. Abouleish (✉)
NatureTex, El Salam City, Egypt
e-mail: konstanze.abouleish@sekem.com

J. Kronenberg and E. T. Lammerts van Bueren (eds.), *On the Earth We Want to Live*, World Sustainability Series, https://doi.org/10.1007/978-3-031-98758-8_34

Fig. 1 Logo NatureTex

1 Introduction

NatureTex is a subsidiary of Sekem in Egypt.[1] Sekem was founded in in 1975, when Dr. Ibrahim Abouleish visited his home country of Egypt after living in Austria for many years. The country had changed a lot, many problems had arisen and the desire to get involved was born. From this grew a variety and multitude of initiatives, companies and working groups. NatureTex as a response to a growing problem in the textile industry in Egypt was one of them (Fig. 1).

2 NatureTex—An Answer to the Pesticide Problem in Egyptian Cotton Production

NatureTex emerged from the fact that the Egyptian state could no longer export cotton due to the high pesticide residues nde. After Sekem was able to prove for two years that cotton also grows well without pesticides, the application of pesticides by aeroplane was stopped throughout the country and the quantities of pesticides used up to that point were significantly reduced.

The quantity of cotton that was now available as a result of these trials later led to the foundation of the company Conytex, now NatureTex. Located in Sekem, the external conditions were favourable. It started with an existing room, 3 or 4 sewing machines, an employee who was familiar with the cotton value chain process and a manager. In the beginning, everything was produced for export, later the local market was added. NatureTex employees, like all Sekem employees, are surrounded by art and culture. Attention is always paid to beauty, both in the landscaping and in and around the buildings where work is carried out. There are spokespersons who are elected and who mediate on a social level, contribute new ideas, help organise an upcoming festival or other things. Optimisations in production, for example, come most effectively from the production itself, so the employees are part of the whole and the whole ultimately consists of them, of the people.

In the meantime NatureTex employs around 200 people, in the subsidiaries of the Sekem initiative still overtaken by Isis[2] with around 600 employees. The company has moved twice to expand and optimise, and is celebrating its 30th anniversary this year. There were ups and downs, which could be mutually

[1] SEKEM is presented in Chapter "SEKEM Group—Sustainable Development Since 1977".

[2] Isis belongs to SEKEM Holding and is the leading tea producer in Egypt. All teas are produced biodynamically.

equalised together with the other parts of Sekem—an associative idea. The export share of NatureTex has always been higher than the supply of the Egyptian market, but the latter is currently being built up and should soon account for at least 25% of the total volume.

The Sekem products sold should be able to answer four questions from the buyer: (1) what impact does the production of this product have on nature, (2) what impact does it have on the people who produced it, (3) did each link in the added value chain receive a fair price, and (4) what impact does the product have on the buyer himself? More and more people are asking about the background and want to know the answers to these questions. The buyer as a determinant of the world market—this gigantic possibility and responsibility is slowly dawning on us, and it is astonishing how our consciousness so easily "pushes it away" and still reaches for the cheaper or more easily available offer.

Change will certainly come, the only question is when. Existence "between the times", i.e. between what was and what wants to become, requires an alertness that is not readily available to us. The practical path of practice, which Rudolf Steiner left us in such abundance and variety, will occupy us for several centuries to come! (Fig. 2).

Fig. 2 Cotton harvest NatureTex

3 Applying Anthroposophy and Sustainable Development

The four dimensions of ecology, economy, social life and culture, with which NatureTex and SEKEM work, emerged from the social threefold concept[3] of anthroposophy on the one hand, and the encounter of the desert north of Cairo on the other. These dimensions determine the definition of sustainable development and characterise the work of all affiliated companies.

The dimension of ecology or nature demands care and attention from people in order to create a framework for a sustainable society, which is what the threefold social order would like to contribute to. The desert, which had to be greened, made it clear that without this appreciation and care of the given, there can be no threefolding at all. The human being is missing from this calculation, because a person cannot live in the desert.

The economic component logically takes centre stage in a company, but this does not mean that earning money is the only goal. It is always about the wholeness that every life requires. The four dimensions exist everywhere, there is no "pure" economy, law or culture anywhere in the world, they always play together. The essential thing is equilibrium, balance. We are currently experiencing a lot of challenges in the world, which are largely due to the gigantic excesses of just one dimension.

Harmonising the four dimensions sounds relatively feasible, but it is not. It is a life task to get to know the other dimensions (usually you are only really at home in one of them) and to learn to understand them. The people who feel at home here or there are often fundamentally different in character and naturally achieve little in the way of true understanding. At some point, however, it can become clear: the certain knowledge that "my" dimension is not the most important and right one, and that it cannot live without the other three. The way there hurts, bites, tears and fires!

The second component that is decisive for any functioning context is the different levels of consciousness in which the human being lives. Anthroposophy distinguishes between the sentient soul, the soul of mind and spirit and the soul of consciousness (Steiner, 1910/1994). Every individual has the opportunity to develop. These concepts provide a lifelong, exciting, infinitely rich and constantly surprising field of practice. And it is a good example of the fact that it is never just about the words and the intellectual learning content, but always about life and action. Without needing the word "anthroposophy", art and culture were celebrated, the various levels of consciousness were cultivated and economic endeavours were promoted. In the background is an idea that came into our world today through anthroposophy and can be brought to life through our actions.

---

[3] The social threefold structure is a framework where society is divided into three dimensions: economic life where solidarity is leading, cultural life where freedom is leading and legal life where equality is leading. More on this in this publication in Chap 4.12 *Earthbound Economics.*

These two principles, the four dimensions and the levels of consciousness, are interwoven in a thousand ways and follow no principle or guideline except one: always wanting to be consciously lived anew. All our structures are like clothes that at some point become too tight, too boring, too colourful or too thin for a living being. The aim is to recognise these questions of time and to deal with them as fruitfully as possible.

4 Conclusion

NatureTex was founded as a subsidiary of SEKEM in Egypt to respond to the pesticide problem in the country's textile production. After an import ban on Egyptian cotton due to excessive pesticide content, NatureTex was able to prove within a few years that it is possible to produce high-quality cotton without pesticides. NatureTex works according to SEKEM's sustainability principle, which is based on the four dimensions of ecology, economy, social life and culture. Biodynamic agriculture forms the basis, while anthroposophical principles create the framework and enable an expanded understanding of sustainable development. In 2024, NatureTex celebrates 30 years of pioneering work for an ecological, socially just and economically fair production method in the textile industry.

Reference

Steiner, R. (1994). *Theosophy: An introduction to the spiritual processes in human life and in the cosmos* (original work published 1904). Anthroposophic Press.

Emergency Pedagogy Without Borders—From Trauma to Developmental Opportunity

Johannes Beurle and Bernd Ruf

Abstract

Turning crises into opportunities for children and young people after traumatic events, helping them to overcome their traumas and avoiding trauma-related disorders—this is the mission of the 'Emergency Pedagogy without Borders' association founded in 2020.

Features:

(a) Year of foundation	2006 (at Freunde der Erziehungskunst Rudolf Steiners); 2019 Foundation of the international network 'Emergency Pedagogy without Borders'; 2020 Foundation of the association 'Emergency Pedagogy without Borders'
(b) Employees	4 full-time employees and 100 trained emergency pedagogues
(c) Location	Parzivalstrasse 1, 76139 Karlsruhe, Germany
(d) Active in which countries	Worldwide in crisis regions and representative offices in 32 countries
(e) Sector	Emergency pedagogical acute interventions in crises, major damage situations, disasters (NFP outreach); emergency pedagogical outpatient services, counselling (NFP outpatient clinic); Seminars, certified training and further education (NFP Academy)

J. Beurle (✉) · B. Ruf
Notfallpädagogik ohne Grenzen e.V., Karlsruhe, Germany
e-mail: j.beurle@nfp-og.org

B. Ruf
e-mail: b.ruf@nfp-og.org

J. Kronenberg and E. T. Lammerts van Bueren (eds.), *On the Earth We Want to Live*, World Sustainability Series, https://doi.org/10.1007/978-3-031-98758-8_35

1 Emergency Pedagogy Without Borders (Notfallpädagogik ohne Grenzen (NFPoG))

Turning crises into opportunities for children and young people after traumatic events, helping them to overcome their traumas and avoiding trauma-related disorders - this is the mission of the 'Emergency Pedagogy without Borders' association founded in 2020. In 2006, Bernd Ruf (Waldorf teacher) was asked by the mayor of Stuttgart to accompany a group of Lebanese children back to their home country, where war had broken out. Since then, the following question has stayed with him: What can we do to help children pass the trials of destiny without being broken by them? As research into psychotraumatology continued, the possibilities of Waldorf education became increasingly clear, with many elements showing a preventative protective effect on extreme psychological situations. The first experiences were soon gained. Individual elements of Waldorf education were applied in a targeted manner in crisis areas. Accompanied by scientific research, a concept was developed as to how educational methods and settings could significantly reduce the risk of developing a trauma-induced disorder. Emergency pedagogy was initially developed within the 'Freunde der Erziehungskunst Rudolf Steiners e.V.'. In 2019, emergency pedagogical intervention teams joined forces to form the 'Emergency Pedagogy without Borders' international network, which now includes teams from 32 countries. Whether it be natural disasters, civilisational catastrophes, terrorist attacks or wars—when children are affected, intervention teams from 'Emergency Pedagogy without Borders' are on hand worldwide to provide help (Fig. 1).

Currently, matters are not looking good for the physical and psychosocial health of our children and young people: Paediatricians are lamenting an explosive

Fig. 1 Logo emergency pedagogy without borders—as provided by the association

increase in psychosomatic illnesses, child and adolescent psychiatrists an enormous increase in emotional disorders, and educational professionals the rapid rise in aggression and behavioural problems ('Donau Universität Krems', 2021; Copsy Study, 2022; Hurrelmann et al., 2023) Paediatricians' surgeries have reached the limits of their capacity, child and adolescent psychiatric clinics are overcrowded and there is a widespread lack of urgently needed therapy places across the country.

What has happened? What causes and backgrounds offer explanatory approaches for the existing distress of children and young people? How can sustainable help be provided?

Against the background of a spiritually expanded understanding of the human being and the world, the following chapter outlines the current psychosocial situation of children and adolescents and illustrates the pedagogical-therapeutic possibilities of sustainable trauma processing through anthroposophical emergency and trauma pedagogy.

2 Trauma: When Nothing is as It Was Before

Trauma is a shock-like paralysation (freeze) in the face of overwhelming feelings of helplessness and powerlessness, triggered by an event that has the effect of threatening one's existence and that cannot be overcome. According to Fischer and Riedesser (2009, p. 84), psychotraumatisation creates a vital discrepancy between a subjectively experienced existential threat and the person's own coping mechanisms. This shatters basic assumptions about the world and one's own self in those affected. The term trauma refers to 'the injury and lasting damage to an existing structure' (Hausmann, 2006, p. 31).

This can affect both the body and the soul. A psychotrauma entails 'an event, usually occurring suddenly, which has a very threatening effect on the person affected and which at the same time appears to be unmanageable. It violates the basic assumptions about the world and one's own person' (Hausmann, 2006, 31).

As long as the trauma-induced shock paralysis does not dissipate, the wound to the soul cannot heal. This often results in painful psychological and psychosomatic reactions. Those affected suffer from their terrible memories, which obsessively assail them and repeatedly trigger deadly fears (flashback). They cannot forget. For others, the experiences are so unbearable that they push the incidents into the subconscious and can no longer remember them at all (amnesia). Rhythm disturbances, as in the relationship between remembering and forgetting, can often occur after traumatisation and manifest themselves in the form of sleep disorders, eating disorders, movement disorders and digestive disorders. In addition, there are concentration disorders accompanied by nervousness and overexcitation. Many children are sad, depressed and paralysed after traumatic experiences. Others are aggressive and display hyperactivity. Still others are numb or feel empty inside. It is understandable that children with terrible experiences try to avoid anything that could evoke memories of the traumatic event (triggers). These can be places, people, smells, objects, colours, sounds and much more. The strategy of avoidance behaviour and irrational feelings of guilt ultimately affect everyday life and put a strain on social relationships.

Trauma is initially an injury and not an illness. Injuries usually heal on their own, provided our resilience is sufficient for the wound healing process. However, wounds can also become infected and then lead to serious courses of illness. The first weeks and months after a traumatisation determine whether the traumatic experience can be processed and integrated into the biography or whether the painful experience is repressed without being processed, often resulting in lifelong, painful trauma-related disorders. Chronic traumas can ultimately throw life completely off course. Severe personality disorders, depression, panic attacks, aggressive or auto-aggressive impulses, violent crime and suicidal behaviour can result and ultimately even lead to a biographical break (Levine & Kline, 2005).

Traumatic congealing impedes further physical, psychosomatic and mental development. Entire areas of the brain can be blocked and then no longer develop. Here, the age at which the trauma occurs is decisive. It makes a big difference whether a developed organism is traumatised or whether the organism is still developing. The younger the child is at the time of traumatisation, the higher the risk of developing serious trauma-induced disorders (Krüger, 2007). Traumas firmly fix the victims to the past and have the potential to prevent the future from unfolding because development may come to a standstill and new things can no longer be absorbed at all. This may also be the cause of complex learning and social disorders in the school context. Traumatisation usually changes lives for a lifetime!

Time and again, children and young people have to suffer the incomprehensible. They experience the loss of caregivers, are involved in serious accidents or have to undergo medical procedures due to illness. They become victims of natural disasters, wars, torture and displacement or suffer neglect, sexual abuse and maltreatment at the hands of their caregivers. When the incomprehensible occurs, the lives of these children and young people are lastingly altered. Here, the consequences of the experiences are all the more severe, the younger the child is and the closer the relationship to the person who caused them is. A child's brain is about 25 per cent developed at birth. Most of the neuronal networks only develop later. According to the current state of research, the imprints of the first two years of life are irreversible despite the incredible plasticity of the brain (Welt-online, 2023b).

3 Ever More Crises—Ever More Traumatisation

Crises, wars and disasters have (massively) increased in quantity and quality worldwide in recent years, at least in the perception of many people. This can also explain, albeit in part, the (explosive) massive increase in traumatisation (Bundestag & Dienste, 2016).

The number of natural disasters worldwide has more than quadrupled over the past 40 years. Millions of children and young people are affected by natural disasters every year (IFRC, Disaster Report, 2012). Disasters pose a particularly high level of stress for children and young people. Many are unable to adequately process their traumatising experiences (Juen et al., 2017; Ruf, 2018).

The number of children forced to suffer in wars has never been as high as it is today. According to the German Association for the United Nations (Deutsche Gesellschaft für die Vereinten Nationen e.V., 2019), around 420 million children under the age of 18 live in a conflict zones worldwide. Almost one in five children is affected by war or conflict. Hundreds of thousands of children die as a result of armed conflicts, millions are traumatised (Shah, 2017; Ruf, 2019a).

According to the UN Refugee Agency, over 100 million people are currently displaced. Around 50 per cent of them are under the age of 18, i.e. children. Their events and experiences as refugees often leave them with deep wounds to the soul. Anxiety, depression, insomnia and long-term psychosomatic illnesses then have a lasting impact on their development (Adam, 2016; Ruf, 2019a; Zimmermann, 2012).

4 The Covid-19 Trauma

The Covid-19 pandemic and the measures taken to contain it have changed the everyday lives of individuals, families, institutions and entire social systems in a previously unimaginable way within just a few weeks. Many people are frightened, unsettled and traumatised. They fear for their health, their economic livelihood and their freedom. This perpetual existential fear is making people ill (Spitzer, 2020)!

The general societal psychosocial effects of the Covid-19 trauma are already highly noticeable and drastic. These are not the traumatic consequences of a virus pandemic, i.e. a natural disaster (type I trauma), but the traumatic effects of the measures taken to combat the pandemic, i.e. traumatic consequences caused by human beings (type II trauma). Scientific studies on the consequences of such 'man-made disasters' show that depression and post-traumatic stress disorder occur more frequently in the first year after the catastrophic event. Longer-term consequences include dysthymia (mood disorders), burnout syndrome, alcoholism and other addiction problems, as well as developmental disorders in children and adolescents. We are increasingly dealing with such symptom formations throughout society (Spitzer, 2020).

The Covid-19 crisis is a collective traumatisation on a previously unimaginable scale. However, it is to be feared that we are only at the beginning of a traumatic process that will present us with unforeseen challenges for society as a whole over the coming decades. After all, the traumas that have emerged or re-emerged in isolation will by no means disappear with the end of the Covid-19 crisis.

5 The Kaspar Hauser Syndrome: The Trauma of the Thwarted Mission

It was on Whit Monday at the end of May 1828, when a 16-year-old adolescent, obviously in need of care, who exhibited the psycho-emotional level of a three-year-old child, was abandoned on the Unschlittplatz in Nuremberg (Wassermann, 1908.; Schötz, 2010).

Already the very first police investigations already pointed to a 'crime against the soul-life of a child' (Feuerbach, 1832). The Nuremberg municipal court physician testified: 'This human being is neither mad nor insane, but has obviously been forcibly removed from all human and social education in the most wretched of ways' (Dr. Preu quoted by Pies, 1966).

The young foundling from Nuremberg, who was also known as the 'Child of Europe' and who finally fell victim to an assassination attempt in Ansbach in 1833, referred to himself as 'Kaspar Hauser'.

Even today, in medicine and psychology, the most severe forms of hospitalism, where human beings have suffered massive damage to their physical, psychosocial and mental development through isolation and abuse, are referred to as 'Kaspar Hauser syndrome' or also as hospitalism or deprivation (Wikipedia: Hospitalism; du Bois & Resch, 2005; Resch et al., 1999); Knölker et al., 2007). The background to Kaspar Hauser syndrome is traumatising extreme stress experiences.

6 Increasing Trauma Vulnerability

The term 'trauma' is on everyone's lips these days. It is currently being used in the media, in everyday language, but also by psychologists and therapists in a seemingly inflationary fashion. Almost everything seems to be able to trigger trauma: Murder and manslaughter as well as the defeat of the home club in the local football league. The inhabitants of entire regions of the world are now considered 'traumatised'.

7 Why Does Traumatisation Seem to be Occurring on an Epidemic Scale Today?

As early as the beginning of the last century, Rudolf Steiner pointed out impending evolutionary changes in the interaction of the human constitutional fabric and described their consequences (Steiner, 1983, GA 184). The firm connection between the vital organisation (etheric body) of the human being and their physical organisation would increasingly loosen. As a result of this change in the overall human constitution, new soul abilities and forms of consciousness as well as a higher sensitivity of the human being would become apparent. 'We are living in a time in which the etheric body is slowly dislodging from the physical body. It is part of the normal development of humanity into the future, that the etheric body gradually leaves the physical body again; (…) We are in the midst of this process, and many of the more subtle symptoms of illness in the present would be understood if we knew this' (Steiner, 1982, GA 113, pp. 131–132).

The altered interaction of the human constitutional fabric, i.e., the loosening of the etheric body from the physical body, forms the anthropological basis for increased psychological vulnerability and a growing risk of traumatisation (Steiner, 1982, GA 113, pp. 131–132).

8 Emergency and Trauma Pedagogy: How Pedagogy Can Help in Processing Trauma

Trauma profoundly impacts the lives of children. They therefore need special support and affection after traumatic experiences. In addition to psychological approaches, concepts for emergency and trauma pedagogical interventions have also been developed in recent years.

Emergency pedagogical interventions take place in the first developmental phases of the traumatic process, in which it is decided whether the stressful experience can be processed or whether a trauma-related disorder will develop. By means of pedagogical methods, the self-healing powers and processing capabilities of the children and adolescents affected are to be stimulated and fostered in such a way, that overcoming the experience is supported and enabled. The aim is to activate resilience and prevent possible subsequent trauma-related disorders. The faster the help, the shorter the suffering (Ruf, 2019b, pp. 9–10).

While emergency pedagogy as secondary prevention at the beginning of the trauma process provides stabilising first aid for the soul, using pedagogical methods in order to avert subsequent trauma disorders, trauma pedagogy works as tertiary prevention in the phases of traumatic illness. In addition to a variety of methods, far more complex organisational structures, multi-professional collaboration and educational-therapeutic services are required here (Ruf, 2019b, p. 9–10).

Parents also require psychoeducational training and counselling as well as support in dealing with their children's trauma-related ways of reacting. In counselling sessions or group settings, they must learn to understand the often conspicuous and usually disruptive behaviours of their children as a failed attempt at self-healing and a cry for help, in order to develop appropriate parenting methods.

Educational professionals in particular have reached their stress limits in the post-Covid phase due to their own traumas, the trauma reactions of children and young people as well as the increasingly conflict-ridden interaction with parents and colleagues. They urgently need specific approaches for dealing with distress and preventing burn-out syndrome. These include targeted exercises to tap into creative and vitalising forces (Ruf, 2018, pp. 225–227). One could also speak of an 'anti-stress programme' for pedagogical professions: Against the background of Rudolf Steiner's anthropology, the physical body is an instrument of spiritual individuality. Without it, self-efficacy of the human being in the physical world is not possible. The idea of the body as an instrument of the 'I' illustrates the responsibility of human beings towards their 'working instrument'. Thus, it goes without saying that hygiene, a healthy diet, sufficient exercise, enough sleep and much more must be taken into account.

Rhythm and time are inextricably linked to life. Rhythm therefore manifests itself as a primordial phenomenon of all life. The whole of human life is characterised by rhythmic elements. Rhythm keeps human life in balance and gives us strength. If the rhythm is disturbed, we lose our balance. Physical and mental illnesses are an inevitable consequence in the long term. Rhythm holds considerable significance for physical health, our attitude to life and the associated social implications. A conscious rhythmisation of everyday life can open up inner, vitalising sources of strength. It is an essential element of a conscious approach to life and constitutes an important aid in dealing with stress.

The harmonious interplay of the fundamental psychological forces of cognition (thinking), emotionality (feeling) and the capacity to act (willing) forms the basis of the inner, psychological health of the human being. The possibilities of creative action are also based on this harmony. Whether eurythmy, music, language, colour or form - all works of art are absorbed and experienced through human feeling. The training of this experience is, alongside the training of the senses and fine motor skills, the task of all artistic practice. Artistic endeavour always appeals to inner activity. Through the inner activity in the artistic process, creative forces can be developed, which in turn have an invigorating and vitalising effect in everyday life.

The importance of imaginative, healing images, mindfulness exercises, religious-spiritual attitudes as well as prayer and meditation in the context of trauma prevention and coping with trauma is well known in psychotraumatology and various schools of trauma therapy.

In the post-Covid phase, indirect educational stabilisation methods are particularly useful for stabilising children and adolescents who have been traumatised due to Covid-19, as they help to activate resilience skills and thus alleviate the painful trauma symptoms and promote the processing of the experience (Ruf, 2022, pp. 283–285; Eckardt, 2005).

Through targeted rhythmic care, the traumatised child's organism is to be harmonised again and its self-healing powers are to be activated. This involves, among other things, structured and rhythmic daily routines as well as regular eating and sleeping times. The ritualisation of table prayers as well as morning and bedtime rituals provide security, support and new orientation. Movement therapy approaches such as eurythmy and Bothmer gymnastics as well as massages, rhythmic embrocations and body geography exercises can help to release trauma-related cramping and to find a new relationship with one's own body. Artistic activities such as painting, drawing, kneading, dancing or making music can also help to give creative expression to what is actually unspeakable and cannot be communicated verbally, thus allowing it to be processed.

Experiential education approaches can, for example through climbing exercises, rebuild the trust in oneself and others that has been lost due to the trauma. They can also train the often severely impaired ability to concentrate, e.g. through thread games, Memory or Mikado, and playfully compensate for the trauma-related loss of social skills as well as develop new social skills. In the telling of age-appropriate fairy tales, stories and biographies, healing images can be used to counteract the images of destruction in the soul caused by traumatisation (Ruf, 2018, pp. 150–152). Puppet shows have also proven to be helpful in emergency pedagogical intervention. By identifying with puppet figures, stressful experiences can be represented and creatively processed through play.

Trauma fixes its victim to the terrible past and blocks their perspectives for the future. The future must therefore be reclaimed with traumatised children and young people. This can be done by planning and realising projects together, such as a joint meal or an excursion. This also helps to overcome trauma-related feelings of powerlessness and helplessness, acquire new behavioural skills and open up experiences of self-efficacy (Ruf, 2019a, 255).

Finally, the factor of joy should not be forgotten when applying all of these methods. Joyful moments increase the organism's disposition to recovery. There are studies by the University of Pittsburgh that predict the correlation between a person's stress level and the likelihood of catching a cold. Stress, anger, resentment or negative memories trigger chaotic heart rhythms for a few minutes, as a result of which the immune system is weakened for about six hours. Immunoglobulins A, which are constantly produced in the mucous membranes and protect against infections, drop significantly after stress, thereby weakening the organism's resilience. This is why after a traumatic experience, there is a significantly increased risk of infection (Ruf, 2018, p. 153). Conversely, joy, the experience of empathy and positive memories lead to heart coherence and an increase in the production of immunoglobulins A and thus to an increase in resilience. Joy stimulates self-healing powers, joy heals (Ruf, 2018, p. 256)!

In times of crisis and in threatening situations, children and young people need stabilisation methods as well as places where they can feel safe and secure (Ding, 2009; Ruf, 2019b, pp. 12–14). In order to achieve such places of safety, educational institutions must now pay particular attention to clear, transparent structures on a physical level (procedures, responsibilities, order, cleanliness, etc.) as well as to aesthetics in the structuring of living, working and communal areas (flowers, picture decorations, table design, etc.). At the beginning of the twentieth century, Rudolf Steiner pointed out the connection between the aesthetics of everyday life at the time and the development of infectious diseases. Aesthetics in external design help to strengthen the immune system. Beauty heals! Shaping our daily routines is also more important now than ever: the structure of everyday life, rhythmisation of the daily and weekly routine. Rituals also strengthen our immune system. Rhythmisation and ritualisation provide stability and orientation! Shaping the relationship level is particularly important: Attention, interest, emotional warmth and closeness as well as community are essential elements of psychosocial prevention. Studies have impressively shown the influence that human affection and social inclusion have on the bio-psychosocial health disposition of human beings. This also applies to the post-Covid era. On the mental-spiritual level, it is ultimately about strengthening the specific identity of the organisation: guiding principles, value standards, objectives and a sense of purpose! This not only strengthens the social organism of the institution, but also has an impact on the resilience of the children and young people as well as the educational staff.

9 Assistance for the Kaspar Hauser Children of Our Time

Kaspar Hauser, the child of Europe, was able to experience a pedagogical maturation and find more and more of his identity and purpose with the support of his patrons: the brilliant humanist court president Feuerbach, who made a significant contribution to the abolition of torture and the establishment of orderly court proceedings in Germany, the gifted pedagogue Friedrich Daumer, who also taught Hegel's sons, and the Ansbach priest Heinrich Fuhrmann. He had an initiation-like experience during Holy Communion at his confirmation on 20 May 1833 at the altar of the Schwanenritterkapelle in Ansbach.

The 'experiment' of spiritual isolation, which ultimately failed with the foundling Kasper Hauser, could today become the civilisational fate of childhood and adolescence, which, forced by the Covid-19 crisis, threatens to become more and more psychosocially, mentally and spiritually captive in a digital world. In its online edition of 13 April 2023, the magazine 'Welt' ran the following headline about the 'media neglect' of children and young people: 'The digital banishment of our children' (Welt-online, 2023a).

In view of the conditions of civilisation and socialisation, more and more children today have to follow the path of Kaspar Hauser. The Covid-19 crisis will drastically accelerate this pre-existing trend. Unresolved traumas tend to be re-staged and re-enacted. Emergency and trauma pedagogy aims to help the Kaspar Hauser children of our time to process and integrate their trauma and try to transform the crisis that has occurred into an opportunity for their biographies.

10 Turning the Crisis into an Opportunity

There is no life without traumatisation, but not every trauma makes you ill! The processing of a traumatic experience depends less on the factors of an event, i.e. duration, severity or outcome, but more on individual factors such as age, temperament, previous experiences, the ability to process them, etc. These individual coping capabilities essentially determine whether the psychological injury can be overcome by means of self-healing powers or whether it will lead to pathological trauma sequelae. Moreover, environmental factors, such as social support, also represent an important influencing factor. This interplay of risk and protective factors ultimately determines the course of the trauma process (Ruf, 2019a, pp. 95–96).

Every crisis is also a biographical opportunity. Psychotraumatological research today impressively illustrates how unresolved trauma can develop destructive potential that throws the victim's life off course. However, research studies show just as impressively how overcoming trauma can build lasting resilience and contribute to personality maturation (Tedeschi et al, 1998). According to studies by Columbia University/USA, this can even be said to be the rule: 60–80 per cent of those who went through traumatisation became happier and more resilient in the long term (Bonanno, 2004; cf. Schopper, 2018). Clinical experience clearly shows that children and adolescents have personal prerequisites that enable them to experience post-traumatic growth after a wide range of traumatisation (Landolt, 2004, 106).

Promoting post-traumatic growth is also a task and goal when working with children and young people who have been traumatised due to Covid-19. Emergency and trauma pedagogy aims to support children and adolescents worldwide in coping with their stressful experiences and thus try to prevent, mitigate or heal possible subsequent trauma-related disorders (Ruf, 2019b, pp. 17–18). Ultimately, it endeavours to transform a painful crisis into a biographical opportunity for personal maturation by overcoming it.

11 The Anthroposophical View of Humankind—A Key to Extended Sustainability

Just as we find a spiritual dimension added in anthroposophy to the concepts of medicine, trauma and education, sustainability can also be understood in this sense. I act sustainably not only when my actions conserve resources in such a way that future generations can still live, but also when I take the purpose of humanity into account. Am I contributing to the advancement of the world through my actions, towards a world where human beings live in love and humanity in harmony with nature and their fellow human beings? Anthroposophy's view of human beings and the world makes an essential contribution to this.

Our work thrives on looking at human beings as a whole. We focus not only on physical and mental health, but also on spiritual goals and tasks. Trauma can prevent human beings from being able to fulfil their mission in life (see Kaspar Hauser). Our work aims to enable human beings to overcome their traumas so that they can realise their very own spiritual impulse on Earth.

Is that sustainable? Absolutely, in every respect. From an economic point of view, the emergency pedagogy approach is extremely interesting. By resolving trauma and reducing and preventing subsequent trauma-related disorders, a very modest effort will save society large sums of money in therapy costs, which often continue to be incurred for years in the case of untreated psychological trauma (Bundestag & Dienste, 2016). Ecological effects are also to be expected. Major environmental destruction is often based on a disturbed relationship between individual human beings and the world. The Earth can no longer be thought of and perceived as an organism in which we are a part with a central function. This alienation from reality can often be the result of a trauma that has not been overcome. Wherever human beings are in harmony with themselves, they are more likely to live in harmony with nature. By helping children to regain a healthy relationship with themselves and the world, our work also contributes to environmental sustainability.

However, the impact of emergency pedagogy is most clearly felt in the area of social sustainability. Disturbed attachment behaviour, which can be the result of many traumas, has a massive impact on social interaction and individual development. How can I respect or love another human being if I cannot love or even really recognise myself? Any contribution to helping children in the early stages of trauma can have a profound effect on their attachment behaviour in the future. What can take years of painstaking therapy to achieve in adulthood in the case of manifest trauma-related disorders, can in some circumstances be achieved through a three-day intervention using pedagogical methods. Emergency pedagogy is understood as prevention of developing a trauma disorder after suffering a shock.

12 Conclusion

The concept of emergency pedagogy combines scientific psychotraumatology with the view of human beings and the world of anthroposophy as well as pedagogical methods. In this way, children who have suffered a shock can be helped to playfully overcome it. This can significantly reduce the risk of developing long-term debilitating post-traumatic disorders. As trauma can have a massive impact on the attachment behaviour of those affected (towards other human beings, oneself and the world) and entails the need for years of therapeutic support, every prevention of a psychological sequelae makes a lasting contribution to the social, economic and ecological organism of our society. We owe this profound effect to the holistic approach of the anthroposophical view of human beings and the world. By expanding our view to include the spiritual dimension of human beings, we are able to provide holistic help (body, soul and spirit). We are also trying to make these successes measurable through increasing scientific monitoring. Some governments in countries where emergency pedagogy teams have been deployed, have become aware of our work and have included emergency pedagogy (based on Waldorf pedagogy) in their national disaster plans.

References

Adam, H. (2016). Krieg, Terror, Flucht – die seelische Belastung der Kinder. *Trauma. Zeitschrift für Psychotraumatologie und ihre Anwendungen, 14*(1) 2016, 24–30

Bonanno, G. A. (2004). Loss, trauma and human resilience: Have we underestimated the human capacity to thrive after extremly adverse events? *American Psychologist, 59*, 20–28.

Deutsche Gesellschaft für die Vereinten Nationen e.V. (2019). Kinder im Krieg: Neue Ausmaße der Gewalt. http://frieden-sichern.dgvn.de/meldung/kinder-im-krieg-neue-ausmasse-der-gewalt. Stand July 05, 2019

Deutscher Bundestag/Wissenschaftliche Dienste. (2016). *Posttraumatische Belastungsstörung. Zahlen sowie Aspekte geschlechtsspezifischer Behandlungsangebote.* https://www.bundestag.de/resource/blob/490504/8acad12ffbf45476eecdfeff7d6bd3f4/wd-9-069-16-pdf-data.pdf. Stand August 18, 2024

Ding, U. (2009). Trauma und Schule. Was lässt Peter wieder lernen? Über unsichere Bindungen und sichere Orte in der Schule. In J. Bausum, L. Besser, M. Kühn, & Q. Weiß (Hrsg.), *Traumapädagogik. Grundlagen, Arbeitsfelder und Methoden für die pädagogische Praxis* (S. 55–66). Juventa.

Donau Universität Krems. (2021). *16 Prozent der Schülerinnen haben suizidale Gedanken.* https://www.donau-uni.ac.at/de/aktuelles/news/2021/16-prozent-der-schuelerinnen-haben-suizidale-gedanken.html. Stand 28.06.2023

Du Bois, R., & Resch, F. (2005). *Kleinische Psychotherapie des Jugendalters. Ein integratives Praxisbuch.* Kohlhammer.

Eckardt, J. (2005). *Kinder und Trauma. Was Kinder brauchen, die einen Unfall, einen Todesfall, eine Katastrophe, Trennung, Missbrauch oder Mobbing erlebt haben.* Vandenhoeck & Ruprecht.

Feuerbach, A. (1832). *Kaspar Hauser. Beispiel eines Verbrechens am Seelenleben des Menschen.* J. W. Dollfuß.

Fischer, G., & Riedesser, P. (2009^4). *Lehrbuch der Psychotraumatologie.* Reinhard Verlag.

Hausmann, C. (2006). *Einführung in die Psychotraumatologie.* Facultas Universitätsverlag.

Hurrelmann, K. (2023). *Partei der Profiteure. Corona, Krieg, Klima: Die Gesellschaft zeigt Anzeichen einer Posttraumatischen Belastungsstörung* (Nr. 167, S. 2). der AFD, in: Süddeutsche Zeitung. Accessed July 22/23, 2023.

International Federation of Red Cross. (2012). *World Disasters Report 2012—Focus on forced migration and displacement.* https://www.ifrc.org/en/publications-and-report/world-disasters-report-2012---focus-on-forced-migration-and-displacement/

Juen, B., Karutz, H., Warger, R., Nindl, S., Walter, H., & Kratzer, D. (2017). Kinder in Naturkatastrophen. In H. Karutz, B. Juen, D. Kratzer, R. Warger (Hrsg.) *Kinder in Krisen und Katastrophen. Spezielle Aspekte psychosozialer Notfallversorgung* (pp. 405–415). Reihe Krisenintervention und Notfallpsychologie, Bd 4. Insbruck.

Knölker, U., Mattejat, F., & Schulte-Markwort, M. (2007^4). *Kinder- und Jugendpsychiatrie und -psychotherapie systematisch.* Uni Med Verlag.

Krüger, A. (2007). *Erste Hilfe für traumatisierte Kinder.* Patmos Verlag.

Landolt, M. A. (2004). *Psychotraumatologie des Kindesalters.* Vandenhoeck&Ruprecht.

Levine, P. A., & Kline, M. (2005). *Verwundete Kinderseelen heilen. Wie Kinder und Jugendliche traumatische Erlebnisse überwinden können.* Kösel-Verlag.

Pies, H. (1966). *Kaspar Hauser. Eine Dokumentation.* Brügel-Verlag.

Resch, et al. (1999^2). *Entwicklungspathologie des Kindes- und Jugendalters. Ein Lehrbuch.* Psychologie Verlag.

Ruf, B. (2018^2). *Trümmer und Traumata. Anthroposophische Grundlagen notfallpädagogischer Einsätze.* Ita Wegman Verlag.

Ruf, B. (2019a). *Krieg-Flucht-Notfallpädagogik. Wie Pädagogik minderjährigen Flüchtlingen helfen kann, traumatische Erlebnisse zu überwinden.* Ita Wegman Verlag.

Ruf, B. (2019b). Seelen-Tsunamis: Wie Notfallpädagogik Kindern und Jugendlichen in Katastrophengebieten hilft. In B. Ruf (Hrsg.), *Trauma. Zeitschrift für Psychotraumatologie und ihre Anwendungen* (H 3. pp. 6–19). Notfallpädagogik. Jg 17/2019.

Ruf, B. (2022^2). Das Corona-Trauma und die therapeutischen Möglichkeiten der Notfallpädagogik. In M. Glöckler, & A. Neider (Hrsg.), Berhardt, C., Hartmuth, T., Hueck, C, Ramm, H., Ruf, B. (2022^2). *Corona—Was uns die Pandemie lehren kann. Diagnosen, Erfahrungen und Kraftquellen für die Zukunft* (pp. 243–309). Akanthos Akademie, Edition Zeitfragen.

Shah, H. (2017). Kinder aus Kriegsgebieten. In H. Karutz, B. Juen, D. Kratzer, & R. Warger (Hrsg.), *Kinder in Krisen und Katastrophen. Spezielle Aspekte psychosozialer Notfallversorgung. Reihe Krisenintervention und Notfallpsychologie* (pp. 439–450), Bd 4. Insbruck.

Schopper, C. (2018). *Trauma überwinden. Ein Handbuch für Therapeuten und Betroffene.* Verlag Urachhaus.

Schötz, H. (2010). *Kaspar Hauser - Kronprinz oder Schwindler?* Verlag und Offsetdruck Eppe GmbH.

Spitzer, M. (2020^3). *Pandemie. Was die Krise mit uns macht und was wir aus ihr machen.* MVG-Verlag.

Steiner, R. (1982⁵/1908/GA 113). *Der Orient im Lichte des Okzidents.* Rudolf Steiner Verlag.
Steiner, R. (1983²/1918/GA 184). *Die Polarität von Dauer und Entwicklung im Menschenleben.* Rudolf Steiner Verlag.
Copsy-Studie. (2022). https://www.uke.de/kliniken-institute/kliniken/kinder-und-jugendpsychiatrie-psychotherapie-und-psychosomatik/forschung/arbeitsgruppen/child-public-health/forschung/copsy-studie.html
Tedeschi, R. G., Park, C. L., & Calhoun, L. G. (1998). *Postraumatic growth: Positive changes in the aftermath of crisis.* Irving B. Weiner.
Wassermann, J. (1908). *Caspar Hauser oder Die Trägheit des Herzens* (6th ed.). Deutsche Verlagsanstalt. Retrieved from Internet Archive
Welt-online (13.04.2023): Mediale Verwahrlosung: Die digitale Verbann7ung unserer Kinder, https://www.welt.de/kultur/plus244678204/Medien-Warum-schicken-wir-unsere-Kinder-so-selbstverstaendlich-in-die-digitale-Verbannung.html, (stand 14.04.2023)
Welt-online (14.04.2023): „Was sich in den ersten zwei Lebensjahren im Gehirn abspeichert, ist irreversibel", https://www.welt.de/wissenschaft/plus244304151/Stefanie-Stahl-Was-sich-in-den-ersten-zwei-Lebensjahren-im-Gehirn-abspeichert-ist-irreversibel.html, Stand 14.04.2023
Zimmermann, D. (2012). *Migration und Trauma. Pädagogisches Verstehen und Handeln in der Arbeit mit jungen Flüchtlingen.* Psychosozial Verlag.

Newton Dee Village—What Sustains Us to Live in Community? (A Harmony of Four Voices)

John Addison, Wolodymyr Radysh, Marjan Sikkel, and Ginny Bayman

Abstract

In a well-established community such as Newton Dee, it might be assumed that the community's contribution to a publication such as this would be well-polished and coherent, and, as such, reflect the state of having completed a successful endeavour in community-building. However, the reality is that an intentional community is not only co-created by all who participate in it but is an act of renewed creation (a re-creation) every time we awake from sleep and step out into the day.

Features:

(a) Founding year	1960
(b) Employees/Villagers/Co-workers/Day placements	Around 250
(c) Location	Newton Dee Village, Aberdeen
(d) Activity of Newton Dee	Schotland
(e) Sector	Intentional Community

J. Addison · W. Radysh · M. Sikkel (✉) · G. Bayman
Newton Dee Village, Aberdeen, Scotland
e-mail: marjan@newtondee.org.uk

J. Kronenberg and E. T. Lammerts van Bueren (eds.), *On the Earth We Want to Live*,
World Sustainability Series, https://doi.org/10.1007/978-3-031-98758-8_36

Fig. 1 Logo Newton Dee Village

1 Opening Commentary

In a well-established community such as Newton Dee (Fig. 1), it might be assumed that the community's contribution to a publication such as this would be well-polished and coherent, and, as such, reflect the state of having completed a successful endeavour in community-building. However, the reality is that an intentional community is not only co-created by all who participate in it but is an act of renewed creation (a re-creation) every time we awake from sleep and step out into the day. In this sense, what is presented here is (to borrow a phrase from Zen) set out with an attitude of "beginner's mind" (Suzuki, 1970)—unfinished, partial, even raw in parts. This attitude is in harmony with Goethe's idea that to be truly human is to be in the process of becoming.

Moreover, to present a perfectly presented document with no flaws, contradictions, or omissions, would be to portray a finished project that, in its perfection, would no longer be capable of further development. Such a presentation would be far from an honest description of the reality of our life here. Despite existing for decades, Newton Dee is a community in which there is still much room for growth and development (and it is most certainly not a perfect community!) In this sense, what follows is an accurate (if partial) snapshot of how life in Newton Dee is currently, perceived through the lens of "sustainability" in the widest sense of the word.

What follows are the testimonies and reflections of four voices from within the Newton Dee community. These three voices speak for themselves as individuals, and none of them represent the views of all the individuals who live in Newton Dee; yet, between them, much of the essence of what it means to sustain our community life is expressed. The structure of the essay is to present the four voices with as little editing as possible, to let them speak for themselves. At the end, there will be another commentary—not to sum up or encapsulate the four voices, but as a fifth voice, answering the other four in the sense of "call and response."

Written in the period up to, and during, the Michaelmas period of 2024, this speaking of souls, heart to heart, is hopefully in harmony, not only with the Michaelic quality of the autumnal season, but also with the "zeitgeist" of these dramatic times.

2 Voice 1

Every initiative will have an inner and outer reason for its beginning.

In Vienna, a doctor by the name of Karl König and a group of young people were interested in anthroposophy and were working on existential questions of life. This was taking place at the same time that nazi oppression was spreading in Europe. It was clear to them that this was not a place to continue working together and vowed to meet up somewhere at some point. Prior to their departure from Vienna, they read the last lecture that Rudolf Steiner had given about the task of the Younger Generation which inspired them immensely (Pietzner, 1990).

In 1939 they regrouped in the north-east of Scotland on the generous offer of the Houghton family on the Williamston Estate in Insch, north of Aberdeen.

The impulse to work with children who had challenges started there by this group. These children at that time were considered in-educable. This was the initial mission of what came to be known as Camphill. The name Camphill only received this label when the Camphill estate was purchased in 1940 just outside Aberdeen in the beautiful Royal Deeside along the Dee River.

In the case of Newton Dee, the outer reason for its origin had to do with socially challenged boys with an overabundance of energy who were not compatible with the work that was being done with the children. When Newton Dee estate of 180 acres came on the market in 1945, in the same vicinity, it was purchased to accommodate the work with these boys who needed a serious farming challenge. The young doctor, by the name of Thomas Weihs, took on the farming in the estate not only to help the boys but to also heal the land through biodynamics.[1]

In 1960 Newton Dee became a 'village community' for adults with additional needs where productive tasks in a workshop, garden or farm brought about dignity to the human being which in itself served as a therapeutic and healing activity. Through life- sharing and a community approach to economics enables biodynamics to be carried and become viable within the context of community life.

Alongside these outer initiatives Karl König wrote the First Memorandum in 1945 which set out a clear 'inner' mission statement for the emerging community based on three 'essentials':

1. that every human being is an eternal, immortal and divine self.
2. helping others depends on our individual self-development.
3. human beings strive to create community as a natural expression of social Encounter (Pietzner, 1990).

This has carried Newton Dee to the present day.

[1] Biodynamic Agriculture arose from Anthrosopohy in 1924 after a course of Rudolf Steiner in Koberwitz, now Poland. In Chap. 4.6 *Iodynamic Agriculture—and the Future of Agriculture* in this publication you can find more information.

Community building and life sharing enhance the social, cultural and spiritual activity whereby a common vision can be carried together despite the most varied individual approaches. All members of Newton Dee review their common 'Guiding Vision Statement' from time to time to see if it is still applicable and if any adjustments need to be made to reflect the current experience.

These elements broaden the meaning of sustainability and offer a holistic and viable lifestyle for those who have the courage to try it, as it comes with its own challenges.

The current approach to sustainability generally relies on regulatory systems to ensure that sustainability is implemented even though there is general agreement this is a good thing to do for the sake of the earth and climate change. Regulation often produces a hesitant, bitty and disconnected implementation often impeded by self-interest and the inability to feel connected to nature and our earth as an organism and on account of our experience that we are essentially separate from it.

The background and approach that is taken in Newton Dee originates from the broad aspects of consequential implementation what is dear to us in community life. This 'big picture approach' allows for a clear appreciation of working out of the wholeness of our existence and appreciation that we as human beings are custodians of the earth planet and the human race as part of a transitional and ever evolving process. It was back in 1945 that Newton Dee was already implementing the 'big picture approach' as organic and biodymanic methods in agriculture were being applied in the gardens and farms. It was not only a matter of getting healthy food for the community but it was and is simply a natural consequence of an inclusive way of thinking. Since the 1970's, well before the green agenda became popular, solar panels were being installed on roofs, and sustainable and natural materials were a vital part of the palette in the architect's office. Car sharing, responsible co-operation with money and different social forms in inclusive supportive households are a core principle of sustainable living and part of an ecological agenda. All this was and is because: 'the universe is part of my being and I need to look after it.'

This begs the question as to the inner orientation that is required of each member of the community who is called upon to turn themselves towards a holistic approach which needs to take on board that our earth has had an evolutionary path to which we inherently belong and cannot isolate ourselves from the ongoing process of change.

Without our experiencing ourselves as united with the universe, decisions and actions become isolated and can become self-serving and egocentric without the ability to take account of the integration of elements which naturally belong together. The interconnection of parts and processes cannot be broken down into isolated events without an understanding which arises out of a cohesive whole. Perhaps this is best exemplified by Scotland's neuroscientist and philosopher, Iain McGilchrist who describes thinking as originating out of two distinct processes: one is purely analytic and quantifiable and the other is purely qualitative and process- driven (McGilchrist, 2009). Our current civilisation and culture rely heavily on the analytic dimension which has the tendency to lead to isolated but nonetheless brilliant thoughts and discoveries, which however do not arise out of a complete wholeness but out of separate and isolated events leading to a limited conclusion. Right hemisphere thinking, on the other hand, arises out of dynamic interrelationships of living processes that respect the qualitative nature of what underpins our life and existence. The right half works with flow, flux, fluidity and interrelationship. It is the intuitive and artistic aspect of perception and experience. It works with wholeness and unity to grasp the interlinking of a complete experience. It reaches to deep experiences and can comprehend the completeness of nature and the cosmos. It can become the force which can focus 'total attention' to perceive the whole. McGilchrist (2009) describes these two dimensions of thinking as arising out of the two hemispheres of the brain and concludes that if we continue to rely wholly on our left hemisphere, we will never truly be able to grasp the significance of the wholeness of the universe and our own integrated activity within it.

Sustainability of the earth will not become a strong enough experience that we wholeheartedly embrace it. Our thinking needs to benefit from the interaction and working together of both brain hemispheres to offer a full understanding of life and nature from the standpoint of a sustainable approach to our planet and humanity. This applies to three areas of building a sustainable community in Newton Dee that arises out of:

- our love for nature, humanity and the universe
- the love of human encounter
- the pursuit of personal development for the good of the world.

When we work out of these three ideals then what flows into daily life enables us to say that this is a sustainable approach out of a universal whole (Fig. 2).

3 Voice 2

The aim of Camphill has always been to create communities in which people of different abilities, ages, cultures and countries live and work together. We work with a strength-based practice: if each contributes to their best ability, then between us we can create a strong and healthy community. By modelling a way of working and living together in a way that is good for the planet and all the living beings on it, we hope to contribute to a sustainable future in the widest sense. In his document on the three pillars of Camphill, Karl König quotes Rudolf Steiner:

> If a man works for another man, he must find in the other the purpose for his own labour; he has to sense and to experience the value, the being and the meaning of it. This is only possible if the community is something other than a more or less undefined sum of single people. Such a community must be permeated by a true spirit in which every one of its members partakes. In such a community, each person should say: 'This community is right, and I want it to be so.' The community must have a spiritual mission, and every individual should have the urge to help so that the mission may be fulfilled. (König, 2018, pp. 122–123)

Fig. 2 Working in the Bakery. Newton Dee Village

This is a very important statement and not something that is easily achieved. We are a community of around 250 people, most of whom are living within the community, some are living outside as day placements and employees. How can we ever agree on anything when we try to live and work as equals? Nobody can tell the other what to do, but somehow, we have to come to a consensus. To complicate matters, people are coming and going, so our vision needs to be reviewed on a regular basis which, of course, we do not manage because we are all very busy with our daily lives.

We hope to model a way of living and working that can radiate out to the wider society. Throughout the ages, there have always been small groups of people who started to live and work in a different way, the results of which seeped only many years later into mainstream society. Steiner speaks in connection with that about preparing for the society of the future.

> We try to cultivate spiritual science in order to overcome materialism (…) We found communities of human beings within which there must be no dogmatic beliefs or any tendency to accept teaching simply because it emanates from one person or another. We found communities of human beings in which everything, without exception, must be built upon the soul's free assent to the teaching. (Steiner, 1992)

In Newton Dee nobody is anybody's boss. We have created a so-called mandate organisation in which different groups of people have received the mandate from the whole group to perform certain tasks. Thus, we have a finance group, building group, cultural group etc. There are clearly defined mandates within which the group can get on with their task. Of course, there are always situations that do not fall within anybody's mandate, and these have to be looked at individually.

In the above quote, Steiner speaks about the importance of overcoming materialism. We have come to believe that the economy should keep on growing endlessly. For many, the meaning of life is to become rich in a material sense. The divide between rich and poor is growing. Does it have to be like this? Can we find meaning and fulfilment in different ways?

We work out of the conviction that we do not only have a physical body, but also a soul and spirit. All three need to be nourished. Our spirit is healthy, regardless of our outer appearance or disability. When we manage to see ourselves as foremost spiritual beings, our whole outlook on life changes. Material goals will become less important.

We do not own anything in our community, but we make use of it whilst we are living there. We are stewards, not owners. This includes the care for the land, property and people. We aim to add to what is there and leave it to those who come after us in a better state.

What is the meaning of our lives? We try to find meaning in life through work, culture and relationships. We take into account that throughout life we go through different life phases in which our needs may change.

By dancing, playing, making music etc. ourselves, we can find an enormous sense of joy and achievement. Culture is as important in Camphill as work and rest. We celebrate the seasonal festivals throughout the year. Many people nowadays struggle with the Christian element of these festivals, but long before Christianity people were celebrating the change of the seasons in our part of the world. So, if we are not Christian, we can also find meaning in them if we see them as nature festivals or as festivals around the changing of the light throughout the year. Next to that, we offer a variety of creative activities for people to take part in. The so called fundamental social law presented by anthroposophy is formulated as:

> The well-being of a community of people working together will be greater when individuals claim less of the income resulting from their own work for themselves. Instead, they should contribute more of these proceeds to their fellow workers, while satisfying their own needs not solely from their own labour but also from the work done by others. (König, 2018, p. 122)

Newton Dee practises this in a particular way. None of the carrying co-workers gets a fixed salary. All the money goes into one pot. We have divided the community in three Neighbourhoods and annually these sit together to make the budget. Each person who is responsible for a household requests the amount they think they will need monthly in the coming year to cover the financial needs of their household. Within the Neighbourhood they need to come to an average allocation which has been set by the finance group depending on the available resources. All requests go back to the finance group who integrate it into the overall budget for Newton Dee.

With this we hope to overcome the trap of the growth economy and never-ending needs. It is not an easy practice. Most of us come from mainstream society where nobody can tell me what I do with my own money. Now, I suddenly need to ask permission from others to buy what I consider to be essential in my life. And of course, someone else can say "No." It can happen that I now no longer like this person! On the other hand, I now suddenly have access to a large pool of money. Will my needs grow accordingly, or will I be able to distinguish between needs and wants?

There is increasing polarisation in society. People have stopped talking with those who do not think similarly. One way out of that is to start meeting again as is for instance practised in deep democracy (Mindell, 2002). In Newton Dee we cannot escape each other. We share so many things and we have to make decisions together. On top of that, we are often short-staffed, buried under paperwork and short of money. Of course, conflicts will arise. We need to learn to see that this is to be expected and part of the human condition. I need to be able to go to the other and try to understand where they are coming from. I also need to self-reflect. If I do not learn to look inside myself and change my expectations and behaviour, I will not last long in the community.

Newton Dee started in 1960. We have moved on from the pioneer phase long ago and are now in the maintenance phase. New generations have joined, and they bring new ideas and needs. As a community we need to keep on changing and evolving. Steiner formulated what he called the fundamental sociological law:

> Mankind strives in the beginning of cultural states for the emergence of social associations; the interest of the individual is first sacrificed to the interest of these associations; further development leads to the liberation of the individual from the interest of the associations and to the free development of the needs and powers of the individual. (Steiner, 1887–1901, p. 255)

This law expresses itself very clearly in Newton Dee. In the beginning all sacrificed everything for the strength and health of the community. More and more over time, the individual needs have come to the foreground. We now need to be aware that this does not go so far that we lose the community element! It is not always easy to live in a community like Newton Dee. We have many strong individuals with their own opinions and ideas. However, overall, we have managed to create a beautiful community in which we can all flourish and grow.

4 Voice 3

My main role in the community is that of a House Coordinator. What motivates me is that I am deeply interested in human connections, relationships, working creatively with people and self-development. The reason I came here was, in the first place, to live in a community.

In my time here, I have had successes, and experienced challenges.

The successes are found in developing relationships, feeling connected with others. Striving, reflecting, through true collaboration—you find mutuality between each other. That makes me feel free. We need each other to develop. Looking at my own inner and outer situation, community living helps me to become free of my limitations.

The challenges come through the process of developing these relationships. I feel that forcing my own will on others is a big mistake, but when it happens I try and transform it by using it as a point for learning and growing. Living in community can often involve a confrontation with my own double. I had to learn that what served me well in the past does not necessarily do so in the present.

In the encounter with others, opportunities for self-reflection and growth become visible. For instance, the specific challenges that one resident presented provoked a deep antipathy within me that I felt I had to address. To move from antipathy to empathy, I looked at their biography and saw them in a different light. Their mum had abandoned them. In my heart, I knew that I could not do the same.

I participated in an anthroposophical mental health course which opened up a different way of looking at the human being. I realised that I had always felt unspecialised and unspecial, and saw this as a drawback. However, through this course I realised that to be human is to hold back from specialisation. This gave me more confidence in myself as a person.

The course gave me insight into the various personality disorders and related them to the zodiac signs. I could see all my own traits in the macrocosm of the Zodiac. I found where I was in life, and who I was becoming. I developed compassion for myself and other people. I began to understand zodiacal clothing, the masks that we put on to protect ourselves. And the wonder to experience the effects on others when they feel accepted, and held in mind, how these tight zodiacal cloths can open and our light can shine through.

I have been deeply moved by Karl König's "Social Farming" and "Creating Islands of Health."

I am drawn to the quantities of Goethean conversation,[2] through the cultivation of which we really meet each other, as individuals living in community (Fig. 3).

5 Voice 4—Closing Commentary

Hearing these three voices, their differences and their similarities, a few things become apparent as underlying patterns. One is that the concept of "sustainability" is not addressed directly but is perceived as an inherent and integral part of the overall practice and discipline of living in community.

In recent years, Newton Dee has been endeavouring to raise to heightened consciousness the principle of sustainability that has been embedded and implicit in

[2] See for instance **Spock, M. (1982)**. *The art of Goethean conversation.* Retrieved from https://westdalechildrensschool.weebly.com/uploads/1/7/7/9/17799545/article_-the_art_of_goethean_conversation.pdf.

Fig. 3 Working on the biodynamic garden in Newton Dee

the entire community enterprise from the outset. A Sustainability Group has been formed and a range of environmentally conscious practices are being established in the daily life of the houses and workshops.

An underlying dynamic behind this is that there can sometimes appear to be a conflict in the so-called "daughter movements" of anthroposophy (the practical applications, such as biodynamics, Waldorf education, and Camphill communities) between two polar attitudes. As a gross generalisation, there are, on the one hand, older, more traditional anthroposophists, and, on the other, younger people who are more critical. Yet the experience in Newton Dee is that these two poles are far from being fixed in a "generation gap" model, and can co-exist (sometimes in the same individual!) Thus, the older attitudes and approaches to community living exist in a creative dynamic tension alongside the newer sensibilities and approaches (such as the heartfelt emphasis on ecological and social justice issues living in the younger generation).

As for this essay, the red thread running through these contributions is one of being grounded. The community's approach to sustainability is grounded in its foundational narratives of the early days of Camphill and the caring for the Earth that these early ideas generated. It is also grounded in the newer approaches and practices of sustainablility, in making them active and effective in daily life (albeit by painfully small increments) rather than resting content in grandiose, but empty, gestures of policy and presentation. And, last, but not least, it is grounded in the daily community life alongside the so-called "disabled" people, without whom our community would not exist.

Above and behind the words four voices expressed here, is the living reality of the spiritual reality, the true and eternal "ground of existence," which is the fertile soil in which the tree of sustainability can take root, grow, blossom and bear fruit, both in time and eternity.

In summary, a short addition on how Newton Dee aims to contribute to sustainability on different levels:

- Socially: each community member is equally part of a complex social organism. We focus on strength instead of disability. Each person matters. All this can help us find meaning in our lives which contributes to our long-term mental health and wellbeing.
- Ecologically: We treat the earth and all that lives and grows on it with respect and aim to give back when we take from her.
- Economically: By decoupling work from monetary reward, we stop seeing money as a goal instead of a means. We value making / buying products that last, sharing resources, reflecting on our needs and wants. All this counteracts over-consumption and waste of resources.
- Spiritually: Our spiritual worldview underpins everything. Once you realize the whole world and its surroundings is one amazing, organic whole, you know that the only way forward is by making sure it is sustainable for everyone and everything.

References

König, K. (2018). *The spirit of Camphill: Birth of a movement*. Floris Books.
McGilchrist, I. (2009). *The master and his emissary: The divided brain and the making of the Western world*. Yale University Press.
Mindell, A. (2002). *The deep democracy of open forums: Practical steps to conflict prevention and resolution for the family, workplace, and world*. Hampton Roads Publishing Company.
Pietzner, C. (Ed.). (1990). *A candle on the hill: Images of Camphill life*. Floris Books.
Steiner, R. (1887–1901). *Collected essays on cultural and contemporary history: 1887–1901* (GA 31). Rudolf Steiner Verlag.
Steiner, R. (1992). *The mystery of death: The nature and significance of Central Europe and the European folk-spirits* (U. Dietler, Trans.). Rudolf Steiner Press (Original work published 1915).
Suzuki, S. (1970). *Zen mind, beginner's mind*. Weatherhill.

Odin Foodcoop—Power to the People and the Planet!

Merle Koomans van den Dries and Jacqueline de Kool

Abstract

Founded in 1983 in the Netherlands, Odin wants to show that a more sustainable, healthy food system is possible when you work together and reconnect with other people and nature. Besides providing the flow of goods, money and information between consumers and producers of organic and biodynamic food, Odin works on more diversity on the field, in the shopping basket and in the teams. The founders already did not opt for personal ownership. Today, Odin has developed into a cooperatively owned retail organisation with over 20.000 members.

Features:

(a) Founding year	September 1983
(b) Employees	926
(c) Location	Headquarters in Geldermalsen (NL). 2 distribution centers, 38 organic shops
(d) Active in which countries	The Netherlands, export to Belgium, import from about 20 countries
(e) Sector	Food

1 Where the Odin Story Starts

In the autumn of 1983, three partners founded Odin. They started in a utility room in Koedijk (North Holland) with experience in the organic and biodynamic sector and a large dose of optimism (Fig. 1). The founders did not opt for personal

M. Koomans van den Dries (✉) · J. de Kool
Odin, Geldermalsen, Netherlands
e-mail: m.koomans@odin.nl

J. Kronenberg and E. T. Lammerts van Bueren (eds.), *On the Earth We Want to Live*, World Sustainability Series, https://doi.org/10.1007/978-3-031-98758-8_37

Fig. 1 Logo Odin

ownership, but for a limited partnership, with a foundation as a silent partner. Later, the limited partnership was transformed in a cooperative.

Healthy food from vital agriculture and a fair economy, that was what Odin was all about from the start. This mission was expressed at the start in the following foundation stone meditation (free translation in English):

May we always remember:
The origin of our endeavour
to cultivate the earth in a healing way
Respecting the wisdom and harmony
of God's creation

To share its revenues in a fair way
Seeking brotherhood
where people depend on each other

To do good from inner strength
Directing our will to God's word that nourishes us in
our food

So that it may become effective:
In material sacrificing light
Between people creative love
In every individual the power of the spirit

2 Fair and Healthy Groceries from Farm to Fork

In the beginning, Odin developed as an international wholesaler for fresh biodynamic and organic food. In addition to fruit and vegetables, products such as dairy, cheese and vegetarian products were added to the range for Dutch buyers.

In the early 1990s, all business activities were concentrated in Geldermalsen, in the middle of the Netherlands. This is also where the idea of a national fruit and vegetable subscription was born. A great way to keep a short line between producer and consumer, to be able to offer producers a purchase guarantee and to provide consumers with super-fresh organic products. In the late 1990s, tens of thousands of people collected a weekly bag with fresh organic and biodynamic fruit and vegetables at one of the hundreds of collection points in the Netherlands. To be able to fill the bags all year round, Odin's buyers drew up cultivation plans with growers in the Netherlands and abroad. A weekly newsletter with background information and recipes was included in the bag. Further, the consumer service dealt with all kinds of complaints and questions on a daily basis and organised excursions and other activities for subscribers. As the demand for additional organic groceries grew, a web shop was built. It was called 'The Jacket Around The Vegetable Bag.' Later, the web shop was expanded with a delivery service that delivers groceries ordered online to family homes (Fig. 2).

In 2000, Odin adopted six organic shops. Being a retailer was new to Odin and therefore, a whole new challenge. In order not to compete with wholesale customers and the vegetable bag pick-up points, the new retail formula was given the name Estafette. Over the years, more and more new Odin owned shops were added, partly by taking over existing organic shops, but also by opening new shops in places where there was no supply of organic groceries at all. One of these is the shop in the Amsterdam department Bos en Lommer, where residents held a petition to convince Odin to open a shop there. In 2016, with about 20 shops by then, it was decided to simply call the own shops Odin from now on. One name, one mission, one clear image and story to the outside world. As of 2024, Odin is running 38 organic shops in the Netherlands.

In 2012, it was decided to convert the limited partnership into a cooperative structure with the following membership categories: participating members, employee members, capital members and, since the end of 2022, financial members. In the participating member category, consumers, sympathisers and stakeholders are offered the opportunity to make an active contribution to realising the ideals of Odin. The employee members have the task of entrepreneurially monitoring the course and impulse of the cooperative and helping to provide direction. Capital members and financial members make sustainably earmarked financial resources available under various conditions for the benefit of business operations. An important capital member is the Sleipnir Foundation. Sleipnir was founded by Odin and plays a role in other steward owned companies (for more information: www.sleipnir.nl). The Odin cooperative started in 2012 with 111 members. The number of members is continuing to grow, counting about 20.000 members in 2024.

In 2018, the activities of Cooperative De Nieuwe Band were added. Also a cooperative that, like Odin, was founded in 1983. Where Odin until then focused on fresh organic products, De Nieuwe Band was a wholesaler and producer specialised in organic dry groceries. With the merger, Odin's range became complete.

In addition to the wholesale and shops, Odin has employed a biodynamic beekeeper since 2010 who takes care of bee colonies on organic and biodynamic farms that supply Odin (for more information: www.odin.nl/over-odin/imkerij). Together with the 26 farmers and a group of volunteers the Odin beekeeper works full-time to create a more healthy natural environment for the bees. The beekeeper also organises courses, farm visits and lectures.

Fig. 2 Home delivery service Odin foodcoop also by Sunrider

Since 2016 Odin also has its own biodynamic farm, De Beersche Hoeve (for more information: www.debeerschehoeve.nl), in the south of the Netherlands. On 20 hectares, the farm crew works on the breeding of open pollinated gmo-free seeds, seed propagation and the production of vegetables. In cooperation with artist and activist Claudy Jongstra (for more information claudyjongstra.com), the Odin farm also grows plants for natural dyes, like woad, sunflower and yellow chamomile.

3 The Vision of Odin Foodcoop Today

Since the start in 1983, Odin wants to contribute to a sustainable, social and fair society by profitably providing the flow of goods, money and information between consumers and producers of organic and biodynamic food. The starting point for Odin is the sustainable connection of consumer demand with the possibilities of producers, with love and care for the earth and natural resources. Creating diversity on the field, in the teams and in the shopping basket, while respecting all lives and maintaining sovereignty in the food system, are pillars under every day choices made in the Odin organisation.

Odin has never been privately owned by a few owners, shareholders or investors since its foundation in 1983. There is no interest or objective to make a profit for owners or shareholders. The term steward ownership is increasingly used for this. This is a fairly unique arrangement of ownership that is rarely found in companies of this size in the Netherlands. You could say the company is 'owned by itself'; all the company's assets (land, capital and means of production) are not freely tradable. As a result, the Odin crew can focus entirely on its mission.

The food system has developed in an unfair and unhealthy system, ruining and poisoning the earth and all life on it. Decline in biodiversity, erosion of fertile soils, water pollution, increase of diseases like cancer, Parkinson and viruses plus a big climate crisis coming our way are some of the results. Multinationals are gaining more and more power about what we sow, the way we grow our food and what we eat. Odin wants to make a visible contribution to the transition to a vital and future-proof food system. To inspire others to also take action, Odin wants to show that a more sustainable food system is possible when you work together and reconnect with other people and nature. Everyone, from farmer to consumer, can join. For that reason, Odin has repeated the following motto as a mantra: With every euro you spend, you help determine what the world looks like.

4 The View on the World and the Human Being

Odin's foundation lies in anthroposophy, inspired by the work of Rudolf Steiner. You can see the connection in the way Odin is organised, but also in the everyday choices made.

First of all, maybe the most easy connection to spot is the amount of biodynamic products in the assortment. More than 20% of all products sold in the Odin shops are certified biodynamic. When you look at the seven ideals of Odin, the first one says that we choose for 100% organic produce, but prefer biodynamic.

Why this preference for biodynamic? Because the biodynamic standards come close to the mission as expressed in the foundation stone meditation: healing the earth, respecting live, being aware that we are all connected, that everything is connected.

Odin has an holistic view on human beings, companies, the food system and the world. The worms, the cows, the plants, the bees and the human beings are all beings between heaven and earth with different roles. It is all connected and there is a mutual dependency.

As human beings, we have three capacities: thinking, feeling and acting. Each of those capacities have a correspondence with the different spheres of the society. In line with that, they are connected to the different activities of Odin.

The thinking is connected to arts, education and plant breeding. It should be free. In Odin, this area is, for example, connected to the plant breeding on our farm or the beekeeping where we work on more biodiversity and do not harvest any honey. Money in this area should not be labelled to an outcome. That is why Odin arranges "free money" for this area, with funds and gifts.

The feeling is connecting to the governance, the way we arrange things. You make appointments about things. In Odin, this area is for example connected to the membership of the cooperative or agreements about loans.

Fig. 3 Odin beekeeper Jos Willemse

The acting is connected to our daily business. In Odin this area is connected to the buying and selling of products. That is where the trucks are driving, people working in shops and warehouses. In this area, you pay for what you ordered. That is a different colour of money than the free money in the thinking area.

The fact that the three spheres are blurring more and more in our food system is part of the problem. Odin is aware of the different spheres and the different approach they need, and keeping them clear helps to work on solutions (Fig. 3).

5 Contributions to Make the World a Better Place

Odin has a deep connection with the Fundamental Social Law of Rudolf Steiner (Geisteswissenschaft und soziale Frage, Steiner, 1905/1906). In this Law he proposes that social wellbeing fundamentally depends upon a relationship of mutuality between the individuals and the community as a whole:

> The well-being of a community of people working together will be the greater, the less the individual claims for himself the proceeds of his work, i.e. the more of these proceeds he makes over to his fellow-workers, the more his own needs are satisfied, not out of his own work but out of the work done by others.

This Fundamental Social Law can be seen as one of the guidelines of Odin. It's translated in the foundation stone meditation of Odin of 1983 and more recently in one of the seven Odin ideals you can find on the Odin website (www.odin.nl) and on signs in the shops: "Live not purely for yourself, but for the good of all of us." This is also the reason why Odin consciously chooses not to own private property or work for shareholders. In the cooperative, anyone can join. With that, Odin can belong to employees, members, customers and suppliers work together for the benefit of the whole. Profits are not for individuals, but are used to invest in order to realise ideals even better.

6 Successes, Challenges and Learning Points

In autumn 2023, Odin celebrated its 40th birthday. On 29 September, Saint Michael's Day, the day Odin was founded in 1983, there was a festive moment with the founders and their partners. The Odin organisation has changed considerably over the years. It developed from a small wholesale company into a consumer supported retail organisation. It has had ups and downs, socially and economically. Balancing between ideals and running a professional and economic healthy business is an everyday challenge. The Odin crew has to deal with the fact that it is not possible to tackle everything at once. And in 2024 still 95% of the Dutch consumers are not into buying organic or biodynamic food.

But thanks to hard work and perseverance, Odin is still there and developing. One of the current challenges is how to combine growing bigger with more shops, more employees and more members with keeping the human scale in the organisation. Another challenge is to be less modest about the urgency of the problems in the environment and the food system. Odin has always been modest about its ideals and results. Odin strives to work with positive energy. But with the ecological and social problems becoming bigger, being positive is not enough. People have to become aware that thanks to monocultures, deforestation and chemicals, the biodiversity, soil fertility and nutritional quality are decreasing rapidly, with serious consequences for our health and the perspective of future generations. More action is required. Therefore, Odin decided for example to support political activities in order to keep organic and biodynamic agriculture free of GMO. Power to the people and the planet, that will be the motto for the coming years! (Fig. 4)

Fig. 4 Odin Ijburg Amsterdam_one of the organic shops of Odin foodcoop

7 Conclusion

For over 40 years, Odin has stood for a sustainable food system from farm to plate. Organic and Biodynamic agriculture are at its core. "Power to the people and the planet" is our latest motto, where, until now, modest positivity is turning into more radical action. Steward-ownership results in all profits flowing back into the development of the organisation and fair prices for the customers all the way throughout the added value chain to the farmers and nature itself. Trust between the consumer and producer marry and result into a true care for the healing of the earth.

References

Steiner, R. (1905/1906). *Geisteswissenschaft und soziale Frage.* Lucifer-Gnosis, (30 & 32).
Steiner, R. (2009). *Verbeter de wereld (en begin samen): De sociale hoofdwet.* Nearchus.

Sativa—Living Seeds, Living Varieties for Today and Our Future—Sativa Is Working on It

Amadeus Zschunke

Keywords

The Founding History of Sativa

Features:

(a) Year founded	1998
(b) Number of employees	80
(c) Locations	Rheinau (CH), Jestetten (DE), Neckarmühlbach (DE), and Pradellona (IT)
(d) Active in which countries	Switzerland and the EU
(e) Sector	Organic Seeds, Biodynamic Plant Breeding

1 The Founding History of Sativa

Ilmar Randuja, the pioneering figure in biodynamic plant breeding and biodynamic vegetable seeds, who passed away in 2024 at the age of 98, laid the foundation for the impulse that Sativa embodies today. He recognised early on that fruitful biodynamic cultivation could only be sustained in the long term if biodynamic seeds and biodynamically bred varieties were also available. After World War II, he encountered anthroposophy and biodynamic agriculture at Schloss Hamborn. The development of biodynamic horticulture became his life's work. With unwavering dedication, he first trained in seed production and plant breeding, which at the time

A. Zschunke (✉)
Sativa AG, Rheinau, Switzerland
e-mail: a.zschunke@sativa.bio

J. Kronenberg and E. T. Lammerts van Bueren (eds.), *On the Earth We Want to Live*,
World Sustainability Series, https://doi.org/10.1007/978-3-031-98758-8_38

Fig. 1 Logo Sativa

could still largely be described as a craft. Later, as a gardener at the Ita Wegman Clinic in Arlesheim, he harvested his own vegetable seeds and began selecting his first varieties. As early as 1963, he published his first list of biodynamic vegetable seeds, which he distributed to other farms and growers (Fig. 1).

In the 1980s, while training as a vegetable gardener at the biodynamic Willmann Garden near Stuttgart, Amadeus Zschunke was struck by the fact that biodynamic farming relied solely on conventional seeds from conventionally bred varieties. Searching for alternatives, he came into contact with Ilmar Randuja, later studying seed production and plant breeding. In 1988, he joined the gardening team led by Ilmar Randuja at the Ekkharthof, a therapeutic institution on the Swiss side of Lake Constance, which had already gained a reputation among insiders for its range of biodynamic seeds.

In 1998, the Fintan Foundation project began in Rheinau, near the Rhine Falls (Fintan Foundation, n.d.). This initiative brought together 13 independent enterprises that closely collaborate across various fields. Amadeus Zschunke seized the significantly improved opportunities provided by this project for advancing Ilmar Randuja's seed impulse and co-founded Sativa Rheinau GmbH with others. Even today, the Fintan project is fundamentally supported by a diverse biodynamic farm (125 hectares), an institution for people with special needs, and Sativa's work.

2 Sativa's Vision

Sativa aims to contribute to ensuring an independent, GMO-free, and patent-free seed supply for organic farming, covering both horticultural and agricultural seeds. We see ourselves as part of a European network and strive to preserve biodiversity while also fostering its creation. Our goal is to offer and make organic seeds available to everyone who seeks them. In doing so, we place particular importance on contributing to the preservation and, where possible, the enhancement of the autonomy and development opportunities of farms. Sativa seeks to promote not only ecological but also social sustainability (see Lammerts van Bueren et al., 2018).

3 Sativa's Work

We are committed to promoting varietal diversity and offer an extensive range of varieties, encompassing both horticultural and agricultural seeds. On the one hand, we preserve many old and traditional varieties, while on the other, we are deeply engaged in biodynamic plant breeding.

Independence and autonomy are central to our work. This is evident in various aspects, starting with our organisational structure: since 2005, we have been a public limited company. Approximately 30% of the shares are held by Sativa employees, and there are currently around 850 shareholders whose subscribed capital enables Sativa to pursue its established goals. Our autonomy is also reflected in the fact that the vast majority of the seeds we sell come from contract production. Currently, around 120 biodynamic and organic seed production farms across various parts of Europe collaborate with Sativa. The varieties in our horticultural seed range are predominantly our own selections or varieties we have bred ourselves.

At the same time, we see ourselves as partners to many initiatives across Europe that share similar goals, and we engage in active exchange with them.

In recent years, Sativa has continued to grow organically. In addition to the main operation in Rheinau, Switzerland, the Sativa family now includes a German trading company handling the distribution of organic seeds outside Switzerland, an operation in Italy specialising in seed production and services for our plant breeding efforts, and, from 2024, a new operation in southern Germany dedicated to plant breeding.

Almost all forms of agriculture depend on seeds. The choice of seeds is critical for organic agriculture in general and biodynamic agriculture in particular. For this type of farming to function effectively today and to continue developing in the right direction in the future, it also requires varieties and seeds with suitable characteristics, such as adaptability to low external system inputs, the ability to suppress weeds, or robust and extensive root systems. These traits do not come from conventional plant breeding.

4 Sativa and Anthroposophy

In biodynamic agriculture, the aim is to understand and develop processes from a human perspective. It is not about minimising human intervention in natural processes or treating humanity as a disruptive factor in an otherwise perfect system. Rather, biodynamic agriculture seeks to enable the co-evolution of humans and nature.

Engaging with the concepts of biodynamic agriculture—such as farm individuality or the organism principle—and exploring what distinguishes biodynamic plant breeding from conventional or organic plant breeding inevitably leads to a confrontation with the principles of anthroposophy.

We find it deeply enriching to address the tasks and challenges of our field in the context of today's world while comparing them with anthroposophical approaches. However, we firmly believe that developing one's own competence and understanding is essential.

Anthroposophy's view of humanity, in general, and the specific insights from the *Agricultural Course*, *Curative Education Course*, and *National Economic Course*, significantly inspire and enrich our work. For us, it is crucial to approach the present situation as we concretely experience it rather than starting from abstract concepts. By incorporating anthroposophical insights, we aim to improve both our understanding and our methods.

To illustrate how productive it can be to first develop one's own insights and subsequently integrate anthroposophical perspectives, we turn to our understanding of plant breeding.

Conventional plant breeding typically has significantly greater resources and access to an extensive array of tools, techniques, and methods—many of which are incompatible with the principles of organic farming and are therefore rejected in organic breeding. But what distinguishes biodynamic plant breeding? Is it more than mere rejection? What are our own strengths?

In collaboration with people both with and without anthroposophical backgrounds, the *Rheinau Theses* 1 and 2 on the rights of plants and plant breeding as dialogue were developed (Sativa Rheinau, 2011; Sativa Rheinau (n.d.)). These theses hold great significance for our practical breeding work. When we understand and take plants seriously as living beings, we are compelled to recognise their rights as well. Plant breeding that acknowledges the rights of plants cannot treat plant material as a genetic Lego kit. Such breeding can only take place through dialogue with the living being that is the plant, respecting intrinsic values like plant integrity (Nuijten et al., 2017). This is breeding as a conversation.

But how can I, as a plant breeder, engage in a conversation with the plants I work with? A dialogue requires two interlocutors: the plant on one side and the human on the other. The course of this dialogue and the resulting plant variety depend not only on the plant itself but also on me as the breeder. As a plant breeder, I must not only excel in genetics and statistics but also engage fully as a human being who is constantly evolving.

In biodynamic plant breeding, the human factor is central. It is not about supposedly neutral processes and techniques, which are increasingly automated today and in the future. Biodynamic agriculture is conceived by humans and can only be understood as such. The same applies to biodynamic plant breeding. As a breeder, I must remain aware of this fact. How I develop myself and my biography must, in turn, be left to the freedom of each individual. Everyone must find their own path to this development. For one, it may be the Goethean approach to observing plants (see Chap. 4.12 Holdrege); for another, it may involve different methods (Fig. 2).

Fig. 2 Sativa headquarters in Rheinau (CH)

5 Sativa and Sustainable Development

A truly sustainable overall development is only possible if we strive to preserve and, where possible, enhance the good in all aspects of life. While progress may vary across different areas, reducing the concept of sustainability to isolated factors such as the share of renewable energy, CO_2 emissions, or working conditions is insufficient to achieve meaningful improvements that advance our overall development. For this reason, we endeavour to make gradual progress in all areas within our means. Some goals are simpler to achieve, while others require more time and effort.

Sustainable development must be understood as a process both inwardly and outwardly. Internally, as a seed company, we aim to make the core production steps—such as seed processing, storage, and packaging—as sustainable as possible. We also prioritise building sustainable, long-term relationships with our seed production partners. To this end, we take responsibility by offering fair prices and increasingly sharing parts of the cultivation risks faced by seed producers.

Our biodynamic plant breeding represents our outward focus. Such breeding, committed to sustainability, considers not only the methods and techniques used but also whether breeding goals genuinely support farms in their tasks. The *systems-based breeding* concept developed by Edith T. Lammerts van Bueren has been a significant aid to our work (Lammerts van Bueren et al., 2018). This is particularly evident in our lettuce breeding. Conventional plant breeding for lettuce relies on monogenic resistances, which are regularly overcome by mildew fungi after a short period. This approach cannot be considered sustainable. At Sativa, we instead focus on broadly supported, more stable field resistance and robustness in our lettuce breeding efforts.

6 Ecological, Social, and Economic Sustainable Development

All three dimensions are equally important and mutually dependent. A business enterprise is first and foremost committed to economic sustainability—this means the business must be managed economically and profitably, as this forms the foundation upon which everything else is built. Concretely: without the economically successful trade of organic seeds, there can be no biodynamic plant breeding. However, we must avoid the mistake of prioritising economic success at the expense of ecological and social sustainability. In line with our vision, we can only achieve economic success if we also operate in an ecologically and socially responsible manner. This principle applies not only to the work within our company but also to our collaboration with seed production partners.

Our efforts are evident in various areas, such as our new company building, which incorporates numerous aspects of ecological sustainability. While Sativa's pioneering phase focused primarily on advancing the company as quickly as possible, in recent years, we have increasingly worked to improve working conditions within the company and involve employees in decision-making processes. We aim for long-term, cooperative relationships with our seed production partners, extending beyond fair compensation for services rendered.

In our plant breeding efforts, we also strive to develop varieties that enable farms to operate in a socially, ecologically, and economically sustainable manner. In the current economic system, people are regarded as a labour force and therefore primarily as a cost factor. To preserve vitality, agricultural production cannot be rationalised without limits. As a result, agriculture generally suffers from the fact that the net added value of its production continues to decline, and farms can only partially compensate for this through larger cultivation areas or increased mechanisation.

Organic and biodynamic farms, in particular, face a growing conflict between the economic constraints of their operations and the ideals and goals of their farming practices. This is because these farms often involve more people in food production. This development increases their economic dependencies and reduces their autonomy, potentially leading them to align decisions solely with external pressures.

As biodynamic plant breeders, we must ask ourselves how our varieties can contribute to giving farms greater autonomy in their operational decisions. For example, we invest significant time in developing sweetcorn varieties capable of fixing nitrogen independently. Such varieties require less fertilisation, allowing

Fig. 3 Harvesting carrots in Sativa fields

farms to use their manure more flexibly elsewhere in the crop rotation, thereby hopefully increasing farm fertility.

We also devote considerable effort to developing pole bean varieties for mixed cultivation with fodder maize. Pole beans and maize share a long history of co-evolution, making them plants that thrive together and mutually support one another. Including pole beans in animal feed increases its protein content, potentially reducing the farm's reliance on external protein sources. This, in turn, enhances the farm's autonomy (Fig. 3).

7 Successes and Challenges

Success is rarely the achievement of an individual; it usually requires more. In recent years, awareness of the critical importance of seeds has increased significantly within the organic movement and among biodynamic farms. "Why biodynamic seeds?"—this is no longer a question anyone asks. This shift is partly due to a growing awareness. Unfortunately, however, much of this awareness stems from a more or less conscious rejection of conventional plant breeding.

The consequences of plant breeding based on a materialistic worldview and embedded in a capitalist context are becoming increasingly apparent to a wider audience. Not only within agriculture but also among consumers, there is growing concern and questioning about hybrid breeding, old and new genetic engineering techniques, and the privatisation of seeds through patents. Interest in the work of

Sativa and other initiatives with similar goals has grown in recent years. However, this interest often arises less from an understanding of what we truly stand for and more from a rejection of developments perceived as threatening. In the long run, this alone is not sufficient to achieve substantial progress and sustainable development.

On a practical level, Sativa can look back on numerous smaller and larger successes. Over 25 years, we have developed many new varieties that enable the production of high-quality biodynamic vegetables. At the same time, we have built a system that extends beyond the pioneering efforts of individuals like Ilmar Randuja. Today, we have an entire team of qualified breeders working with continuously improving infrastructure to further develop and refine our unique breeding approaches. This is already a significant achievement. However, when measured against what is still needed to match the capabilities of conventional breeding systems, there remains a great deal to do.

Although Sativa has grown into a medium-sized seed company, we are still far from achieving true financial independence in our plant breeding. Currently, three-quarters of the costs are covered by external funding, most of which comes from donations. This means our work remains heavily reliant on the goodwill of donors and the ability of society to provide such resources.

8 The Coming Years

Our first priority is to pass on the founding impulse of Sativa to the next generation. We have already begun this process and are on a promising path. A key element of this transition is to detach the foundations of Sativa's work from the acceptance and influence of individual people and instead anchor the underlying values as broadly as possible within the company. This is an ongoing process, and while the groundwork has been laid, we recognise that it will require continuous effort.

At the same time, we observe that people in our society are increasingly unable to rely on traditions as they could 20, 30, or 40 years ago. This has led to a widespread sense of uncertainty and lack of knowledge. In the future, we will need to explain more often—both to our employees and to our customer base—what we do and why we do it. Why don't we choose the seemingly simple and environmentally friendly solutions proposed by modern plant breeding? Why don't we address problems with straightforward genetic manipulation? These explanations will become increasingly challenging in a society where fewer and fewer people have direct experience with gardening or farming. We anticipate that, in addition to our current activities, we will increasingly face "educational responsibilities." We must begin preparing for this now and establish the foundations for addressing these tasks.

Over the past 25 years, Sativa's decisions and actions have always been future-oriented. Plant breeding is a concrete act of shaping the future. Sativa is committed to a vision of the future that we believe is relevant for society as a whole. Seeds are the key to the future of agriculture, and how we shape our agricultural and food systems will determine our collective future.

As agricultural practitioners, we regularly witness the tangible effects of climate change. Seed production has also become more demanding. Extreme weather events are increasing, not only in absolute terms (such as record temperatures or rainfall) but also in the persistence of weather patterns. There are parallels to societal dynamics: polarisation is on the rise, and nuanced perspectives are receiving less attention. Differences are increasingly perceived as irreconcilable, making dialogue and mutual understanding more difficult.

Insights from Rudolf Steiner, as well as modern microbiome research, highlight the connections between environmental health—especially in agriculture and soil—and human health. Healthy humans encompass more than just the absence of physical ailments; health includes mental and spiritual well-being. The old saying, "You are what you eat," has taken on renewed significance. If humanity is to meet the challenges ahead, it will depend on the forces we draw from the food we consume. The food we eat must also support us in perceiving and valuing not only the material but also the spiritual aspects of the world. What responsibilities and future tasks does biodynamic plant breeding face when it is oriented towards these principles?

At the same time, we are witnessing an enormous acceleration of societal developments and processes. Many people are increasingly overwhelmed by this rapid pace. The use of so-called artificial intelligence (AI) is accelerating many processes exponentially, threatening to outpace society's ability to maintain control. How can we keep up with these developments in our work, within our company, with our employees, and in dialogue with our customers? How should we respond, and what must we do?

We do not yet have concrete answers to all these questions, but we will need to seek and find them. One thing is certain: we are called upon to bring as much light and awareness as possible to these developments, wherever we can.

9 Conclusion

Sativa can be regarded as a pioneer in organic and biodynamic seeds and plant breeding. With its origins dating back to the 1960s, Sativa possesses a wealth of experience and a now-systematised approach to its work. The question of an earth-connected, ecological agriculture begins with seeds and plants. However, according to Sativa, this awareness is still underrepresented in debates about the future of agriculture and, consequently, the future of humanity's food systems.

The isolation of agriculture into a rootless, technocratic system offers no path forward. A sustainable, future-oriented solution cannot emerge from the mechanistic assembly of isolated measures. Instead, we need living seeds that enable a vibrant and life-affirming agriculture, pointing the way toward a future grounded in connection to the Earth.

References

Fintan Foundation. (n.d.). *Home*. Retrieved January 14, 2025, from https://www.fintan.ch

Lammerts van Bueren, E. T., Struik, P. C., van Eekeren, N., & Nuijten, E. (2018). Towards resilience through systems-based plant breeding: A review. *Agronomy for Sustainable Development, 38*(42). https://doi.org/10.1007/s13593-018-0522-6

Nuijten, E., Messmer, M. M., & Lammerts van Bueren, E. T. (2017). Concepts and strategies of organic plant breeding in light of novel breeding techniques. *Sustainability, 9*(1), 18. https://doi.org/10.3390/su9010018

Sativa Rheinau AG. (n.d.). *Rheinauer Thesen zu Rechten von Pflanzen* [PDF]. Retrieved January 14, 2025, from https://www.sativa-rheinau.ch/dateien/Zuechtung/Rheinauer_Thesen_I.pdf

Sativa Rheinau AG. (June 2011). *Rheinauer Thesen zur Ökologischen Pflanzenzüchtung* [PDF]. Retrieved January 14, 2025, from https://www.sativa-rheinau.ch/dateien/Zuechtung/Rheinauer_Thesen_II.pdf

SEKEM—Sustainable Development Since 1977

Christine Arlt

Keywords

sustainability • Egypt • holistic development • potential unfolding • ecology • economy • cultural life

Features:

(a) Founding year	1977
(b) Employees	2000
(c) Location	EG, Belbeis, Sharkeyya
(d) Countries	Worldwide
(e) Sector	Ecology, Economy, Culture and Societal Life
(f) Other	Right Livelihood Award—Alternative Nobel Prize in 2003, Gulbenkian Prize for Humanity 2024, United Nations Champion of the Earth 2024

C. Arlt (✉)
Member of the SEKEM Future Council and Head of SEKEM-Friends Germany, SEKEM-Friends, Cologne, Germany
e-mail: Christine.Arlt@Sekem.com

J. Kronenberg and E. T. Lammerts van Bueren (eds.), *On the Earth We Want to Live*, World Sustainability Series, https://doi.org/10.1007/978-3-031-98758-8_39

1 The SEKEM Initiative: Origins, Sources of Inspiration and Vision for the Future

Known to many as a "miracle in the desert," the Egyptian SEKEM Initiative evokes a variety of images. Some think of a flourishing biodynamic farm in the middle of the desert, others of holistic educational institutions for rural communities or of companies producing organic products. Still others picture a sustainably living community of Christians and Muslims, Europeans and Egyptians. The name SEKEM is derived from an ancient Egyptian hieroglyph meaning "the power of the sun."

What SEKEM truly represents in reality is difficult to capture in just a few words. Nearly 50 years after its founding in 1977, SEKEM is undoubtedly a realised miracle—the vision of a holistically conceived initiative for sustainable development brought to life. Yet at the same time it remains far from perfect or fully realising its lofty vision. This apparent contradiction is, in truth, a defining characteristic and perhaps even a secret of SEKEM. It is at once utopian and a living example of countless practical concepts that coexist and evolve in dynamic harmony. But how does such a system function in reality? What are its sources of inspiration? And how is SEKEM evolving now that its founder Ibrahim Abouleish passed away in 2017 (Fig. 1)?

The creation and implementation of the SEKEM vision are closely tied to the biography and spiritual inspirations of Ibrahim Abouleish. Throughout his life, SEKEM's founder was able to integrate the diverse streams and impulses that shaped his journey, forging synergies between East and West, Egypt and Europe, Islam and Christianity, anthroposophy and philosophy, spirituality and science. "I lived in two worlds that I experienced as opposites: the spiritual stream of the East from birth and that of Europe to which I felt an elective affinity. Increasingly, I experienced moments in which these two streams united in my soul. (…) The two fundamentally different cultures began to dissolve in their opposition within me and merge into a new, third element, so that I was neither entirely one nor the other," wrote the Egyptian-born Abouleish, who lived studied and worked in Austria for 20 years, in his autobiography (Abouleish, 2016, p. 39).

SEKEM

Sustainable Development since 1977

Fig. 1 Logo SEKEM Group

Today, SEKEM provides employment for approximately 2000 people, involves around 40,000 individuals in its activities and serves tens of thousands more in Egypt and beyond with its products and services. Subsidiaries and initiatives associated with SEKEM operate in areas such as food processing, pharmaceuticals and textiles. Educational institutions range from kindergarten and school to vocational training, therapeutic education and a university. Additionally, there is a day clinic and affiliated farmers' associations such as the Egyptian Biodynamic Association—Demeter Egypt. All these initiatives and activities contribute to a lived model of sustainable development.

2 Lived Sustainable Development

Abouleish's vision of a holistically lived sustainable development emerged when, after two decades in Europe, he returned to his homeland of Egypt and was shocked by the dire state of the environment, society, culture and economy. The population had grown dramatically (from 25 million in 1960 to over 115 million today), infrastructure and the education system were inadequate and cultural life was almost nonexistent. Looking toward the future—both recognising what we create for the future and what comes toward us from the future—he articulated his vision:

> Sustainable development for a future in which every human being can realise their individual potential; where humanity lives in social forms that reflect human dignity; and where all economic activities align with ecological and ethical principles. (SEKEM Vision 2057)

He transformed barren desert land using biodynamic agricultural methods, established companies to process agricultural products and created transparent value chains where all participants received a fair share and opportunities for cultural and individual growth. With the profits from these enterprises, he funded schools for the people in neighbouring villages and built an international network of like-minded initiatives and ventures (Fig. 2).

3 SEKEM's Vision 2057 for Egypt and the World

After all this became a reality—turning a desert into an oasis where people live, work and learn together—and after Ibrahim Abouleish departed from earthly life in 2017, 40 years after SEKEM's founding, a pivotal question arose: How can what has been achieved and practically tested at SEKEM become relevant on a larger scale? How can models be developed and multiplied from SEKEM's oasis and lighthouse project that bring systemic relevance to Egypt and contribute to real change in the country?

Fig. 2 On the farm in the oasis Wahat EL-Bahareyya SEKEM is greening desert soils and creating a community that supports people in their individual potential unfolding

In response, the SEKEM community formulated a vision for the next 40 years: the SEKEM Vision 2057, with 16 vision goals. Within this framework, 16 prototypes are being developed that can be easily scaled and subsequently established. For example, in the field of ecology SEKEM aims to establish sustainable and biodynamic agriculture nationwide. In Egypt's education system, the focus is to place the individual potential of learners at the centre. In the economy, SEKEM envisions an "Economy of Love," a concept it promotes to serve people equitably and fraternally. Lastly, Egyptian society is to undergo a transformation that encourages its citizens to responsibly co-shape the country's development.

In the current situation, where Egypt faces significant challenges across all sectors, these goals may initially seem unpromising. However, the SEKEM vision would never have become a reality if its founder had not unwaveringly held onto his vision despite all obstacles and continuously sought to implement his ideas. Fifty years ago, biodynamic farming in the desert seemed as absurd as a fully sustainable Egypt might seem today. The advantage that the SEKEM community has now compared to then is the wealth of practical experience it carries, supported by the spiritual inspirations of the original vision (Fig. 3).

Some initiatives are already underway. One example is the "Economy of Love" (EoL) model, which certifies a new form of economy while simultaneously incentivising farmers to transition from conventional to sustainable biodynamic farming methods (see Chapter "Biodynamic Agriculture—And the Future of Agriculture").

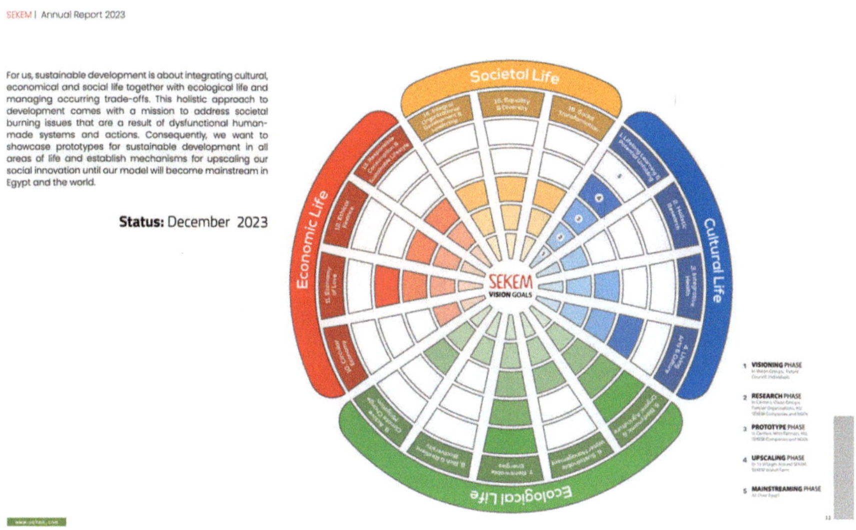

Fig. 3 SEKEM's sustainability flower and vision goals for 2057

Between 2020 and 2024, this model has already reached over 20,000 farmers, making the vision of a biodynamic Egypt suddenly seem much less far-fetched.

In the realm of social and cultural life, SEKEM is focusing on the transformative potential of surrounding communities. This involves providing financial support, sharing knowledge and conducting cultural activities. At the same time, it is continuously developing the structures of its own educational institutions and programmes to focus increasingly on unlocking individual potential, ensuring a long-term positive impact on society. For instance, alternative examination methods are being developed, and innovative teaching methods are being implemented outside the traditional classroom setting.

A significant contribution to this mission is made by the Heliopolis University for Sustainable Development, founded under SEKEM's patronage. The university currently educates over 2000 students in fields such as pharmacy, engineering, business, organic agriculture and physiotherapy. Its holistic approach ensures that students across all disciplines are required to take courses in art and culture. This supports their individual development while fostering creativity, innovation and a sense of social responsibility (Fig. 4).

4 The Spiritual Dimension of Sustainable Development in SEKEM

These are just some examples of the approaches rooted in SEKEM's history and future vision. Its development emerged against many odds, inspired by the founder Ibrahim Abouleish, who firmly believed: "I do not think even the smallest part of

Fig. 4 SEKEM's 2023 key figures. Annual Report 2023

our vision could be realised without a daily, concrete connection to the spiritual world" (Abouleish & Arlt, 2022). These spiritual inspirations for holistically lived sustainability came from diverse sources. They drew from Islam as well as anthroposophy, from Egyptian culture as much as European culture, and from figures like Goethe or Beethoven as well as from Arabic mystics.

The fundamental idea of development, which lies at the heart of SEKEM's vision, can, for example, be found in the Quran. The first revelation of the Prophet Muhammad (Surah, 96) can be interpreted as a call to development. The Archangel Gabriel urges the Prophet: "Read!" For Muhammad, who was illiterate, this call to read likely referred to the pursuit of understanding and engagement with oneself and the world rather than the act of reading written words. In addition to this personal task of development, humanity has also been entrusted with a responsibility for the world: "Indeed, we offered the Trust to the heavens and the earth and the mountains, and they declined to bear it and feared it; but man [undertook to] bear it. Indeed, he was unjust and ignorant" (Surah, 33:72). The SEKEM Initiative seeks to take this world responsibility seriously, creating a framework that enables people to discover their unique tasks of development within this context.

A significant realisation regarding the importance of development in all areas of daily life at SEKEM came later through Ibrahim Abouleish's engagement with anthroposophical views. Recognising the developmental stages of human beings has become a fundamental element of all activities within the SEKEM vision. Daily efforts are made to address the varying needs of the Sentient Soul, the Intellectual Soul and the Consciousness Soul, as this is essential for genuinely sustainable action. It is also worth mentioning that the SEKEM vision does not

Fig. 5 One of SEKEM's circles

place itself above the developmental possibilities of individuals but continually seeks to evolve itself in a living, dynamic way (Fig. 5).

5 Conclusion

The SEKEM Initiative and the vision that sustains it are far more than a flourishing desert, a thriving Economy of Love or children learning freely and joyfully. Like its founder, the SEKEM Vision seeks, through spiritual inspirations from Egypt and Europe, East and West, Islam and Christianity, philosophy and science, to contribute to a synergy that creates something new—a world where, as Ibrahim Abouleish envisioned, "people can realise their individual potential, live in social forms that reflect the dignity of every individual and where all economic activities align with ecological and ethical principles."

References

Abouleish, I. (2016). *Die SEKEM-Symphonie: Nachhaltige Entwicklung für Ägypten in weltweiter Vernetzung*. Frankfurt am Main: Info 3.

Abouleish, H., & Arlt, C. (2022), *SEKEM Inspirationen: Impulse für einen zukunftsfähigen Wandel*. Frankfurt am Main: Info 3. Quran. (n.d.). Surah Al-Ahzab [33:72]. Sahih International. Retrieved December 13, 2024, from https://quran.com/al-ahzab/72

Sonett: Washing and Cleaning Agents that Honour Water as the Carrier of All Life

Beate Oberdorfer and Gerhard Heid

Features:

(a) Founding year	1977
(b) Employees	120
(c) Location	DE 88603 Deggenhausertal
(d) Active in which countries	More than 40 countries worldwide
(e) Sector	Ecological detergents and cleaning agents
(f) Other	German Sustainability Award 2022 and 2024

1 Starting Point and Impulse

The impetus for the development of Sonett detergents and cleaning agents can be traced back to research carried out by the natural scientist Johannes Schnorr at the Institute for Flow Sciences in Herrischried at the end of the 1960s. A method was developed there to assess the quality of bodies of water and drinking water, the so-called drop picture method [Tropfbildmethode], which, in addition to the chemical analysis of water for contamination by heavy metals, herbicides, pesticides, fertilisers, etc., makes the quality of the water visible as a holistic image. The drop picture method can provide information about the quality of water as a carrier of life and as food. Using this method, Johannes Schnorr was able to detect

B. Oberdorfer (✉) · G. Heid
Sonett GmbH, Deggenhausertal, Germany
e-mail: beate.oberdorfer@sonett.eu

G. Heid
e-mail: gerhard.heid@sonett.eu

J. Kronenberg and E. T. Lammerts van Bueren (eds.), *On the Earth We Want to Live*, World Sustainability Series, https://doi.org/10.1007/978-3-031-98758-8_40

sonett
ÖKOLOGISCH KONSEQUENT

Fig. 1 Logo Sonett

non-degradable or poorly biodegradable surfactants (washing-active substances) in drinking water (Wilkens et al., 1995. P25).

In the 1960s, household washing machines became widespread in Germany, and with them came new synthetic detergents that washed ‘whiter than white’ and whose effects on the environment and water had not been taken into account. These surfactants were washed into the sewage treatment plants with the wastewater and could not be broken down there, or not quickly enough. Via reintroduction into the water and the extraction of drinking water from it, they then found their way back to human beings. In Ullmann’s Encyclopaedia of Technical Chemistry, Volume 24, 1983 edition, chapter ‘Detergents,’ you can read about the development of detergents from simple soap to the synthetic, washing machine-suitable surfactant tetrapropylene benzene sulphonate (TPS), which already covered 65% of the entire demand for washing raw materials in the western world in 1959. This surfactant was only 20–30% biodegradable. Residues of the detergent surfactant were found in the groundwater, but also in lakes and rivers and led to huge mountains of foam at dams and sluices. As a result of this catastrophe, the first detergent law was passed, which came into force in 1964, stipulating that anionic surfactants had to be at least 80% biodegradable. It was not until later, in 1975, that the Detergents and Cleaning Agents Act—WRGM—was passed, which stipulated more comprehensively that detergents and cleaning agents ‘may only be placed on the market in such a way that after their use any avoidable impairment of the waters is prevented.’[1]

This shocking contamination of drinking water inspired Johannes Schnorr to develop a new environmentally friendly detergent. The approach for this new detergent was: (1) Rapid and 100% complete biodegradability of all detergent ingredients (2) the greatest possible saving in the use of raw materials. (3) Use of soap as almost the only surfactant that already starts to biodegrade on the way to the sewage treatment plant (Fig. 1).

[1] Original in German «*Sonett, Mittel für Waschen und Reinigen, die das Wasser achten als Träger alles Lebendigen“*

2 This Was the Birth of the Sonett Modular Principle

Detergent, softener and stain remover/bleach complex as separate components that can be optimally dosed to save resources, depending on the water hardness and the degree of soiling of the laundry. Sonett's starting point was therefore the care and responsibility for water as the carrier of all life. This starting point is still the centre of our production and product development today and is expressed in the Sonett slogan "Sonett, detergent for washing and cleaning, which honour water as the carrier of all life."

Washing and cleaning is a cultural achievement that, while benefiting human beings, also inevitably pollutes water. We see it as our task to harmonise washing and cleaning with nature, human beings and especially water in a considerate and constructive manner. Today, this task can no longer be carried out in isolation, because in the living organism Earth, which comprises human beings, nature and the cosmos, everything is interconnected and everything has an effect on everything. This applies not only to our work at Sonett, but basically to all work that is not carried out for our own needs and well-being. In other words, what is needed is: (1) A new way of dealing with raw materials, with the Earth as a whole and with water in particular; (2) new social structures: cooperation and partnership, and (3) a new way of dealing with company property, capital and profits. Only the simultaneity of these three fields of organisation is sustainable, lively and therefore future-proof in the true sense of the word.

3 Anthroposophy—The Starting Point and Basis of Our Work at Sonett

One of Sonett's ideological principles is the realisation and the impulse that an understanding of the living world is necessary for a proper approach to nature, the environment and human beings, for which responsibility is assumed. Only if we understand the whole Earth as a living being, i.e. the Earth itself with all the beings of nature, and human beings as beings who want to develop towards freedom and responsibility, will there be a chance of overcoming the crises of the present:

Instead of exploiting nature—the main cause of the climate catastrophe—cooperating with nature and promoting the forces of life.

Instead of exploiting people and disregarding human rights—the main cause of the refugee disasters—willingness to help and cooperation in partnership.

Instead of selfish appropriation and economic supremacy—a cause of the extreme inequality in the distribution of wealth and income—orientation towards the common good and cooperation. These motives are not just 'reform' ideas, but represent a fundamental paradigm shift in our relationship with nature, human beings and the world as a whole. It is a reversal of consciousness that is exemplified by biodynamic agriculture: not exploitation of soil and plants, but support and care based on their own living conditions. The result is healthy food that we do not take from nature, but receive as a 'gift.' This paradigm shift means that nature is no longer there for us, we are there for nature.

For us, anthroposophy represents this paradigm shift. Anthroposophy is the science of the living. Furthermore, it is about new vitality in all areas and not limited to life in nature. Life exists equally on a soul level, on a social level and on a spiritual level. The paradigm shift towards a new vitality in the individual/spiritual realm is the development from self-centred utilitarian thinking towards creative and substantive thinking. In the social/entrepreneurial sphere, it is the shift from wanting to have to wanting to give, and with regard to nature and the environment, it is the shift from the highest possible level of exploitation to taking responsibility for nature and its further development. Anthroposophy is therefore all-pervading. For us, anthroposophy is identical with individual and entrepreneurial responsibility for human beings and nature.

4 Sonett—Quality

With regard to Sonett products, this responsibility is expressed in Sonett's 3 quality levels:

1. Responsible selection of components:
   - 100% biodegradability of all raw materials;
   - No petrochemical ingredients: no synthetic surfactants, fragrances, colorants or preservatives;
   - No genetic engineering, nanotechnology and microplastics.
2. As many raw materials as possible from organic cultivation. All the oils that we saponify, such as olive oil, sunflower oil, rapeseed oil, coconut oil and palm oil are 100% organic. All our essential oils come from controlled organic cultivation or wild harvesting.
3. Development and use of rhythmic processes and dynamised preparations to stimulate life forces that have a restorative effect in the water via our products. For instance, we rhythmise our balsamic detergent preparations in an oloid mixer, swirl the process water for Sonett's liquid products in twelve egg-shaped glass swirlers and sensitise our summer and winter mistletoe preparations in a fluidic oscillator. We not only want to utilise nature 'at our service,' but also give something back that is important for its future development.

With these principles and processes, Sonett has been a major and successful supplier of ecological detergents and cleaning agents in the wholefood market worldwide for almost 50 years. Sonett is the market leader in the German wholefood trade with a share of over 40%. In addition to quantitative market leadership, we also see Sonett as a quality leader in ecology and efficiency. This is documented, for example, by the fact that the Sonett modular system—the Sonett original product since 1977—was recognised as the 'Best Detergent' in the overall assessment in the ÖKO TEST 2019 test magazine compared to 23 conventional

Fig. 2 Production hall Sonett

(including all well-known supermarket brands) and 5 ecological detergents in terms of washing power and ingredients (Fig. 2).

5 Sonett—Culture

In the social field, we start from an image of human beings that sees them as developmental beings predisposed to freedom and responsibility. This means that we build on the will for insight and the capacity for insight, whereby insight is not based on personal opinion or desires, but on the spirit and nature of the task at hand. Understanding Sonett in terms of its essence means understanding the company itself as a developmental being. In this sense, every company is intrinsically and in its development dependent on the human beings who bear responsibility for it. We see ourselves as its 'stewards,' working towards the immanent development of this being. Based on this attitude, we always rely on shared insight and consensus rather than majority decisions or separate areas of responsibility. This applies first and foremost to the leadership of Sonett itself. We have been managing Sonett as partners for over 30 years and always reach agreement on all key decisions. Experience shows that this is possible if those responsible for management orientate themselves to the company's development principles. For us, the principle of partnership involves the cooperation of two very different human beings who agree in the spirit of the cause. We also practise this in the management of the departments and project teams at Sonett. This requires a sufficient number of employees who not only do their work for a living, but see it as a vocation and a life's mission. We cultivate the principle of 'diversity' in our workforce. We employ people from over 20 nations, from academics with PhDs to people with assistance needs. The proportion of male and female employees at our company has always been very balanced—even without quotas. This also applies to the management levels.

We have an intensive culture of dialogue at Sonett, which means that hierarchies can be remarkably flat. In the regular development meetings with our employees, the focus is not on the function holder, but on the human being. However, we have found that it is always very unusual for employees to ask themselves what they really want in life, and even more so to be asked this question.

Our error and conflict culture also often takes some getting used to for new employees. What matters to us here is not that a mistake or misconduct is reprimanded or even sanctioned. Rather, it is important that the mistake is analysed objectively and that it is not just a matter of affirming good intentions that this should not happen again; instead, a very individual lesson is drawn from the error analysis and concrete consequences are formulated in writing. However, we also repeatedly receive the feedback that taking oneself and others 'seriously' not only proves its worth in the workplace, but also has a positive effect on colleagues' private lives.

6 Sonett in the Public Eye

Our partnership-based management model and our corporate form as a foundation company were publicly recognised, most recently in 2022 when we were awarded the German Sustainability Award in the 'Transformation Field - Society'. We are also delighted to have won the German Sustainability Award in 2024, this time for the detergents and cleaning products sector, making Sonett the most sustainable producer of detergents and cleaning agents in Germany.

Sonett collaborates closely with the neighbouring Camphill Workshops Lehenhof. There, 36 people with assistance needs work for Sonett in two workshops. In addition, three 'Lehenhofers' work directly in our company. We have also worked closely with the 'Sieben Zwerge' substance abuse healing centre for many years. A number of former 'Zwerge' are now in permanent employment with us, some of them in management positions.

Through the Sonett charitable foundation, we support cultural, artistic and social initiatives. We support two anthroposophical curative education centres, not only financially but also through the involvement of two members of the Sonett Foundation's Board of Directors in the work of these organisations. The Sonett charitable foundation is a 100% shareholder of 'Sonett Vermögensverwaltung GmbH' (Sonett Asset Management Ltd), which is the owner of Sonett's fixed assets, land, buildings, machinery and equipment. It leases these to the Sonett operating company, Sonett GmbH. This is where the sales and profits are generated. Every year, Sonett

GmbH transfers funds to the Sonett Foundation, depending on the income situation. This continuously increases the foundation's assets, and the foundation can use the interest income to carry out charitable activities.

We still see a large area for development with regard to our foundation. We see the allocation of donations as part of Sonett's operational corporate task and do not outsource it to a foundation committee. Our desire is to develop completely new forms of cooperation between donors and recipients—beyond 'charity' and 'patronage.' We still find ourselves at the very beginning here. There is hardly any experience or role models for this approach.

7 What Does Sustainability Mean to Us?

For us, sustainability means working organically in a threefold manner. Ideational, social and ecological-economic developments belong together and form an organism. Rudolf Steiner's concept of the threefold structure of the social organism is fundamental for us. We are not only concerned with identifying the three different constituent elements that can be distinguished in every social structure, but above all with the task of shaping organic relationships. A quote from Rudolf Steiner's National Economic Course provides orientation: 'You cannot organise an organism, an organism grows.' Organics always work from the future: the whole tree is already laid out in the seed. In social terms, this means that leadership responsibility requires an existential, life-defining decision and responsibility. This is a major task that stands in stark contrast to today's common concepts and ideas of life, which are primarily determined by individual desires and advantages.

For many people, the astonishing experience of consciously dealing with the living is that everything is reversed here in relation to habitual ways of thinking: this is exemplified in a quote by IPV Troxler (1828, p. 27): "Man can only come to himself, by going beyond himself. And he only gets beyond himself by entering into himself." This means that human beings can only find themselves and their purpose in life, i.e. what they really want, if they do not focus their activities and endeavours primarily on themselves, but instead 'go beyond themselves,' i.e. place themselves at the service of an ideational task. Experience shows that your own development and happiness in life then come to you indirectly, like a gift, to an extent that you could never have achieved yourself.

8 What Do We Understand as the 'Sonett-Task', Now and in the Future?

The central concept of our activity at Sonett is the 'living': Water is the carrier of everything 'living,' that is our motif and working theme. The living cannot be understood with 'opposite' concepts. You can only do justice to the living if you 'immerse' yourself in it. I can only grasp the living in a living way. Bringing thinking to life is therefore the first task. My thinking comes to life when I connect with the content and processes of my senses and thinking. In other words, it comes to life when every thought, every perception becomes a task, an 'obligation', i.e. when I connect with it, stay with it and feel responsible for my thoughts and everything that enters my consciousness. However, this has the dimension of a life decision; it is about entering a spiritual stream of development and life that cannot be limited in terms of content or time, because the living is unlimited; it is infinite. The fact that this attitude to life must be anchored in the Sonett management, describes the major task of succession planning. That is because this approach to life is the fulfilment of human freedom, sovereignty and responsibility.

We are facing major challenges for the future:

1. **The consistent maintenance and further development of our quality standards**

This is by no means a matter of course given the rapidly changing market conditions, uncertain supply chains and unpredictable fluctuations in commodity prices. We are anticipating extreme swings and rollercoaster rides in all areas of life and work in the future. Forward planning and healthy development conditions are becoming increasingly uncertain. When making decisions, we will therefore increasingly have to tentatively feel our way forward step by step. We are currently experiencing the stabilising effect of our foundation model on the company's resilience. The fact that no profits are withdrawn by shareholders at Sonett, but instead remain in the company, means that we can withstand fluctuations such as those caused by Covid-19 and the consequences of the war in Ukraine, although we are also experiencing limits here.

2. **A second major challenge is company succession**

Here, too, our foundation form provides us with optimal conditions, as we can shape the succession purely from within, free from investors' and inheritance claims. Nevertheless, taking over the management of the company and continuing the principle of partnership-based management based on a spiritual understanding of the company, remains a major development task. This task likewise includes the fundamental question of how the connection to the predecessors can live on after the transfer of responsibility—even beyond death.

3. **The third future field we see is the new cooperation between economically value-generating companies and initiatives of spiritual life**

Organisations in the spiritual life, such as schools, homes, cultural and art initiatives, etc., cannot create economic value and are dependent on surpluses from economic life. Here we are looking for a relationship of equals and mutual interest, including relationships of responsibility—beyond charitable aid or generous

Fig. 3 Team Sonett

patronage. It is very unusual for recipients of funds to realise that they also bear responsibility for where the money they live and work with comes from. On the other hand, it is just as unusual for donors to realise that donations are the most 'productive' use of money. After all, where do education, skills and innovation—the sources of all economic value creation—come from (Fig. 3)?

9 Conclusion

Sonett is only successful where customers, wholesalers or shops are already asking for detergents and cleaning agents that harm neither nature nor health. Our task is not to proselytise for our products, but to answer questions and offer or develop solutions for customers who are looking for ecologically, ethically and socially acceptable and responsible quality detergents and cleaning agents. This means that we are fundamentally and from the outset dependent on resonance; experience has shown that strategic pushing through marketing measures hardly works for us. We are, however, noticing that more and more people are developing a new appreciation for life-sustaining and life-affirming quality in the face of increasing environmental disasters and pollution. Sooner or later, they will also turn to ecological detergents and cleaning products. Basically, it's about thinking ahead and not running behind developments. The division in society will continue to grow, and we have to reckon with the fact that even ecology and organic farming could come under attack in the future. The same applies all the more to idealistic and spiritual issues. Our 'niche' could become very small, although only a short time ago it seemed as if reason, ecology and responsibility had arrived 'in the centre of society.' At the same time, it is noticeable that there are fewer and fewer personalities with the leadership qualities and willingness to take on responsibility as 'lone fighters.' The good thing about this 'shortage,' however, is that the complexity of the future can only be mastered in partnership. However, partnership-based cooperation can neither be decreed nor organised; partnership must arise from an existential inner need—on both sides! Achieving this is a spiritual, emotional and physical life event and, in our opinion, currently the greatest challenge.

References

Troxler, I. P. V. (1944). *Naturlehre des menschlichen Erkennens* [Natural theory of human knowledge] (1st ed., originally published 1828, p. 27). Troxler Verlag.

Willkens, A., Jakobi, M., & Schwenk, W. (1995). *Wasser verstehen lernen* [Learning to understand water] (Special vol., p. 25). Verlag Verein für Bewegungsforschung.

Stockmar (Hans Stockmar GmbH & Co. KG)—Dimensions of Sustainable Entrepreneurship

Inke Kruse

Features:

(a) Founding year	1922
(b) Employees	29
(c) Location	Kaltenkirchen, Germany
(d) Active in which countries	Together with our sister company Mercurius in over 50 countries
(e) Sector	Colours and products for educational and artistic needs

1 Introduction

Art holds societal relevance. Through engaging in artistic activities, human beings experience themselves as creative, productive beings. However, "art" should not be understood solely in terms of formative creation. Humanity prepares the ground upon which what is fruitful for the world may unfold. This applies to agriculture as well as society. STOCKMAR stands for creating, inspired by the core principles of healthy economic practices, where ecological agriculture serves as a source of inspiration, and for fertilising the societal sphere through its activities. STOCKMAR products stimulate creative and formative forces, thereby contributing to a sustainable world. Art makes sense. It provides access to meaningful future development and is an integral component of sustainable development (Fig. 1).

I. Kruse (✉)
Hans Stockmar GmbH & Co. KG, Kaltenkirchen, Germany
e-mail: inke.kruse@stockmar.de

J. Kronenberg and E. T. Lammerts van Bueren (eds.), *On the Earth We Want to Live*,
World Sustainability Series, https://doi.org/10.1007/978-3-031-98758-8_41

Fig. 1 Logo Stockmar

Who shapes the future? What does a human being need to be able to co-create the future? They need an awareness of their self-efficacy and a guiding principle for the future, along which they can develop creatively. At all times, in every challenging period, aligned with the demands of their generation. A creation in the present. STOCKMAR's mission is to nurture the formative powers of the developing human being through creative products that make their self-efficacy tangible. This is our foremost contribution to a "good future". Consistently following this is the manner in which we achieve it.

2 Back to the Future

Hans Stockmar was born on 17 March 1890 in Sydney and spent his early school years in Cairo, San Francisco, London and Montreux. He abandoned his commercial apprenticeship in Bremen following his ideals and inner impulses and likely experienced, for the first time, a sense of spiritual "home" at a drama school in Berlin. Together with his future wife Vilma, he emigrated to New Zealand after a brief stint at an agricultural school, where he established a large-scale apiary. Life with the bees became his life's purpose. At the same time, Stockmar encountered anthroposophy, and later the family returned to Germany to enable their children to attend a local school and to found Stockmar in 1922.

In working with bees, impulses became evident early on that remain central to STOCKMAR today: vitality, community and transformation. These processes are also reflected in the later logo, the honeycomb. This emblem thus serves as a kind of company seal, symbolising openness and a readiness for inner transformation. From the production of wax and candle-making—which remains a part of STOCKMAR's manufacturing today—there developed, over time, an orientation towards creative, artistic processes and products. The impulse for wax colours came in no small part from his son Anselm, who was active as a teacher.

Early on, Stockmar's life threads began weaving an increasingly complex tapestry of connectedness to nature, craftsmanship, pedagogy and creative forces, which also consciously engage with social questions. The wax workshop and family home became a cultural source in Kaltenkirchen. Community, in the STOCKMAR sense, is understood as a resonant space "where what is fruitful for the world may unfold." It is thus fitting that the company introduced profit-sharing for employees as early as the 1950s, and in the 1970s, initiated the process of neutralising corporate capital, which later took place in association with the Neuguss

Group. The seed of what we now call sustainable development was deeply rooted in STOCKMAR's work from the very beginning.

3 Localisation in the Here and Now and a Sustainable Development

That what is fruitful for the world may unfold—this image leads us foremost to agriculture. Humanity works on and with the soil, so that the seeds may sprout. There can hardly be a greater orientation towards the future. But for this future to succeed sustainably, humanity must care for the soil. It must not demand too much from it and must give something back to it time and again. This is a circular economy in the strictest sense. In this regard, sustainably designed agriculture is to be viewed as the foundation of all (sustainable) economic practices. It inspires us, on the one hand, to keep both feet on the ground but also to prepare it so that we can trust in our ability to withstand uncertain times, to have the courage to embrace the new.

STOCKMAR strives to shape such an economy for the future in three dimensions, implementing a more sustainable future as a manufacturer of art supplies:

Dimension: Products

The natural material beeswax, the starting point of the STOCKMAR company, and the question of responsible use of it, inevitably lead us towards diverse landscapes and agriculture and a sustainable approach with respect to biodiversity. The STOCKMAR brand is renowned worldwide for its colours and creative products, which hold a significant place in Waldorf schools, among others, and are firmly rooted in their educational approach. Their quality enables a truly profound experience. Our products create a space where children and those engaged in artistic activities can experience themselves as creative beings. Each person can shape: relationships, education, society—the future. STOCKMAR products sow this seed through artistic experience from the outset. For us, this is the essential impulse towards creating a future worth living. Humanity, as the shaper of the future, is central.

People, especially children, should be supported in experiencing themselves as self-effective and creative. The sensory experience of the products can, through engaging with them, become something akin to a "home for the soul"—the space from which, according to anthroposophy, thinking, feeling and willing arise. It is a grounding in the present but also a connection to the world and, in doing so, a piece of human identity. STOCKMAR is part of the "Waldorf brand." Integrated within the Waldorf school system, this supports a way of life with perspectives on the world that aim towards a respectful way of relating to and caring for one another.

Dimension: Collaboration

Economics is shaped by people. Therefore, we place people at the centre by focusing on fostering successful cooperation: Humanity as the shaper of relationships is central. Collaboration does not simply mean working together; it speaks to the way this togetherness is structured. It is about living in partnership and, as much as possible, in a spirit of fraternity within the company. To be capable of community, one must preserve and develop one's independence. This requires a community that strengthens and develops synergies. The individual needs the presence of others to evolve and also needs friction to grow both individually and collectively. As we know, friction also generates warmth. In this sense, we live our partnerships and are associatively connected with one another.

STOCKMAR maintains and cultivates a worldwide network of diverse partnerships. We are in active and close exchange with suppliers, service providers, sales partners and people in the fields of education and art. STOCKMAR stands for dialogue and the recognition that something arises from flexibility and a shared process of movement and breath. From this community of encounter and knowledge, we become creatively active. Our collaboration is a lived association, striving to consistently implement Steiner's idea of "transferring profits from economic life into the spiritual." For us, the guiding principles are love, light and warmth—as capacities to enable relationship-supportive cooperation.

Dimension: Financial Flows

What can a business enterprise contribute to a sustainable future? The company is, first and foremost, made up of people, and it is people who "do" and thus make an impact. This is why we firmly believe that companies cannot be sold. Through the structure of our overall organisation—the non-profit Neuguss Foundation—we implement this conviction by ensuring that STOCKMAR remains unsellable through this arrangement.

Money must serve humanity. Therefore, we are constantly challenged in every aspect of our financial management to direct financial flows in meaningful ways. We strive to contribute through innovative concepts, not only to our own enterprise but also to the improvement of everyone's living conditions. Public good over self-interest. Thus, we include the Gemeinnützige Treuhandstelle (Charitable Trust Office) in Hamburg in an agreement to share our profits and, through donations and grants embedded firmly within our corporate reality, we actively support the fields of culture and education. Furthermore, our surplus profits are distributed to Neuguss, where, together with GLS Treuhand and other institutions, we continuously contribute to their socially impactful, society-relevant activities.

To evaluate the effectiveness of our actions, we have been producing a common good balance sheet since 2019. The overall balance arising from this evaluation guides us in further aligning our actions with our principles. It certifies us in the best sense of our founding thought: Art makes sense (Fig. 2).

Fig. 2 Stockmar production of wax crayons

4 What Responsibility Must Businesses, Above All, Assume in Relation to the Future?

STOCKMAR views economics as a tool that serves to help shape a future worth living. We do this by actively directing our financial flows through the qualities of gift money, donation money and other innovative concepts. At STOCKMAR, this perspective is reflected in the way we, as part of the Neuguss Group, put our generated profits to social use. Here too, we consistently direct our actions towards creating enabling spaces for humanity. We comprehensively prepare the ground for the creative human being.

Furthermore, we do this by broadly integrating the spheres of human life into our activities—agriculture, education, society and art, ecology, business management and collaboration—as areas of action that we also understand as part of our economic activities. Humanity is central, as a responsible shaper of financial flows. We thus view sustainable development as an incentive for awareness, encouraging us to regard all dimensions of production, collaboration and interaction with the Earth as a unified whole, in our case, in the service of a creative human being.

For the quality of our products, we use carefully selected raw materials, which we apply judiciously based on experience and scientific insights. STOCKMAR products are harmless to the health of humans, animals and the environment, and are regularly tested by certified institutions.

5 Love, Light and Warmth—A Vision

> Just the process of creating beeswax—a material born of light, warmth and love. That holds a great fascination for me. – Inke Kruse

We work with bees and incorporate their precious beeswax into our products. Bees form unique communities that follow the principles of togetherness and mutual care. They are highly sensitive beings and respond immediately to external influences. Despite the greatest care and dedication, it is not always possible to bring a bee colony through the winter. These creatures therefore also represent the mystery of creation and serve as a metaphor for the unavailability of life. They give us what they produce from light, warmth and love, serving as a model for a way of working that can also be sustainable in economic contexts.

Love
In 2022, STOCKMAR celebrated its hundredth anniversary, and in preparing for the occasion, the company posed the question of what exactly there was to celebrate about 100 years of STOCKMAR. Naturally, such occasions first prompt a look back at what has been achieved. Yet, the company's guiding impulse is actually directed toward the future. Out of tradition, it is committed to the well-being of current and future generations. When we ask who will shape the world of tomorrow, our thoughts quickly turn to the child. STOCKMAR products are designed in such a way as to support the experience of 'I can create!' This follows an impulse rooted in the past century from Waldorf schools and anthroposophy. Beyond its products, however, the company also understands itself as a responsible part of this world. Therefore, all perspectives are directed towards the child—that is, towards the future—and to acting responsibly for the common good. This can be seen as the fundamental gesture of the STOCKMAR company.

The founder, Hans Stockmar, entrusted all future generations with a mission: to keep alive the creative impulse inherent in the products. Alongside this, he imparted an attitude towards how and for what purpose we engage in business.

Light
Light is the prerequisite for all living things and stood at the origin of the story of creation. We seek light not only externally but also within ourselves, where it can be kindled through a loving dedication to what we do and how we do it, and through an empathetic attention to others. However, light is primarily perceived in contrast to darkness— its counterpart. Its effect is amplified when it reaches us in places where we do not expect it. These contrasts create a spatial effect, opening a space that suddenly feels three-dimensional rather than flat. The gradation also creates both colour depth and transparency, bringing "light into the darkness." Goethe, too, explored this in his Theory of Colours (which he considered his most important work), focusing on the poles of light and dark (von Goethe, 1810). Starting with the three primary colours—yellow (light), red, and blue (dark)—all other colours on the colour wheel arise between these poles, through the mixing

of primary colours. Or, in other words, diversity through composition. Colour has an effect. These insights are part of STOCKMAR's DNA and permeate the very substance of its products.

Transparency, including corporate transparency, also brings light into the darkness. STOCKMAR's self-concept aligns with a triad of economy, ecology and social responsibility, placing it squarely within the modern pursuit of sustainable development. Thus, it is natural that a portion of the profits is allocated to charitable purposes via the Neuguss Administration Society. The company is committed to biodiversity and the preservation of ecological agriculture. It also establishes and continuously cultivates an aesthetic approach to management that is flexible enough to respond anew to the individual needs of all employees and partners while remaining economically successful.

Warmth

Cold and warmth play an essential role in the production of wax products. The wax must first be melted in order to be shaped as desired. Warmth liquefies, makes malleable and flexible. Cold hardens, stabilises and gives form. This technique is fundamental to the origins of wax casting and is still applied today in STOCKMAR's manufacturing processes. At the balanced centre of these processes stands the human being. The natural material beeswax is subject to subtle fluctuations, making it essential for people to guide the production processes. Balancing the right moment, texture and colour is the responsibility of the person at the machine.

Wax, as a material, is also to be understood as a creative mandate. Working with the product requires the quality of warmth. It requires the (creative) will of the creator in order to take shape. The living materials of wax and colour serve as teachers for entrepreneurial and social processes alike. Artistic, creative practice has a long-standing tradition at STOCKMAR. It is reflected in the focus on processes and products, it lives in partnerships with training institutions and universities, and it radiates inward as well.

Through not only its products but, above all, its ethos, STOCKMAR seeks to help art and meaning come together, thereby generating healing impulses for society. Art can create meaning and can even be understood as an integral component of sustainable development through the future-creating forces of art itself (Fig. 3).

6 Senses and Forms

Hans Stockmar did not only leave behind a truly unique enterprise; he also lent his name to this company, its products and the values associated with it. STOCKMAR has become a personality in its own right or even an adjective—a symbol, an idiom, a value expressing a way of life characterised by respect, appreciation, dialogue and a future-oriented creative will. Hans entrusted all future generations with a mission: to keep alive the creative impulse inherent in the products along with an attitude towards how and for what purpose we conduct business. By transferring the company to Neuguss, he and his descendants "set it free"—and now, after over

Fig. 3 Stockmar wax crayons

100 years, we look back on generations shaped by this very moment of freedom who have contributed to the company STOCKMAR through their participation.

The name STOCKMAR is Hans's legacy, but it has itself become a form of art. Hans Stockmar was born as Maximilian Alexander Beheim Schwarzbach but adopted his mother's maiden name and henceforth called himself Hans Stockmar. The name "Hans" was, according to tradition, given to him by his partner and wife Vilma. She also ensured the survival of the business in its early years, while Hans contributed to the family income as a steward on immigrant ships.

In the STOCKMAR gesture of connecting nations, genders and generations through the experiential space of art, a link to the beginnings and to the formative figures and groundbreakers of the past remains alive today—a presence in all phases of the company's history. This gesture of connection arises through a grateful reflection on what has become and also through an ongoing connection with the founding impulses. Thus, looking back is not about preserving the past but rather about dedicating ourselves to keeping it alive and furthering its development.

Community in the STOCKMAR sense is understood as a resonant space. This journey can be described as extending from its origins in Kaltenkirchen to the schools, children's rooms and studios around the world. It tells a story of continuity, innovation and the enduring search for beauty in simplicity. STOCKMAR not only prepares the ground for artistic expression but also contributes to a creative spirit of the future, embodying an aesthetic approach to corporate leadership and a consistent connection to the world around us.

7 Conclusion

Where human beings can shape responsibly for sustainable economic practices, the creative, artistic moment becomes an integral part of sustainable development. We see enabling this foundational experience through our products as a cultural contribution in the broadest sense. In a narrower sense, nature-connected and community-oriented production is deeply aligned with what we call sustainability today.

Reference

von Goethe, J. W. (1810). *Theory of colours* (Vols. 1–2). Cotta.

Studio Claudy Jongstra: Weaving History and Our Common Future—Earthbound Art and Design Towards Resilience and Regeneration

Claudy Jongstra

Features:

(a) Founding year	1995
(b) Employees	30
(c) Location	Spannum, Netherlands
(d) Active in which countries	US, The Netherlands, Belgium, Spain, Turkey, UK
(e) Sector	Art, Design, Textile, Fashion

1 Founding Story

Claudy Jongstra began her Studio as a fashion designer, completing her studies at University of the Arts Utrecht in the context of the emerging fast-fashion industry. However, upon entering the professional world, she faced a pivotal moment of disillusionment. The stark contrast between the creative ideals she developed in art school and the realities of mass production led her to completely reconsider her future work. This period of introspection culminated in a life-changing experience at an exhibition focused on wool (Fig. 1).

Fascinated by the material's inherent genius, its historical significance, and the rich traditions surrounding it, Jongstra decided to entirely shift her focus. For two years, she immersed herself in the experimental exploration of wool, eventually

C. Jongstra (✉)
Studio Claudy Jongstra, Spannum, Netherlands
e-mail: claudyjongstra@claudyjongstra.com

J. Kronenberg and E. T. Lammerts van Bueren (eds.), *On the Earth We Want to Live*, World Sustainability Series, https://doi.org/10.1007/978-3-031-98758-8_42

CLAUDY JONGSTRA

Fig. 1 Logo studio Claudy Jongstra

dedicating her career to it. Over the past three decades, she has worked with this material, not only elevating its artistic value but also advocating for its cultural and ecological importance. With the foundation of Studio Claudy Jongstra in 1995 soon a team was set up to support Jongstra in her endeavours. A biodynamic farm with her own flock of sheep and heritage dye plant garden became part of the Studio and provided direct access to all levels of the value chain.

In the early years, Jongstra observed how wool—sheep herds once were a dominant feature of Northern European landscapes—had become undervalued and disconnected from public consciousness. The realisation of the loss of affinity for a material so central to history compelled her to bring it back into the spotlight. Collaborating with high-profile partners and leveraging these relationships as a platform, she sought to create awareness and renewed appreciation for wool. Over time, her work with wool evolved into a tool for broader societal transformation, particularly when paired with natural pigments. Together, the richly coloured wool and vibrant plant pigments became symbols of hope, diversity, and change.

Through her endeavours, Jongstra discovered that plant-based pigments have profound cross-cultural significance, creating a universal narrative that transcends the regional limitations of wool. Pigments have historically played a central role in communities worldwide, serving as a bridge between cultures, as a means of expression and as carriers of identity. By weaving these elements together, her creations tell stories that connect people, inspire action, and emphasise the importance of preserving cultural and ecological heritage.

2 The Work

Jongstra's primary body of work includes besides autonomous works of art, commissioned pieces for institutions, universities, embassies, and hospitals. Each project is tailored to its context—ranging from artworks with political resonance to those with a healing quality. In recent years, she has also embarked on creating independent works aligned with her mission, characterised by their activist message and nomadic nature.

Three such works stand out: Woven Skin, Guernica de la Ecología, and Tangible Transformation No. 1. These pieces are designed to travel, fostering engagement through events such as symposia, panel discussions, and debates. Their purpose is to inspire audiences to take action. For instance, Woven Skin highlights the issues of industrial farming and humanity's treatment of the Earth and animals. Exhibited

Fig. 2 Claudy Jongstra, Guernica de la Ecologia, 2021, courtesy Iona Stichting

at events like New York Climate Week, it has sparked vital conversations about sustainability.

Guernica de la Ecología addresses ecocide, focusing on the degradation of landscapes, soil, and the far-reaching consequences of human intervention in nature. This work is currently displayed in Valencia, a city profoundly impacted by environmental disasters, making its presence particularly resonant.

The third piece, Tangible Transformation No. 1, shifts the focus toward hope. It explores the potential of collective action in urban or rural settings—planting seeds, fostering biodiversity, and cultivating community involvement. Through this work, Jongstra emphasises the healing impact of such initiatives, inspiring audiences to reconnect with their environments (Fig. 2).

3 Studio Claudy Jongstra, Anthroposophy and Biodynamic Agriculture

Jongstra's artistic process is deeply rooted in biodynamic agriculture, which serves as both inspiration and foundation for her work. Recently, her studio received a Demeter certification for textiles—the first of its kind globally after the cotton initiative of Sekem in Egypt. This milestone highlights the alignment between her work and the principles of vitality, interconnectedness, and responsibility within biodynamic farming.

By integrating biodynamic practices, Jongstra's work embodies a holistic worldview brought by anthroposophy. Human and earth development, inner and outer transformation come together as augmented understanding of sustainable development. From soil vitality to ethical transport, every aspect reflects a commitment to sustainability and interconnectedness. This philosophy is evident in the appearance of her pieces, which transcend words, touching audiences on multiple sensory and emotional levels.

Jongstra describes her process as connected to the thought of "liberating colour from the plant", a concept that underscores the transformative power of biodynamic agriculture. Her approach also serves as a pioneering model for integrating agriculture, art, and sustainability—including sustainability in inter human working relationships. This whole field is gaining recognition through academic studies, such as those highlighting the measurable impact of biodynamic farming on soil health, carbon sequestration and biodiversity.

4 Defining Sustainable Development

Jongstra views sustainable development as an ongoing process rooted in resilience and adaptability. For her, the essence lies in the ability to conserve knowledge, explore new possibilities, and respond effectively to challenges. Sustainability goes beyond preservation; it is about maintaining the agility to adjust course and take responsibility for the future.

She emphasises the importance of weaving patterns—both literal and metaphorical—that reflect and adapt to societal needs. Social and ecological dimensions, especially the health of the soil and the inclusion of marginalised communities, are central to her work. In today's polarised world, fostering inclusivity and mutual recognition is more vital than ever (Fig. 3).

5 For the Future

Looking ahead, Jongstra is committed to promoting earth-based education. Her studio is developing a holistic curriculum that emphasises interdisciplinary learning and sustainability. By documenting traditional recipes, processes, and techniques, she aims to create accessible resources for artists, educators, and communities worldwide.

The curriculum will combine theoretical and practical elements, such as producing natural oil paints or cultivating dye plants from seed to pigment. Through partnerships with universities and international speakers, the initiative seeks to inspire a new generation of artists and change makers to embrace sustainable practices.

Ultimately, Jongstra's vision is rooted in cultivating sensitivity, empathy, and reciprocity. She believes that fostering these qualities can accelerate change and create a more harmonious and resilient world.

Fig. 3 Claudy Jongstra. Tangible Transformation No. 1. 2024

6 Conclusion

Claudy Jongstra's work reflects her unwavering dedication to sustainability, heritage, culture, and community. By blending art, agriculture, and activism, she has created a body of work that transcends aesthetics, inspiring action and nurturing hope for a more balanced future. Grounded in biodynamic agriculture, her work serves as a testament to the transformative power of creativity and the importance of reconnecting with the natural world.

Voelkel—Contributions of the Demeter Pioneer Voelkel to Sustainable Development

Boris Voelkel and Johannes Steinke

Features:

(a) Year of Establishment	1936
(b) Employees	360 staff members
(c) Location	Pevestorf, Lower Saxony, surrounded by the Elbe River biosphere reserve
(d) Active Regions	Primarily in Germany, with suppliers and partners worldwide
(e) Other	Beverage production (organic and Demeter-certified juices, lemonades, kombucha, syrups, and ginger shots) (Fig. 1)

1 Voelkel's History

Voelkel GmbH operates in the picturesque village of Pevestorf, surrounded by the Elbe River Biosphere Reserve, and is currently producing a variety of beverages in its fourth generation. Four brothers are actively involved in the company: Jurek Voelkel oversees marketing and sales, David works in technical operations, Jacob is responsible for production, including quality assurance (QA) and quality management (QM), and Boris manages procurement. Their parents, Grita and Stefan Voelkel, also remain active in the business, where 360 employees work diligently

B. Voelkel (✉) · J. Steinke
Voelkel GmbH Fruit and Vegetable Juices, Pevestorf, Germany
e-mail: boris.voelkel@voelkeljuice.de

J. Kronenberg and E. T. Lammerts van Bueren (eds.), *On the Earth We Want to Live*, World Sustainability Series, https://doi.org/10.1007/978-3-031-98758-8_43

Fig. 1 Logo Voelkel

from morning to evening to produce juices, lemonades, kombucha, syrups, and ginger shots.

In the 1930s, great-grandparents Margret and Karl Voelkel were drawn to the rural area through their involvement with the Wandervogel movement. War-weary and seeking to escape economic constraints, they settled there and began cultivating their own fruit and vegetables. In 1936, they founded the Voelkel cider mill in an old dairy on the Höhbeck in Lower Saxony.

Soon, they encountered anthroposophy, which would later become their central life theme alongside their work in what was then still a very small cider mill. The book *Journey to the Land of Earth* (Fahrt ins Erdenland) by von Koschützki (1952) had a groundbreaking influence and contributed to establishing anthroposophy as a guiding principle in their lives and work. A close friendship with Koschützki developed, and this naturally led to their encounter with biodynamic agriculture, which became a compass for socially, economically, and ecologically responsible business practices.

2 The Products: Market Leader in Organic and Demeter Juices in Germany

Today, Voelkel is the market leader in the production of organic and Demeter-certified juices in Germany. The pioneer offers around 250 different products in its range, prioritising Demeter quality and biodynamic agriculture wherever possible. Additionally, private-label products are manufactured for large retailers to enhance efficiency within the company.

Most food manufacturers specialise in producing specific items, such as apple juice or vegetable juices; others mix various juices together, while yet another group of specialists handles bottling. Voelkel, however, has chosen to specialise in diversity, producing all components in-house where it makes sense. From aronia berries and apples to beans, nettles, cranberries, dandelions, ginger, carrots, beetroot, and even zucchini, Voelkel creates, mixes, and bottles an impressive variety of juices across four filling lines. Although this approach is less efficient and results in higher costs compared to the concept of specialisation, it offers remarkable stability, flexibility, and adaptability.

At Voelkel, purchasing naturally involves close collaboration with suppliers, farmers, and manufacturers, both locally and globally. Over decades, we have witnessed how climatic, social, demographic, and logistical crises have intensified. By

Fig. 2 Some Voelkel Juices

now, one could almost speak of a "multiple crisis" that feels like a kind of showdown. One political or natural crisis follows the next—everything is escalating rapidly (Fig. 2).

3 New Approaches to Business—Healthy Food from Healthy Structures

For a long time, we have maintained that the classical, conventional market mechanisms of the neoliberal economy, operating under the guise of a social market economy, are inadequate for addressing or adapting to these changes. Voelkel has long worked under the motto "Healthy Food from Healthy Structures," providing farmers with stable economic support. Rather than constantly buying from the cheapest sources, we cultivate long-term, empathetic relationships, understanding the challenges and needs of the farmers as they understand ours. This mutual support enables us to align and assist one another when necessary, embodying what we consider to be the social and economic sustainability of the future.

These relationships allow farmers to focus on preparing for the future instead of frantically rotating through volatile markets. For instance, when farmers pursue initiatives such as agroforestry or efficient irrigation, we offer them financial support. More importantly, we ensure a stable pricing framework. In 2022, for example, there was a massive surplus of processing apples in the market. A strong harvest combined with poor quality for the fresh market led to high rejection rates, sending these apples to juice production. In the Altes Land region alone, 8000–10,000 tonnes of apples were left without a market. Prices collapsed, and no one wanted these apples, even in Naturland, Bioland, and Demeter quality. With significant effort, Voelkel acquired 90% of these surplus apples and stored them in its own and external tank facilities, maintaining stable prices throughout. This was a tremendous undertaking.

The following year, demand for apples in association-certified quality surged while the harvest was weaker. Thanks to loyalty from farmers and the previous year's stock, Voelkel was able to meet the high demand effectively, seizing the available potential. Other cider mills, overwhelmed by similar requests for association-quality products, missed these opportunities entirely.

Additionally, suppliers frequently approach us with urgent surplus produce. For example, a supplier might call to say, "We have plums that must be gone by tomorrow, or they will spoil; I don't have a buyer." Voelkel steps in whenever possible, integrating these goods into existing products or developing new items. Without this flexibility, suppliers often face dilemmas, as static markets are insufficiently agile to respond to the realities of farmers who depend on nature and weather conditions.

Thanks to our network of long-term suppliers worldwide, with whom we maintain warm and empathetic connections, Voelkel is able to swiftly implement nearly any desired innovation, even for private-label customers.

4 Social and Economic Warmth for a Healthy Earth

It is becoming increasingly clear that a new, empathetic, and long-term relationship-based approach to economics must be developed across the entire value chain to produce food sustainably and economically. Numerous examples demonstrate how conventional market mechanisms in agriculture are failing, such as in the cases of orange juice or forestry. Here, anthroposophy plays a vital role, offering profound insights through the National Economics Course (Nationalökonomischer Kurs 1922) (Steiner 2013) and the Agriculture Course (Landwirtschaftlicher Kurs) (Steiner, 1924).

The strength required to embed this type of practice into typically business-oriented relationships with individual farmers arises from the ideal that such relationships should also be rooted in emotional and spiritual components. Progressing toward an empathetic and loving interaction must be understood as the foundation for fostering deeper connections between us.

Living this ideal in practice demands constant attention and, in some cases, considerable energy. In our increasingly polarised world, this form of relationship-building does not always bear fruit. Many farmers are often the "weakest" link in the long chain of value creation, enduring significant hardship and disappointment in this role. A respectful approach and the cultivation of trust in these relationships are therefore essential foundations for sustainable, long-term growth. This is particularly necessary for developing a farm organism and working harmoniously with nature.

When relationships lack engagement on emotional and spiritual levels, trust is often lost, and this can hinder bold future planning. Many farms facing generational transitions yearn for security—the security that efforts toward investments or the orientation of farming practices can take place within a fair and sustainable partnership. They seek assurance that their partner will listen and comprehend the often-complex conditions of working with natural elements. This sense of security emerges from trust. Building and maintaining trust is a core aspect of our work in procurement. There is a palpable longing for this type of collaborative growth, and it is deeply rewarding when these relationships begin to flourish.

Rudolf Steiner's principles concerning the development of these soul-level relationships serve as helpful guides and orientation tools in a world that often perplexes us and prioritises short-sighted market mechanisms. Short-lived success followed by abrupt crisis mode characterises these mechanisms. It has become increasingly evident that these approaches cannot work in the long run and are destined to fail in numerous ways. Over three generations, ecological farming has been pioneeringly brought to the forefront of society. Large retailers and associations have taken up the task of refining and continually improving this movement. Now, the next pioneering challenge is to establish a new way of conducting business. Like all pioneering work, this will be arduous and demanding, which is intrinsic to its nature.

Rudolf Steiner has provided us with invaluable tools in the National Economics Course, the human image he outlined, and his principles of organisation and community. These are not rigid formulas but rather reflections of inherent human qualities that simply need to be awakened and brought into resonance.

If we can learn to focus on the well-being of others rather than constantly watching to ensure we gain the greatest advantage for ourselves, we can approach the immense challenges of our time and the future with courage and optimism. Sustainable and resilient land-use systems will have greater opportunities in the coming years and are likely to prevail. However, developing these systems requires courage—a courage that can be strengthened and advanced significantly through trust-building and relationship work.

5 Conclusion

Since 1936, Voelkel GmbH has stood for the integration of sustainable agriculture and high-quality beverage production. As Germany's market leader in organic and Demeter-certified juices, the company sets benchmarks in social, ecological, and economic responsibility. Guided by the values of anthroposophy, Voelkel pursues a holistic approach based on long-term partnerships, stable pricing, and a broad, diverse product range. With a clear focus on biodiversity and resource-efficient production, Voelkel demonstrates how sustainable business practices can succeed in practice. The company embodies its conviction that healthy food arises from healthy structures, taking a pioneering role in shaping a sustainable and empathetic agricultural future.

English quotations in the text referring to works in German, Spanish, or Dutch listed in the reference section were translated by the editor.

References

von Koschützki, R. (1952). *Fahrt ins Erdenland: ein Menschenschicksal*. Verlag Urachhaus.

Steiner, R. (1922). *Nationalökonomischer Kurs: Aufgaben einer neuen Wirtschaftswissenschaft, Band I (GA 340)*. Rudolf Steiner Verlag.

Steiner, R. (1924). *Geisteswissenschaftliche Grundlagen zum Gedeihen der Landwirtschaft: Acht Vorträge, gehalten in Koberwitz bei Breslau vom 7. bis 16. Juni 1924 (GA 327)*. Rudolf Steiner Verlag.

Steiner, R. (2013). *Rethinking economics: Lectures and seminars on world economics (CW 340–341)*. (P. Clemm, Trans.). SteinerBooks.

Weleda

Nadine-Hélène Santiago and Tina Müller

Features:

(a) Founding year	1921
(b) Employees	2500
(c) Location	Arlesheim, Switzerland
(d) Active in which countries	Over 50 countries worldwide
(e) Sector	Natural and organic cosmetics, pharmaceuticals

1 Our Founding Story, Vision and Work

Weleda was founded in 1921 by Dr. Ita Wegman, Dr. Rudolf Steiner, and Oskar Schmiedel. The company emerged from a shared vision to create products that support the health and well-being of individuals while respecting the environment. The founders were influenced by anthroposophy, the spiritual science developed by Rudolf Steiner, which emphasises the interconnectedness of all life and the importance of holistic approaches to health, including sustainability as a key element of development.

Today, Weleda's vision remains rooted in that of its founders. We are the global leader in certified natural cosmetics and anthroposophical medicines, promoting health, beauty, and well-being through products that are in harmony with nature

N.-H. Santiago (✉) · T. Müller
Weleda, Arlesheim, Switzerland
e-mail: nsantiago@weleda.de

J. Kronenberg and E. T. Lammerts van Bueren (eds.), *On the Earth We Want to Live*, World Sustainability Series, https://doi.org/10.1007/978-3-031-98758-8_44

Fig. 1 Logo Weleda

Fig. 2 Harvesting roses for Weleda Morocco

and the human being. Our commitment to sustainability is reflected in every aspect of our business, from sourcing raw materials to manufacturing processes and community engagement.

Weleda produces a wide range of natural and organic cosmetics and pharmaceuticals. Our products are designed to support and regulate the body's natural healing processes and enhance overall well-being. We believe that true beauty and health come from within and that our products should strengthen body, mind, and spirit. Our commitment to sustainability drives us to use only the highest quality natural ingredients, sourced ethically and sustainably (Fig. 2).

2 The Role of Anthroposophy

Anthroposophy is the foundation of our understanding of the human being as a holistic entity, encompassing physical, emotional, and spiritual dimensions while actively considering the planet and the cosmos. Our approach to product development ensures that our products support or regulate the human organism. Anthroposophy also influences our commitment to social and environmental responsibility, as we recognise the interconnectedness of all life and the need to care for the Earth and its inhabitants.

Anthroposophy highlights that humans and nature possess different fields of functioning, which extend beyond mere physical or material existence to include sensory perception, thought, and a spiritually self-conscious dimension. Because humans partially share these aspects with nature, there is a resonance between the human organism and natural substances. When we take natural remedies or apply natural substances, these can stimulate processes in our organism that regulate specific aspects of the human being. In this way, humans unfold beauty and health with the support of nature. Anthroposophy invites a holistic perspective that allows us to contemplate and understand these facets of the world and humanity.

Anthroposophy makes us aware that humans are constantly evolving beings. It provides a path and tools for individual development. Personal development is never isolated but always occurs within the context of social surroundings and nature. In the human heart lies the source of creativity and love for people and all beings, as well as the bridge to the wisdom of nature and the cosmos.

From this consciousness, humans can take responsibility for nature and the planet. This forms the foundation of forward-looking sustainability. Weleda's medicines and cosmetics acknowledge and integrate this interplay of material, soul, and spiritual forces. Weleda carefully considers the entire value chain in terms of ecological, economic, and social sustainability.

3 Our Definition of Sustainable Development

For Weleda, sustainable development means creating products and practices that are environmentally, socially, and economically responsible. We define sustainability as the ability to meet the needs of the present without compromising the ability of future generations to meet their own. This involves a commitment to organic and biodynamic farming, fair trade practices, and reducing our environmental footprint through innovative manufacturing processes.

Weleda's approach to sustainability encompasses three dimensions: social, ecological, and economic.

- **Social Sustainability**: We are committed to fair labour practices, community engagement, and supporting the well-being of our employees and partners. We work closely with our suppliers to ensure fair wages and safe working conditions, and we invest in community development projects around the world.
- **Ecological Sustainability**: We prioritise the use of organic and biodynamic ingredients, which are grown without synthetic pesticides or fertilisers. Our manufacturing processes are designed to minimise waste and energy consumption, and we continuously seek ways to reduce our carbon footprint.

- **Economic Sustainability**: We aim to create long-term value for our stakeholders by operating a financially stable and resilient business. This includes investing in research and development, maintaining high standards of quality, and fostering innovation to meet the evolving needs of our customers.

• **Contributions of Anthroposophy to Sustainable Development**
Anthroposophy provides a framework for understanding the interconnectedness of all life and the importance of holistic approaches to sustainability. At Weleda, this framework guides our commitment to ethical sourcing, environmental stewardship, and social responsibility. By integrating anthroposophical principles into our business practices, we aim to create products that not only benefit individuals but also contribute to the well-being of the Earth.

• **Leading Concepts and Terms**
Several key concepts guide Weleda's work:

- **Holistic Health**: The belief that true health encompasses physical, emotional, and spiritual balance.
- **Biodynamic Agriculture**: A farming method that emphasises the health of the soil, plants, and animals, considering the farm as a self-sustaining ecosystem.
- **Ethical Sourcing**: Ensuring that our raw materials are sourced in a way that respects the environment and the rights of workers.
- **Environmental Stewardship**: Taking responsibility for the impact of our operations on the environment and striving to minimise our ecological footprint.

• **Successes, Challenges, and Learnings**
Over the years, Weleda has achieved significant milestones in sustainability. We have been recognised with numerous awards for our commitment to environmental and social responsibility. However, we also face challenges, such as navigating complex supply chains and adapting to changing market conditions. Through these experiences, we have learned the importance of continuous improvement, collaboration, and staying true to our core values.

• **Future Steps and Societal Developments**
Looking ahead, Weleda is focused on several key initiatives:

- **Innovation in Sustainability**: We will continue to invest in research and development to create new products and processes that further reduce our environmental impact and meet consumer needs.
- **Community Engagement**: We aim to deepen our partnerships with communities and support initiatives that promote social and environmental well-being.
- **Education**: We will work to raise awareness about the importance of sustainable business practices, biodiversity, and healthy soil.

Fig. 3 Harvesting calendula for Weleda

• Living and Shaping the Future

Weleda is committed to living our values and shaping a sustainable future. Our new strategy, "#_growth that matters", involves not only creating products that are in harmony with nature and the human being but also fostering a culture of responsibility and care within our organisation. We believe that by staying true to our principles and continuously striving for improvement, we can make a positive impact on the world.

Key questions guiding our work are: How can we further reduce our environmental footprint while maintaining the highest standards of quality?; How can we support the well-being of our employees, partners, and communities in a rapidly changing world?; and how can we continue to innovate and lead in the field of certified natural cosmetics and anthroposophical pharmaceuticals? (Fig. 3).

4 Conclusion

Weleda's commitment to sustainability is deeply rooted in the principles of its founders. By integrating these principles into every aspect of our business, we aim to create products that support the health, beauty, and well-being of individuals while respecting the environment. We believe that true sustainability requires a holistic approach that considers the interconnectedness of all life and the need to care for the Earth and its inhabitants. As we look to the future, we remain dedicated to our mission of promoting health, beauty, and well-being through products that are in harmony with nature and humankind (*In Harmonie mit Mensch und Natur*).

9graden Architectuur—Architecture: A Road Toward the Earth

Yaike Dunselman

Abstract

9graden architectuur is an ambitious and international architecture firm with offices in Amersfoort (NL), Oldenburg (D) and Zurich (CH). We represent an organic, functional, ecological, and modern architecture in which human health takes center stage. Building with and for nature and with ecological materials is important to us. A building with a positive impact on people's vitality is an investment that is meaningful and contemporary—in a human, ecological and economical way. Our goal is to create a truly human architecture in every respect. In doing so, we aim to include not only the physical, but also the emotional and spiritual aspects of the human being in the design process. The coherence between these aspects forms the foundation for organic and functional architecture. We aim to integrate buildings into existing urban and landscape structures in a contemporary way, with respect for the past—to further develop existing qualities and to enrich them with new ideas.

Features:

(a) Founding year	2003
(b) Employees	24
(c) Location	Amersfoort (NL), Oldenburg (D), Zürich (CH)
(d) Sector	Architecture

Y. Dunselman (✉)
Architect, co-founder and director, 9graden architectuur, Amersfoort, The Netherlands
e-mail: dunselman@9graden.net

J. Kronenberg and E. T. Lammerts van Bueren (eds.), *On the Earth We Want to Live*,
World Sustainability Series, https://doi.org/10.1007/978-3-031-98758-8_45

(e) Features — BDA-Preis des Landes Niedersachsen, AIT Award, Fritz Höger Preis (honourable mention), AIA Architecture Prize, Hedy d'Ancona Prize (honourable mention), Iconic Awards, Deutscher Ziegelpreis (honourable mention), Bremer Design-Preis, Best architects 23, Stadbouwprijs Amersfoort (NL), Nationale Houtbouwprijs 2022 (NL), Häuser des Jahres 2024 and several nominations

1 Starting Point and Impetus

After completing a joint architecture study (1993–1999) at the Alanus Hochschule für Kunst und Gesellschaft in Alfter (Germany), Lars Frerichs (Germany) and Yaike Dunselman (Netherlands) founded the Dutch-German architecture firm 9graden architectuur in 2003. The firm formulated its vision for a sustainable future through four key thematic pillars: ideals (1), culture (2), working method (3), and result (4). 9graden architectuur developed this perspective from a profound connection with anthroposophy, initially with little work experience but with a clear inner will to carrying anthroposophically inspired architecture into a new century in an individual and centemporary way (Fig. 1).

"Our goal—our ideal (1)—is to create human-centred architecture in every respect. In doing so, we strive to incorporate the physical, emotional, and spiritual dimensions of human beings into the design process. The connection between these aspects forms the foundation of an organic functional architecture. Architecture, without a doubt, is always a reflection of its time and the consciousness of the people living within it. We see it as a challenge to integrate buildings into existing urban environments and landscape structures in a contemporary way that honours the past, further develops existing qualities, and enriches them with new ideas for the future.

Fig. 1 Logo 9graden architectuur

In our design process, we maintain an awareness—our culture (2)—of the phenomena at work in nature, such as rhythm, polarity, dynamism, and metamorphosis. This approach allows architecture to have an experiential, inspiring, and supportive effect on human beings. We recognise that every architectural decision affects human living spaces. Therefore, we treat every individual architecture project with the utmost care, always aiming to create architecture of the highest quality with social value, even when working within small budgets. Without question, good design can positively influence both living and working environments.

Through dialogue—our working method (3)—with the client, we aim to articulate ideas and aspirations for the future structure, thereby developing a shared design motive. This process enables us to create individualised designs with which future users can personally identify.

The question of sustainability and the ecological aspects of building—our result (4)—is self-evident to us. For us, a building that positively influences human life represents a meaningful and contemporary investment in human, ecological, and economic terms. The overall environment—including the architectural aspects of space, shape, colour, material, and proportion—affects not only the health of our physical bodies but also our emotional and spiritual lives. This approach to design establishes new principles that influence and shape human beings on multiple levels. From this arises the necessity for both clients and architects to build consciously and responsibly for people."

These core themes have evolved within 9graden architectuur through its connection with anthroposophy, exposure to the passage of time, the questions posed by clients, and the realities of its projects. While, over 20 years ago, this was still largely a theoretical approach for the founders, it has since become an integrated inner attitude that manifests in their external designs. According to 9graden architectuur, it is intrinsic to anthroposophy that we connect with the spiritual level on a conceptual level, engage outwardly with the questions of our time, and have the courage to merge inner and outer dimensions into a unified, connecting, and practical approach.

2 Organic

The term 'organic architecture' is used in various ways both within and outside anthroposophy. It is often characterised by external elements, with the outer form reflecting either the formal language of nature or the underlying natural laws. For 9graden architectuur, however, it represents primarily an attitude—an inner attitude marked by an awareness of the relationship between human beings and architecture, and an inner connection with the world around us.

Human beings and architecture are intrinsically related. The built environment profoundly affects our physical, mental, and spiritual well-being. We move through buildings, constantly absorbing sensory impressions emanating from this environment. Physical aspects such as mass, the play of forces, direction, movement, and balance are internally mirrored within us. These impressions from our surroundings can either support or disturb our physical well-being. Beyond these physical perceptions, sensory impressions also evoke experiential values, such as moods and feelings, shaped by elements like design, light, colour, and warmth, which determine these inner sensations.

We also engage with the built environment through our thinking. Thoughts, intentions, and considerations underpin its creation, reflecting themselves subconscious in the design. These elements influence people's moral lives, their sense of truth, and their capacity for social harmony.

We perceive our built environment as an external world, yet it is an 'artificial' construct arising from our inner world. We maintain a relationship with it, internally mirroring this external world we have created. This relationship forms the basis for the development of our perceptive organs. In this sense, architecture can be understood as a man-made 'organ' external to our organism, yet one that directly impacts our organism and, consequently, our health. Viewed in this way, organic architecture, for 9graden architectuur, holds no merely external meaning; instead, it expresses an inner attitude.

3 Development and Connection

Facing the challenges of our time, 9graden architecture observes that humanity is currently handling all available materials on Earth in an extremely polluting manner. Never before have there been so many possibilities; in the West, we live in immense prosperity, with unparalleled opportunities. We travel across the globe, transporting food and building materials worldwide, using them indiscriminately, and discarding them when they no longer suit us—placing them 'away' and out of sight. Never before have future generations been confronted with so many challenges as they are today. Never before has the world exhibited such stark contrasts, and never before has our built environment been so disrupted, polluted, and outwardly brutal.

This situation comes along with its own responsibilities. There is the responsibility to treat our Earth with care, to consider its origins and future, to reflect on demand and use, on before and after, on construction and demolition, and on the reality of finiteness. 9graden architectuur sees it as its mission to delve into the questions underlying each individual task. We aim to design by engaging with what arises from these questions, enabling the future to shape our buildings.

We refuse to be governed by a fascination with unconscious, often outdated images of the past. Instead, we seek to engage with these questions through artistic awareness, allowing us to develop meaningful answers in which beauty is expressed. This is the kind of beauty that fosters connection, enabling people to feel at home on this Earth.

4 A Fragile Earth

The connection to our Earth remains a central concern for 9graden architectuur. We, as human beings, seem intent on conquering the Earth, shaping it according to our own laws. But should we not instead learn to love it and understand its laws? Should we not strive to live in harmony with the Earth rather than forcing it to adapt to us? This approach requires recognising and experiencing the Earth's vulnerability. To preserve its liveability, we must see, feel, and comprehend this fragility. Architecture has the potential to embody this awareness, serving as a (pre)image of these values. Through the use of natural materials and environmentally-conscious designs, we encourage people to reflect on the Earth's natural phenomena.

The passage of time can be read in our buildings. Beyond their material aspects, sustainability must also encompass immaterial qualities: the connection to context, an awareness of the Earth as a living organism, and a sense of 'attachment' to the beauty of the built object. For the future of our planet, it is vital that people connect with their surroundings, making it a natural part of their consciousness: "This is the world in which and with which I live, which I care for, and for which I take responsibility."

From this awareness of a fragile Earth, an Earth as a living organism, 9graden architectuur designs projects that make this vulnerability visible. One example is the Lichtenberg House, realised in Amersfoort (Netherlands). With this design, 9graden architectuur sought to consciously minimise the use of raw materials while making the world more beautiful, softer, and more 'touchable.' The house was allowed to be vulnerable, just as our Earth is. Just as people are. From the understanding that our survival depends on whether and how we address this vulnerability, the house was designed with refined detailing, craftsmanship, and natural materials such as wood and reed.

The starting point was to create a house that was specific to its location—one that could not simply be replicated elsewhere. The design responded to the unique characteristics and qualities of the site. To harmonise with its surroundings, a sculptural building mass was created, with reed as the defining material. Reed, a natural product, was historically used in the richly detailed buildings from the 1930s in the area. This design crowns clean, contemporary architecture with traditional craftsmanship, namely thatching. Respecting the past, the new building integrates into its surroundings, further developing existing qualities and enhancing them with contemporary ideas (ideals (1)).

Traditional wickerwork was used to carefully design various edge details, eliminating the need for metal eaves and waterslaps. The refined craftsmanship of the thatcher is evident in the building's termination, where the sculptural thatched facade becomes tactile and the traditional skills remain visible (culture (2)).

The house was constructed using timber-frame techniques with high insulation values, reducing the need for complex installations. The interior features natural materials, including wood and loam paint. The insulating thatch on the facade is combined with wooden facade frames, panels, and slats, all of which are designed

Fig. 2 Residential house Lichtenberg in Amersfoort, The Netherlands

to age naturally over time. The passage of time can be read in the building, allowing us, as human beings, to reflect on the natural phenomena of the Earth (result (4)). This design achieves a reduction in raw materials while making vulnerability both visible and experiential (Fig. 2).

5 Organism and Health

The awareness of the Earth as a living organism aligns with Rudolf Steiner's architectural impulse. In Steiner's approach to architecture, design reflects an artistic search for balance between polar processes—architecturally expressed as bearing and burdens. The design motif appears to emerge in the encounter and balance of "support and load", but also between other polar processes such as inner and outer worlds, contraction and expansion, and so forth. This balance is the artistic phenomenon of 'health.'

In 9graden architectuur's design for the new timber construction of the Klinik Arlesheim (H4H), a connection was sought with Rudolf Steiner's 1924 design of the wooden house for Ita Wegman on the clinic grounds of the Klinik Arlesheim. 9graden architectuur sought to engage with history, linking past and future in the new building. The Ita Wegman house illustrates how polar form language facilitates a meeting point—where "support and load", "—and how this encounter gives rise to design. It is at the meeting of "support and load"," that balance seems to emerge, even to take form. Indeed, to form! Balance, which Rudolf

Steiner describes as something divine, forms the foundation of health and is an essential element in organic design.

An initial conceptual motif ("support and load",) develops into architecture. This architecture expresses a polar language of forms in a newly created balance. It envelops and defines interior space, opens up, and enters into a relationship with its surroundings. In this design, we experience inhalation and exhalation, openness and closure, roundness and straightness, activity and rest. It is never one-sided. These are the phenomena of the healthy organism!

9graden architectuur has worked with these phenomena of health in the design of the Klinik Arlesheim, translating them into a tangible design that integrates seamlessly with its surroundings. In the clinic's patient rooms, emphasis is placed on support, warmth, and a sense of physical envelopment. In the facade, an enveloping and tender gesture is created. Simultaneously, this enveloping form language in the facade connects to its surroundings. Through the horizontal termination of the eaves, an open and connecting gesture is formed—a gesture of interest. These are polar forms: a concentration on the interior space and an expansion into the exterior space, coming together in a newly discovered balance.

In the facade, an individual and newly found gesture becomes observable: movement (world interest) transitions to rest (inner concentration), and rest returns to movement. Inner space connects with outer space. The design seeks to bring architectural principles—support, load and balance—into a living, experiential language of form that aligns with the building's function. Here, the phenomena of health have been translated into an experiential reality, expressed through an architectural formal language (Fig. 3).

6 Sustainability and Circularity in Our Culture

The demand for sustainable and ecological aspects of building is fundamental to 9graden architectuur. In material selection, we prioritise ecologically sound materials that are natural, renewable, reused or reusable wherever possible. When choosing specific materials, we pay attention to their stylish, contemporary, and sustainable qualities. For 9graden architectuur, sustainability also involves the lifespan of a building; nothing is more polluting than having to demolish and rebuild every thirty years. Too often, discussions around sustainability focus solely on the material aspects of a building.

Circular construction entails the use of renewable, reused, and/or reusable materials. Renewable materials grow in nature (such as wood and flax), forming part of a natural circular chain. Reused materials have already served a previous purpose and are reprocessed for incorporation into new buildings. Reusable materials are designed to be demountable, allowing them to be repurposed in the future. However, many current circular building projects are only designed to be demountable but rely on new and finite materials, thereby shifting the responsibility for disassembly—and the project's circularity—onto future generations.

Fig. 3 Klinik Arlesheim, in Switzerland

9graden architectuur is committed to advancing circular thinking. We firmly believe that building should rely exclusively on renewable or reused materials. By taking this approach, we assume circular responsibility from the outset of each project.

To make sustainable material choices, 9graden architectuur employs a circularity matrix. Beyond construction costs, we consider environmental and maintenance costs, as well as factors such as flexibility during the usage phase, health, and lifespan. This provides a comprehensive understanding of the building's entire life cycle and its specific components, enabling us to make conscious, sustainable choices that benefit both people and the environment. 9graden architectuur consistently seeks opportunities to preserve or reuse as much of the existing structure and materials as possible.

We also emphasise low-tech sustainability, prioritising the building envelope and orientation to achieve a passive, energy-efficient, and healthy indoor climate. Our goal is to create comfortable buildings with healthy climates using minimal technology. Instead of relying on complex, maintenance-intensive installations that require replacement every 20 years, we invest in durable, healthy materials with long lifespans. 9graden architectuur champions low-tech sustainability—an approach that is healthy for both people and the environment!

7 Shaping Sustainability

In the developments of contemporary architecture, we find ourselves in a challenging field of tension: on one hand, an awareness of the climate crisis, and on the other, the emerging possibilities of artificial intelligence. Awareness of the climate crisis is transforming architecture. Buildings are increasingly designed to be demountable, suitable for future reconstruction, and recyclable. However, the result of these new approaches is that architecture is becoming increasingly temporary. At 9graden architectuur, this is where the danger of one-sidedness arises. If buildings are reduced solely to their material resources, architecture risks becoming a technocratic exercise, devoid of memory or history.

The ability to remember a building—how it was built, what it communicated to us personally, and what it meant in our lives—is a fundamental human experience. This experience forms part of our identity and sense of belonging on Earth. We build to orient ourselves, to remember, to understand ourselves, and even to find our place on this planet. Yet, in a one-sided pursuit of sustainability, we risk erasing this essential meaning that architecture has historically provided. Memory connects us, and is not connection a central, intangible aspect of sustainability?

In demountable and circular construction, the focus often shifts to the economic value of materials at the end of a building's life cycle. Materials, which were once a means to an end, are increasingly treated as the central purpose. As a result, architecture itself risks becoming a 'disposable item,' stripped of meaning and reduced to something merely temporary.

On the other hand, the possibilities of artificial intelligence evoke imaginative forms: plastic, seemingly alive, even organic, and highly seductive. A world of previously unimagined or unthought forms has become possible. However, the question arises: can the essential experiences of architecture still be realised in these seductively shaped forms? Or, put another way, do we not have the responsibility to develop an awareness of what is truly alive, truly organic, rooted in original ideas, and individually artistic?

Both directions outlined here are fascinating. Yet fascination, for 9graden architectuur, is not synonymous with beauty. Fascination often has the potential to become one-sided, precisely because it is captivating. In both directions—circularity and artificial intelligence—there is a tendency for technology to dominate, leading to one-sidedness and the diminishing of the human element. Our individual and creative human nature is at risk of being lost amidst this growing one-sidedness. Technology increasingly dictates the creative process, redefining our human and architectural experiences.

The connection between humans and the Earth—a tangible, touchable, fragile environment shaped by human hands—is in jeopardy. On one side, abstract and lifeless thinking causes form and meaning to disappear into matter. On the other, machines conjure unprecedented forms in which the meaning of matter remains elusive. Is it not humanity's task, through art, to connect form and matter, giving them meaning? 9graden architectuur sees it as its responsibility to engage with both approaches while preserving the human and artistic quality as a connecting balance. This also requires defining what is truly human and artistic—a monumental and substantive task to which 9graden architectuur is dedicated.

How do we preserve humanity as the foundation for sustainability in our built environment? Through a love of spirit and physical reality, we tap into the truly creative potential of humanity—the creativity that fosters love for the Earth. Recognising that we are still at the very beginning of this journey, our task is to lovingly dedicate ourselves to the spiritual in every design process. By focusing our efforts in this direction, practising skills, and engaging deeply, we can bring motifs to life within us and through us.

We connect, through concentration, with the spiritual content of our designs. We familiarise ourselves with this world. At the same time, we connect our hands to the Earth, to the material. In this connection, something new often flows toward us—something transformed by our hands and given form on Earth. What was once in the inner space of the spirit is transformed into the outer space of the Earth through our inner and outer work. Our hands serve as a bridge, connecting us to the spiritual and linking cosmos and Earth.

This process is essential: we connect with both motif and material, and through artistic transformation, the material evolves. What flows through our hands rises into the sphere of the ideal, becoming beauty. Only human beings can achieve this! Beauty, in this sense, becomes the foundation of sustainability.

8 Conclusion: Awareness of Vulnerability

Life—or survival—on our Earth depends on how we, as human beings, address its fragility. For 9graden architectuur, this awareness is fundamental. The technical progress of past centuries appears to have led humanity on a path of alienation from the Earth—a road 'away from the Earth.' 9graden architectuur seeks to contribute to a future shaped by an inner and outer attitude that connects with the living organism Earth. This is an individual path, a 'road toward the Earth,' a journey that can only be taken with spiritual awareness of the delicate and fragile living organism, Earth.

Conclusion

Concluding Remarks and Outlook

Johannes Kronenberg and Edith T. Lammerts van Bueren

1 Introduction

This concluding section summarises the main themes and contributions of anthroposophy to sustainable development as explored in the 45 texts of this compendium. Parts 1 and 2 examined the relationship between anthroposophy and sustainable development, while Part 3 introduced multi- and transdisciplinary perspectives, offering insights into the anthroposophical worldview and its implications for various fields engaged in sustainable development. Part 4 then presented 26 enterprises from 12 countries across five continents—referred to as "Sustainable Natives"—who shared their experiences of integrating anthroposophy into their work and sustainability efforts. The compendium aimed to bridge conceptual and practical approaches to anthroposophy and sustainable development.

As mentioned in the introduction, this work serves as a collection and an initial sketch of anthroposophical contributions to sustainable development rather than a comprehensive overview. It follows the earlier work of McKanan (2017), which first examined the relationship between anthroposophy and the environmental movement over the 20th century.

The central themes covered in this compendium range from anthroposophy's historical engagement with social and ecological challenges to its broader understanding of development. McKanan (Chap. 2) introduced four key contributions

J. Kronenberg (✉)
Research Associate Section for Agriculture, Goetheanum - Freie Hochschule für Geisteswissenschaft, Dornach, Switzerland
e-mail: johannes.kronenberg@goetheanum.ch

E. T. Lammerts van Bueren
em. Prof. Dr. Wageningen University and Louis Bolk Institute, Organic Plant Breeding, Utrecht, The Netherlands
e-mail: edith@bioacademy.nl

J. Kronenberg and E. T. Lammerts van Bueren (eds.), *On the Earth We Want to Live*, World Sustainability Series, https://doi.org/10.1007/978-3-031-98758-8_46

("gifts") that anthroposophy offers to the socio-ecological challenges, including a meta-perspective on human and planetary development, emphasising the profound impact of human activity and its responsibility for the Earth's future. This perspective extends sustainable development beyond concepts such as efficiency, sufficiency, consistency, and the traditional three pillars of sustainable development. Thus, sustainable development is not merely a desirable option or a directive to report on, but rather forms the very backbone of our common future.

In order to introduce anthroposophical principles, an inquiry into the nature of "who the human being is" (Chap. 4) and "the Earth as a living organism" (Chap. 5) was provided. The role of the human "I" (Ego) was presented as a fundamental theme, emphasising its potential as a free yet responsible co-creator in Earth's development (Chap. 3). This concept is not to be mistaken for a destructive anthropocentrism but rather for a potentially constructive recognition of humanity's role in shaping the future of the Earth, indicated as an "appropriate anthropocentrism."

The evolution of sustainable development, with its broad definitions and potential to emerge as a meta-discipline (Ametepey et al., 2023), was examined with particular attention to its deeper cultural and spiritual dimensions (Part 2). The socio-ecological crisis was presented as a cultural-spiritual crisis, a perspective already touched upon by Rudolf Steiner and later explored by initiatives such as the Earth Charter (2000), Pope Francis 'Laudato Si' (2015), The Club of Rome (2018), and Paul Kingsnorth (2018). The current socio-ecological crisis can also be approached through the framework of the three divides proposed by Scharmer and Kaufer (2013), which describe the disconnection between humanity and the Earth (ecological divide), between people (social divide), and within oneself (spiritual-cultural divide). The inner dimension of sustainable development has gained recognition in recent years, as reflected in the Inner Development Goals (Eleftheria & Suarez, 2024), which position inner transformation as a prerequisite for achieving Sustainable Development (Goals). Aligning with Steiner's perspective on humanity's co-creative role with the Earth, Scharmer & Kaufer (2013) framed this as "Seeing and Acting from the Whole." Steiner even envisioned the purpose of humanity as transforming the Earth into a "spirit-filled work of art" (Steiner, 1989), with humanity being an inherent and directive part of the Earth's organism. Viewing the Earth as a living organism in development rather than a broken machine in need of repair, is a fundamental premise of anthroposophy and represents a significant contribution to the broader sustainable development discourse.

This expanded perspective provides an initial outline of what we introduced as Integral Sustainable Development, which aligns the development of the individual, the collective, and the Earth in a deeper, more interconnected way. This framework suggests that before engaging with ecological, social, or economic sustainability, three dimensions should be considered: World-Viewing—recognising the Earth as an interconnected living organism; World-Sensing—moving from a mindset of control and exploitation to one of co-creative participation; and World-Acting—integrating humans, animals, plants, and entire ecosystems (Earth) into ethical, responsible action, forming the foundation of a cultural-spiritual approach to sustainable development. A connection to Social Threefolding (or the Threefold Social Order) given by Steiner (2022) was discussed as well and linked to ecological, social, and economic sustainable development.

This "Threefold Social Order" offers a distinct perspective on sustainable development by complementing the ecological, social, and economic pillars through principles of economic solidarity, legal equality, and cultural freedom. Unlike conventional economic models driven by perpetual economic growth, profit maximisation, market competition, or strict regulation, Social Threefolding proposes an associative economy rooted in solidarity and civil society engagement. Dunlop and Stein (1937), as early Social Threefolding adopters, already emphasised that the economy is not separate from the Earth but inherently part of it, requiring a shift in consciousness towards recognising the planet as a living organism. This perspective suggests that sustainable development extends beyond resource management and social justice to encompass moral responsibility for the Earth's development as well. In this view, human activity is not merely about fulfilling basic human needs but also about co-creation with the Earth in a way that fosters both planetary and human health and evolution.

Part 3 explored how anthroposophy has influenced disciplines such as medicine, agriculture, pedagogy, economics, and natural sciences, reconnecting the ecological, social, and spiritual dimensions of sustainable development. Key themes exemplified include Biodynamic Agriculture (Chap. 11) and Anthroposophic Medicine (Chap. 13), both of which have, for over a century, sought to reintegrate human and Earth development rather than viewing the planet as a mere stage for random human activity where only the fittest survive. Agriculture, in this view, is not only responsible for feeding humanity but also for regenerating and rejuvenating soil organisms in reciprocity with the realms of plants, animals, and vital eco-organisms. Similarly, anthroposophic medicine approaches "health as more than just the absence of disease"—it fosters an active relationship between physical, emotional, and spiritual health, integrating modern medical advances with natural remedies and self-regulation principles.

Finally, Part 4 examined 26 enterprises that exemplify the practical application of anthroposophical principles. These Sustainable Natives embody the deep connection between grand vision and concrete action, fostering long-term sustainable development. Companies such as Stockmar (Chap. 41), Voelkel (Chap. 43), and Weleda (Chap. 44) have, for over a century, integrated ecological, social, and economic sustainability from a cultural-spiritual perspective—and therefore can be seen as sustainable native. For these enterprises, fundamental questions such as "Who is the human being?", "What is the Earth?", and "What does development mean?" are not abstract concerns but core elements of their business models. As Engelsman (Chap. 23) pointed out, true "profit" is only meaningful if both humanity and the Earth benefit as an integral part of any economic activity.

It also became evident that there is no single "anthroposophical" approach to sustainable development. The anthroposophical movement, initiated by Rudolf Steiner just a century ago, remains in its early stages of integration into diverse fields of work and life. Additionally, definitions and understandings of sustainable development continue to be plural. However, it is important to recognise that among the listed Sustainable Natives, sustainable development is not merely a directive to report on but an intrinsic part of their work and responsibility.

In the following sections, we present a synthesis of key contributions, a critical discussion, and an outlook for the future, followed by our final conclusion.

2 Synthesis of Key Contributions

This section provides a synthesis of the key contributions of anthroposophy to sustainable development as explored in this compendium. The synthesis is structured around the three main dimensions identified in the introduction: conceptual contributions (related to Parts 1 and 2), multidisciplinary perspectives (related to Part 3), and practical applications (related to Part 4).

Conceptual Contributions: Anthroposophy's Perspective on and Extension of (an Integral) Sustainable Development (Related to Parts 1 and 2)
Anthroposophy can offer an integral and holistic understanding and extension of sustainable development. It contributes a perspective that extends beyond conventional definitions of SD on several levels: in addressing the well-being and social justice of current and future generations without exceeding planetary boundaries, anthroposophy not only emphasises Brundtland's aim (1987) of satisfying basic human needs in harmony with the carrying capacity of the Earth, but recognises the potential co-creatorship and responsibility of each individual and society in the larger development of the Earth as a living organism. A technical sustainable development approach focused on efficiency, sufficiency, and consistency is complemented with what Broman and Karl-Robert (2017) call "meaning-making" as a Sustainability Principle. Herewith, "a new narrative" for our global society, as the Club of Rome called for (2018), can be formulated. McKanan (2017) called this one of four gifts of anthroposophy to the environmental movement. This was introduced as "planetary transmutation," where the human being should not only protect, conserve, and refrain from harming "nature" to secure its own future but actively heal (regenerate) and evolve (generate) the very substance of the Earth. This is what Steiner called for, expressed in his statement that "Our mother Earth has hardened through pain. Our mission is to spiritualise her again, by reworking her through the power of our hands into a spirit-filled work of art" (Steiner, 1909, p. 113). Paul Kingsnorth (2018) repeated this finding independently after decades of environmental activism of all kinds against Big Industry, destroying Earth's living organisms. He stated that:

> The crisis which we experience, the by ourselves created downfall of nature, and the likely collapse of our modern civilisation in this century, is in essence a spiritual crisis. Because this must be the first civilisation in history where nothing is sacred anymore. There is no central faith, but moreover: no notion anymore of something greater than ourselves. Something to have reverence for. According to the myth of progress, we have transcended the

> outdated notion of faith and sacredness we find primitive. And this, I think, this lack of the divine is the cause of the destruction we create. (…) We have become individuals that devour the world. (Kingsnorth as cited in VPRO Tegenlicht, 2018)

For Kingsnorth, the old narrative, the myth of progress, lost the higher sense and reason of human existence and, therefore, the relationship with the Earth. He, therefore, declares the socio-ecological crisis as a spiritual crisis of humanity. Creating a new narrative is not meant as another anthropocentric replacement of the myth of progress or green growth but as a necessary transition to an "appropriate anthropocentrism" as another "gift" of anthroposophy to the environmental movement (McKanan, 2017). This will at the same time be the main point of tension and critique, especially in the environmental movement, which has already deemed the human being as the main evil for the Earth and warns against any human-centred colonial-suspicious thought—of course, with justified reasons. More on this point will be discussed in section "6.3 Critical Discussion."

To overcome this spiritual crisis, inner and outer (sustainable) development are encouraged to become more interrelated. The Inner Development Goals (Eleftheria and Suarez, 2024), briefly touched upon in this publication, align with anthroposophy's threefold anthropology of the human condition (thinking, feeling, willing) and its understanding of development. These point towards another extension of traditional SD. In the anthroposophical approach, outer action and development always relate strongly to inner development and vice versa. Scharmer and Kauffer (2013), who draw on anthroposophy as well, identified the socio-ecological crisis as a spiritual crisis too. They state that modern humanity is challenged by an ecological divide (with the Earth), a social divide (with others), and a spiritual divide (with oneself). Reconnecting with the Earth, others, and oneself might sound like self-help psychology, but it will need to form the backbone of a new understanding of development in the century to come, and it was already an integral part of many cultures and civilisations before. The attribution of a Human "I" to each individual, as introduced by Lavecchia (Chap. 3), is the very faculty of the human being that can direct and train the inner part of oneself. This shapes another cornerstone of the anthroposophical approach, referred to as the inner path of development.

The proposed Integral Sustainable Development approach, therefore, advocates for relating inner and outer (sustainable) development more closely. This means that sustainable development is not just a governance question as the highest authority and leverage, but a deeply rooted cultural-spiritual affair, touching the very core values and worldviews of societies, organisations, and individuals. The proposed integral approach can be seen and utilised as a tool to make World-Viewing, World-Sensing, and World-Action visible before analysing ecological, social, and economic sustainability challenges, strategies, and measurements. This means that, on the one hand, World-Viewing openly asks what concept of "nature," the Earth, plants, animals, human beings, ecosystems, etc., one holds. On the other hand, it stimulates seeing the Earth as a living organism on a path of development, where one can define one's own place as a co-creative actor. World-Sensing works similarly—on the one hand, openly asking what one's attitude towards one's surroundings is. On the other hand, it addresses the transition from a "human as ruler" attitude to a "human as participant and co-creator" of a common path of development. Finally, World-Acting states that all action follows, consciously or unconsciously, a certain attitude (World-Sensing) and worldview (World-Viewing). Therefore, World-Acting culminates in integrating the four dimensions of earthly life: (1) anthropocentric (only the human being is considered ethically relevant); (2) zoocentric (humans and sentient animals are considered ethically relevant); (3) biocentric (all living organisms are considered ethically relevant); and (4) ecocentric (here, beyond individual organisms, also species or ecosystems/organisms are considered ethically relevant). This will not necessarily make the whole discipline and act of sustainable development less complex, but it tries to expand its scope to a larger understanding of development at large and the common responsibility for it.

Worldview & Multidisciplinary Perspectives: Anthroposophy as a Bridge Between Disciplines (Related to Part 3)
This part of the publication shows that both sustainable development and anthroposophy have in common that they can be perceived as meta-disciplines. This means that both can provide understanding and orientation to, for instance, the addressed fields of agriculture, medicine, pedagogy, social development, environmental work, and beyond. Anthroposophy can offers an integral worldview essential for sustainable development, according Van Egmond and de Vries (2011).Their research results of a survey amongst 40.000 people in the Netherlands about their values and worlview, showed paralels to Steiners 12 worldviews (Steiner, 2015) and aim to find an equilibrium or middle ground, in the quadrant between materialism and spiritism, and collective and self.

Whereas sustainable development is still more understood as a tool to fix problems, anthroposophy is intended as "a path of knowledge to guide the Spiritual in the human being to the Spiritual in the universe" (Steiner, 1973, p. 13). This seems to create more distance instead of common ground. However, the multidisciplinary contributions in Part 3 showcase how anthroposophy contributes to sustainable development by becoming more integrated into the backbone of society. Examples include the redefinition of the task of agriculture, wherein Biodynamic Agriculture, from 1924 onward, does not only perceive its task as feeding the world but also as healing (regenerating) and evolving (generating) the very substance of the Earth and life on it. Rejuvenating aging and degraded soils and ecosystems thus becomes a completely new understanding of the task of agriculture. Of course, it is crucial to link this core task with ensuring the satisfaction of basic human needs at all times. Biodynamic Agriculture, in this respect, emphasises the long-term stability of satisfying the human need for food by ensuring that soils and ecosystems improve rather than degrade. One can recognise a shift in understanding agriculture (relating to World-Viewing), the emphasis on a co-creative task between farmer and Earth (relating to World-Sensing), and finally the production of human- and earth-dignified food and "land management" (relating to World-Acting). Regarding the discipline of medicine, the definition of its core task of ensuring health is also expanded by the anthroposophical approach. This practice defines health as "more than the absence of disease" and focuses on an active relationship between physical, emotional, and spiritual health, integrating modern medical advances with natural remedies and self-regulation principles. The transdisciplinary relation and collaboration between agriculture and medicine thus receive renewed attention.

Goethean science, as part of anthroposophical scientific work (Chap. 16), complements natural sciences by integrating thinking and perception, bridging the dualities between the living and non-living, objective and subjective, and transcending sympathy and antipathy. The understanding of metamorphosis—the movement in a certain development—becomes central to the scientific work. This forms a foundation for other disciplines, including pedagogy: how can a child's development be seen as an unfolding metamorphosis, progressing from one stage to the next, embedded in a broader context and path of development? The same understanding can be applied to a farm (known in Biodynamic Agriculture as the farm organism) and even to the Earth as a larger organism in an even broader context. Ultimately, this understanding contributes to an expanded definition of development in relation to sustainability.

The concept of sustainability is etymologically rooted in the Latin sustinere/sustineri, which combines tenere ("to hold") with sub- ("under, up from below"), conveying meanings of support, endurance, and maintenance. In this sense, sustainability can be understood as the capacity to uphold or carry ecological, social, economic, and cultural systems over time. When paired with "development," the term signifies not merely growth but the ability to foster and "uphold" progress in a manner that ensures long-term viability. The institutionalisation of sustainable development as a global framework represents a significant milestone, offering a shared language for addressing planetary challenges. However, critiques persist regarding its Western, human-centered origins, neocolonial tendencies, and the limitations imposed by economic paradigms such as the Triple Bottom Line (people, planet, profit). While these concerns are valid, dismissing sustainable development prematurely overlooks its potential for adaptation, inclusivity, regeneration, and "generating." Some might reject the framework of sustainable development outright. The contribution of anthroposophy could be to promote a more constructive approach and to refine and expand its scope, ensuring that it serves as common ground for a shared future.

Practical Applications: Enterprises and Sustainable Practices (Related to Part 4)

The 26 enterprises from 12 countries, 5 continents, and operating worldwide showcase decades—and for some, more than a century—of pioneering experience in what we today call sustainable development. They are all founded, linked to, or inspired by anthroposophy and have maintained a connection to it, some explicitly, some more implicitly. A key observation is that a strong sense of responsibility for social justice and "Earth dignity" forms a golden thread through all Sustainable Natives. A significant portion explicitly states that their work with anthroposophy contributed to their understanding of carrying responsibility for socio-ecological development at large. For Alnatura (founded in 1984), a leading organic supermarket in Germany and a system-relevant enterprise, this is expressed in their understanding of sustainability:

> Our understanding of sustainability encompasses the dimensions of environment, society, economy, and culture. In doing so, we expand the conventional sustainability model to include a unifying spiritual-cultural dimension. This reflects how we think about humanity and the Earth and embodies our belief that sustainable action stems from sustainable thinking. (Alnatura, n.d.)[1]

Other enterprises such as Eosta (founded in 1990), SEKEM (founded in 1977), and the Biodynamic Federation Demeter International (founded in 2020, with their Demeter certification founded in 1928 as the first organic certification standard) also showcase similar understandings of their work in relation to anthroposophy and sustainable development (also see Sect. 7 in Chap. 6, Examples of Implementing Integral Sustainable Development).

The practical contribution of anthroposophy to sustainable development can, in this aspect, be understood as both pioneering and extending its definition, as already indicated by McKanan (2017), as well as having numerous practical applications in many fields of work and life. SEKEM in Egypt, for instance, was awarded both the Gulbenkian Prize for Humanity 2024 and the UN Environmental Programme's Champion of the Earth 2024 laureate for their contribution to Biodynamic Agriculture as a climate solution and environmental innovation—both prizes being among the highest recognitions in their field on a global level. Although enterprises working with anthroposophy form a small part of the global economy and cultural life, they seem to have reached the stage of global discourse concerning the socio-ecological challenges of our time and their contributions to finding new narratives and pathways for a shared future.

Finally, what today is called "purpose" in a business model could be seen as the translation of the anthroposophical approach and practices in economic, social, and cultural life. By integrating a cultural-spiritual dimension into their work, long-term horizons, meaning, and deeply rooted commitment can emerge, as seen in the Sustainable Natives.

3 Critical Discussion

As mentioned in the introduction of this publication, the aim of this compendium was to collect diverse but not exhaustive perspectives of both a conceptual and practical nature, in order to explore anthroposophy's contributions to sustainable

[1] Translated by the editor from the Alnatura website https://www.alnatura.de/de-de/ueber-uns/nachhaltigkeit-bei-alnatura/.

development. Given that both anthroposophy and sustainable development are multidisciplinary and extensive fields, a full and systematic review or overview was not intended to be covered. The absence of a systematic comparative study of the enterprises and their work remains an area for further research.

One of the desired outcomes of this compendium, however, is to facilitate engagement with anthroposophy and its related enterprises, disciplines, and research in public, academic, and broader sustainability discourses. This will need to be an ongoing effort.

A key critique of anthroposophy's contributions to sustainable development is likely to be its perceived human-centred approach, a concern already embedded in the etymology of anthroposophy itself, where "ánthrōpos" means "human" and "sophía" means "wisdom" (thus signifying "human wisdom"). While it is necessary to acknowledge that anthroposophy fundamentally aims to cultivate human wisdom in relation to the broader development of the Earth and even the cosmos, we argue against the interpretation that it is inherently and destructively anthropocentric. McKanan (2017) suggests that anthroposophy presents an "appropriate" or "constructive anthropocentrism," which does not result in human dominance over other life forms but instead emphasises humanity's immense and decisive responsibility for the trajectory of Earth's development. This is currently evidenced in the negative sense by the destructive impact of human activity on the planet's habitability. A crucial task for sustainable development discourse is therefore to shift from a destructive anthropocentrism to a constructive one. Of course the protection of "nature" and vital eco-organisms is a very important and an essential part of einvironmental work being understood as part of sustainable development. However currently shutting off certain areas for human activity, indirectly justifies misuse and deploitation in non-protected areas. An appropriate anthropocentrism therefore aims for a new understanding and narrative of the human being as inherent part of the Earth.

In the current global context, where geopolitical paradigms are shifting dramatically and (national) "survival" has become a central topic of discussion globally, it seems increasingly challenging to establish a shared understanding and foundation for a common future. Whether anthroposophy and sustainable development can contribute meaningfully to this state of the world remains an open question. However, we advocate for an open and constructive collaboration among all individuals, initiatives, and share- and stakeholders in striving to overcome the present socio-ecological crisis through commitments to social justice and Earth dignity, grounded in a renewed cosmopolitan awareness. Since the COVID-19 pandemic and of climate emergency, it has become evident that human societies must be addressed as a whole, transcending national and ideological boundaries.

By expanding the understanding and definition of sustainable development, the intention is not necessarily to criticise current sustainability efforts. Instead, we hold the conviction that this field is still in the early stages of its existence and development. While some authors, such as Jhagroe (2024), criticise sustainable development as being neocolonial, Western-centric, one-sided, and misleading, we acknowledge these concerns without fully endorsing them. It is crucial to recognise the relative juniority of sustainable development as a discipline and to focus on its growth rather than prematurely dismantling it. We encourage a constructive discourse on the development of sustainable development itself and the expansion of its definition, as also advocated in the 2018 Report to the Club of Rome (Wijkman & Weizsäcker, 2018).

4 Outlook and Future Directions

Given the findings of this compendium, further research is needed to explore anthroposophy's role in sustainable development in greater depth. This includes comparative studies on enterprises working with anthroposophy, as well as empirical assessments of its contributions to sustainability practices across different disciplines. Of course, always following the goal of practical application into daily life.

There is significant potential for deeper collaborations between anthroposophy and global sustainable development initiatives. As seen in various case studies throughout this publication, enterprises and organisations inspired by anthroposophy have already achieved meaningful contributions in fields such as agriculture, medicine, education, and social development. Strengthening dialogues between anthroposophical initiatives and broader sustainability efforts could help bridge knowledge gaps and encourage a more integrative approach to sustainable development.

Future pathways for integrating the Integral Sustainable Development framework into academic, business, and policy spheres should also be explored. The framework's ability to combine ecological, social, economic, and cultural-spiritual dimensions provides an opportunity to refine sustainable development strategies and ensure long-term resilience. As sustainability challenges continue to evolve, inter- and transdisciplinary approaches that embrace cultural-spiritual perspectives may prove increasingly valuable to built a new narrative of human and earth development.

Finally, emerging opportunities for applying anthroposophical insights to pressing environmental and social challenges should be further investigated. This includes amongst others exploring the relevance of Goethean science in ecological research, the role of biodynamic agriculture in climate resilience, and anthroposophical medicine's potential contributions to holistic health approaches. By continuing to refine and apply these insights, anthroposophy may offer valuable approaches for shaping sustainable futures.

5 Conclusion

The exploration of anthroposophy's contributions to sustainable development highlights the potential of an integral, holistic approach that extends beyond conventional frameworks. Sustainable development shifts from being a mere "nice-to-have" or directive to report on, towards becoming the very backbone of society's future. New narratives and approaches for our common future are articulated. Both on a conceptual and practical level, anthroposophy contributes to sustainable development in a pioneering way, starting more than a century ago. A late example is SEKEM, which has been working on sustainable development since 1977 in Egypt, and was recognised as UNEP's Champion of the Earth 2024 laureate for their contributions to Biodynamic Agriculture as a climate solution and innovative practice.

While not without critique or limitations, anthroposophy provides valuable perspectives that can enrich contemporary sustainable development discourse, particularly in relation to how the relationship between humanity and the Earth could be understood. The human being appears as a co-creator in relation to Earth's development, resulting in an "appropriate" or "constructive anthropocentrism." The task is not only to heal (regenerate) soils, eco-organisms, and landscapes, but also to "generate" and rejuvenate them. These are definitions of sustainable development that stem from anthroposophy. This entails, in addition to "nature" protection and conservation, also, and in particular, an active involvement and understanding of the Earth as a living organism, with the human being as an integrated responsible actor. These "outer" actions and transformations are strongly related, within the anthroposophical approach, to inner development, independently related to and known from the recently introduced Inner Development Goals as prerequisites for (outer) Sustainable Development (Goals). If the geopolitical paradigm shift we are experiencing at the time of writing allows the United Nations to continue their shared effort for a renewed Sustainable Development Agenda after 2030, we envision a collective effort to integrate these inner and outer sustainable development approaches in order to rebalance the relationship between humanity and the Earth.

This compendium serves as a contribution to ongoing conversations in the field, offering a collection of perspectives that illustrate both conceptual and practical applications of anthroposophy in sustainable development. By fostering a dialogue between anthroposophy and sustainability efforts, this work seeks to encourage further research, collaboration, and innovation.

Ultimately, the necessity of an integrated and holistic approach to sustainable development has never been so pressing, given the many social, environmental, and planetary boundaries and tipping points that are either almost reached or already exceeded.

In the end, the outcomes of the twenty-first century will rest on the understanding of the ancient question: why are we inhabiting the Earth as human beings, what are our tasks, and how should we respond and live accordingly? Anthroposophy, therefore, could define, complementary to Brundtland, that sustainable development points towards the very essence of human existence and the main task of the human being on an Earth in development. Or as Steiner himself proclaimed: "Our mother Earth has hardened through pain. Our mission is to spiritualise her again, by reworking her through the power of our hands into a spirit-filled work of art" (Steiner, 1992, p. 113 [translated by the editor]).

References

Alnatura. (n.d.). *Sustainability at Alnatura.* Retrieved February 25, 2025, from https://www.alnatura.de/de-de/ueber-uns/nachhaltigkeit-bei-alnatura/

Ametepey, S. O., Aigbavboa, C. O., Ansah, S. K., Gyadu-Asiedu, W., & Boamah, L. (2023). *Meaning, evolution, principles, and future of sustainable development: A systematic review* (Preprint No. 2023020158). Preprints. https://doi.org/10.20944/preprints202302.0158.v1

Broman, G., & Robèrt, K.-H. (2017). A framework for strategic sustainable development. *Journal of Cleaner Production, 140*, 17–31. https://doi.org/10.1016/j.jclepro.2015.10.121

Brundtland, G. H. (1987). *Our common future.* Oxford University Press.

Dunlop, D. N., & Stein, W. J. (Eds.) (1937). *World survey* (Vol. 1. 1). World Power Conference.

Eleftheria, E., & Suarez, M. C. (Eds.) (2024). *Inner development goals: Stories of collective leadership in action.* De Gruyter.

Francis, P. (2015). *Laudato Si': On care for our common home* [Encyclical letter]. Vatican Press.

Jhagroe, S. (2024). *Voorbij duurzaamheid: Op weg naar een ecorechtvaardige samenleving.* Mazirel Pers.

McKanan, D. (2017) *Eco-Alchemy: Anthroposophy and the history and future of environmentalism.* University of California Press.

Scharmer, C. O., & Kaufer, K. (2013). *Leading from the emerging future: From ego-system to eco-system economies.* Berrett-Koehler Publishers.

Steiner, R. (2015). *Der menschliche und der kosmische Gedanke* (7th, ed., GA 151). Rudolf Steiner Verlag.

Steiner, R. (2022). *Die Kernpunkte der sozialen Frage in den Lebensnotwendigkeiten der Gegenwart und Zukunft* (7th ed., GA 23). Rudolf Steiner Verlag.

Steiner, R. (1973). *Anthroposophical leading thoughts: Anthroposophy as path of knowledge.* Rudolf Steiner Press.

Steiner, R. (1989). *Die Impulsierung des weltgeschichtlichen Geschehens durch geistige Mächte* (3th ed). Rudolf Steiner Nachlassverwaltung. Rudolf Steiner Verlag.

VPRO Tegenlicht. (2018, December 16). *Paul Kingsnorth: Portrait of a recovering environmentalist* [Video]. VPRO.

Van Egmond, K., & De Vries, H. J. M. (2011). Sustainability: The search for the integral worldview. *Futures, 43*(8), 853–867.

Wijkman, A., & von Weizsäcker, E. U. (2018). *Come on! Capitalism, short-termism, population, and the destruction of the planet.* Springer.

Accountability and Research Approach

Johannes Kronenberg and Edith T. Lammerts van Bueren

1 Accountability in the Context of This Compendium

Within the framework of this compendium, accountability manifests in the attempt to integrate a diverse and large body of perspectives on anthroposophy and sustainable development, while simultaneously acknowledging the limitations and interpretative nature of the presented research. Yet, maintaining a balanced level ranging from academic quality to professional practitioners' experience determines this work.

This accountability requires recognising the complex and interdisciplinary nature of sustainable development and anthroposophy. As outlined in Part 3, the integration of anthroposophy into various disciplines—such as medicine, agriculture, pedagogy, and economics—and their contribution to sustainable development demands an integral approach that considers both practical and conceptual implications. The inclusion of case studies of enterprises engaging with anthroposophical principles and contributing to sustainable development (Part 4) illustrates how this accountability extends beyond conceptual discourse into daily applications. In doing so, an even broader view of the main question was sought.

J. Kronenberg (✉) · E. T. Lammerts van Bueren
Research Associate Section for Agriculture, Goetheanum—Freie Hochschule für Geisteswissenschaft, Wageningen, The Netherlands
e-mail: johannes.kronenberg@goetheanum.ch
URL: https://www.sektion-landwirtschaft.org/en/

E. T. Lammerts van Bueren
e-mail: edith@bioacademy.nl

Wageningen University and Louis Bolk Institute, Organic Plant Breeding, Utrecht, The Netherlands

J. Kronenberg and E. T. Lammerts van Bueren (eds.), *On the Earth We Want to Live*, World Sustainability Series, https://doi.org/10.1007/978-3-031-98758-8_47

We recognise that we could not involve everyone we would have wished to contribute. At the same time, it may be the case that certain individuals or organisations who should have been part of this publication will only hear of it after its release, which we, in those cases, regret. An open call was sent out in early 2024, alongside an active search for contributors. Therefore, there was a window in which we welcomed all who wished to submit their proposals.

2 Research Approach: Integrative and Multi-perspective Inquiry

The research approach employed in this compendium is inherently multi- and transdisciplinary, integrating insights from different fields to offer a comprehensive understanding of anthroposophy's contributions to sustainable development. This methodology reflects an effort to bridge conceptual exploration with experience-based examples, facilitating a dialogue between theory and practice.

A key methodological consideration in this research was the necessity of contextualising anthroposophy within the broader sustainability discourse, as explored by scholars and thinkers in the realms of anthroposophy and/or sustainable development (Part 1). While anthroposophy offers a unique perspective on ecological, social, and economic sustainability, it was crucial to examine its alignment with established sustainability frameworks. This task was undertaken by the editors themselves, given their professional background in both sustainable development and anthroposophy (Chap. 1 and Part 2).

Furthermore, the research approach acknowledges the historical evolution of anthroposophy's engagement with sustainable development. Rather than presenting anthroposophy as a singular, definitive solution, the compendium recognises its position within an ongoing discourse. This awareness of historical and conceptual fluidity underscores the necessity of an open, iterative research process that invites further scholarly inquiry and practical experimentation.

Another methodological aspect of accountability in this research is the incorporation of diverse voices. The compendium does not claim to present a single, unified anthroposophical or solely academic approach but rather highlights a spectrum of interpretations and applications, encompassing both academic and professional practitioner texts. This pluralistic approach ensures that anthroposophy's contributions to sustainability are represented in a nuanced and balanced manner, acknowledging both its strengths and areas for critical reflection.

Following a round of invitations for multidisciplinary perspectives on anthroposophy's worldview and its relationship to sustainability (Part 3), twelve texts were submitted, the majority authored by people engaged in academic discourse. We provided an open guideline on how to write these essay-like contributions, framed around the research question of what anthroposophy's contributions to sustainable development are in relation to their respective disciplines.

The 27 Sustainable Natives (Part 4) were selected through both an open call and a targeted search for a wide range of enterprises actively working with anthroposophical principles and contributing to sustainable development. We aimed to include different cultural and geographical backgrounds, resulting in a selection encompassing twelve countries across five continents and diverse cultural contexts. In addition to inviting enterprises with over 100 years of experience, we also sought to include younger organisations.

Finally, the two editors of this compendium represent different generations and backgrounds in anthroposophy, sustainable development, and Plant Sciences, particularly Organic Plant Breeding.

3 Editors' Translations of Cited Quotations

A part of the chapters were originally submitted in German and have been translated into English by Till Höffner. In several instances, he also provided his own translations of quoted material, as official English versions were either unavailable or inaccessible. All chapteres where this applies, are provided by a footnote.

4 Ethical Considerations and Reflexivity

Given the intersection of anthroposophy with cultural, spiritual, and ecological dimensions, reflexivity played a crucial role in shaping this compendium. Researchers and contributors remained aware of their own perspectives and potential biases, striving to present their findings in a manner that respects both anthroposophical discourse and contemporary sustainability science.

Furthermore, accountability in research necessitates a commitment to accessibility and knowledge dissemination. By structuring the compendium to engage both academic and practitioner audiences, this work aims to foster a broad and inclusive dialogue on anthroposophy and sustainable development. The integration of Sustainable Natives as case studies further ensures that conceptual discussions are grounded in lived experiences, reinforcing the principle that sustainability must be actively practiced.

MIX
Papier aus verantwortungsvollen Quellen
Paper from responsible sources
FSC® C105338

If you have any concerns about our products,
you can contact us on
ProductSafety@springernature.com

In case Publisher is established outside the EU,
the EU authorized representative is:
Springer Nature Customer Service Center GmbH
Europaplatz 3, 69115 Heidelberg, Germany

Printed by Libri Plureos GmbH
in Hamburg, Germany